Applied Nonparametric Statistical Methods

Nonparametric statistical methods minimize the number of assumptions that need to be made about the distribution of data being analysed, unlike classical parametric methods. As such, they are an essential part of a statistician's armoury, and this book is an essential resource in their application. Starting from the basics of statistics, it takes the reader through the main nonparametric approaches with an emphasis on carefully explained examples backed up by use of the R programming language.

Key features of this fully revised and extended fifth edition include the following:

- An introductory chapter that provides a gentle introduction to the basics of statistics, including types of data, hypothesis testing, confidence intervals and ethical issues
- An R package containing functions that have been written for the examples in the text and the exercises
- Summary bullet points at the end of each section to enable the reader to locate important principles quickly
- A case study from medical research to demonstrate nonparametric approaches to the data analysis
- Examples fully integrated into the text, drawn from published research on contemporary issues, with more detail given in their explanation
- Extensive exercises along with complete solutions that allow the reader to test their understanding of the material
- Articles used in the examples and exercises carefully chosen to enable readers to identify up-to-date literature in their field for research, publications and teaching material
- Numerous historical references throughout the text, from which to explore the origins of nonparametric methods

Applied Nonparametric Statistical Methods, Fifth Edition, is a comprehensive course text in nonparametric techniques suitable for undergraduate students of mathematics and statistics. It assumes only basic previous experience of statistics, and with algebra kept to a minimum, it is also ideal for quantitative methods modules delivered to undergraduate or postgraduate students in science, business and health service training. It is an invaluable resource for researchers, medical practitioners, business managers, research and development staff, and others needing to interpret quantitative information. Suitable for self-directed learning in continuing professional development, it also acts as a handy accessible reference manual.

CHAPMAN & HALL/CRC

Texts in Statistical Science Series

Recently Published Titles

Applied Linear Regression for Longitudinal Data: With an Emphasis on Missing Observations
Frans E.S. Tan and Shahab Jolani

Fundamentals of Mathematical Statistics
Steffen Lauritzen

Modelling Survival Data in Medical Research, Fourth Edition
David Collett

Applied Categorical and Count Data Analysis, Second Edition
Wan Tang, Hua He, and Xin M. Tu

Geographic Data Science with Python
Sergio Rey, Dani Arribas-Bel, and Levi John Wolf

Models for Multi-State Survival Data: Rates, Risks, and Pseudo-Values
Per Kragh Andersen and Henrik Ravn

Spatio–Temporal Methods in Environmental Epidemiology with R, Second Edition
Gavin Shaddick, James V. Zidek, and Alex Schmidt

A Course in the Large Sample Theory of Statistical Inference
W. Jackson Hall and David Oakes

Statistical Inference, Second Edition
George Casella and Roger Berger

Nonparametric Statistical Methods Using R, Second Edition
John Kloke and Joesph McKean

Generalized Linear Mixed Models: Modern Concepts, Methods and Applications, Second Edition
Walter W. Stroup, Marina Ptukhina, and Julie Garai

Analysis of Categorical Data with R, Second Edition
Christopher R. Bilder and Thomas M. Loughin

Applied Nonparametric Statistical Methods, Fifth Edition
Nigel C. Smeeton, Neil H. Spencer, and Peter Sprent

For more information about this series, please visit: https://www.routledge.com/Chapman--HallCRC-Texts-in-Statistical-Science/book-series/CHTEXSTASCI

Applied Nonparametric Statistical Methods

Fifth Edition

Nigel C. Smeeton, Neil H. Spencer, and
Peter Sprent

CRC Press
Taylor & Francis Group
Boca Raton London New York

CRC Press is an imprint of the
Taylor & Francis Group, an **informa** business
A CHAPMAN & HALL BOOK

Fifth edition published 2025
by CRC Press
2385 NW Executive Center Drive, Suite 320, Boca Raton FL 33431

and by CRC Press
4 Park Square, Milton Park, Abingdon, Oxon, OX14 4RN

CRC Press is an imprint of Taylor & Francis Group, LLC

Second edition published by CRC Press 1993
Third edition published by CRC Press 2001
Fourth edition published by CRC Press 2007

ISBN: 9780367344894 (hbk)
ISBN: 9781032449104 (pbk)
ISBN: 9780429326172 (ebk)

DOI: 10.1201/9780429326172

Typeset in Palatino
by codeMantra

Contents

Preface

Applied Nonparametric Statistical Methods first appeared in 1989. Major developments in computing, especially for exact permutation tests, inspired the revised edition that followed in 1993. The third edition in 2001 reflected not only further advances in computing but widening developments in nonparametric and distribution-free methods. With the fourth edition, published in 2007, the scope was broadened to target those with an understanding of basic statistics but only a limited acquaintance with nonparametric methods. It began with a summary of relevant general statistical concepts and a first-principles introduction to the concepts of nonparametric or distribution-free methods. In addition, designed experiments, including those with factorial treatment structures, became the focus of an entire chapter. Coverage of the analysis of survival data and the bootstrap method was expanded. A new final chapter focused on large sample methods and computer-intensive applications.

Having established a widely respected text in applied statistics, Peter Sprent was keen to see the book developed further. At the time of his death, he was actively considering the possibility of a further edition. This work brings his aspirations to fruition and combines the best of the previous editions with new material that reflects the unabated evolution in the application of statistical methods.

Greater emphasis has been given to the importance of statistical software in statistical analysis. Extensive use has been made of the R programming language. An R package dedicated to the edition has been specifically developed for the examples in the text and exercises that follow each chapter.

Another major change is in the format of the text. This has been substantially redesigned to make the book free flowing, thereby increasing its readability. The chapter summaries have been replaced by summary bullet points at the end of each section. Examples have been fully integrated into the text, and rather than having a distinct section describing fields of application within each chapter, areas of application are illustrated by the examples, many of which draw on real-world studies. The exercises have a greater emphasis on the analysis of published data from a wide range of fields. Apart from the explorative questions that follow the case study, detailed solutions are available for all the exercises.

Several chapters have been significantly expanded and some rewritten. A new Chapter 1 provides a gentle introduction to the basics in statistics. It discusses types of data, hypothesis testing, confidence intervals and ethical issues. Material on nonparametric statistics in Chapter 1 has been relocated to Chapter 2 to create an expanded introduction to nonparametric tests. This chapter now contains an analysis of data on the benefits of citrus fruit consumption in the context of scurvy taken from the historic study conducted by Lind in the 1700s. Chapter 3 has been reordered to start with the straightforward concept of the sign test. Chapters 4–8 provide new contemporary examples. In Chapter 7, the reader is reacquainted with Lind's citrus fruit study from a different perspective. The first section of Chapter 9 has been expanded to describe an early study of life expectancy and give a more detailed introduction to survival curves. The examples used in Chapters 11–14 feature contemporary issues. In Chapter 11, regression methods are addressed in the context of the recreational use of video games by young people. Further areas of current interest are encountered in Chapter 12 (public and patient involvement in research studies) and

Chapter 13 (engagement in physical activity by young people). Chapter 14 revisits recreational gaming.

Chapter 15 is completely new. Based on a case study on pre-operative diet and the surgical removal of the gall bladder, it starts with an explanation of the formation of gallstones using non-medical language. Having introduced the study, nonparametric approaches that can be used in the data analysis are compared. The final section of this chapter assesses the future for nonparametric statistical methods.

This book has been written for a wide audience. It can be used as an introductory course text in nonparametric methods for undergraduate students following mainstream statistics courses and in quantitative methods modules for students majoring in other disciplines at either undergraduate or postgraduate level. In addition, it has been designed with researchers in medicine and allied professions, social sciences, business and economics in mind, along with medical practitioners, business managers, research and development staff and others needing to interpret quantitative information. The format adopted makes this book suitable not only as a student class text but also for self-directed learning in, for example, continuing professional development. This book can be used as an accessible reference manual.

The numerous summary points allow the time-pressured reader to locate the most important principles quickly. The articles used in the examples and exercises have been chosen so that, along with their associated cited and citing papers, readers will be able to identify up-to-date literature in their own field of work for their research, publications and teaching material.

Statistical ideas have developed in parallel rather than in series, like threads woven into cloth rather than stations along a railway line. Instead of having a short section on the history of nonparametric methods, historical comments have been placed in the most relevant chapter. Conventional wisdom has been clarified or corrected where necessary. The historical references include some unfamiliar early publications and form a good starting point for those intending to investigate the origins of nonparametric methods.

The pioneering contributions to nonparametric statistics made by the late Professor EJG Pitman have not been overlooked. Peter Sprent was one of Pitman's students, and Nigel Smeeton has been involved in the restoration of Pitman's Notes on Non-Parametric Statistical Inference. Pitman's highly cited teaching material from the 1940s, arguably the first collection of nonparametric methods, is now an online resource at the University of North Carolina.

We are grateful to the reviewers and readers of the earlier editions for their constructive feedback on the content and structure of the text. In terms of the fourth edition, Roger Newson provided detailed comments and Nick Cox meticulously checked the text for typing errors. Ian Goudie sent us several papers on capture-recapture methods that were proving difficult to track down. Mike Lewis provided us with the data for the case study, and Angela Madden checked the case study chapter from a clinical point of view.

Nigel Smeeton

Neil Spencer

Authors

Nigel Smeeton commenced his career as the Statistical Assistant at Leeds University in the Department of Statistics, teaching statistics to undergraduate and postgraduate students, and running the statistical advisory service. He then joined the General Practice Research Unit at the Institute of Psychiatry, London, focusing on the modelling of episodes of mental illness, classification of mental illness and repetition of attempted suicide. Moving to the United Medical and Dental Schools (now part of King's College London), he specialized in stroke and asthma epidemiology. Having established the undergraduate (BDS) course in dental statistics, he wrote the introductory text *Dental Statistics Made Easy*, which has run to three editions. With Peter Sprent, he co-authored the third and fourth editions of *Applied Nonparametric Statistical Methods*. He has been Editor of the applied statistics journal *The Statistician*.

He joined the Centre for Research in Public Health and Community Care, University of Hertfordshire in 2013. Having worked in the areas of adolescent behaviour and epilepsy risk, he is currently part of a UK National Institute for Health and Care Research (NIHR) funded team evaluating public health interventions. His statistical expertise and interests include capture-recapture methods, proportional hazards regression, kappa statistics and the historical development of statistical methods.

Neil Spencer is Professor of Applied Statistics and Head of the Business Research Unit and Statistical Services and Consultancy Unit in Hertfordshire Business School at the University of Hertfordshire. He completed a BSc in Applied Statistics at the University of Reading before moving to the University of Southampton for an MSc in Social Statistics and then to Lancaster University for a PhD in Applied Statistics. He then became a lecturer at Staffordshire University, where he first became involved in consultancy, and moved on to the University of Hertfordshire. He is a Chartered Statistician.

He has undertaken research in a variety of subjects, from Victorian censuses to value-added school league tables, paramedics treating patients whilst wearing protective equipment, health behaviour surveys of school children, compassion in education, creation of family memories and surveys of gig economy workers across 13 European countries. His consultancy work has included testing the randomness of National Lottery machines (for over two decades), parish plan surveys, appearing as an expert witness, methods for in-service testing of utility metres and others, for a range of clients large and small. He is the author of the books *SAS Programming: The One-Day Course* and *Essentials of Multivariate Data Analysis*.

Peter Sprent started out as Tutor, then Lecturer in Mathematics at the University of Tasmania, where he taught undergraduate students and provided statistical support within the University and to government departments and agencies. A sabbatical at the renowned British agricultural institution, Rothamsted Experimental Station, resulted in an appointment at East Malling Research Station, where he focused on the application of statistical methods to agricultural science. Moving to the University of Dundee, he developed the provision of teaching in statistics and headed up the statistical consultancy service within the university. His research areas included statistical regression, the analysis of small experiments, and the mathematics of size and shape. He was appointed to a personal Chair and continued working at the University of Dundee until his retirement when he was awarded an Emeritus Professorship.

He devoted his later life to writing texts, mostly on nonparametric statistics. In addition, determined to raise the understanding of statistics in society as a whole, he published the books *Taking Risks: The Science of Uncertainty* and *Understanding Data*. With his wife, Janet, a botanist, he co-produced the text *Nitrogen Fixing Organisms: Pure and Applied Aspects* and a guide to the mountains of north-west Scotland.

1

Basic Concepts of Statistical Inference

1.1 Setting the Scene

The use of statistical methods is an important component in the accumulation of knowledge about a *population* through what has been described as the cycle of research. A research question is formulated, and a relevant population for its investigation is identified. The question will be concerned with particular characteristics of the population, for instance the mean height of adult men. It is rarely feasible either in terms of staff resources or money to include every relevant member of the population (e.g., all adult males), so a *sample* is chosen instead. Statistical analyses are conducted on the sample selected, the results of which are used to deduce information about the original population. This part of the process is referred to as *statistical inference*. As the sample generally forms only a small proportion of the population, the information obtained, for instance about mean height for adult males, will not be exact. However, the sample should give some indication of plausible values. Results for mean male height might indicate, for instance, that the population value could reasonably be between 170 and 180 cm. Once this study has been completed, more information can be found out about this or other populations by conducting further studies in the cycle: population → sample → analyse → deduce → population, etc.

Almost always, characteristics take values that differ between individuals, so they are generally referred to as *variables*. Noting down the values that a variable takes for a series of individuals will give an indication of which values are most likely. Values can be illustrated graphically as a histogram, giving a profile of relative frequencies. As the sample size increases, this profile becomes smoother and approaches the shape for the relative frequencies within the population. This shape, which can generally be described concisely using algebra, is known as the *probability density function* of the variable. A well-known example is the bell-shaped *normal* distribution for which most individuals take a value close to the mean with just a few relatively high and relatively low values, although technically any value, positive or negative, is possible. Many variables, such as height and weight, cannot take negative values, so their distribution cannot be exactly normal, although in certain situations, it can take a close approximation. Another example is the *uniform* distribution, for which values are equally likely within a specific interval (e.g., between 0 and 100), but none occur outside of the interval.

One issue that individuals unfamiliar with statistics might find off-putting is the use of algebra. Introductory texts are available within which the use of algebra is avoided. This raises the question as to whether algebra can be dispensed with altogether. The truth is that in mainstream statistics, algebra is not only useful but also necessary. When introducing statistical ideas, it may be helpful to refer to the sample mean in words, but for written

DOI: 10.1201/9780429326172-1

documents, it is more convenient to indicate the mean by the symbol $\overline{x}$. For the process of calculating the mean (adding up all the values and dividing by the number of observations), using an algebraic
equation is much more succinct:

$$\overline{x} = \frac{\sum_{i=1}^{n} x_i}{n} \tag{1.1}$$

where n is the number of observations, $x_1, x_2, x_3, \ldots, x_n$ are the n observations and Σindicates adding up the values of the n observations.

As a further example, consider the variance for the sample, a measure of variation. The procedure can be explained in words as follows: subtract the sample mean from each observation to create a set of differences, square each difference and add them together, then divide by the sample size minus one. In algebra, this is written as:

$$s^2 = \sum_{i=1}^{n} (x_i - \overline{x})^2 / (n-1) \tag{1.2}$$

The benefit of algebra in statistical work rapidly becomes clear. An additional case for the use of algebra is that it is transferrable between languages, facilitating international communication between researchers.

In general, algebra referring to a population uses elements of the Greek alphabet; sample characteristics are represented by Latin symbols. For instance, the population mean is indicated by the symbol μ, the sample mean by $\overline{x}$; the variance is shown by σ^2 for a population and by s^2 for a sample. For the standard deviation (the square root of the variance), the symbols are σ (population) and s (sample), respectively.

Readers for whom statistical concepts, methods and analysis are either unfamiliar or a distant memory will benefit from working through a clearly laid out general text (e.g., Spiegel and Stephens, 2017) along with a book on statistical distributions such as Bury (1999).

SUMMARY POINTS:

- The statistical analysis of data from a sample for the purpose of making deductions about that population is known as statistical inference.
- The distribution of the relative frequencies for a variable is known as the probability density function.
- The use of algebra is essential in all but the most basic statistical work.

1.2 Populations and Samples

When making deductions about a population based on a sample, a common assumption is that the observations constitute a *random sample* from the population. This means that for sampling without replacement, each sample of a particular size has an equal probability of being selected. For sampling with replacement, each member of the population has the same probability of being selected each time an item is taken. If random sampling is not the case, the sample is likely to be *biased* relative to the population, and it is difficult to generalize the findings obtained from the sample to the population as a whole.

How far the findings can be generalized depends on the context of the study. Suppose that in a study of the prescription of medication by general practitioners, a sample of the general practices in an English town is studied. A useful variable to record would be the proportion of patients attending each practice who are issued with a prescription. Strictly speaking, the findings would apply only to the practices within that town, although it might be reasonable to assume that other towns with a similar age distribution and socio-economic characteristics have similar patterns of prescription. However, it would be unwise to generalize the deductions to large cities or thinly populated rural communities. The findings are unlikely to be true in areas of Britain having different criteria for free prescriptions (e.g., Wales) and in countries with different forms of health care such as the United States or Germany.

The abstract notion of random sampling from an *infinite* population (implicit in most inferences based upon the population taking a normal distribution) often works well in practice but is never completely true, as some individuals will be easier to identify than others. In addition, individuals selected in the future may differ in their characteristics from those currently available even within the same population. At the other extreme, cases may be so small in number that the sample constitutes the whole population, for instance with an extremely rare disease in which patients are highly motivated to participate in research.

Another important assumption necessary in applying many statistical procedures is that the observations are *independent* of each other. This implies that the value taken by any one observation is uninfluenced by the values of other observations. Avoiding this difficulty may require careful planning, particularly in community studies for which individuals within the same family have similar views about participation or nonparticipation in the study.

SUMMARY POINTS:

- A common assumption made in research is that the observations form a random sample from the population. In practice, this is often not the case.
- If the method of sampling is not random, the sample is likely to be biased relative to the population.
- Many statistical procedures require the assumption that the observations are independent of each other.

1.3 Types of Variable

1.3.1 Binary

A variable is described as being binary if it can only take two values (e.g., outcome of tossing a coin: head, tail). One of the outcomes (in coin tossing typically a head) is referred to as a "success". The distribution of the number of successes obtained from a sample of independent observations drawn from a binary variable leads to the commonly encountered *binomial* distribution (Bury, 1999).

1.3.2 Nominal

Observations for a nominal variable can be divided into more than two categories, but any numbers associated with them cannot be ordered in a meaningful way; they have no numerical significance (e.g., country of residence: England, France and Germany). Nominal variables can be summarized by the mode, the category containing the highest number of observations, which is a measure of *central tendency.*

1.3.3 Ordered

A variable is described as being ordered (or ordinal) if it can take more than two values and the categories can be ordered (e.g., reported level of health: excellent, good, fair and poor). Here, there is a natural sense of downward ordering in reading the categories from left to right.

For any ordered variable, an appropriate measure of central tendency (average) is the median. Put simply, the median is the value relative to which 50% of the observations in the sample are lower and 50% are higher. The sample median (M) is found by arranging the observations in order of size. The middle value is taken if the number of observations is odd, and the mean of the middle pair of observations is calculated if the number of observations is even.

In a similar way, the lower quartile (Q_1) is the value relative to which 25% of the observations in the sample are lower and 75% are higher, and the upper quartile (Q_2) is the value relative to which 75% of the observations are lower and 25% are higher. For an ordered variable, the difference between the two quartiles or interquartile range ($Q_2 - Q_1$), and the semi-interquartile range, $½(Q_2 - Q_1)$, are appropriate measures of *spread.*

The median, quartiles and deciles (points based on multiples of 1%) are examples of what are known as *percentiles.*

1.3.4 Quantitative

A quantitative variable is one for which differences between categories have numerical meaning as with weight and height. Quantitative variables can consist of whole number values only (e.g., number of visits to a doctor within a period of 12 months) or be continuous with accuracy restricted only by the method of measurement (e.g., pH value).

The arithmetic mean ($\bar{x}$) is an appropriate average for symmetrical quantitative variables, but not for those with a clear lack of symmetry (asymmetric or skewed distributions). Such distributions are referred to as being *positively skewed* if the asymmetry exists only because of high extreme values, and *negatively skewed* if the asymmetry is solely due to low extreme values. For positively skewed distributions (e.g., annual income in most societies), the mean is greater than the median, and for negatively skewed distributions (e.g., length of gestation at birth), the mean is less than the median. With the mean being susceptible to extreme values (also known as outliers), where they are evident, the median (M) is preferred. Alternative approaches include adjusting outliers to make them more plausible or removing them altogether (Section 2.5).

For a symmetrical quantitative variable, the standard deviation (s) is a suitable measure of spread; the interquartile range ($Q_2 - Q_1$) is preferred for skewed distributions. The range (highest value – lowest value) is seriously inflated by a single outlier and is not recommended. The mean should be reported with the standard deviation; the median and interquartile range should be given together.

Observations from quantitative variables can be labelled in order of size to give what are known as *order statistics* (Chapter 2). This procedure is known as *ranking*. The lowest value is given a rank of 1, the next lowest a rank of 2, etc. The highest value has a rank equal to the sample size. If two or more observations share the same value, the average of the ranks that would have been assigned to those observations had they differed slightly is given to each.

SUMMARY POINTS:

- Variables may be binary, nominal, ordered or quantitative.
- In choosing summary measures for a distribution, for instance in deciding between the mean and the median, the shape of the distribution needs to be taken into account.

1.4 The Normal Distribution

1.4.1 Symmetry

For the normal distribution, a particular example can be distinguished from others by the values of its mean μ and standard deviation σ. Entities such as these are known as *parameters*. However, whatever the values of these parameters, a normal distribution is continuous and symmetrical. Around 95% of the distribution lies within two standard deviations of the mean. Given the mean and standard deviation of a normal distribution, it is straightforward to identify outliers.

It is sometimes possible to show that a distribution is non-normal from the values of the mean and standard deviation. If the mean minus twice the standard deviation is negative and the data must be positive (e.g., lengths of inpatient admissions in days), the assumption of normality leads to a contradiction. In order to accommodate the lower limit of zero, the data must be positively skewed. The relatively high value for the standard deviation compared to the mean is most likely due to the presence of large outliers.

Example 1.1

In a study of alcohol consumption by men living in Belfast and France, Ruidavets et al. (2010) found that for the 1,229 men in Belfast who were regular drinkers, daily alcohol consumption (g) had a mean of 40.2 and a standard deviation of 39.9. To a good approximation, the mean minus two standard deviations = $40 - (2 \times 40) = -40$.

For drinkers, daily alcohol consumption must be greater than 0 so an observation of -40 is impossible. Therefore, alcohol consumption (g/day) cannot follow a normal distribution.

1.4.2 The Central Limit Theorem and the Standard Error

Making deductions from data is facilitated by what is known as the *central limit theorem*. If large samples (often of size >30) of independent observations are taken from a population having any distribution, then as samples continue to be drawn:

a. the sample means will have an approximately normal distribution
b. the mean of the sample means will be approximately the same as the true population mean
c. the standard deviation of the sample means (*standard error* or se) can be calculated by dividing the population standard deviation by the square root of the sample size.

SUMMARY POINTS:

- A specific normal distribution is defined by two parameters, the mean and the standard deviation (Section 1.4.1).
- The distribution is always symmetrical and continuous (Section 1.4.1).
- The central limit theorem states some of the characteristics of the sample mean in larger samples (Section 1.4.2).

1.5 Hypothesis Testing

Statistical inference plays an important part in ascertaining the plausibility of statements made regarding a specific population, which may or may not be true. An assertion of this nature, which should be made before the data are collected (*a priori*), is called a *null hypothesis* and denoted by H_0. This should be linked to an *alternative hypothesis* H_1 regarding the population. This is considered should there be little evidence to support the null hypothesis. The statement H_0 is called the null hypothesis because it is assumed initially that the asserted value is correct. Formally, this is stated as:

$$\text{Test } H_0 : \mu = \mu_0 \text{ against } H_1 : \mu \neq \mu_0 \tag{1.3}$$

The onus is on the investigator to demonstrate to a cautious audience that there is evidence against the null hypothesis. The rationale behind the null hypothesis follows the philosophy of Karl Popper, who suggested that most hypotheses can never be proved and knowledge increases through the falsification of currently held ideas (Popper, 1980). Growing evidence against a widely held view eventually leads to it being rejected in favour of an alternative, which may in turn be cast aside by future discoveries.

Although null hypotheses are typically applied in the comparison of groups, such as individuals resident in the United Kingdom, France and Germany, to keep things simple, a population consisting of a single group will be considered in this chapter. The null hypothesis could be that the mean of the population has a particular value μ_0, with an alternative that it is not equal to μ_0. The alternative hypothesis is referred to as two-tailed since a difference from μ_0 is of interest whatever the direction (i.e., whether $\mu > \mu_0$ or $\mu < \mu_0$). In contrast, with a one-tailed test, only a difference in one direction is of interest (e.g., $\mu > \mu_0$).

Generally, a two-tailed test is preferred. Consider an investigation into the proportion of males who access health services at least once per year. Suppose that the null hypothesis value to be tested is 0.4 despite the researchers being almost certain that this is an overestimate. Using a one-tailed test with the alternative hypothesis based on their prior impression (the proportion of males is less than 0.4) would imply that a finding in the opposite direction (a proportion of more than 0.4) would not be regarded as evidence against H_0, however great the difference.

Once the necessary observations have been collected, they need to be analysed and interpreted in the light of the null hypothesis. Suppose that for a population of adult males the mean height is claimed to be 175 cm. Even if the null hypothesis is true, it is unlikely that a sample would have precisely this value for the mean. Some difference from the true value can be expected due to random variation between the heights of individual males. The important issue here is the size of the difference between the sample mean and the null hypothesis value for the mean (175 cm) that makes it reasonable to cast doubt on the null hypothesis. If out of a sample of 100 males, the mean is, say, 176 cm, this does not necessarily indicate much evidence against the null hypothesis. However, a sample mean of 170 or 185 cm would point more strongly towards the existence of a genuine difference.

Apart from a few exceptions (e.g., if one case of measles is found in a village when it has been assumed that everyone is measles-free), it is not possible to disprove a null hypothesis, only to say that there is strong evidence against it. Intuitively, the evidence against the null hypothesis should be weak if the differences between an assumed population mean and an observed sample mean are small and strong if they are large. The measure that is used to quantify this is known as the *P-value*.

The *P*-value is the probability of obtaining *at least* the difference observed between the sample and null hypothesis values of a measure (e.g., the mean) *assuming that the null hypothesis is true*. Usually, it is the magnitude of the difference (regardless of whether it is positive or negative) that is important.

Pearson, who developed the concept of the *P*-value (Pearson, 1900), tabulated these values to assist in their interpretation. However, Fisher (1925) favoured the use of *P*-value thresholds to indicate levels of *statistical significance* and it became standard practice to describe *P*-values of 0.05, 0.01 and 0.001 as 'significant', 'highly significant' and 'very highly significant', respectively. If the *P*-value exceeded 0.05, the result was described as *not significant* ($P > 0.05$) and H_0 accepted. Tables commonly found in texts on statistics and research methods from the 1920s onwards facilitated the use of these arbitrary benchmarks.

It is preferable to state *the actual P*-values and subjectively interpret the *strength of evidence* against H_0. For instance, a *P*-value of 0.002 represents *strong evidence* against the null hypothesis, whereas a value of 0.2 would indicate *weak evidence* against the null hypothesis, with values around 0.05 indicating moderate evidence. Figure 1.1 compares the traditional 'fixed-level' approach to significance, with the more flexible assessment based on strength of evidence using a *P*-value. Rigid rules based on thresholds should be avoided; it is illogical that similar *P*-values (e.g., 0.049, 0.051) are interpreted in entirely different ways. In addition, it is tempting to conclude (erroneously) that if $P > 0.05$ the null hypothesis must be true, particularly if $P = 1$.

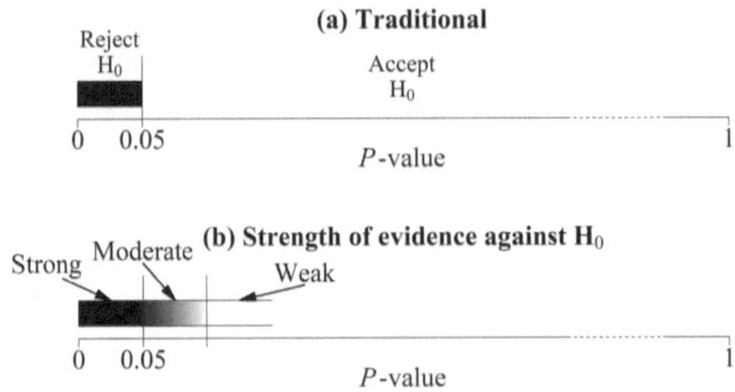

FIGURE 1.1
(a) A fixed-level approach to significance testing and (b) assessment based on strength of evidence where some flexibility in interpretation may be allowed for *P*-values close to 0.05.

Example 1.2

A university runs a range of 3-year undergraduate degree courses. Most students enter at 18 years, so the mean age across the course as a whole might be thought of as around 20. However, due to the presence of mature students, aged 25 or more, the overall mean age is actually equal to 23. An external investigator who is unaware that the courses admit mature students specifies a null hypothesis H_0 that the mean undergraduate student age is 20.0. A sample of 50 undergraduates is selected and found indeed to have a mean age of 20.0. With the sample mean equal to the null hypothesis mean, the *P*-value for the test is equal to 1, the greatest possible value. Nonetheless, H_0 is clearly untrue.

A common misconception is that a low *P*-value indicates a departure from the null hypothesis that is of *practical importance*. For larger samples, a relatively small divergence from the null hypothesis can be associated with low *P*-values; in describing findings, it is important to consider the size of any divergences and not solely report *P*-values.

Whether or not the null hypothesis is true, errors can be made in deciding between the null and alternative hypotheses. These errors are most clearly explained through the traditional benchmark approach to hypothesis testing. If the alternative hypothesis is chosen when the null hypothesis is true, this is known as an *error of the first kind* or Type I error. The probability of a Type I error occurring, denoted by α, is equal to the *P*-value. If the criterion for rejection is $P < 0.05$, then $\alpha = 0.05$. If the null hypothesis is retained when the alternative hypothesis is true, an *error of the second kind* or Type II error has been made. The probability of a Type II error occurring is denoted by β. This type of error is more likely to occur if the null and alternative hypotheses are similar and/or the sample size is small.

The probability of correctly selecting the alternative hypothesis $(1 - \beta)$ is known as the *power* of the test. This is likely to be high if the sample size is large, if the *P*-value for rejection is relatively high (e.g., 0.05 rather than 0.01) and if there is a clear difference between the null and alternative hypotheses. Power is also influenced by how well the observations match the assumptions made in performing the statistical methods involved.

In designing a study, the desired power is generally chosen in advance and the appropriate sample size is calculated. Generally, the power of the test is chosen as 0.8 or 0.9 (the Type II errors then being 0.2 and 0.1, respectively). The value 0.95 is rarely selected as a study having such a power can require a sample size that is too large to be practical given the resources available. In addition, rejecting a correct null hypothesis is considered to be a more serious error than failing to choose a correct alternative hypothesis. A correct alternative may be identified by a subsequent study, whereas a rejected null hypothesis is unlikely to be revisited.

A serious difficulty remains. This is because with small data sets, one may never observe sufficiently small *P*-values to justify rejection of H_0 even when it is not true.

Example 1.3

Suppose a coin is tossed five times to test the hypothesis

$$H_0 : \textit{the coin is fair,}$$

against

$$H_1 : \textit{the coin is biased.}$$

i.e., is either more likely to fall heads or more likely to fall tails.

In five tosses, the strongest evidence against H_0 is associated with the outcomes 5 heads or 5 tails. Under H_0, the probability of obtaining 5 heads = 0.5 × 0.5 × 0.5 × 0.5 × 0.5 = 0.03125; the probability of obtaining 5 tails is the same. Hence, the strongest evidence against H_0 occurs with a probability of 0.03125 × 2 = 0.0625. This is the smallest attainable *P*-value when $n = 5$, so H_0 is never rejected at a conventional $P = 0.05$ level whatever the outcome of the five tosses – even if the coin has two heads. That the experiment is too small is the only useful information given by the *P*-value in this example.

This weakness of hypothesis testing, together with the perpetration of myths, such as equating acceptance of a hypothesis to prove that it is true, has led to justified criticism of what is sometimes called the *P*-value culture. Krantz (1999) and Nelder (1999) both highlight dangers arising from inappropriate use of, and misunderstandings about, the meaning of a *P*-value.

SUMMARY POINTS:

- A null hypothesis regarding a population needs to be made before the data are collected.
- The null hypothesis generally represents currently held ideas. The role of the investigator is to search for evidence against the null hypothesis.
- If evidence against the null hypothesis is sufficiently strong, it may be rejected in favour of an alternative hypothesis.
- A two-tailed null hypothesis is generally preferred.
- A *P*-value is based on the assumption that the null hypothesis is true.
- Where possible, the actual *P*-values should be reported.
- The probability of correctly selecting the alternative hypothesis is known as the power of the test.

1.6 Estimation

1.6.1 Confidence Intervals

The sample mean is generally an appropriate estimate of the population mean. A measure of the *precision* of this estimate is required as the sample mean varies between samples. The confidence interval is one such measure. For samples of a reasonable size, there is a simple relationship between the standard error of the mean and the 95% *confidence interval* for the *population* mean. From the central limit theorem, the sample mean will approximately have a normal distribution. The standard deviation of the sample means or standard error represents the variation in the values of the sample mean when samples of a particular size are taken. From the properties of a normal distribution, 95% of the values of the sample mean will be within 1.96 standard errors of the population mean. From this, it can be deduced that if an interval is constructed from 1.96 standard errors below the sample mean to 1.96 standard errors above the sample mean, it will contain the population mean with a probability of 95%.

A 95% confidence interval for a mean is often (correctly) defined as a range of values within which one can be 95% confident that the true mean lies. If the confidence interval contains the population mean indicated in the null hypothesis, there is no strong evidence against the null hypothesis. On the other hand, a value specified in the null hypothesis that lies well outside the confidence interval indicates strong evidence against H_0.

A common interpretation of a 95% confidence interval is in terms of the property that if such intervals are formed for repeated samples, then in the long run, 95% of these intervals would contain (or cover) the true but unknown mean.

Confidence intervals are useful because their width/narrowness give a feel for the imprecision/precision of the estimate around which they are based. In addition, they assist in deciding whether a significant result is likely to be of practical importance.

In more rigorous language, a useful way of looking at the distinction between hypothesis testing and estimation is to regard testing as answering the question:

- Given a hypothesis H_0: $\theta = \theta_0$ about, say, a parameter θ, what is the probability (P-value) of getting a sample as or less likely than that obtained if θ_0 is indeed the true value of θ?

Estimation using a confidence interval answers the question:

- Given a sample, what values of θ are consistent with the sample data in the sense that they lie in the confidence interval?

1.6.2 Statistical Significance and Practical Importance

Prior to the study being conducted, the lowest difference from θ_0 of practical importance should be defined. Assume here that the null hypothesis value of θ_0 is equal to 0; similar reasoning applies for other values of θ_0. The confidence interval may include values that represent a difference of practical importance but may also include zero (indicating no evidence against the hypothesized value for the population mean). On the other hand, the confidence interval may not contain zero, but all the likely values for the difference might be too small to be of practical importance. Findings such as these can give unclear messages. Unless the 95% confidence interval contains only values having practical importance, the findings have limited application.

Example 1.4

Drugs A and B for treating hypertension are considered effective if they reduce mean systolic blood pressure by a minimum (in terms of practical importance) of 15 mm Hg. Suppose that for these drugs the 95% confidence intervals for mean blood pressure reduction when applied in similar populations are:

Drug A: (−3, 35)

Drug B: (21, 25)

A two-tailed hypothesis H_0: *drug does not change blood pressure* would not be rejected at a 5% significance level for Drug A (the confidence interval contains the null hypothesis value zero). However, the finding is of both statistical significance and practical importance for Drug B, as the confidence interval values exclude not only 0 but 15 mm Hg as well.

The 95% confidence interval for Drug A is very wide and therefore imprecise; one can be 95% confident that the true reduction in mean lies between −3 (an increase) and 35 mm Hg. For Drug B, the 95% confidence interval is much shorter and has much greater precision. Despite the upper limit of the 95% confidence interval being higher for Drug A than Drug B, the finding for Drug B is of much greater value as it indicates practical importance.

The art of designing experiments is to take enough observations to ensure a good chance of detecting with reasonable precision departures from H_0 of practical importance, but to avoid wasting resources by taking so many observations that trivial departures from H_0 provide strong statistical evidence against it. Practical design of experiments is best done with guidance from a trained statistician, although many statistical software packages include programs giving recommendations in specific circumstances.

SUMMARY POINTS:

- Confidence intervals contain possible values for the summary measure in the population (not the sample) (Section 1.6.1).
- A 95% confidence interval for a summary measure is defined as the range of values within which one can be 95% confident that the population value for the measure lies (Section 1.6.1).
- It is not usually sufficient to demonstrate statistical significance. Differences from the null hypothesis value should be of practical importance to a professional in the relevant subject area (Section 1.6.2).

1.7 Ethical Issues

1.7.1 Research Ethics

Ethical considerations are important both in general working practices (Herring, 2022) and in the planning and conduct of investigations (Hutton, 1995). The main principles are respect for autonomy, nonmaleficence, beneficence and justice. Most research proposals need to be assessed by ethical committees before being approved. This applies particularly in medicine, but increasing attention is being given to ethical issues in environmentally sensitive fields like biotechnology and the social sciences, where questions of legal rights or civil liberties may arise. Williamson et al. (2000) discuss the role of statistics and statisticians in research ethics committees. The CONSORT Statement for good practice in the design, execution and reporting of clinical trials is described by Schulz, Altman and Moher (2010).

It is unacceptable to study some issues by allocating individuals to possible groups at random. For instance, in a study of the effects of smoking on health, one could not ask individuals to smoke or to abstain from smoking. This disregards the autonomy of the study participants. An individual's choice of whether to smoke or not must be respected, and an alternative type of study should be planned to make this possible.

When planning a new study, a systematic review should be performed to identify what is currently known about the subject under investigation. It is unethical to conduct research that ignores previous relevant work because time, money and scarce resources will not be used to the best effect. Nevertheless, results of literature searches need to be interpreted with caution. Studies with interesting findings are more likely to appear in print, leading to publication bias [see Easterbrook et al. (1991)]. When there is little relevant prior knowledge, it may be prudent to conduct an initial pilot study to highlight potential problems that might arise. Results from a pilot study can also be helpful in choosing an appropriate number of participants for the main study. A sufficiently large number should be involved at the pilot stage to have a reasonable chance of finding the expected difference between the two groups if it really exists. The intended method of statistical analysis also influences the sample size requirement. Small studies often fail to yield useful findings and are thus a poor use of resources. On the other hand, resources are wasted if more participants are recruited than needed. In either situation, more individuals are recruited than is strictly necessary. Careful planning should consider the composition of the sample with respect to age, sex, ethnic group, etc., as this will enable problems under investigation to be answered more effectively.

It is generally good practice to incorporate early stopping rules into a study. If it becomes clear at an early stage of an investigation that a new treatment is much better than the established alternative, the study should be closed. A similar decision should be made if the new treatment quickly shows a highly increased risk of harmful side effects. Continuing a study could deprive patients of a more effective treatment or expose participants to unnecessary risks from side effects. Early stopping recommendations are reached through a small number of planned interim analyses that take place throughout the period of the study. Cannistra (2004) discusses ethical issues around the application of early stopping rules.

Data collected in studies should be kept confidential. In the United Kingdom, for example, data records held in whatever form should adhere to the principles laid down in the General Data Protection Regulation 2018. Data used for statistical purposes should not contain sensitive information such as participants' names or addresses.

Limited availability of a treatment for experimental use may create ethical problems. Suppose that there were high hopes that a new drug might greatly relieve suffering in severe cases, but only enough doses were available to treat six patients. The principles of beneficence and justice suggest that the six patients to receive the drug should be those with the most severe symptoms. In a situation like this, the drug may reduce suffering, but such patients may still, after treatment, be ranked as less well than patients receiving an alternative treatment because, although their condition may have improved, their symptoms may still be more severe than those of patients receiving the alternative treatment. In this situation, any statistical analysis should be based on some measure of "degree of improvement" shown by each patient. At the other extreme, an experimenter might allocate the new drug to the patients with the least severe symptoms. From a research point of view, this is misleading, as even if it were ineffective or no better than an existing treatment, these patients may still show less severe symptoms. However, if it is likely that only patients in the early stages of the disease will benefit, it is ethically more appropriate to give the new drug to these patients.

Even with random allocation, and strong evidence for abandoning a hypothesis of no treatment effect, a statistically significant outcome based on $P < 0.05$ may be of no practical importance, or there may be ethical reasons for ignoring it. A doctor might have concerns about offering a new treatment if it only prolonged by a few days the life expectation of terminally ill patients in considerable distress, but may, from the principle of beneficence, prescribe it if it substantially improved survival prospects and quality of life.

A researcher who performs a number of competing tests, each producing a different *P*-value, but only publishes those results with 'interesting' *P*-values ignoring the other findings, is guilty of unethical suppression of evidence. Ethical considerations may influence not only how an experiment is carried out (the experimental design) but also what inferences are possible and how these should be made.

1.7.2 Ethical Considerations Regarding Study Participants

In medical investigations, each potential participant should receive a written information sheet outlining the main points about the study. The autonomy of the patient should be respected, and the patient should only decide on whether to enter the trial following careful consideration of the information provided. Ideally, they should be allowed at least 24 hours to absorb the information and clarify issues. This is particularly important with tests for inherited diseases that only become evident in later life. A positive finding may cause distress, have serious implications for family members and prejudice applications for life assurance. On a more positive note, someone may agree to participate in the belief that future patients might benefit from their contribution to the research.

If after due consideration a patient is willing to be involved in the investigation, they should give witnessed informed positive written consent (where willingness to participate is not assumed until this is received) to the investigator(s) before being entered and they should be allowed to withdraw from the investigation at any time.

All available information about the possible efficacy and side effects of treatments involved in the study should be given to the patient. In practice, not all patients will understand, or even wish, to receive details beyond those given in the information sheet, particularly in a sophisticated trial. In this situation, the patient should be given a choice about what information is supplied. Once a trial has been completed, patients who feel that they have received an effective treatment for their health problem may wish to continue with it. Financial constraints and/or the concerns of the patient's general practitioner may prevent long-term use of the treatment; this should be discussed in advance as part of the patient information. Ethical problems may preclude the use of the same patients to compare two treatments. For instance, in a comparison of two methods of treating oral cancer, both of which involve radiation, the estimated dose of radiation from the combined treatments may be considered unacceptably high for the patient.

Patient and public involvement (PPI) in research enables individuals from the study population to contribute to the design and conduct of the investigation. Instead of being viewed simply as participants, those who become engaged in PPI provide input based on their own personal experiences and so develop a degree of ownership in the research. Involvement may include commenting on patient information leaflets, acting as project advisors and being co-applicants in research projects. These activities should incorporate the provision of feedback between PPI representatives and researchers (Mathie et al., 2018). Some research groups have established an ongoing PPI team for the provision of input into new projects as they evolve. Co-production, in which researchers, practitioners and the public work together, sharing responsibility throughout the project is considered best practice. Increasingly, funding bodies require evidence of a high level of PPI in applications for research support.

SUMMARY POINTS:

- It is unethical to ignore previous relevant work (Section 1.7.1).
- A sample should be large enough to detect true differences of practical importance, bearing in mind that the resources required should be kept to a minimum (Section 1.7.1).
- Early stopping rules should be incorporated into a study to address large differences between treatments and the occurrence of unexpected harmful side effects (Section 1.7.1).
- Data collected in studies should be kept confidential (Section 1.7.1).
- Participants should give informed positive written consent prior to entering an investigation (Section 1.7.2).
- Appropriate patients and members of the public should be involved in the design and conduct of research studies (Section 1.7.2).

1.8 Exercises

1.1 In a group of 145 patients admitted to hospital with a stroke, the number of cigarettes smoked per day had a mean of 10 and a standard deviation of 15. Explain why their cigarette use does not follow a normal distribution. Is this finding surprising?

1.2 Following a television campaign about the risks of smoking tobacco, the cigarette consumption of a group of 50 smokers decreases by a mean of 5 cigarettes per day with a standard deviation of 8. Explain why the reasoning of Exercise 1.1 cannot be used to show that this distribution is not normal.

1.3 In Section 1.5, it was explained that five tosses of a coin would never provide evidence against the hypothesis that a coin was fair (equally likely to fall heads or tails) at a conventional 5% significance level. Show that just one further toss would resolve this difficulty, stating the calculated P-value.

1.4 A national chain of supermarkets sells a particular pineapple juice product. Following an increase of 10% in the price of the pineapple juice, mean weekly sales per store based on a representative sample of stores fell from 41 to 35 cartons per week, with a 95% confidence interval for the decrease from 2.5 to 9.5. Give the null hypothesis that would be appropriate for this study and interpret the 95% confidence interval.

1.5 Explain the concept of negative consent in the context of a clinical trial.

2

Fundamentals of Nonparametric Methods

2.1 Parametric and Nonparametric Methods

2.1.1 Approaches to Inference

A probability density function, as described in Chapter 1, can be used to represent the distribution of a population. As hinted in Section 1.4.1, distributions are associated with parameters, and a particular distribution has a specific shape uniquely determined by the value(s) of its parameter(s). For instance, a normal distribution must have a mean equal to zero and a standard deviation equal to 1 in order for it to take the shape of a standard normal distribution. Other combinations of values for the mean and standard deviation lead to different specific shapes that have the general normal form, giving rise to the *family* of normal distributions. A particular distribution is often indicated by the notation $N(\mu,\sigma^2)$. Other families of distributions include the uniform (Section 1.1), binomial, Poisson and exponential distributions (Bury, 1999). Statistical deduction based on estimating the values of one or more parameters is known as *parametric inference.*

It may be reasonable based on past experience or current knowledge to assume that observations come from a particular family of distributions. Many measurements, including those for height and weight, generally have a distribution that, apart from a few extreme values, is close to normal. For larger samples, the central limit theorem justifies the use of the normal distribution in making deductions about a population from a sample, even if the population is non-normal. These are described as *asymptotic approximations* as the accuracy of the approximation increases as the sample size becomes larger.

Parametric inference may not always be appropriate, and in some situations, it is impossible. For example, records of school examination results may only give candidate attainment in the form of banded and ordered grades designated Grade A, Grade B, Grade C, etc. Given two different schools and information on the grade frequencies for each, it may be of interest to determine whether there is a genuine difference in performance that might be attributed to different methods of teaching or to the ability of one school to attract more able students. There is no obvious family of distributions associated with these data, and there are no clearly defined parameters about which inferences can be made. Two terms are in common use for the type of inferences that can nevertheless be made. They are *"nonparametric"* and *"distribution-free"*. The description "nonparametric data" is incorrect as these terms refer to the method of analysis.

The use of the terms nonparametric and distribution-free is not consistent, and they are often interchanged. This is of no great consequence in practice, and to some extent, it simply reflects historical developments. There is not even universal agreement about what constitutes

DOI: 10.1201/9780429326172-2

a parameter. Quantities such as μ and σ^2 appearing in the density functions for the normal family are unquestionably parameters. The term is often used more widely to describe any population characteristic within a family such as a mean, median, quantile or range.

Situations can arise where observations are composed of a deterministic and random element, and constants occurring in the deterministic element are to be estimated. Such constants are also sometimes called parameters. In nonparametric, or distribution-free, methods, inferences about parameters in this wider sense can be made.

In this situation, the name distribution-free is more appropriate if interest is in parameters in the broader senses mentioned above. Some procedures are both distribution-free and nonparametric in that they do not involve parameters even in the broader use of that term. The above example involving examination grades falls into this category.

Historically, the term nonparametric was in use before distribution-free became widespread. There are procedures for which one name is more appropriate than the other, but as in many areas of statistics, terminology does not always fit procedures into watertight compartments. A consequence is the spawning of hybrid descriptions such as *asymptotically distribution-free* and *semiparametric* methods. There is even some overlap between descriptive statistics and inferential statistics, evident in a practice described as *exploratory data analysis*. As shown in Section 2.4 and elsewhere, sensible use of exploratory data analysis may prove invaluable in selecting an appropriate technique, parametric or nonparametric, for making statistical inferences.

Designating procedures as distribution-free or nonparametric does not mean that they are assumption-free. In practice, nearly always some assumptions are made about the underlying population distribution. For example, a distribution may be assumed to be continuous and symmetric. These assumptions do not restrict reference to a particular family such as the normal, but they exclude both discrete and asymmetric distributions. Given some data, exploratory data analysis will often indicate whether an assumption such as one of symmetry is justified.

Many nonparametric or distribution-free procedures involve, through the test statistic, distributions and parameters (often the normal distribution). This is because the terms refer not to the test statistic but to the fact that the methods can be applied to samples from populations having distributions only specified in broad terms, e.g., as being continuous, symmetric, identical, differing only in medians, means, etc. The distribution of the appropriate test statistic is the same no matter what the population distribution may be, providing only that it satisfies the broad-term specification. There is a grey area between what is clearly distribution-free and what is parametric inference. Some of the association tests described in Chapters 12 and 13 fall in this area.

2.1.2 The Use of Nonparametric Methods

Some parametric tests do not depend critically on the correctness of an assumption that samples come from a distribution in a particular family. These tests are described as *robust*, and findings are similar whether or not the underlying assumptions are satisfied.

Nonparametric methods are usually more robust than their parametric counterparts. They are widely used with data that can be ranked and are often the only ones available for data consisting of the number of individuals in various categories.

In most statistical problems, no matter whether parametric or nonparametric methods are appropriate, what can be deduced depends on what assumptions can validly be made. An example illustrates this.

Example 2.1

Two machines produce metal rods. For each, 2.5% of all rods produced have a diameter exceeding 30 mm. This condition is met if the first machine produces items having a normal distribution with mean 27 mm and standard deviation 1.53 mm. This is because, for any normal distribution, 2.5% of all items have a diameter at least 1.96 standard deviations above the mean (Section 1.6.1), so 2.5% exceed $27+1.96\times1.53\approx30$ mm. The condition is also met if the second machine produces items with diameters uniformly distributed between 20.25 and 30.25 mm (i.e., with mean diameter 25.25 mm). This follows because any interval between 20.25 and 30.25 mm of width 0.25 mm (in particular from 30 to 30.25 mm) contains a proportion 1/40 (i.e., 2.5%) of total production.

This uniform distribution is unlikely to be met in practice in this context, but the example shows that the proportion of defectives may be the same in two populations, yet each has a different mean and their distributions do not even belong to the same family.

If an additional assumption is made that distributions of diameters for each machine differ, if at all, only in their means, then if the proportion over 30 mm in samples of, say, 200 from each machine is known, it would be possible to test whether the means could reasonably be supposed to be identical. Doing this would not make best use of the data. It would be better to measure the diameter of each item in smaller samples and, then, use an appropriate test.

Means and medians are widely used to indicate where distributions are centred. Both are formally described as *measures of location*. Not all distributions have a mean, but all have a median. For a symmetric distribution, if the mean exists, the mean and the median have the same value. Their values usually differ for asymmetric, or skew, distributions. The Cauchy distribution is a well-known example of a symmetric distribution that has no mean. It has a well-defined median, this being zero for the standard Cauchy distribution.

Tests and estimation procedures are often concerned with location measures, e.g.,

- Is it reasonable to suppose that a sample comes from a population with a prespecified mean or median?
- Do two samples come from populations whose means differ by at least 10?
- Given a sample, what is an appropriate estimate of the population mean or median? How good is that estimate?

Variation or *spread* or *dispersion* is often of interest also. Manufacturers of new cars or computers need not only a good average performance but also consistency, i.e., not too much variation in performance from item to item of the same brand. Each buyer expects his or her purchase to perform as well as those of other buyers of that model. Success of a product often depends upon personal recommendations, so mixed endorsements – some glowing, others warning of niggling faults – are not good publicity.

Dispersion is often assessed by *variance* or *standard deviation* (Section 1.1), but these measures may not exist for all distributions. Moreover, they are not well suited on their own to describe, or compare, the spread of skew distributions. There are both parametric and nonparametric methods for assessing spread or variability. In other situations, how well data conform to some hypothesized population distribution function may be of interest, the appropriate test being one for *goodness of fit*. Tests of association or correlation are also of considerable interest.

Nonparametric techniques may be the only ones available when the information available is limited. For instance, in testing that weights of a large batch of items have a median equal to 2 mg, the only information available might be the number of items in a sample weighing more than 2 mg. If it were difficult, expensive or impossible to get exact weights, an available nonparametric approach may be cost effective. Simple nonparametric methods are also useful when data are in some sense incomplete, like those in Example 2.2.

Example 2.2

In medical studies, the progress of patients is often monitored after treatment, but this may be for a fixed period only due to resource constraints. Ling et al. (2017) reported the survival times in years for 14 retired footballers following a diagnosis of dementia. Assuming that follow-up is restricted to exactly 15 years, the precise survival time is unknown for the longest surviving participant as he was alive at that point; this observation is referred to as being censored.

Survival time (years) were:

5 6 7 8 9 9 9 9 10 11 14 14 14 15*

The asterisk denotes the censored observation.

Is it reasonable to suppose that these data are consistent with a median survival time of 12 years? Censored observations cause problems in many parametric tests, but in Example 2.4, a simple nonparametric test is used to show there is no strong evidence against the hypothesis that the median is 12. For that interpretation to be meaningful and useful, it must be reasonable to assume that the data are a random sample from some population of patients with the disease.

To confirm that the median might well be 12 without additional information is not particularly helpful. It would be more useful to have a 95% confidence interval for the median survival.

SUMMARY POINTS:

- A particular distribution has a shape that is uniquely determined by the value(s) of its parameter(s) (Section 2.1.1).
- In nonparametric inference, no assumption is made prior to the analysis that the samples are associated with any family of distributions (Section 2.1.1).
- A test is robust if it does not depend critically on the correctness of an assumption that samples come from a distribution in a particular family (Section 2.1.2).
- Nonparametric methods are usually more robust than analogous techniques that require assumptions about parameters (Section 2.1.2).
- Nonparametric techniques may be the only ones available if the information is limited (Section 2.1.2).

2.2 Permutation Tests

There is a long history of studies that produce data suitable for analysis by *permutation tests*. The earliest known example dates from around 600 BC. The book of Daniel in the Bible (Daniel 1 vv. 3-16) records that on the orders of Nebuchadnezzar certain favoured children of Israel were to be specially fed on the king's meat and wine for 3 years. Reluctant to defile himself with such luxuries, Daniel pleaded that he and three of his brethren be fed instead on pulse for 10 days. After that time, the four were declared "fairer and fatter in flesh than all of the children which did eat the portion of the king's meat". This indicates that the health of the study "participants" could be ranked (Section 1.3.4), which is needed for a permutation test to be performed. Be warned though that the common-sense conclusion may not be justified here. Allocation to the two groups was anything but random. Daniel and his three brethren may already have been "fairer and fatter" before the experiment began. In addition, the method used for assessing the condition of the participants is not indicated. Given the lack of detail available from this study, data from another historic experiment will be used to illustrate these procedures.

Example 2.3

Before the nineteenth century, scurvy was a common condition suffered by sailors on long voyages of more than a few weeks. Little was known about the factors associated with scurvy and how the disease could be controlled or cured. To address this issue, a study of sailors who had contracted scurvy was conducted in May/June 1747 onboard the British vessel HMS *Salisbury* (Lind, 1753).

A group of 12 sailors suffering from scurvy was selected. The men were divided into six pairs, each pair receiving a different form of treatment. One pair received a diet that included citrus fruits in the form of two oranges and one lemon daily. Lind, as the ship's surgeon, made clinical notes on each participant throughout the subsequent 2 weeks.

In assessing the benefits or otherwise of consuming citrus fruit, an appropriate analysis would be a comparison of the pair in the "citrus group" with the 10 participants who received a noncitrus diet. Using the clinical notes made by James Lind, it is possible to rank the participants in terms of health following 2 weeks of intervention from the least sick (rank 1) to the most sick (rank 12). Of those receiving citrus fruits, one was fit for duty after 6 days and the other was employed in nursing the others, all of whom remained ill. These sailors could therefore be assigned the ranks 1 and 2. Following the publication of this study, most of the general public accepted this as clear evidence that citrus fruits should be included in the supplies when preparing a ship for a long voyage at sea. However, could Lind's results have been purely down to chance?

The reasoning involved in a permutation test is as follows: With no association between the course of scurvy and citrus fruit consumption, it is unlikely that either sailor would have been ranked as 1 or 2. Nevertheless, the possibility of the pair receiving ranks 1 and 2, as observed by Lind, cannot be ruled out. Assuming independence of the twelve observations, the probability of a specific outcome is based on the number of ways in which two individuals can be selected from a group of twelve.

There is a choice from 12 for the first individual, then 11 individuals to select from for the other member of the pair. First impressions might suggest that 12×11 or 132 pairings are possible. However, the order of selection is not of importance here, only the individuals that make up the pair. The number of *distinct* pairings is therefore equal to 132/2 or 66.

If all six interventions are equally effective/ineffective, the pair of ranks associated with the sailors receiving citrus fruits is equally likely to be any of the 66 pairs of numbers listed in Table 2.1. Thus, there is only 1 chance in 66 that the pair showing greatest improvement (ranked 1 and 2 in order of condition) are the two sailors allocated to the citrus fruit diet. This casts doubt on the claims of cynics, of which there were plenty in the eighteenth century, that citrus fruits are not beneficial.

From a hypothesis testing perspective, the study consists of a group of two treated with a diet that includes citrus fruits and a group of ten (the remainder) given no citrus fruits, a two independent sample experiment. The most favourable evidence for the use of citrus fruits would be that those receiving them are ranked 1 and 2 in health; the least favourable would be that they are ranked 11 and 12.

Consider a test of:

H_0: *citrus fruit consumption has no effect*

against the two-sided alternative

H_1: *citrus fruit consumption has an effect (beneficial or deleterious)*

Given the lack of information on the treatment of scurvy prior to Lind's study, a two-tailed approach would be perfectly reasonable.

In order to perform a hypothesis test, an appropriate statistic is required with possible values that reflect the degree of divergence from what might be expected were the null hypothesis to be true. An intuitively reasonable choice is the sum of the two ranks, denoted by S. Table 2.2 shows the possible pairs along with the summed ranks.

Counting the number of times that each rank sum occurs in Table 2.2 gives the distribution of the test statistic (Table 2.3).

TABLE 2.1

Possible Selections of Two Individuals from Twelve Labelled 1–12

1, 2	2, 3	3, 5	4, 8	5, 12	7, 12
1, 3	2, 4	3, 6	4, 9	6, 7	8, 9
1, 4	2, 5	3, 7	4, 10	6, 8	8, 10
1, 5	2, 6	3, 8	4, 11	6, 9	8, 11
1, 6	2, 7	3, 9	4, 12	6, 10	8, 12
1, 7	2, 8	3, 10	5, 6	6, 11	9, 10
1, 8	2, 9	3, 11	5, 7	6, 12	9, 11
1, 9	2, 10	3, 12	5, 8	7, 8	9, 12
1, 10	2, 11	4, 5	5, 9	7, 9	10, 11
1, 11	2, 12	4, 6	5, 10	7, 10	10, 12
1, 12	3, 4	4, 7	5, 11	7, 11	11, 12

TABLE 2.2

Possible Selections of Two Individuals from Twelve Labelled 1–12 with the Sum of the Labels (Ranks) in Parentheses

1, 2	(3)	2, 3	(5)	3, 5	(8)	4, 8	(12)	5, 12	(17)	7, 12	(19)
1, 3	(4)	2, 4	(6)	3, 6	(9)	4, 9	(13)	6, 7	(13)	8, 9	(17)
1, 4	(5)	2, 5	(7)	3, 7	(10)	4, 10	(14)	6, 8	(14)	8, 10	(18)
1, 5	(6)	2, 6	(8)	3, 8	(11)	4, 11	(15)	6, 9	(15)	8, 11	(19)
1, 6	(7)	2, 7	(9)	3, 9	(12)	4, 12	(16)	6, 10	(16)	8, 12	(20)
1, 7	(8)	2, 8	(10)	3, 10	(13)	5, 6	(11)	6, 11	(17)	9, 10	(19)
1, 8	(9)	2, 9	(11)	3, 11	(14)	5, 7	(12)	6, 12	(18)	9, 11	(20)
1, 9	(10)	2, 10	(12)	3, 12	(15)	5, 8	(13)	7, 8	(15)	9, 12	(21)
1, 10	(11)	2, 11	(13)	4, 5	(9)	5, 9	(14)	7, 9	(16)	10, 11	(21)
1, 11	(12)	2, 12	(14)	4, 6	(10)	5,10	(15)	7, 10	(17)	10, 12	(22)
1, 12	(13)	3, 4	(7)	4, 7	(11)	5, 11	(16)	7, 11	(18)	11, 12	(23)

TABLE 2.3

Number of Occurrences for Each Sum of Ranks of Two Items from Twelve

Rank sum (S)	3	4	5	6	7	8	9	10	11	12	13	14	15	16	17	18	19	20	21	22	23
Occurrences	1	1	2	2	3	3	4	4	5	5	6	5	5	4	4	3	3	2	2	1	1

The outcomes 1, 2 (rank sum = 3) and 11, 12 (rank sum = 23) are extremes with a total associated probability of $P=2/66 \approx 0.0303$ if H_0 is true.

In classic hypothesis testing terms, H_0 is rejected at an exact 3.03% significance level if either of these extreme outcomes is observed. This small P-value provides strong evidence that the citrus fruit diet has an effect.

To find what other outcomes would be consistent with a P-value not exceeding 0.05, a region in each tail is selected (since H_1 implies a two-tailed test) with a total associated probability not exceeding 0.025. From Table 2.3, selecting in the lower tail $S=3$ and $S=4$ gives an associated probability of $\Pr(S \le 4)=2/66 \approx 0.0303$. This exceeds 0.025, so the lower tail critical region should be $S \le 3$ giving $P=1/66 \approx 0.0152$. By symmetry, the upper tail region is $S \ge 23$ also with $P \approx 0.0152$. Thus, for a two-tailed test, the largest symmetric critical region with $P \le 0.05$ is $S=3, 23$ and the exact $P=2/66 \approx 0.0303$. This critical region consists only of the two extreme values, showing the paucity of information available from a small study.

Some statisticians suggest choosing a critical region with probability as close as possible to a target level such as $P=0.05$ rather than the more conservative choice of one no larger. In this example, adding $S=4$ and the symmetric $S=22$ to the critical region gives a two-tailed $P=4/66 \approx 0.0606$. This is closer to 0.05 than the size (0.0303) of the region chosen above. It is best to quote the exact P-value where possible. The practical argument (though there are further theoretical ones) for quoting nominal sizes such as 0.05 is that many tables give only these. For example, Gibbons and Chakraborti (2020) give relevant exact P-values for many sample size combinations and different values of S. Computer programs giving exact P-values may remove the need to use such tables.

In the preliminary testing of drugs for treating a rare disease, the population may be in a strict sense the only cases available. However, if these patients are typical of all who might have the disease, it is not unreasonable to assume that findings from this small experiment may hold for any patients with a similar condition providing other factors (nursing attention, supplementary treatments, consistency of diagnosis, etc.) are comparable. When an experiment involves what is effectively the whole population, and the only data are ranks, a permutation test is the best test available. Random allocation of treatments is essential for the test to be valid; this may not always be possible in the light of some ethical considerations as discussed in Section 1.7.1. For a small study, a marginal P-value associated with what looks to be an intuitively encouraging result may indicate that a larger experiment might give more convincing positive findings.

Tests based on permutation of ranks or on permutation of certain functions of ranks (including the original measurements on a continuous scale when these are available) are central to many nonparametric methods. They are called *permutation* or *randomization* tests. The latter term applies when the permutation process is based on the randomization procedure used to assign treatments to units. That was the situation in Example 2.3, the permutations giving all possible assignments. These tests have an intuitive appeal and comply with well-established theoretical criteria for sound inference. Hettmansperger and McKean (2011) summarize this theoretical basis for many different procedures.

SUMMARY POINTS:

- A permutation test considers the number of ways in which subgroups can be selected from the overall sample.
- Permutation tests can be applied to ranked data to investigate whether higher ranks are concentrated in a particular subgroup.
- When an experiment involves what is essentially the whole population, and the only data are ranks, a permutation test is the best test available.

2.3 Binomial Tests

2.3.1 Number of Ways of Selecting a Sample from a Larger Group

In Example 2.3, it was demonstrated that the number of ways in which a pair can be selected from 12 individuals is equal to 12×11/2=66. Another way of representing this is as:

$$\frac{12\times 11\times 10\times 9\times 8\times 7\times 6\times 5\times 4\times 3\times 2\times 1}{(2\times 1)\,(10\times 9\times 8\times 7\times 6\times 5\times 4\times 3\times 2\times 1)}$$

This can be written more concisely as:

$$\frac{12!}{2!\times 10!}$$

The term 12! is often referred to as "12 factorial", and other numbers written in this form are described in a similar way. Highlighting the fact that 2 objects are being selected from a group of size 12, the above quantity can be written as:

$$\frac{12!}{2!\times(12-2)!} \text{ or } \binom{12}{2}$$

In general, the number of ways in which r observations can be selected from a larger group of n observations is:

$$\frac{n!}{r!\times(n-r)!} \text{ or } \binom{n}{r}$$

2.3.2 Independent Events

An assumption required for most of the methods described in this text is that the observations are independent. If this is the case, the multiplication rule of probability can be applied. This states that for independent events (e.g., head and tail outcomes when a coin is tossed repeatedly), the probability of obtaining the series observed is given by multiplying together the probabilities of the individual outcomes. In particular, the probability of obtaining n successes in a series of length n is given by the probability of a success (p) raised to the nth power, i.e., p^n.

In a somewhat light-hearted style, the satirical writer John Arbuthnot (1710) observed that in each of the 82 years from 1629 to 1710, the number of males christened in London exceeded the number of females. By considering the outcomes of throws with two-sided dice, he noted that if male and female births were equally likely, the observed finding would have an extremely low probability of 0.5^{82}. This finding, he reasoned, demonstrates strong evidence against the commonly believed assumption. Note that he did not distinguish between births and christenings; his conclusion is only true if the probability of an infant being christened was equal for males and females.

2.3.3 An Application of the Binomial Test

One observation was censored in the data for the survival of retired footballers in Example 2.2. Despite its exact value being unknown, it could be shown that it was reasonable, given that data, to retain a hypothesis that the population median was 12. An appropriate test to justify that conclusion is now considered.

Example 2.4

Survival times (years) in order of retired footballer enrolment were:

9 9 8 5 9 6 14 10 14 9 11 7 15* 14

The asterisk denotes the censored observation.

The appropriate null hypothesis here is that the median, θ, of survival times for the population from which the sample was obtained is 12 against the alternative of some other value:

$$H_0 : \theta = 12 \text{ against } H_1 : \theta \neq 12 \tag{2.1}$$

A simple test needs only a count of the number of sample values exceeding 12. In the data above, values over 12 are replaced by a "plus", while other values being replaced by a "minus" to give the sequence consisting of 4 "plus" and 10 "minus" signs:

− − − − − − + − + − − − + +

Under the null hypothesis, since H_0 is concerned with the median, the probability of the next observation being a plus is 0.5. As only two outcomes are possible, the probability of the next observation being a minus is also 0.5. Assuming that the 14 retired footballers form a random sample, and order of enrolment is taken into account, the probability of observing the sequence above is obtained by multiplying the individual probabilities together giving 0.5^{14}.

Of more relevance for testing H_0 is the probability of obtaining a combination of 4 plus and 10 minus signs irrespective of ordering. This gives the overall chance of observing 4 values in excess of the median in a sample of size 14. The probability is calculated by multiplying the probability of obtaining a specific sequence of plus and minus signs (0.5^{14}) by the number of ways in which a sample of 4 observations can be selected from a group of 14:

$$\frac{14!}{4! \times 10!} \text{ or } \binom{14}{4}$$

Writing the number of pluses as X, the probability of 4 pluses in a sample of size 14 can be written as:

$$\Pr(X = 4) = \binom{14}{4} \times 0.5^{14}$$

It follows that the probability of obtaining *r* pluses from 14 observations is:

$$\Pr(X = r) = \binom{14}{r} \times 0.5^{14}$$

Probabilities for the possible values of *r* are given below (Table 2.4).
Note that the number of pluses takes a binomial distribution B(*n*, *p*):

$$\Pr(X = r) = \binom{n}{r} p^r (1-p)^{n-r}$$

The sample size $n=14$, the probability of a positive outcome (plus sign) of $p=0.5$, the probability of a negative outcome (minus sign) of $(1 - p)=0.5$, the number of pluses being *r*. Hence, this distribution is denoted by B(14, 0.5).

If the null hypothesis is true, the expected number of plus signs is 14×0.5 or 7. From the definition of a *P*-value, the relevant tail probabilities are those for $r\leq4$ pluses and $r\geq10$ pluses. These add up to $(0.0001+0.0009+0.0056+0.0222+0.0611)\times2=0.1798$.

Hence, there is no strong evidence against H_0. This implies that departures from the expected number of plus signs, 7, as large, or larger, than that observed will occur in slightly more than one-sixth of all samples when H_0 is true. This simple test, called the *sign test*, is discussed more fully in Section 3.2. The test is *distribution-free* because no assumption about the form of the continuous distribution of the underlying observations is made. Formulation and testing of hypotheses are only concerned with possible values of the population median.

For most parametric tests (e.g., the *t*-test), all values of *P* between 0 and 1 are possible. However, for many nonparametric methods *P*-values are limited to a finite number of discrete values. These do not always correspond to conventional thresholds for statistical significance.

This is the case for the sign test. In this example, for a two-tailed test, the four smallest are as follows: $P=2\times0.0001=0.0002$ corresponding to 0 or 14 plus signs; $P=2\times(0.0001+$

TABLE 2.4

Probabilities of Obtaining *r* Pluses from 14 Observations under H_0

Number of+	0	1	2	3	4	5	6	7	8
Probability	0.0001	0.0009	0.0056	0.0222	0.0611	0.1222	0.1833	0.2095	0.1833
Number of+	9	10	11	12	13	14			
Probability	0.1222	0.0611	0.0222	0.0056	0.0009	0.0001			

0.0009) = 0.002 corresponding to no more than 1 or at least 13 plus signs; $P = 2\times(0.0001 + 0.0009 + 0.0056) = 0.0132$ corresponding to no more than 2 or at least 12 plus signs; and then, $P = 2\times(0.0001 + 0.0009 + 0.0056 + 0.0222) = 0.0576$ corresponding to no more than 3 or at least 11 plus signs. Next comes the observed $P = 0.1798$ (no more than 4 or at least 10 plus signs). In all cases, probabilities have been rounded to four decimal places.

The statistic used in this example – the number of plus signs – has a discrete distribution. This means that, as in Example 2.3, there is no direct way of obtaining a critical region of exact size 0.05 for a two-tailed test; the only choice is between regions of size 0.0132 or 0.0576.

Once they are recognized, and the consequences appreciated, discontinuities in possible *P*-values do not cause serious interpretational problems in the analysis of a particular data set. However, these discontinuities do lead to some theoretical difficulties in comparing performance of competing tests. A device called a *randomized decision rule* has been proposed with the property that in the long run an error of the first kind has, in repeated testing, a probability at a pre-chosen nominal level, e.g., at 5%. Gibbons and Chakraborti (2020) describe how a randomized decision rule works (pp. 24–25). They commented that such devices may seem artificial and are "probably seldom employed by experimenters". Hence, it is better, when known, to use exact levels, rather than worry about nominal arbitrary levels. When there are discontinuities, however, there is a case for forming a tail probability by allocating only one-half of the probability that the statistic equals the observed value to the "tail" when determining the size of the "critical" region. This approach has many advocates. It is not used in this book, but if it is used, this should be done consistently. The sign test provides a basis for forming a confidence interval for the population median.

Example 2.5

In Example 2.4, when using a sign test for a median with a sample of 14, in a two-tailed test at the 1.32% level, H_0 is retained if between 2 and 12 plus signs are observed.

Consider again the data in that example ordered by size, i.e.,

$$5 \quad 6 \quad 7 \quad 8 \quad 9 \quad 9 \quad 9 \quad 9 \quad 10 \quad 11 \quad 14 \quad 14 \quad 14 \quad 15^{*}$$

where the asterisk represents a censored observation. There are between 2 and 12 plus signs if the median specified in H_0 has any value greater than 6 but less than 14. This implies that the interval (6, 14) is a $100(1 - 0.0132) = 98.68\%$ confidence interval for θ, the population median survival time. Since any H_0 that specified a value for the median greater than 6 but less than 14 would be retained, there is considerable doubt about the population median value. It is stating the obvious to say the estimate lacks precision.

2.3.4 A Binomial Distribution with Unequal Success and Failure Probabilities

The next illustration involves a more general binomial distribution.

Example 2.6

Inchley et al. (2020) found that for 15-year-old boys in England, 75% brush their teeth frequently (more than once daily). Hence, the number who brush their teeth frequently, S, in a sample of 10 independent boys aged 15 years has a binomial B(10, 0.75) distribution. Here, the probabilities for the various values, r, of the statistic S, where r takes integer values between 0 and 10, are given by:

TABLE 2.5

Probabilities of Obtaining r Individuals Who Brush Their Teeth More Than Once Daily Out of 10 Boys Aged 15 under H_0

Number	0	1	2	3	4	5	6	7	8
Probability	0.0001	0.0000	0.0004	0.0031	0.0162	0.0584	0.1460	0.2503	0.2816
Number	9	10							
Probability	0.1877	0.0563							

$$\Pr(X = r) = \binom{10}{r} \times 0.75^r 0.25^{n-r}$$

The relevant probabilities are (Table 2.5):

If based on data from previous years, it was thought that the proportion should be higher than 0.75 (e.g., it has never been less than 0.9 before), the relevant test would be:

$$H_0 : p = 0.75 \text{ against } H_1 : p > 0.75$$

for which the smallest P-value for testing is in the upper tail and is associated with $r=10$, i.e., $P=0.0563$. Applying the convention $P\le 0.05$ as sufficiently strong evidence to discredit H_0, such values are never obtained.

This is not a problem for the one-tailed test of H_0: $p=0.75$ against H_1: $p<0.75$, since, in the appropriate lower tail, $P=\Pr(S\le 4)=0.0162+0.0031+0.0004 = 0.0197$.

This example also shows a logical difficulty associated with a rule suggested in the literature that the appropriate level for a two-tailed test is twice that for a one-tailed test, for if the outcome is $S=4$, the two-tailed test level based on this rule is $2\times 0.0197=0.0394$. This presents a dilemma, for there is no observable upper tail area corresponding to that in the lower tail. This means that if a two-tailed test is appropriate, departures from the null hypothesis can only be detected in one direction. There may well be a departure in the other direction, but if so, detection is highly unlikely at the conventional level $P\le 0.05$. Even if a departure were to be detected, it would be for the wrong reason. This is not surprising when, as shown above, the appropriate one-tailed test must fail to detect it; for generally, a one-tailed test at a given significance level is more powerful for detecting departures in the appropriate direction than is a two-tailed test at the same level.

An implication is that in this example a larger sample is needed to detect departures of the form H_1: $p>0.75$. Again, the unconvincing P-value associated with the possible critical region for the one-tailed test only indicates that the sample is too small. The stipulation that the study participants be independent is important. If the boys included members from the same household (e.g., twins), it is quite likely that their teeth brushing habits would be similar if not identical. Situations of this kind are considered in more detail in Agresti (2013). There is no universal agreement that one should double a one-tail probability to get the appropriate two-tail significance level – see, for example, Yates (1984) and the discussion thereon. An alternative is that once the exact size of a one-tail region has been determined, one should, for a two-tailed test, add the probabilities associated with an opposite tail situated equidistant from the mean value of the test statistic to that associated with the observed statistic value. In the symmetric case, as already pointed out, this is equivalent to doubling the probability, but it seems inappropriate with an asymmetric distribution. In Example 2.6, the region $r \leq 4$ is appropriate for a lower tail test. The mean of the test statistic (the binomial mean np) is here 7.5. Since $7.5-4=3.5$, the corresponding deviation above the mean is $7.5+3.5=11$. Because $\Pr(r \geq 11)=0$, the two-tailed test based on equidistance from the mean would have the same exact significance level as the one-tailed test.

SUMMARY POINTS:

- The number of ways in which r observations can be selected from a larger group of size n may be calculated using a formula based on factorial terms (Section 2.3.1).
- If events are independent, the probability of obtaining the series of outcomes observed is given by multiplying together the probabilities of the individual outcomes (Section 2.3.2).
- For many nonparametric methods, P-values are limited to a finite number of discrete values (Section 2.3.3).
- If a binary outcome has unequal success and failure probabilities, the distribution of the number of successes in a sample is asymmetric. This can cause problems in defining appropriate critical regions for two-tailed tests (Section 2.3.4).

2.4 Order Statistics and Ranks

Many nonparametric procedures are based on the ordering, or ranking, of detailed observations. In Example 2.4, ranks were not used, but ordering was inherent in the procedure. Order was considered in determining the number of survival times that exceeded the hypothesized median.

Ordering data is important in more general statistical contexts, both parametric and nonparametric. There may be interest in the distribution of the largest or smallest observations in a sample to answer questions such as:

- On the basis of maximum flood levels recorded in a river over a number of years, what is the probability of the level exceeding, say, 5 m, in future?
- Given a sample of times to first breakdown of a certain brand of computer, what is the probability of a first breakdown being observed within 6 months in one machine in a production run of 1,000 machines?

In a parametric context, such questions are often answered using families of distributions called *extreme value distributions*. A simple example of the role of order statistics in a parametric context is given in Exercise 2.8. Greatest and least values in samples are just two examples of *order statistics*. The sample median is also an order statistic.

Consider a sample of n observations $x_1, x_2, \ldots, x_n$ from a continuous distribution. Continuity implies that there should be no ties, and thus, observations may be uniquely ordered from smallest to largest. Denote the smallest observation by $x_{(1)}$, the second smallest by $x_{(2)}$ and so on, finally the largest by $x_{(n)}$. It follows that,

$$x_{(1)} < x_{(2)} < \ldots < x_{(n)}.$$

The $x_{(i)}$, $i=1, 2, \ldots, n$ are called the *order statistics*. The minimum order statistic, $x_{(1)}$, is relevant to the study of minimum extremes such as the distribution of shortest times to a machine breakdown or minimum survival times after some treatment. The largest, $x_{(n)}$, is relevant to the study of floods or maximum time to failure of a certain type of light bulb.

The measures introduced in Sections 1.3.3 and 1.3.4 can be written algebraically using order statistics. For a sample of n, the median is $x_{[(n+1)/2]}$, if n is odd, and is usually defined as $[x_{(m)}+x_{(m+1)}]/2$ if $n=2m$ is even. A possible measure of dispersion is the sample range $x_{(n)}-x_{(1)}$. A more satisfactory measure is the *interquartile range*. One reason for preferring the latter is that the extreme order statistics are often strongly influenced by suspect observations associated with the terms *outliers* (Section 1.3.3) and *dirty data*.

SUMMARY POINTS:

- The ith order statistic in a sample is the observation having a rank of i.
- The largest and smallest values in a sample are examples of order statistics that can be modelled using extreme value distributions.

2.5 Exploring Data

The addition of nonparametric or distribution-free methods to the procedures for making statistical inferences widens the choice of techniques appreciably. An invaluable first step in selecting an appropriate technique in any given situation is to use *exploratory data analysis* (EDA). Some basic tools of EDA are:

- Descriptive statistics.
- Boxplots.
- Histograms and frequency curves.
- Empirical and theoretical cumulative distribution graphs.

Descriptive statistics are commonly presented in lists or tables. The other tools above are graphical by nature. Commonly met descriptive statistics that summarize key features of a group of observations are the sample mean, median, maximum value, minimum value, standard deviation and quartiles.

When questions of robustness arise due to the presence of extreme values, one approach is to remove or adjust the outliers. With the *trimmed mean*, the most extreme observations (typically amounting to 10%) are removed and inferences are based on the remaining observations. For the *Winsorized mean*, small outliers are replaced by the value of the smallest plausible observation and large outliers are replaced by the value of the largest plausible observation.

Statistics such as the mean, median and standard deviation are often referred to as *secondary data* to distinguish them from the original raw or observational data called *primary data*. Most general statistical software packages have a facility for computing a wide range of descriptive statistics.

A study of relevant descriptive statistics may give a quick indication of, for example, whether an assumption of normality appears to be seriously invalidated, or whether it is reasonable to suppose the sample comes from a symmetric or a skew distribution; and if the latter, whether the long tail is to the left or right. Typically, the five descriptive statistics presented in the order *minimum, first quartile, median, third quartile, maximum* (or *five-number summary*) are used for this purpose.

Such basic characteristics may be explored more fully by graphical techniques. For instance, the descriptive statistics from a five-number summary can be used to construct a *boxplot* or a *box and whisker plot*. These plots can be used to assess how well samples reflect population features.

Example 2.7

Appendix 1 gives four small data sets indicating how they were collected. Table 2.6 gives a set of descriptive statistics useful for summarizing and comparing these data for each of the four sets. The first row gives the number of data for each set. The small sample of 13 observations for the *McDelta* clan might be expected to be less informative than the sample of 59 *McAlphas*.

The sample mean for *McDelta* is markedly lower than that for the other clans. There may be interest in whether this indicates a shorter average life expectancy for that clan, or whether the difference represents some sampling quirk that might disappear if a larger sample was available.

The medians are all appreciably higher than the means, suggesting that the distributions of ages are asymmetric and skewed to the left (Section 1.3.4). This follows because samples are expected to reflect broadly the population characteristics, and for symmetric population distributions, the mean and median coincide.

The abbreviation *SD* is used in the table for the standard deviation, a commonly encountered measure of spread that in the case of a sample from a normal population is an appropriate estimator of the parameter σ. Once again, the clan *McDelta* is the odd one out.

SE mean is an abbreviation for *standard error of the mean*. If the sample standard deviation is denoted by s, then the standard error of the mean is computed as $s/\sqrt{n}$. Thus, the standard error decreases with sample size for a given standard deviation.

TABLE 2.6

Descriptive, or Summary, Statistics for Badenscallie Data Given in Appendix 1

Clan	McAlpha	McBeta	McGamma	McDelta
Number	59	24	21	13
Mean	61.8	61.1	62.9	48.1
Median	74.0	67.5	77.0	65.0
SD	27.52	24.92	26.77	33.45
SE Mean	3.58	5.08	5.84	9.28
Minimum	0	0	13	1
Maximum	95	96	88	87
First Quartile	44.0	41.5	33.0	13.0
Third Quartile	81.00	78.75	83.50	80.00
Range	95	96	75	86
IQ Range	37.00	37.25	50.50	67.00

The maximum and minimum ages at death indicate at least one case of infant mortality for each clan except *McGamma* and at least one nonagenarian survivor for two of the clans.

The *quartiles* divide each ordered sample into four groups of equal size. If the median is considered as dividing the sample into two groups of equal size, the first quartile is in effect the median of the group of lower values and the third quartile is the median of the group of higher values. More formally, the first quartile is the median of $x_{(1)}, x_{(2)}, \ldots, x_{((n-1)/2)}$ if n is odd and is the median of $x_{(1)}, x_{(2)}, \ldots, x_{(n/2)}$ if n is even, with corresponding definitions for the third quartile. The *second quartile* is the sample median. While the third quartiles are similar for all clans, the first quartile is much smaller for *McDelta*.

The first quartiles raise the issue of whether these differences can be accounted for by a quirk of the relatively small sample or whether it represents a different age distribution from that of the other clans. Range and interquartile range, the latter abbreviated in the table to *IQ range*, are, respectively, the differences *maximum–minimum* and *third quartile–first quartile*. Each is a measure of spread alternative to standard deviation. Of the two, the interquartile range is preferred because range depends only on the two observations x_1 and x_n, either of which may represent some unusual, or even a rogue, observation. However, the interquartile range covers an interval containing the central 50% of the observations. Intuitively, this may be expected to be a more stable estimate of general variability. As an alternative to the interquartile range, the *semi-interquartile* range is occasionally used. As its name implies, it is obtained by dividing the interquartile range by 2. For the clan data, the striking differences in interquartile range might be an aspect of the data requiring further analysis.

Figure 2.1 gives boxplots for each clan for the Badenscallie data based on five-number summaries presented in Table 2.6. The labels attached to the boxplot for the *McDelta* clan apply to any boxplot and indicate that the box section extends from the first to the third quartile. The vertical line dividing this box into two portions represents the median. The horizontal line outside the boxes extends from the minimum to the maximum.

Including box and whisker plots for all four clans on the one diagram enables useful comparisons of the kind outlined above to be made very easily. In particular, remembering that half the observations lie at or above the median and half lie at or below the median, it is seen that for all clans the distribution of ages at death is skewed to the left or lower tail. This is made very clear by the median in all cases being nearer to the third quartile than to the first quartile. Recall that the quartiles are effectively the medians of the lower and upper halves of the data, respectively.

Histograms are another widely used graphical device to exhibit key data characteristics. Figure 2.2 is a histogram based on the clan McAlpha data for ages at death with a class interval of 10 years. The long tail to the left is evident. There is also an indication of a mixture of distributions, with a smaller portion of the data indicating *infant mortality* or death before reaching adulthood, while the larger portion represents a more normal (in the physiological but not necessarily in the statistical sense) lifespan peaking around 80 years.

As the size of a random sample increases, it mirrors the population characteristics ever more closely. Modern statistical software packages allow one to draw random samples of any chosen size from a wide range of distributions. For reasonably large samples, i.e., those of at least 50 observations, constructing appropriate histograms and superimposing these on the relevant population distribution frequency function gives a good impression of how effective these matches are.

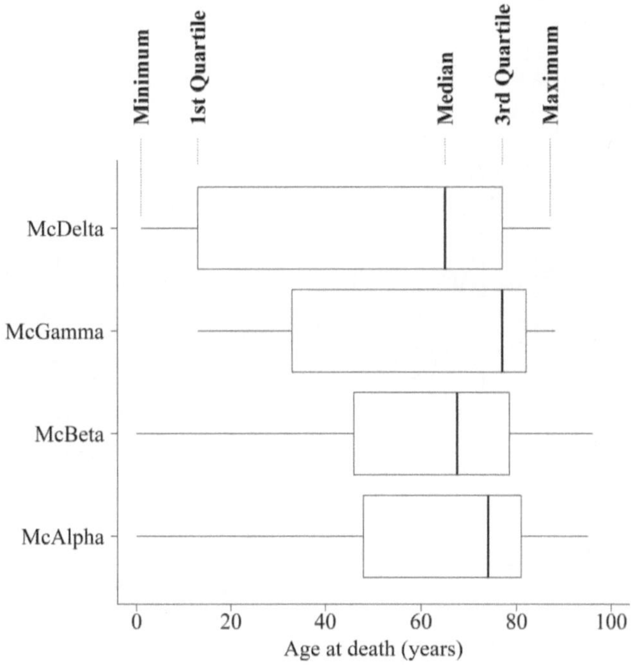

FIGURE 2.1
Boxplots for Badenscallie data given in Appendix 1.

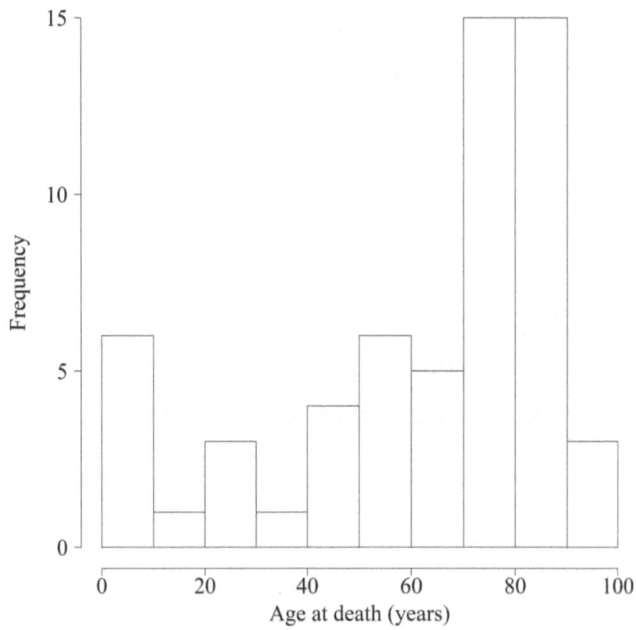

FIGURE 2.2
Histogram for clan McAlpha data given in Appendix 1.

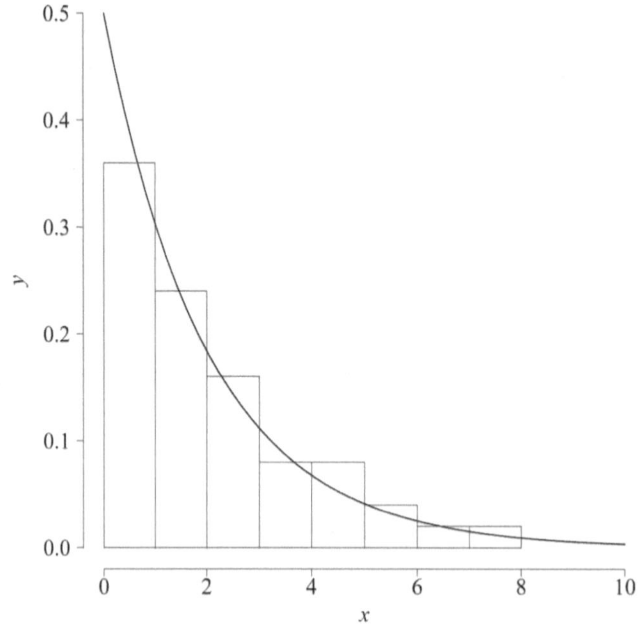

FIGURE 2.3
Histogram for a sample of 50 from an exponential distribution with mean 2. The fitted curve is that of the probability density function.

Example 2.8

A computer-generated sample of 50 observations from an exponential distribution with mean 2 using Minitab gave the histogram in Figure 2.3. All sample values were less than 10, and 20 of them lay in the interval [0, 1), 14 in the interval [1, 2), 6 in the interval [2, 3), and so on. The curve superimposed on the histogram is that of the *frequency function* or *probability density function* of the exponential distribution with mean 2, which has the form:

$$f(x) = ½\, e^{-x/2}, \quad x \geq 0.$$

Statisticians would regard the closeness of the curve to the histogram as an indication that the data might be a sample from this distribution.

For samples smaller than 50, the grouping required to form a histogram may result in a rather poor fit to the population frequency function. However, even for small samples, the sample distribution step function usually lies fairly close to the population cumulative distribution function.

Example 2.9

A computer-generated sample of eight from an exponential distribution with mean 2 gave the values:

0.25 0.53 0.91 0.94 1.56 1.73 4.71 5.50

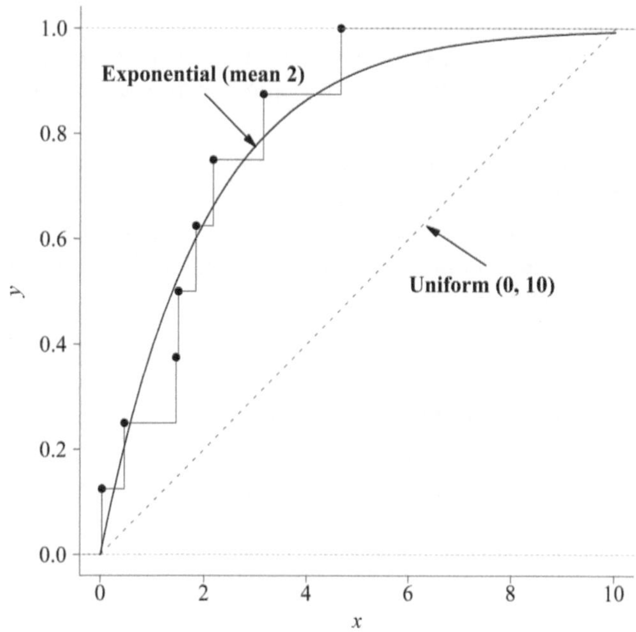

FIGURE 2.4
Sample cumulative distribution function (stepped) for a sample of eight from an exponential distribution with mean 2. The curve is the population cumulative distribution function, and the straight line is that for a uniform distribution over (0, 10).

where these have been arranged in ascending order. Figure 2.4 shows the sample cumulative distribution function for these data (stepped function) and the cumulative distribution function for an exponential function with mean 2. This takes the form:

$$F(x) = 1 - e^{-x/2}, \quad x \geq 0.$$

The step function lies close to this population cumulative distribution function. For illustrative purposes, the straight line joining the points (0, 0) and (10, 1) on the graph is the cumulative distribution function for a uniform distribution over (0, 10). It is almost self-evident that the sample was not taken from that distribution.

More sophisticated EDA methods include the so-called P–P and Q–Q plots, abbreviations for plots of probabilities and of quantiles, respectively, associated with two distributions or with a hypothesized distribution and a sample believed to be from a population having that distribution. Thas (2010) gives a description of how these are used and interpreted.

The examples in this section only touch on the potential of an EDA approach. Further examples are given throughout this book.

SUMMARY POINTS:

- Before choosing the statistical methods to be applied, samples should be investigated using EDA.
- Descriptive statistics in common use include the sample mean, median, maximum value, minimum value, standard deviation and quartiles.

- The boxplot provides a graphical representation of the data based on the five-number summary (minimum, first quartile, median, third quartile and maximum).
- Histograms can be used to investigate characteristics of continuous data grouped into intervals.
- Superimposing a sample histogram on a population probability density function gives an indication of how well the data fit the proposed model.

2.6 Efficiency of Nonparametric Procedures

As discussed in Section 1.5, the power of a test depends upon (i) the sample size, n, (ii) the choice of the largest P-value to indicate significance (usually denoted in power studies by α), (iii) the magnitude of any departure from H_0 and (iv) whether assumptions that are needed for validity hold. Most intuitively reasonable tests have good power to detect a true alternative that is far removed from the null hypothesis providing the data set is large enough. Tests are sometimes required to have as much power as possible for detecting alternatives close to H_0 even when these are of no practical importance. This is because such tests are usually also good at detecting larger departures, a desirable state of affairs.

A statistic is regarded as efficient (put simply) if the sample size required to obtain low probabilities for the Type I and Type II errors is "small". *Efficiency* is usually described in terms of one statistic relative to another. If α is the probability of a Type I error, and β is the probability of a Type II error (the power is $1-\beta$), then the efficiency of a test T2 relative to a test T1 is the ratio n_1/n_2 of the sample sizes needed to obtain the same power for the two tests with these values of α, β. In practice, α is usually fixed at some P-value appropriate to the problem at hand. Then, β depends on the particular alternative as well as the sample sizes. Fresh calculations of relative efficiency are required for each particular value of the parameter or parameters of interest in H_1 and for each choice of α, β.

Pitman (1949) introduced the concept of *asymptotic relative efficiency* for comparing two tests (p. 54). He considered sequences of tests T1 and T2 in which α is fixed. The alternative in H_1 is then allowed to vary in such a way that β remains constant as the sample size n_1 increases. For each n_1, the value n_2 is determined such that T2 has the same β for the particular alternative considered.

Increasing sample size usually increases the power for alternatives closer to H_0. Therefore, for large samples, Pitman studied the behaviour of the efficiency, n_1/n_2, for steadily improving tests for detecting small departures from H_0. He showed under very general conditions that in these sequences of tests n_1/n_2 tended to a limit as $n_1 \to \infty$. More importantly, this limit, which he called the asymptotic relative efficiency (ARE), was the same for all choices of α, β. A detailed discussion of ARE is given by Mood (1954) and Gibbons and Chakraborti (2020, Chapter 13).

Bahadur (1967) proposed an alternative definition that is less widely used, so for clarity and brevity, Pitman's concept is here referred to simply as the *Pitman efficiency*. The concept is useful because, when comparing two tests, the small sample relative efficiency is often close to, or even higher, than the Pitman efficiency.

The Pitman efficiency of the sign test relative to the t-test when the latter is appropriate is a rather low $2/\pi \approx 0.64$. Lehmann (2006) showed that for samples of size 10 and a range of values of the median θ relative to the value θ_0 specified in H_0 with α fixed, the relative efficiency exceeds 0.7. For samples of 20, it is nearer to, but still above, 0.64. Here, Pitman efficiency gives a pessimistic picture of the performance of the sign test at small sample sizes.

When it is relevant and valid, the *t*-test is the most powerful test for any mean specified in H_0 against any alternative. When the *t*-test is not appropriate, other tests may have higher efficiency. Indeed, if a sample comes from the double exponential distribution, which has much longer tails than the normal, the Pitman efficiency of the sign test relative to the *t*-test is 2. That is, a sign test using a sample of n (at least for large samples) is as efficient as a *t*-test applied to a sample of size $2n$. There are, however, situations where ARE may give an unduly optimistic picture of small sample behaviour.

SUMMARY POINTS:

- A test is regarded as efficient if the sample size needed to obtain low probabilities for the Type I and Type II errors is small, relative to other available tests.
- For tests T2 and T1 having the same probability of a Type I error, the efficiency of T2 relative to T1 is the ratio n_1/n_2 of the sample sizes needed to obtain the same power for the two tests.
- As the sample size for T1 is increased, the ratio for T2 relative to T1 approaches the ARE.

2.7 Computers and Nonparametric Methods

Computer software packages suitable for nonparametric analysis fall into three main categories. The first is specialist menu-driven packages that use exact permutation or related methods for small- to medium-sized samples and provide Monte Carlo and/or asymptotic tests for larger samples. The second category consists of the mainstream menu-driven statistical software packages that allow exact inferences for some, but by no means all, widely used nonparametric tests, or are user-friendly in the sense that they allow the user to write macros to carry out such procedures. The final category consists of versatile interactive statistical packages that have a variety of options, or tools, to perform various data manipulations and statistical operations. These are not menu driven. The user combines relevant tools, often with further self-designed options, to achieve some desired objective. Such programs are by their nature generally less user-friendly than menu-driven packages, but they are often more powerful.

In the first category, StatXact is perhaps the most widely used. It gives exact permutation *P*-values for small samples together with Monte Carlo estimates of these, for a large range of tests. Large sample, or asymptotic, results are also given, and there are facilities for computing confidence intervals and also the power of some of the tests for assigned sample sizes and specified alternative hypotheses. Some of the tests in StatXact are also available in SAS. There are also specialized programs dealing with particular aspects of the broad fields of nonparametric and semiparametric inference. These include LogXact, which is especially relevant to logistic regression.

General statistical packages such as SAS, Minitab, SPSS and Stata include some nonparametric procedures. In some of these exact tests are given, but many rely heavily on asymptotic results, sometimes with little warning about when, particularly with small or unbalanced sample sizes, these may be misleading. Facilities for creating Monte Carlo approximations to exact *P*-values, or for bootstrap estimation, are often also available in these standard packages.

In the third category, the open source R is dominant. A number of exact and asymptotic nonparametric tests are available via standard packages that are maintained by the R Core Team, while others are available through packages created by R users. Packages such as R and SAS have a versatility that makes combining of approaches such as EDA and more formal analyses quick and easy.

Users should test nonparametric procedures in any package programs they use with examples from this book and other sources. In some cases, the output will be different, being either more or less extensive than that given in the source of the examples. For instance, output may give nominal (usually 5% or 1%) significance levels rather than exact *P*-values. Sometimes, the convention of doubling a one-tailed *P*-value may be used to obtain a two-tailed test value, but as indicated in Example 2.6, this may not always be appropriate. Particular care should be taken to check whether exact or asymptotic results are given.

This book is largely about well-established methods, although powerful computing facilities are needed for the application of some of them. Solutions to examples or illustrations in this book largely use the ANSM5 package in R, written to accompany the book. In many cases, it would be equally appropriate to use other well-known packages such as SAS, SPSS, and Stata, providing these packages contain relevant programs. In creating the ANSM5 package, we do not claim to have produced software which is necessarily the most computationally efficient nor the most comprehensive. At the time of writing, there are other R packages which also undertake nonparametric statistical analyses and which the reader may wish to investigate. We do not list them here because over time various packages may cease to be maintained and other packages may emerge. However, the ANSM5 package has been designed to cover all the techniques covered in this book and be dependent only on other R packages which are maintained by the R Core Team. It can be downloaded from CRAN and associated mirror sites, or from the "neilhspencer/ANSM5" repository on GitHub. The package also contains data used in the book. Readers who replicate the analyses in the book which use the ANSM5 package should obtain the same results as presented in the examples and exercises. A possible exception to this is where Monte Carlo methods are used. These are based on repeatedly taking random samples of data, and of course, the random samples chosen by the readers' computers may be different from those chosen by the authors' computers. As a result, the Monte Carlo results may differ to a small degree.

Developments in statistical computer software are rapid, and mention should also be made of the programming language Python which is hugely flexible and has packages that can perform some nonparametric tests.

2.8 Further Reading

Hollander, Wolfe and Chicken (2014), Conover (1999), Gibbons and Chakraborti (2020), Higgins (2004) and Desu and Raghavarao (2004) give, in some cases, more background for some of the procedures described here. Each book covers a slightly different range of topics, and at varying depths, but all are suitable references for those who want to get a broad picture of the many aspects of basic nonparametric methods. Daniel (2000) is a general book on applied nonparametric methods.

The text by Hettmansperger and McKean (2011) gives a moderately advanced mathematical treatment of the theory behind nonparametric methods. Likewise, Randles and Wolfe (1991) and Maritz (1995) are recommended for covering the theory at a more advanced mathematical level than that used here. A classic is the book by Lehmann (2006). This book repays careful reading for those who want to pursue the logic of the subject in more depth without too much mathematical detail. Applications in the social sciences are covered by Leach (1979). Siegel and Castellan (1988) and Corder and Foreman (2014) are readable introductory texts.

Noether (1991) takes a nonparametric approach to introduce general statistical concepts. Although dealing basically with rank correlation methods, Kendall and Gibbons (1990) gives an insight into the relationship between many nonparametric methods. Rayner and Best (2001) gives a wide-ranging treatment of many standard and a few specialist procedures using methods based largely on partitioning of the chi-squared statistic. Wasserman (2006), despite its title, mainly deals with more advanced modern topics in nonparametric statistics, a few of which are described in Chapter 14. The author gives a lucid introduction to those topics addressed. Li and Racine (2006) is an advanced text that gives detailed coverage of modern nonparametric methods in the context of econometrics. Agresti (2010, 2013, 2019) and Everitt (1992) give detailed accounts of various parametric and nonparametric models used in categorical data analysis.

Manly (2006) gives a sophisticated treatment of randomization tests with emphasis on biological applications. Randomization and permutation tests are covered by Good (2005), Edgington and Onghena (2007) and Berry, Johnston and Mielke (2014). Hájek, Šidák and Sen (1999) gives the theory behind rank tests. Texts dealing with the bootstrap include Efron and Tibshirani (1993), Davison and Hinkley (1997), Efron and Hastie (2016), Chernick and LaBudde (2011) and Dickhaus (2018).

2.9 Exercises

2.1 A new type of intensive physiotherapy is developed for individuals who have undergone spinal surgery. Due to limited hospital resources, it can only be given to three out of ten patients. The patients are aged:

15 21 26 32 39 45 52 60 70 82

Explain how a permutation test could be used to investigate whether use of the physiotherapy is related to patient age (i.e., whether there is a policy to give the treatment to younger as opposed to older groups or *vice versa*). If the patients aged 15, 26 and 32 have the intensive physiotherapy, find the P-value for a two-tailed test of an appropriate null hypothesis. Comment on your findings.

2.2 Suppose that a new drug under test has all the ingredients of a standard drug at present in use and an additional ingredient that has proved to be of use for a related disease, so that it is reasonable to assume that the new drug will do at least as well as the standard one, but may do better. Explain why a one-tailed test might be justifiable here and formulate appropriate hypotheses for the test. If 9 patients are involved in the study and the post-treatment ranking of the patients receiving the new drug is 1, 2, 4, 6 (a low rank representing a better outcome), assess the strength of the evidence against the relevant null hypothesis. Would a two-tailed test give similar findings?

2.3 An education authority is responsible for 114 primary schools. A random sample of 12 schools is selected to test the hypothesis that the median number of children, θ, in all 114 schools is 225. In the sample of 12, it is found that 3 schools have less than 225 children. Does this justify retention of the hypothesis that $\theta=225$? What would be an appropriate alternative hypothesis? What is the largest critical region for a test with $P\leq 0.05$ and what is the corresponding exact *P*-value?

2.4 The numbers of children in the sample of 12 schools in Exercise 2.3 were:

126 142 156 228 245 246 370 419 433 454 478 503

Find a confidence interval at a level not less than 95% for the median θ.

2.5 In Section 1.6.1, a confidence interval was associated with a two-tailed test. As well as such two-sided confidence intervals, one may define a one-sided confidence interval composed of all parameter values that would not be rejected in a one-tailed test. Use this argument to obtain a confidence interval at a level not less than 95% based on the sign test criteria for the 12 schools given in Exercise 2.4 relevant to a test of H_0: $\theta=\theta_0$ against a one-sided alternative H_1: $\theta>\theta_0$.

2.6 From 6 consenting patients requiring a medical scan, 3 are chosen at random to undergo positron emission tomography (PET), while the others receive magnetic resonance imaging (MRI). Image quality is ranked in order by a hospital consultant from 1 (best) to 6 (worst). Describe how you would test H_0: *scan quality is unrelated to scan method against* (i) H_1: *PET scans are better* (ii) H_1: *the scans differ in quality depending on whether they are from PET or MRI.* Interpret the finding that the consultant rates the three PET scans as the three highest quality images.

2.7 In Example 2.6, it was remarked that a situation could arise where the rejection of the null hypothesis might be for the wrong reason. Explain how this is possible in that example.

2.8 A sample of 12 observations is taken from a continuous uniform distribution over the interval (0, 1). What is the probability that the largest sample value exceeds 0.95? (Hint: Determine the probability that any sample value exceeds 0.95. The condition is met if at least one value exceeds 0.95.)

2.9 A sample of 24 observations is known to come either from a uniform distribution over the interval (0, 10) or else from a symmetric triangular distribution over the same interval (0, 10). The sample values are:

4.17 8.42 3.02 2.89 9.77 6.06 2.72 5.12 6.00 4.78 2.62 7.20
1.61 5.92 7.25 8.01 4.76 5.36 5.34 7.59 0.66 7.27 3.39 1.40

Use appropriate graphical or other EDA techniques to obtain an indication as to which of these distributions is the more likely source of the sample.

3

Exploring Averages for Single Samples

3.1 Continuous Data Samples

This chapter is mainly devoted to inferences about location measures for a single sample of continuous data, typically measurements. These provide simple illustrative examples of key nonparametric concepts, although in most real-world problems two or more samples are compared. The methods then used are often described as extensions of, or developments from, those for single samples. Single samples are also useful for showing how to deal with certain real-world complications such as the effect of tied data values, censored observations and rogue or outlying values.

Choosing the right inference procedure depends on both the type of data and the information required. Even with complete measurements on a continuous scale, analyses that use only part of the information contained in that data may be more reliable, and more informative, than analyses using the original data. For example, an analysis based on a ranking of the order statistics will in many cases have better power and higher Pitman efficiency than tests based on the original measurement data. These analyses may be more reliable than using a parametric procedure of doubtful validity.

In Examples 2.4 and 2.5, the presence of a censored observation precluded direct application of a *t*-test. Instead, these data were investigated using a simple procedure based only on a count of the numbers of observations exceeding the median value hypothesized in the relevant null hypothesis. Detailed consideration of the important topic of censored data in survival analysis is postponed until Chapter 9.

In this chapter, several nonparametric tests are developed and their performance is compared using simple data sets. Sections 3.2–3.4 show how to compute and interpret some exact *P*-values and confidence intervals. In practice, even the best currently available software gives exact *P*-values only for small- to medium-sized samples. Sections 3.2.3 and 3.3.3 present some asymptotic results that often provide good approximations for larger samples. If programs for exact tests are not available, a sensible compromise for small- to medium-sized data sets is to use Monte Carlo sampling to get good estimates of exact *P*-values.

3.2 The Sign Test

3.2.1 A Closer Look

In Section 2.3, the sign test was introduced in the context of the survival of retired footballers diagnosed with dementia. Use of the sign test does not require an assumption of distributional symmetry or even one of continuity. Indeed, providing all the observations

DOI: 10.1201/9780429326172-3

are mutually independent, they need not all have the same distributions, but they then must be from distributions with identical medians.

Example 3.1

The severity of seizures experienced by people with epilepsy can be assessed using the Chalfont Seizure Severity Scale (Duncan and Sander, 1991). The scale consists of ten items each relating to a particular characteristic of the seizure, which are scored either in terms of presence/absence or level of severity. For instance, the duration of seizure is scored as <10 seconds (0); 10 seconds to 1 minute (1); 1 to 10 minutes (4); and more than 10 minutes (16). Item scores are summed to give an overall score that can take a value between 1 (least severe) and 178 (most severe). For a sample of ten patients who participated in the Epilepsy Risk Awareness (ERA) study (Braun et al., 2017), the seizure severity scores arranged in ascending order of magnitude were:

Sample I	1	11	35	41	50	62	104	104	151	161

A larger sample of ERA study participants (20 individuals, none from the previous group) had seizure severity scores of:

Sample II	2	10	27	35	38	40	47	50	58	60
	68	70	79	103	108	151	157	158	162	166

Use these samples with a sign test to investigate a population median of 110.

There is no reason to confine alternatives to one direction, so the chosen null and alternative hypotheses are H_0: $\theta = 110$ and H_1: $\theta \neq 110$, respectively.

If the population median is equal to 110, then any observation in a random sample is equally likely to be either above or below 110. Consider the first sample. A plus sign is associated with values greater than 110. If the hypothesis holds, the number of plus signs has a binomial distribution with $n = 10$ observations and a success probability $p = 0.5$, i.e., a B(10, 0.5) distribution. For the alternative, H_1: $\theta \neq 110$, a two-tailed test is appropriate.

There are 2 values, 151 and 161, greater than 110, implying 2 plus signs. From Table 3.1, the probability of 2 or less or 8 or more plus signs is $P = 0.109$. Hence, there is no convincing evidence against H_0.

Output from the ANSM5 package in R can be obtained using `sgn.test(ch3$sampleI, 110)`.

For the second sample, there are 5 values greater than 110 (151, 157, 158, 162 and 166) each given a plus sign. The relevant binomial distribution is for a sample size of 20 and a success probability of 0.5. For this distribution, the probability of 5 or less or 15 or more plus signs is $P = 0.041$, showing moderate evidence against the null hypothesis H_0.

TABLE 3.1

Probabilities of Obtaining r Pluses from 10 Observations under H_0

Number of+	0	1	2	3	4	5
Probability	0.0010	0.0098	0.0439	0.1172	0.2051	0.2461
Number of+	6	7	8	9	10	
Probability	0.2051	0.1172	0.0439	0.0098	0.0010	

Output from the ANSM5 package in R can be obtained using `sgn.test(ch3$sampleII, 110)`.

If one or more sample values coincide with the population median θ specified in H_0, it is unclear as to how to assign plus or minus signs to these observations. They may be ignored, and the sample size reduced by 1 for each such value. For instance, if with the first sample in Example 3.1 the null hypothesis is H_0: $\theta=62$, the problem is treated as one with 9 observations giving 4 plus signs. Doing this rejects evidence that strongly supports the null hypothesis, but which is uninformative about the direction of possible alternatives. Another proposed approach is to toss a coin and allocate a plus sign to a value equal to θ if the coin falls heads and a minus sign if it falls tails: yet another is to assign a plus or minus sign to such a value in a way that makes rejection of H_0 less likely. The coin-toss approach usually makes only a small difference and has little to commend since it only introduces additional randomness, being in a sense more "error". The last approach is ultraconservative. Lehmann (2006, p. 144) discussed the pros and cons of these choices in more detail. Rayner and Best (2001, Section 2.2) discussed these and some more sophisticated options for dealing with this problem.

3.2.2 Confidence Intervals

Example 3.2

Obtain approximate 95% confidence intervals for the sets of seizure severity scores in Example 3.1 and determine the exact confidence levels.

The confidence interval required contains values for the population median θ that would not be rejected in a two-tailed test at a 5% significance level.

Boxplots for the samples are shown in Figure 3.1. For the first sample, the observations are 1 11 35 41 50 62 104 104 151 161. From Table 3.1, a critical region consisting of 0, 1, 9 or 10 plus signs has size $2\times(0.0010+0.0098)=0.0216$, and this is the largest exact size less than 0.05. Thus, H_0 is rejected if it leads to 1 or fewer, or 9 or more, plus signs. The null hypothesis H_0 is retained if between 2 and 8 plus signs are found. It is clear from the sample values above that there are between 2 and 8 plus signs if θ lies in the open interval from 11 to 151, i.e., if $\theta>11$ and $\theta<151$. This implies an exact confidence level of $100(1-0.0216)=97.84\%$ for the interval (11, 151). For any θ not in this interval, the hypothesis that it is the population median would be rejected at the $100-97.84=2.16\%$ significance level.

Output from the ANSM5 package in R can be obtained using `sgn.test(ch3$sampleI)`.

The second sample has values of 2 10 27 35 38 40 47 50 58 60 68 70 79 103 108 151 157 158 162 166. A critical region consisting of 0, 1, 2, 3, 4, 5, 15, 16, 17, 18, 19 or 20 plus signs has size=0.0414,

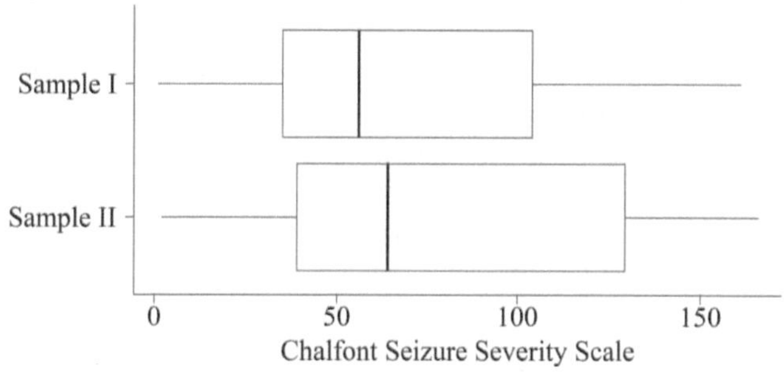

FIGURE 3.1
Boxplots for Samples I and II in Example 3.1.

and this is the largest exact size less than 0.05. Thus, H_0 is rejected if it leads to 5 or fewer, or 15 or more, plus signs. The null hypothesis H_0 is retained if between 6 and 14 plus signs are found. It is clear from the sample values above that there are between 6 and 14 plus signs if θ lies in the open interval from 40 to 108, i.e., if $\theta > 40$ and $\theta < 108$. This implies an exact confidence level of $100(1-0.0414)=95.86\%$ for the interval (40, 108). For any θ not in this interval, the hypothesis that θ is the population median would be rejected at the $100-95.86=4.14\%$ significance level. This confidence interval is noticeably shorter than the one for the first sample.

Output from the ANSM5 package in R can be obtained using `sgn.test(ch3$sampleII)`.

The sample medians (here, equal to 56 and 64, respectively) are appropriate point estimators of the population median when no assumption is made regarding symmetry. The rationale for this choice is that the sample median gives equal numbers of plus and minus signs, the strongest supporting evidence for H_0.

Note that the observations include a possible cluster of values of 151 or more, well above the respective medians. Inspection of the Chalfont Seizure Severity Scale questionnaire (maximum possible score 178) reveals why this could be the case. The last item, given particularly high weight, is "time to return to normal from onset": less than 1 minute (0); 1–10 minutes (5); 10–30 minutes (20); 30–60 minutes (30); 1–3 hours (50); and greater than 3 hours (100). The predominance of this item means that patients with severe epilepsy who take a long time to return to normal are likely to form a separate subgroup.

3.2.3 Asymptotic Approximations for the Sign Test

For sufficiently large samples, the binomial distribution can be approximated by a normal distribution. In particular, this applies to the sign test as it is equivalent to a binomial test with success probability $p=0.5$. If S is the number of successes (here, the number of values above the hypothesized median), then

$$Z = \frac{S - 0.5n}{0.5\sqrt{n}} \tag{3.1}$$

broadly follows a normal distribution with a mean of 0 and a standard deviation equal to 1.

Denoting $1-p$ by q, this approximation is regarded as appropriate if both np and nq are at least 5. The approximation is improved by a continuity correction replacing S by $S+0.5$ if S is less than the value expected under the null hypothesis and by $S-0.5$ otherwise.

Example 3.3

For the samples in Example 3.1, use the asymptotic approximation for the sign test to investigate the hypothesis that $\theta=110$ against the alternative that $\theta \neq 110$.

The first sample consists of $n=10$ observations, and the number of values above the hypothesized median (S) is equal to 2. The proportion of values above the median (p) is therefore $2/10=0.2$, so $q=0.8$. Hence, $np=2$ and $nq=8$, so the use of the normal approximation might cause problems. Since S is lower than the value expected under H_0 ($10\times 0.5=5$), for the continuity correction, its value of 2 is replaced by 2.5.

Applying the continuity correction:

$$Z = \frac{2.5 - (0.5 \times 10)}{0.5\sqrt{10}} \approx -1.58$$

equivalent to $P=0.114$.

Output from the ANSM5 package in R can be obtained using `sgn.test(ch3$sampleI, 100, do.asymp = TRUE)`.

Without the continuity correction:

$$Z = \frac{2-(0.5\times 10)}{0.5\sqrt{10}} \approx -1.90$$

which is equivalent to $P=0.058$.

Output from the ANSM5 package in R can be obtained using `sgn.test(ch3$sampleI, 100, do.asymp = TRUE, cont.corr = FALSE)`.

For the second sample, the number of observations is $n=20$, with $S=5$ values above the hypothesized median. The proportion of values above the median (p) is therefore $5/20=0.25$, so $q=0.75$. Hence, $np=5$ and $nq=15$, and the use of the normal approximation can be considered appropriate. Since S is lower than the value expected under H_0 ($20\times 0.5=10$), for the continuity correction, its value of 5 is replaced by 5.5.

Applying the continuity correction:

$$Z = \frac{5.5-(0.5\times 20)}{0.5\sqrt{20}} \approx -2.01$$

equivalent to $P=0.044$.

Output from the ANSM5 package in R can be obtained using `sgn.test(ch3$sampleI, 100, do.asymp = TRUE)`.

Without the continuity correction:

$$Z = \frac{5-(0.5\times 20)}{0.5\sqrt{20}} \approx -2.24$$

which is equivalent to $P=0.025$.

Output from the ANSM5 package in R can be obtained using `sgn.test(ch3$sampleII, 100, do.asymp = TRUE, cont.corr = FALSE)`.

The exact P-values for the first and second samples are 0.109 and 0.041, respectively. Despite the small samples, the use of the asymptotic method with the continuity correction gives adequate approximate P-values, leading to similar conclusions. If the continuity correction is not used, the method performs less well, overestimating the strength of evidence against each null hypothesis. With small samples, inappropriate reliance on asymptotic methods can lead to misleading findings.

SUMMARY POINTS:

- No assumption of symmetry is required for the validity of the sign test (Section 3.2.1).
- The sign test statistic is the number of observations above or below a median specified in the null hypothesis (see, e.g., Example 3.1, Section 3.2.1).

- Significance is determined using binomial distribution probabilities with the success probability p equal to 0.5 (Section 3.2.1).
- For hypothesis testing, modifications are required to deal with values equal to the median specified in the null hypothesis (Section 3.2.1).
- Confidence intervals can be obtained using binomial distribution probabilities (Section 3.2.2).
- Normal theory approximations (Section 3.2.3) may be used for sample sizes such that both np and nq are at least 5.

3.3 Inferences About Medians Based on Ranks

A procedure for hypothesis testing and estimation using a permutation test applied to the ranked magnitudes of deviation from a hypothesized median is now examined. This test, proposed by Wilcoxon (1945), has many attractive features.

3.3.1 The Wilcoxon Signed-Rank Test

This procedure has some of the characteristics of the sign test, but rather than just considering differences as positive or negative, the size of each difference is taken into account. Suppose that a sample of size n is drawn from a symmetric continuous distribution with an unknown median θ. The null hypothesis H_0: $\theta=\theta_0$ is to be tested against the two-sided alternative H_1: $\theta\neq\theta_0$. The assumption of symmetry implies that under H_0:

- For each observation in the sample, the sign of its difference from θ_0 is equally likely to be positive or negative.
- A difference of any particular magnitude is equally likely to be positive or negative.

As noted previously, for a symmetric population distribution, the mean coincides with the median. In those circumstances, the test described here may also be formulated in terms of means. Tests against a one-sided alternative are also possible.

To carry out the test, proceed as follows:

- Determine the *magnitude* of the deviation of each data value from the hypothetical median θ_0 specified in H_0 without regard to whether the deviation is positive or negative.
- Arrange these absolute deviations in increasing order of magnitude and assign ranks to these in ascending order (1 for the smallest, sample size n for the largest).
- Attach a negative sign to each rank that corresponds to a negative deviation (i.e., to ranks assigned to data values below θ_0) and a positive sign to each rank that corresponds to a positive deviation. The resulting ranks with attached signs are referred to as *signed ranks*.

If θ_0 is indeed the population median, a near-equal scatter of positive and negative ranks would be expected. This implies that under the null hypothesis the sum of all positive ranks and the sum of all negative ranks should not differ greatly. A high sum of the positive (negative) ranks relative to that of the negative (positive) ranks implies θ_0 is unlikely to be the population median.

The appropriate permutation test is described here for a sample size of $n=7$, making the assumption that no deviations have the same magnitude. The ranks therefore sum to $1+2+3+4+5+6+7=28$. Denote by S_+ the sum of the ranks associated with positive deviations from the median θ_0 specified in H_0 and by S_- the sum of ranks associated with negative deviations. If all sample values are above θ_0, all ranks will be positive, so $S_-=0$ and $S_+=1+2+3+4+5+6+7=28$. Similarly, if all sample values are below θ_0, then $S_+=0$ and $S_-=28$. These outcomes represent the strongest possible evidence against H_0. Intermediate values of S_+ or S_- cast less doubt on H_0.

For any allocation of signs to ranks 1–7, $S_++S_-=28$ and $S_+=28-S_-$. This implies that any of the statistics S_+ and S_- or the difference $S_d=|S_+-S_-|$ can be used as a test statistic since there is an ordered linear relationship between them.

When $n=7$, signs may be allocated randomly to ranks in $2^7=128$ different and equally likely ways. The distribution of the statistic S_+ depends only on the sample size, 7, since any data set gives rise to the rank magnitudes 1, 2, 3, 4, 5, 6 and 7.

The distribution of S_+ may be obtained by recording the value of S_+ for all possible sign allocations. For example, with $k=2$, there is only one possible combination of ranks (when just rank 2 has a positive sign), whereas for $k=3$, there are two possible combinations of ranks (rank 3 with a positive sign, ranks 1 and 2 each with a positive sign). The probabilities for each k are obtained by dividing the number of times a sum occurs by 128. It is left for the reader to complete the relevant sums and associated probabilities in Table 3.2.

TABLE 3.2

Exact Probabilities That S_+ (or S_-) Equals a Value k in a Wilcoxon Signed-Rank Test for Sample Size $n=7$ When H_0 Is True

K	Probability	*K*	Probability
0	1/128	15	8/128
1	1/128	16	8/128
2	1/128	17	7/128
3	2/128	18	7/128
4	2/128	19	6/128
5	3/128	20	5/128
6	4/128	21	5/128
7	5/128	22	4/128
8	5/128	23	3/128
9	6/128	24	2/128
10	7/128	25	2/128
11	7/128	26	1/128
12	8/128	27	1/128
13	8/128	28	1/128
14	8/128		

Exact P-values for one- or two-tailed tests on a sample of $n=7$ observations with no tied values may be obtained from Table 3.2. For example, in a one-tailed test where a small lower-tail value of S_+ is relevant, the exact P-value corresponding to a sum of positive deviations $S_+=3$ is:

$$P = \Pr(S_+ \le 3) = (1+1+1+2)/128 = 5/128 \approx 0.039.$$

This probability is doubled for a two-tailed test.

Example 3.4

The heart rate (beats per minute) when standing was recorded for seven members of a tutorial group. It is assumed that in the population of such students the distribution of heart rates when standing is symmetric. Given observations of 73, 82, 87, 68, 106, 60 and 97, test the hypothesis H_0: $\theta=70$ against the alternative H_1: $\theta>70$ using the Wilcoxon signed-rank test.

Symmetry implies that under H_0 sample values should have a near-symmetric scatter about $\theta=70$, so the magnitudes of the sums of positive and of negative deviations from 70 (denoted by d_i) should not differ greatly. The one-sided alternative, H_1, implies that when it holds there are likely to be more positive deviations from 70 than there would be under H_0 and that these positive deviations will tend to be larger than the negative deviations. These properties carry over to the signed ranks of the deviations.

For the given data, deviations from 70 are $73-70=3$ and $82-70=12$, and similarly, those remaining deviations are 17, –2, 36, –10 and 27. The corresponding signed ranks derived from allocating signs to the ranks of the absolute differences $|d_i|$ are 2, 4, 5, –1, 7, –3 and 6. Under H_0, the sum of the positive and negative signed ranks, S_+ and S_-, should be close to equal, whereas under H_1, it is likely that S_+ will be appreciably larger than S_-. If the alternative had been the two-sided H_1: $\theta\neq 70$, small and large values of S_+ would support H_1. For these signed ranks, $S_+=2+4+5+7+6=24$. Table 3.2 shows that $\Pr(S_+\ge 24)=(2+2+1+1+1)/128=7/128\approx 0.055$.

Output from the ANSM5 package in R can be obtained using `wilcoxon.signedrank(ch3$heartrates1, 70, "greater")`.

When $P=0.055$, the evidence against H_0 is sufficiently strong to perhaps warrant a similar study with a larger group of students.

This example shows that, despite the warning in Section 1.5 that small experiments may never produce P-values indicating strong evidence against H_0, an interesting effect may sometimes be suggested even with very small samples. This is encouraging, given the current emphasis on small group work in teaching, where data concerning individuals in the group are often used for illustrative purposes. A teaching advantage of using data from small groups is that tedious arithmetic that may hide the logic of the problem is avoided. However, this example draws attention to a major limitation of a conventional hypothesis test. While it indicates that there are some doubts about the wisdom of retaining H_0, it provides no indication as to what other hypotheses might be plausible. A confidence interval is needed to give this information. This is a problem considered in Section 3.3.2.

Modifications for a two-tailed test, when that is appropriate, are straightforward.

The test statistic S_+ was chosen for this example, but S_-, the sum of the negative deviations, could have been used instead. This is because, as already pointed out, the sum of the unsigned ranks remains constant for all possible allocations of signs. Computer software sometimes calculates the smaller of S_+ and S_-, but each leads to the same relevant P-value, as does $S_d= |S_+-S_-|$.

Following the classic convention of requiring $P \leq 0.05$ before asserting significance would not have led to a rejection of the null hypothesis in a formal significance test at the 5% level. There is no merit in following such an arbitrary rule in this example. It is preferable to report that the case for retaining H_0 is not strongly supported, even though 0.055 slightly exceeds 0.050. The normal theory *t*-test result for this example gives $P=0.051$ for a one-tailed test.

Manual calculation of probabilities associated with all values of S_+ or S_- for all but small samples is time-consuming and error-prone. The number of different values of the rank sums increases with increasing sample size. The differences between the discrete probabilities associated with each sum become smaller, especially in the tails. Also, the distributions of S_+, S_- approach that of a normal distribution as the sample size increases. When $n=12$, there are $2^{12}=4{,}096$ possible associations of signs with ranks, and the possible signed-rank sums range from 0 to 78. Symmetry means that it is only necessary to compute the probabilities for each sum between 0 and 39.

Table 3.3 gives sufficient information to calculate the complete permutation test distribution of S_+ or S_-. From Table 3.3, adding up the probabilities between $k=0$ and $k=13$ inclusive gives $\Pr(S_- \leq 13)=87/4{,}096 \approx 0.021$. Similarly, summing the probabilities between $k=0$ and $k=14$ inclusive gives $\Pr(S_- \leq 14)=107/4{,}096 \approx 0.026$. For two-tailed tests, these probabilities are doubled giving 0.042 and 0.052. Thus, values for which the smaller of the signed-rank sums S_+ and S_- do not exceed 13 indicate significance at level 4.2% or less, i.e., below a conventional 5% level. Similarly, from Table 3.3, $\Pr(S_- \leq 7)=19/4{,}096 \approx 0.0046$, while $\Pr(S_- \leq 8)=25/4{,}096 \approx 0.0061$. Doubling these probabilities for the two-tailed tests gives 0.0092 and 0.0122. Thus, values for which the smaller of the signed-rank sums S_+ and S_- do not exceed 7 indicate significance at level 0.92% or less, i.e., below a conventional 1% level.

Figure 3.2, based on Table 3.3, shows the frequency function of S_+ or S_- for a sample of size 12 in the form of a bar chart. The shape resembles that of the familiar probability density function for the normal distribution. In Section 3.3.3, important implications of this approach to normality are considered.

TABLE 3.3

Exact Probabilities That S_+ (or S_-) Equals a Value k in a Wilcoxon Signed-Rank Test for Sample Size $n=12$ When H_0 Is True

k	Probability	*K*	Probability	*k*	Probability
0	1/4,096	14	20/4,096	28	89/4,096
1	1/4,096	15	24/4,096	29	94/4,096
2	1/4,096	16	27/4,096	30	100/4,096
3	2/4,096	17	31/4,096	31	104/4,096
4	2/4,096	18	36/4,096	32	108/4,096
5	3/4,096	19	40/4,096	33	113/4,096
6	4/4,096	20	45/4,096	34	115/4,096
7	5/4,096	21	51/4,096	35	118/4,096
8	6/4,096	22	56/4,096	36	121/4,096
9	8/4,096	23	61/4,096	37	122/4,096
10	10/4,096	24	67/4,096	38	123/4,096
11	12/4,096	25	72/4,096	39	124/4,096
12	15/4,096	26	78/4,096		
13	17/4,096	27	84/4,096		

For $k>39$, by symmetry, $\Pr(S_+=k)=\Pr(S_+=78-k)$.

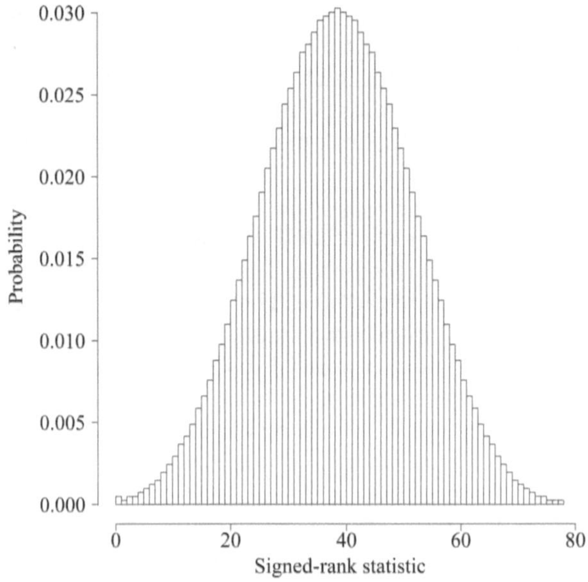

FIGURE 3.2
Bar chart illustrating the distribution of the Wilcoxon signed-rank test when $n = 12$.

Example 3.5

For a group of 12 female students, the changes in heart rate (beats per minute) when standing up from lying down are:

–2 4 8 25 –5 16 3 1 12 17 20 9

If population symmetry is assumed, the Wilcoxon signed-rank test may be used to investigate H_0: $\theta = 15$ against the alternative H_1: $\theta \neq 15$. The deviations from 15 are arranged in ascending order of magnitude and ranked, associating with each rank the sign of the corresponding deviation. The lesser of the sum of positive and negative signed ranks is calculated.

Subtracting 15 from each sample value gives the deviations:

–17 –11 –7 10 –20 1 –12 –14 –3 2 5 –6.

These rearranged in increasing order of magnitude while retaining signs become:

1 2 –3 5 –6 –7 10 –11 –12 –14 –17 –20

whence the signed ranks are:

1 2 –3 4 –5 –6 7 –8 –9 –10 –11 –12.

The sum of the positive ranks, $S_+ = 1 + 2 + 4 + 7 = 14$, is less than the sum of the negative ranks, S_-. From Table 3.3, the one-tailed P-value corresponding to $S_+ = 14$ is 107/4,096 = 0.026.

Output from the ANSM5 package in R can be obtained using `wilcoxon.signedrank(ch3$heartrates2, 15, "less")`.

By symmetry, the two-tailed *P*-value corresponding to $S_+=14$ is therefore $P=2\times 107/4{,}096=0.052$. The evidence against H_0 is not very strong, but the *P*-value is sufficiently small for a further study involving a larger group to be considered.

Output from the ANSM5 package in R can be obtained using `wilcoxon.signedrank(ch3$heartrates2, 15)`.

Doctors often assume that heart rate measurements are symmetrically distributed. In this example, the median, 8.5, is close to the mean, 9.0, indicating that symmetry is possible. There is, however, some evidence of a departure from normality because there is no particular cluster of values around the mean of 9.0; instead, several values are close to zero, and there is another group of values around 17. Given the likely degree of symmetry in the distribution, the Wilcoxon test should not be seriously misleading.

The two clusters of values may reflect two distinct groups of students: those who are heavily involved in sports such as hockey or tennis and those who rarely take part. Females who are physically active may not experience a large increase in heart rate on standing compared with those who follow a more sedentary lifestyle. In addition, one must be cautious about generalizing the above results to the community at large. For several reasons, female students are probably not typical of young adults in general.

The asymptotic *P*-value for these data for a two-tailed test without using a continuity correction described in Section 3.3.3 is $P=0.050$ but increases to $P=0.0546$ when the continuity correction is used. In the light of the approximate normality of the Wilcoxon statistic evident in Figure 3.2, this good agreement with the exact test value is not surprising. For these data, the two-tailed *t*-test value is $P=0.046$. This does not differ greatly from the exact permutation test value, but gives a timely warning on the dangers of adhering to a rigid significance level such as $\alpha=0.05$ as a decision-making basis for accepting or rejecting H_0.

There are several tests for symmetry when only the sample values are given. One proposed by Randles et al. (1980) was described by Hollander, Wolfe and Chicken (2014, Section 3.9) and also by Siegel and Castellan (1988, Section 4.4), but caution is needed in assessing significance if there are fewer than 20 cases where the asymptotic result is on the borderline of significance. However, for these data, the test statistic has a value well below any that would suggest asymmetry. The required computations are tedious rather than difficult, and the reader should refer to either of the above references for details.

3.3.2 Confidence Intervals Based on Signed Ranks

For nonparametric methods, including straightforward techniques such as the sign test (Section 3.2), the more important aim of calculating confidence intervals is not as simple as hypothesis testing. However, the procedures, once mastered, often require only routine calculations that are not difficult to program by writing a macro (a single command that expands automatically into a set of instructions) if no package program is available.

The theory is now outlined, but a reader prepared to take this on trust may move directly to Example 3.6 to see an application. Assume that a sample of n observations with no ties has been arranged in ascending order $x_1, x_2, \ldots, x_n$. The null hypothesis for the mean or median is specified as H_0: $\theta=\theta_0$. In practice, θ_0 will usually lie between the lowest and highest values, but this is not a theoretical requirement.

A deviation $x_i-\theta_0$ will have a negative signed rank for any x_i less than θ_0.

Since the x_i values are in ascending order, among all negative ranks, if there are any, that of the greatest magnitude (disregard the negative sign) will be associated with x_1, that of the next greatest magnitude with x_2 and so on. Similarly, all observations greater than θ_0 have positive signed ranks. Of these, x_n will have the highest rank, x_{n-1} the next highest and so on.

Consider the paired averages of x_1 with each of $x_1, x_2, \ldots, x_n$. Suppose that the signed rank of the deviation associated with x_1 is $-p$. Each of the averages $(x_1+x_q)/2$ will be less than θ_0 when the deviation associated with x_q has either a negative rank or a positive rank less than p. Since rank ties are assumed to be absent, no observation has a positive rank p (since the rank p is already associated with x_1). If x_q is the smallest observation associated with a positive rank greater than p, this implies that $|x_q-\theta_0| > |x_1-\theta_0|$. It follows that the average of x_1 and x_q is greater than θ_0. Then, because the data are in ascending order, so is the average of x_1 and any $x_r>x_q$. Thus, the number of averages involving x_1 that are less than θ_0 is equal to the negative rank associated with x_1.

Similarly, if for any x_i less than θ_0 paired averages are found with each of $x_i, x_{i+1}, \ldots, x_n$, the number of these averages less than θ_0 will equal the (negative) rank associated with x_i. When x_i has a positive signed rank, none of the averages with itself or greater sample values will be less than θ_0. The complete set of averages $(x_i+x_j)/2$ for $i=1, 2, \ldots, n$ and all $j \geq i$ are called *Walsh averages*, which were proposed by Walsh (1949a, b). By inspecting these averages, it is seen that the number of Walsh averages less than θ_0 equals S_- and the number of Walsh averages greater than θ_0 equals S_+. Walsh averages may be used to calculate the test statistic S for a given H_0 and to obtain confidence limits for a population mean or median in the way illustrated in the following example.

Example 3.6

Consider again the data on changes in heart rate for the female students in Example 3.4, viz.:

$$-2 \quad 4 \quad 8 \quad 25 \quad -5 \quad 16 \quad 3 \quad 1 \quad 12 \quad 17 \quad 20 \quad 9.$$

Obtain an estimate of the population median with confidence intervals at levels at least 95% and 99%, using Walsh averages.

Walsh averages are used along with the principle explained in Section 1.6.1 that the $100(1-\alpha)\%$ confidence interval contains those values of the parameter that would be accepted in a significance test at the $100\alpha\%$ significance level.

Table 3.4 shows a convenient way to set out Walsh averages. Both the top row and the first column (in italics) of the table give the sample values in ascending order. The triangular matrix forming the body of the table gives the Walsh averages corresponding to the relevant top row and first column values.

An intuitively reasonable point estimator of the population median is the median of the Walsh averages since it follows from the arguments given before this example that for this estimate the sums of both the negative and positive rank deviations are equal, both being equal to the number of Walsh averages above (or below) this median. The median of the Walsh averages may be determined by inspecting Table 3.4. Entries in any row increase from left to right, and those in any column increase from top to bottom, so that the smallest entries are at the top left of the table and the largest entries are at the bottom right of the table. For a sample size of n, there are $n(n+1)/2$ Walsh averages, i.e., $12\times13/2$ or 78 in this example. Thus, the median of these lies between the 39th- and 40th-ordered Walsh

TABLE 3.4

Walsh Averages for Changes in Heart Rate from Lying to Standing

−5	−5.0	−3.5	−2.0	−1.0	−0.5	1.5	2.0	3.5	5.5	6.0	7.5	10.0
−2		−2.0	−0.5	0.5	1.0	3.0	3.5	5.0	7.0	7.5	9.0	11.5
1			1.0	2.0	2.5	4.5	5.0	6.5	8.5	9.0	10.5	13.0
3				3.0	3.5	5.5	6.0	7.5	9.5	10.0	11.5	14.0
4					4.0	6.0	6.5	8.0	10.0	10.5	12.0	14.5
8						8.0	8.5	10.0	12.0	12.5	14.0	16.5
9							9.0	10.5	12.5	13.0	14.5	17.0
12								12.0	14.0	14.5	16.0	18.5
16									16.0	16.5	18.0	20.5
17										17.0	18.5	21.0
20											20.0	22.5
25												25.0

averages. Inspection of the table establishes that both equal 9.0, so this is the appropriate estimate of the population median (or mean if it exists).

To establish a 95% confidence interval, select as end points values of θ that will just be acceptable if $P=0.05$. In Example 3.4, P was found to be only slightly more than 0.05 if the sum of the lesser of positive or negative ranks is 14.

Thus, if a value less than the 14th smallest Walsh average is chosen, H_0 is rejected. Similarly, if a value greater than the 14th largest Walsh average is chosen, H_0 is also rejected since in either case the test statistic S would be such that $S \leq 13$. Thus, for $n=12$, a confidence interval at not less than a 95% confidence level (i.e., implying no more than 2.5% in each tail) is the interval (14th smallest Walsh average, 14th largest Walsh average). From Table 3.4, by counting from the most extreme value, this is seen to be the interval (2.5, 16.0). From Table 3.3, using $S \leq 13$ as a criterion for significance corresponds to an exact significance level of 4.25% in a two-tailed test. It is intuitively reasonable to regard the interval (2.5, 16.0) as an actual 95.75% confidence interval.

Output from the ANSM5 package in R can be obtained using `wilcoxon.signedrank(ch3$heartrates2)`.

Owing to discontinuities in the distribution of the test statistic and the fact that the end points of the confidence interval often coincide with observed values, values outside this interval must be regarded as ones indicating rejection at a 4.25% significance level. In addition, if the exact level is critical, the end values must be tested to see whether they would be accepted or rejected at this significance level. Exact tests show evidence would be found against the hypothesized means of 2.5 and 16.0. Thus, in this case, all points in the open interval $2.5<\theta<16.0$ would be acceptable, but the closed interval [2.5, 16.0] is a conservative 95% interval, having a confidence level exceeding 95%. In practice, it is usual to cite a confidence interval simply as the interval (2.5, 16.0) without being specific as to whether it is open or closed, bearing in mind that discontinuities inherent in the Wilcoxon test statistic distribution may imply (and it will vary from case to case) that the end points may or may not be included as acceptable values at a nominal significance level such as 5%. One should not get concerned about subtleties due to discontinuity at the end point of intervals. If computer programs are used, it may be worthwhile checking the actual confidence level associated with any quoted interval.

From Tables 3.3 and 3.4, a confidence interval at not less than a 99% confidence level (implying no more than 0.5% in each tail) is (8th smallest Walsh average, 8th largest Walsh average) giving the closed interval [0.5, 18.0].

An appropriate point estimator of the population median is therefore 9.0, and actual 95.75% and 99.08% confidence intervals are, respectively, (2.5, 16.0) and (0.5, 18). This implies that if values of θ outside these intervals are specified in the null hypothesis, there will be evidence against these at levels given by the relevant *P*-values, i.e., $P=0.0425$ and $P=0.0092$.

Output for these intervals from the ANSM5 package in R can be obtained using `wilcoxon.signedrank(ch3$heartrates2)` and `wilcoxon.signedrank(ch3$heartrates2, CI.width = 0.99)`.

A possible alternative to the median of the Walsh averages as a point estimator is the mean of all sample values. It is unbiased in the sense that if samples were taken repeatedly and the mean of the sample values taken as the estimate in all cases, the mean of these estimates would converge to the population mean providing that mean exists. It does not, for example, exist for the Cauchy distribution. However, the median of the Walsh averages is, in a technical sense, not discussed here, generally closer to the population median. This estimator is called the Hodges–Lehmann estimator, which was proposed by Hodges and Lehmann (1963).

If only confidence limits are required, just a few Walsh averages in the top left and bottom right of the triangular matrix in Table 3.4 need to be found. If a computer program is available to form this matrix, there is no problem in computing all the averages. If the matrix has to be formed manually, once the Walsh averages in the first row have been calculated, those in the second row may be obtained by adding the same constant, namely half the difference x_2-x_1, to the average immediately above it. To verify this, denote by x_{ij} the average of x_i and x_j. This is the entry in row i and column j of the matrix of Walsh averages. Now, for any k greater than 2:

$$x_{2k} = \tfrac{1}{2}(x_2 + x_k) = \tfrac{1}{2}(x_2 - x_1 + x_1 + x_k) = \tfrac{1}{2}(x_2 - x_1) + \tfrac{1}{2}(x_1 + x_k) = \tfrac{1}{2}(x_2 - x_1) + x_{1k}.$$

The idea generalizes for all subsequent rows, each entry in row $r+1$, say, being found by adding the same constant $(x_{r+1}-x_r)/2$ to the entry immediately above it.

In Example 3.5, testing H_0: $\theta=15$ gave $S_+=14$. In Table 3.4, there are 14 Walsh averages exceeding 15.0.

3.3.3 Asymptotic Approximations for the Wilcoxon Signed-Rank Test

Using software for exact permutation tests removes the need to use asymptotic results if the number of observations is modest. Asymptotic results may not be reliable unless samples are moderately large. The point at which the size is adequate depends partly upon the procedure being used. In the case of the Wilcoxon signed-rank test, Figure 3.2 suggests that the distribution of the test statistic (S_+ or S_-) approaches normality when $n=12$. It can be shown that for sufficiently large n:

$$Z = \frac{S - E(S)}{\sqrt{\mathrm{Var}(S)}}. \tag{3.2}$$

takes approximately a standard normal distribution with:

$$E(S) = n(n+1)/4 \text{ and } \mathrm{Var}(S) = n(n+1)(2n+1)/24.$$

The formula (3.2) can be improved by a *continuity correction* that allows for the fact that a continuous distribution is being approximated with a discrete distribution where S can take only integer values. If S is the smaller of the sums of the positive or negative ranks, the appropriate correction is to replace S by $S+1/2$. If S is the larger sum, it is replaced by $S-1/2$.

If the statistic $S_W = |S_+ - S_-|$, equation (3.2) reduces to:

$$Z = \frac{S_W}{\sqrt{\mathrm{Var}(S_W)}},$$

where $\mathrm{Var}(S_W) = n(n+1)(2n+1)/6$. The continuity correction is now to subtract 1 from the numerator if $S_+ < S_-$ and to add 1 if $S_+ > S_-$, since changing either S_+ or S_- by 1 changes the value of S_W by 2.

Example 3.7

For the sample of 12 female students in Example 3.5, use the asymptotic approximation to test the hypothesis that $\theta = 15$ against the alternative that $\theta \neq 15$.

Assume that a sample of 12 is sufficiently large to justify using (3.2). From Example 3.5, it is seen that $S_+ = 14$ when $n = 12$ and the use of the formulae given above shows that:

$$E(S) = (12 \times 13)/4 = 39 \text{ and } \mathrm{Var}(S_+) = (12 \times 13 \times 25)/24 = 162.5,$$

Whence using (3.2) with a continuity correction gives:

$$Z = \frac{14.5 - 39}{\sqrt{162.5}} \approx -1.92.$$

This is equivalent to $P = 0.055$, as shown in the output from the ANSM5 package using `wilcoxon.signedrank(ch3$heartrates2, 15, do.asymp = TRUE)`. Without the continuity correction, (3.2) gives $Z \approx -1.96$ equivalent to $P = 0.050$, as shown using `wilcoxon.signedrank(ch3$heartrates2, 15, do.asymp = TRUE, cont.corr = FALSE)`.

Here, as expected from the earlier discussion based on Figure 3.2, there is close agreement with the permutation test result. The continuity correction here makes little practical difference unless one places undue weight upon a magic cut-off point for significance at $\alpha = 0.05$.

It is possible to obtain confidence limits for the values of S_+ and S_- that arise with acceptable hypothesized values of θ for some chosen α. One may use these in association with a table of Walsh averages to determine a confidence interval for the true value of θ.

3.3.4 Wilcoxon with Ties

In theory, when sampling from a continuous distribution, the probability of tied observations, or of obtaining a sample value exactly equal to the population mean or median, is zero. Real observations never have a distribution that is strictly continuous. This may be because of their nature, rounding or limited measurement precision.

Lengths may be measured to the nearest centimetre or millimetre; weights may be measured to the nearest kilogram, gram or milligram. Numbers of words in a sentence must be an integer. These realities may produce data with rank ties or zero departures from a hypothesized mean or median. If they do, the exact distribution of S, the lower of the positive or negative rank sums, requires fresh computation for different numbers of ties and ties in different positions in the rank order.

Before the advent of software to compute exact P-values for permutation distributions with ties, a common procedure was, after adjusting the scoring method based on ranks to allow for ties, to use the asymptotic normal approximation introduced in Section 3.3.3 after certain small adjustments. This usually works well for a few ties if n is reasonably large, but may have bizarre consequences for small n. Another approach for smaller samples was to use tables for the no-tie case assuming that the relevant value of S would be little affected by ties. This need not be so.

When two or more ranks are equal in magnitude, but not necessarily of the same sign, the most widely used procedure, and the one adopted here, is to replace them by their mid-rank, an idea easily explained by an example. If 7 observations are 4, 4, 8, 8, 11, 11 and 11, for a Wilcoxon signed-rank test with H_0: $\theta=6$, the relevant signed differences from 6 are –2, –2, 2, 2, 5, 5 and 5. The magnitude of the smallest difference is 2, and four differences have that magnitude. The mid-rank rule assigns to these ties the mean of the four smallest ranks 1, 2, 3 and 4; i.e., these observations are allocated the signed ranks –2.5, –2.5, 2.5 and 2.5. The remaining three differences are all 5, to which each is given the mean of the remaining ranks 5, 6 and 7; i.e., each is ranked 6. The exact permutation test is based on the signed mid-ranks –2.5, –2.5, 2.5, 2.5, 6, 6 and 6. An appropriate test statistic is $S_-=5$, the sum of the negative mid-ranks.

Testing the null hypothesis using the permutation distribution for the no-ties case given in Table 3.2 gives $\Pr(S_-\leq 5)=(1+1+1+2+2+3)/128=10/128\approx 0.078$. However, with ties, if the mid-rank procedure is used, the distribution of S_- under H_0 is not that in Table 3.2. The exact distribution with ties depends on both how many ties there are and where they lie in the rank sequence. Exact significance levels with ties can be calculated for small- to medium-sized samples.

In each case, the distribution of the test statistic is symmetric (Tables 3.2 and 3.5), which is why the test may be based on either S_+ or S_-. However, the distribution with no ties has

TABLE 3.5

Exact Probabilities That S_+ (or S_-)$=k$ in a Wilcoxon Signed-Rank Test for Sample Size $n=7$ with the Null Hypothesis True and Mid-Ranks of 2.5 (Four Times) and 6 (Three Times)

k	Probability	k	Probability
0.0	1/128	14.5	12/128
2.5	4/128	16.0	3/128
5.0	6/128	17.0	18/128
6.0	3/128	18.0	1/128
7.5	4/128	19.5	12/128
8.5	12/128	20.5	4/128
10.0	1/128	22.0	3/128
11.0	18/128	23.0	6/128
12.0	3/128	25.5	4/128
13.5	12/128	28.0	1/128

only one mode with values of the statistic confined to integer values increasing in unit steps, while that in Table 3.5 is heavily multimodal with S taking unevenly spaced, and not necessarily integer, values. Discontinuities are also more marked.

From Table 3.5, or by using the ANSM5 package in R and the command `wilcoxon.signedrank(ch3$withties, 6, "greater")`, we obtain $\Pr(S_- \le 5) = (1+4+6)/128 = 11/128 \approx 0.086$, which is close to the no-tie probability $10/128 \approx 0.078$ found from Table 3.2. Despite many ties in this example and the differences in the distribution of the test statistic in the two cases, it may appear that ties have little effect on the conclusion. This is not always true.

Example 3.8

Ties often result from rounding. The data 4, 4, 8, 8, 11, 11 and 11 considered above could arise from rounding to the nearest integer in either of the sets:

3.9	4.1	8.2	8.3	10.9	11.0	11.1
3.9	4.0	7.8	7.9	10.9	11.0	11.1

For a Wilcoxon test with H_0: $\theta = 6$, the signed differences from 6 are:

−2.1	−1.9	2.2	2.3	4.9	5.0	5.1
−2.1	−2.0	1.8	1.9	4.9	5.0	5.1

with signed ranks:

−2	−1	3	4	5	6	7
−4	−3	1	2	5	6	7

There are now no tied ranks in either set, and for the first set, $S_- = 3$, while for the second set, $S_- = 7$. Table 3.2 gives $\Pr(S_- \le 3) = 5/128 \approx 0.039$ and $\Pr(S_- \le 7) = 19/128 \approx 0.148$. Thus, a one-tailed test at a level not exceeding 5% would indicate an implausible hypothesis in the first case, but not in the second. Rounding to the nearest integer may greatly affect the strength of evidence against the null hypothesis.

The above example is a simple illustration of difficulties of interpretation with ties due to rounding, a process that makes small perturbations or changes to most observations. Many statistical analyses are sensitive to such changes. The impact of ties due to rounding in this example is more extreme than one is likely to meet in practice, since it results in only two different magnitudes for ranks. For larger samples and not too many ties, the effect on P-values is less marked.

Example 3.9

The age in months of 12 inpatients selected at random from a children's hospital were:

6 18 24 26 37 40 42 47 69 69 87 130

Use the Wilcoxon signed-rank test to test for the population median θ the hypothesis H_0: $\theta = 30$ against the alternative H_1: $\theta > 30$.

To justify the Wilcoxon test, the population is assumed to be symmetric. The deviations of observations from 30 are ranked with due regard to sign, using mid-ranks for ties. The lesser of S_+ and S_- is then found.

The signed deviations from 30 are:

$$-24 \quad -12 \quad -6 \quad -4 \quad 7 \quad 10 \quad 12 \quad 17 \quad 39 \quad 39 \quad 57 \quad 100$$

and the corresponding signed ranks, using mid-ranks for the tied magnitudes at 12 and 19, are:

$$-8 \quad -5.5 \quad -2 \quad -1 \quad 3 \quad 4 \quad 5.5 \quad 7 \quad 9.5 \quad 9.5 \quad 11 \quad 12$$

giving $S_-=16.5$. A one-tailed test is required for the null hypothesis. Output from the ANSM5 package in R using the command `wilcoxon.signedrank(ch3$ages, 30, "greater")` gives the exact probability of observing $S_- \leq 16.5$ with this pattern of tied deviations when $n=12$ as $P=0.0405$. There is moderate evidence against H_0.

Even without software to give the exact distribution of the test statistic when there are ties, the advice to resort to asymptotic (large sample) results of a type described below has little justification for small samples. The data values 69, 69, 87 and 130 throw doubt on the validity of the symmetry assumption. Had these observations been replaced by, say, 55, 55, 60 and 67, the Wilcoxon rank-sum test would give the same result, but then, there would have been much less reason to doubt the symmetry assumption. In this case, but not in all such situations, the test is said to be *robust* against the outliers.

It is a useful exercise to generate exact distributions for a number of tied situations for small sample sizes to develop a feeling for the effect of ties on standard results.

Example 3.10

An extreme case arises when all ranks of deviations are tied apart from signs. Consider the observations 4, 4, 8, 8, 8, 8 and 8 with a null hypothesis for the median of $H_0=6$. The deviations from the null hypothesis value (6) are –2, –2, 2, 2, 2, 2 and 2. These are equal in magnitude, so using the mid-rank approach, each observation receives the rank $(1+2+3+4+5+6+7)/7=28/7=4$. The signed ranks are therefore –4, –4, 4, 4, 4, 4 and 4. Table 3.6 shows the exact permutation distribution for S_+.

As an alternative approach, with such extreme tying, one could attach a plus sign to each positive rank and a minus sign to each negative rank. The distribution for the number of pluses would be binomial (details in Example 2.4):

The distributions shown in Tables 3.6 and 3.7 are equivalent, and hence, the Wilcoxon test reduces to the sign test if all deviation magnitudes are tied at the same value.

TABLE 3.6

Exact Probabilities That S_+ (or S_-)$=k$ in a Wilcoxon Signed-Rank Test for Sample Size $n=7$ When H_0: $\theta=6$ Is True and All Mid-Ranks Are Equal to 4

k	Probability
0	1/128
4	7/128
8	21/128
12	35/128
16	35/128
20	21/128
24	7/128
28	1/128

TABLE 3.7

Probabilities of Obtaining *k* Pluses from 7 Observations under H_0

k	Probability
0	1/128
1	7/128
2	21/128
3	35/128
4	35/128
5	21/128
6	7/128
7	1/128

When there are ties, the asymptotic test requires modification to the formulae given in Section 3.3.3. This involves the variance, and the precise modification depends upon the pattern of ties. Several texts, including Hollander, Wolfe and Chicken (2014, Section 3.1), give a general formula to cover all possible patterns of ties. An alternative procedure, which has the advantage that it extends to many other situations covered in this book, is to regard each rank or tied rank as a *score*, denoting the score associated with the sample value x_i by s_i. In the Wilcoxon procedure and several others to be considered, tests are based on the sum S of the positive (or negative) scores. Under the null hypothesis, a key feature is that each score s_i is equally likely to have a positive or negative sign. It follows that any such statistic S has mean value $\sum_{i=1}^{n}|s_i|/2$ and variance $\sum_{i=1}^{n} s_i^2/4$. It follows from the central limit theorem that for large n the distribution of:

$$Z = \frac{S - \frac{1}{2}\sum |s_i|}{\frac{1}{2}\sqrt{\sum s_i^2}} \tag{3.3}$$

is approximately standard normal. Equation (3.2) is a special case of (3.3). With computer software, it is straightforward to use (3.3) given any set of scores such as those for the mid-ranks in Example 3.9.

Several alternative ways of handling ties in ranked data occur in the literature, but the mid-rank procedure is the one most commonly used in practice. A critical review of procedures can be found in Gibbons and Chakraborti (2020, Section 5.6).

With ties in signed ranks, the method for obtaining confidence intervals needs modification. If computer software for confidence intervals in these circumstances is not available, one might use the Walsh average method described in Section 3.3.2 and then use an available computer program in a trial-and-error manner to study appropriate end-point adjustments. Such adjustments are usually small unless there is heavy tying. If one or more sample values equal the mean hypothesized under H_0, the associated deviations are zero. Statisticians are divided on how best to proceed in this case. Some argue such observations should be omitted from the analyses because they are uninformative about any alternative hypothesis. The test is then applied to a reduced sample.

An intuitively sensible alternative is to rank all deviations initially giving the rank 1 to any zero difference. If there is more than one zero difference, the appropriate tied rank is given to each (e.g., 1.5 if there are two zero differences). After all signed ranks are allocated, those associated with a zero difference are changed to 0, leaving the other ranks as allocated. To obtain exact tests in the latter situation, a suitable computer program is needed for the exact distribution of the test statistic. Alternatively, asymptotic results using the form (3.3) with the adjusted ranks as scores can be applied providing n is reasonably large.

Deleting zero differences is the simplest procedure if no suitable program for exact probabilities of S for the modified rank statistic is available, but the method recommended above has the theoretical advantage of providing a more powerful test.

SUMMARY POINTS:

- For the Wilcoxon signed-rank test, a suitable test statistic is S, the sum of the lesser of the positive, S_+, or negative, S_-, signed-rank differences from the mean or median specified in the null hypothesis (Section 3.3.1).
- Tables or, more satisfactorily, appropriate computer software may be used for significance tests.
- Walsh averages (Section 3.3.2) provide an alternative hypothesis testing procedure, and they are particularly useful for estimation and calculating confidence intervals.
- Open intervals exclude the end points and are indicated by parentheses, and closed intervals include the end points and are indicated by square brackets (Section 3.3.2).
- An assumption of population symmetry is needed for the validity of test and estimation procedures.
- Asymptotic normal approximations (Section 3.3.3) work well for sample sizes of at least 20 if there are no (or very few) ties.
- Ties require special treatment (Section 3.3.4), and suitable computer software is then needed to determine exact permutation distributions.
- Modification of the denominator in the asymptotic test statistic is required for numerous ties, even for relatively large sample sizes.

3.4 Use of Alternative Scores

3.4.1 Raw Data Permutation Tests

The scoring system for the Wilcoxon test is based on ranks or, more specifically, signed ranks of deviations. These became the relevant base for a permutation test.

Historically, an earlier test was proposed by Fisher (1935) who described a randomization test for hypotheses about the mean or median of a symmetric distribution given a random sample of observations from a population that has that type of distribution. Rather than using the ranks of the deviations, Fisher's approach involves the values of the deviations themselves. The Fisher raw data permutation test is generally known as either the

Fisher–Pitman test or, most commonly, the *Pitman test* (Holt and Sullivan, 2023). Pitman himself attributed the test to Fisher (Pitman, 1949, pp. 8–10).

This test is described briefly mainly because of its historical interest. Raw data permutation tests are seldom used in practice because, as shown in Section 3.5, they lack robustness. Prior to the availability of computer packages, the calculations involved were prohibitive except for trivially small samples. However, illustrating the test with a simple numerical example assists in understanding its main features. A comparison with other tests highlights some weaknesses.

The assumptions for the validity of the Pitman test are the same as those for the Wilcoxon signed-rank test. Instead of signed ranks of deviations from the hypothesized median θ_0, it uses the signed deviations themselves to form a test statistic. The appropriate test statistic is either the sum of the positive deviations or the sum of the negative deviations. A simple example suffices to highlight both the strengths and the weaknesses of the test.

Example 3.11

Consider the data set for heart rate (beats per minute) used in Example 3.4 with the same assumptions. The observations were 73, 82, 87, 68, 106, 60 and 97. The null hypothesis H_0: $\theta=70$ is tested against the alternative H_1: $\theta>70$ using the Pitman test.

Symmetry implies that under H_0 sample values should have a near-symmetric scatter about $\theta=70$, so the magnitudes of the sums of positive and of negative deviations from 70 should not differ greatly. The one-sided alternative H_1 implies that when it holds there are likely to be more positive deviations from 70 than there would be under H_0 and that these positive deviations will tend to be larger than the negative deviations.

The deviations from 70 are 3, 12, 17, –2, 36, –10 and 27. The sum of the positive deviations (i.e., differences from 70) is an intuitively reasonable statistic to help assess the strength of the evidence against H_0 in considering the specified one-sided alternative H_1. If the alternative had been the two-sided H_1: $\theta\neq70$, small and large sums of positive deviations would also have supported H_1.

For a randomization test based on S_+, the sum of the positive deviations, the appropriate randomization distribution is formed in a manner similar to that in Example 3.4 for the statistic S used there. The key to obtaining the randomization distribution of S_+ when H_0 holds is that under that hypothesis the sign of each deviation is equally likely, on account of the symmetry assumption, to be positive or negative. Thus, any combination of plus and minus signs attached to a set of deviations having the observed magnitudes is equally likely. In the current example with 7 observations, there are $2^7=128$ possible allocations of sign (2 to each of the 7 differences). Each allocation is equally likely under H_0. Then, determine how many of the equally likely allocations of signs give an S_+ greater than or equal to that for this sample, i.e., $S_+=3+12+17+36+27=95$. The greatest possible value of S_+ occurs when all the deviations have positive signs, and then, $S_+=3+12+17+2+36+10+27=107$. The next highest sum, 105, is attained when only the smallest deviation, 2, has a negative sign. In order of decreasing magnitude, other sums greater than the observed $S_+=95$ occur when negative signs are attached to 3 only (sum 104), both 2 and 3 (sum 102) and 10 only (sum 97). If a negative sign is attached to 12 only, or to both 10 and 2, as above, the sum is 95. Thus, there are seven sums greater than or equal to the observed $S_+=95$. Since all are equally likely, the relevant P-value for assessing the strength of evidence against H_0 is $P=7/128\approx0.055$. Therefore, the evidence against H_0 when $P=0.055$ is sufficiently strong to perhaps warrant a similar study with a larger group of students. The conclusion is identical to that reached in Example 3.4.

Output from the ANSM5 package in R can be obtained using `pitman(ch3$heartrates1, 70, "greater")`.

Unfortunately, while computer programs enable a Pitman hypothesis test to be carried out with relative ease, it is harder, though not impossible, to obtain an exact confidence interval for the true θ. A practical alternative would be to use software to obtain a bootstrap confidence interval, which is shown as (68, 97) for a 95% interval in output from the ANSM5 package in R, obtained using `pitman(ch3$heartrates1)`.

In this example, S_+ was chosen as the test statistic, but the sum of the negative deviations, S_-, could equally well have been chosen. This is because the sum of the magnitudes of the deviations remains constant for all possible allocations of signs (in this example, always having the value 107). The statistic has a symmetric distribution because for any of the 128 sign permutations $S_+ + S_- = 107$ or $S_- = 107 - S_+$. Computer software sometimes calculates the smaller of S_+ and S_-, but each leads to the same relevant P-value. Another statistic that is sometimes used is $S_d = |S_+ - S_-|$. For this example, $S_d = |2S_+ - 107|$, so there is a one-to-one ordered relationship between S_d and S_+. One may even use the one-sample t-test statistic instead of S_+ or S_- because there is a one-to-one ordered relationship between the values of t and S_+. Remember though that the distribution of this statistic t is not that given by the usual t-tables that assume normality but is identical to that for S_+.

The approach has a strong intellectual appeal because it uses all the information in the data, but it lacks robustness against departures from symmetry. This is explained in Section 3.5.

Because the possible values of S_+ depend upon the values of the deviations that, in turn, depend upon the data values, it follows that the S_+ corresponding to a given P-value varies between data sets even for samples of the same fixed size. The test is called a conditional test because it is conditional upon the data values. A practical implication of this is that the P-value associated with H_0 must be computed afresh for each data set.

Although it does not take long for a computer to calculate an exact P-value, an asymptotic approximation can also be obtained that is strictly valid only for large samples. In effect, it uses the signed deviations as scores in (3.3). For these data, the asymptotic P-value for the one-tailed test is $P = 0.051$, which, despite the small sample size, is close to the exact P-value, but such close agreement is not always obtained for small samples. The normal theory t-test for this example also gives $P = 0.051$.

To estimate P by Monte Carlo sampling, one would randomly allocate, each with probability $p = 0.5$, a plus or minus sign to every difference for, say, 10,000 samples, and record the number, r, of these samples giving values of S_+ greater than or equal to 95. Then, $P^* = r/10{,}000$ is an estimate of the exact one-tailed P-value.

3.4.2 Normal Scores

In essence, the transformation of a sample of n continuous ordered data to ranks replaces the sample values with something like the values expected when sampling from a uniform or rectangular distribution over $(0, n)$ – more precisely, the rank r corresponds to the $r/(n+1)$th quantile of that distribution. Might further transformation of these ranks increase the Pitman efficiency or have other advantages?

In many parametric analyses, a normality assumption is a key feature, and when it can validly be made, it often leads to mathematical and computational simplicity. Exact Pitman tests are only practical for small to medium sample sizes even with the best software. It is not unreasonable to hope that by starting with data that resemble a sample from a normal distribution, asymptotic or large sample results will approach exact results more rapidly, in the sense that they may be reasonable for not very large samples.

Such hopes inspired a study of transformations to give data more like those from a normal distribution as a basis for exact tests while retaining desirable properties such as the test being unconditional on the actual data. This is the situation for the Wilcoxon signed-rank test (but not for the Pitman test). A possible transformation is suggested by the fact that ranks correspond to quantiles of a uniform distribution. Why not transform ranks to values, i.e., *scores,* corresponding to quantile values for a normal distribution? More specifically, why not choose the *standard normal distribution*. For then, it is easy to make the transformation using tables of the standard normal *cumulative distribution function*. For example, for the very non-normal samples 2, 3, 7, 21 and 132, first replace these values by the ranks 1, 2, 3, 4 and 5. Then, transform these ranks to normal scores. These are the 1/6th, 2/6th, 3/6th, 4/6th and 5/6th quantiles of the standard normal distribution, obtainable from standard normal cumulative distribution function tables. Many software programs compute these as one of a set of possible data transformations.

The normal score corresponding to the $1/6=0.1667$ quantile is the x value such that the standard normal cumulative distribution function, commonly denoted by $\Phi(x)$, is $\Phi(x)=0.1667$. Tables or software give $x=-0.97$. For the $2/6=0.3333$ quantile, $x=-0.43$, and for the 3/6th (mean or median), $x=0$. By symmetry, the remaining quantiles are $x=0.43$ and $x=0.97$. These quantile scores are sometimes called *van der Waerden scores,* which were proposed by van der Waerden (1952, 1953).

Alternatives to the above scores is discussed by Conover (1999, Section 5.10) and others. These include expected normal scores where the ith-order statistic $x(i)$ is replaced by the expectation of the ith-order statistic for the standard normal distribution. Fisher and Yates (1963, Table IX_I) gives expected normal scores corresponding to ranks for $n \leq 50$. In practice, using either van der Waerden or expected normal scores usually leads to similar conclusions.

Both need adapting for use in the one-sample situation. Only van der Waerden scores are considered here. Unfortunately, the direct application of van der Waerden scores in an analogue to the Wilcoxon signed-rank test is not possible because the rationale of the test demands that signs are allocated to magnitudes of ranks, but van der Waerden scores are of equal magnitudes but opposite signs.

This difficulty may be partly overcome by obtaining positive scores to which appropriate signs may be allocated, for instance, by using only some of the positive quantiles of the standard normal distribution as scores. Appropriate signs, i.e., those for the Wilcoxon signed ranks, are then attached to these. Recommended quantiles corresponding to the ranks 1, 2, 3, …, n are the $(n+2)/(2n+2)$th, $(n+3)/(2n+2)$th, $(n+4)/(2n+2)$th, …, $(2n+1)/(2n+2)$th quantiles. Effectively, the procedure makes use of the quantiles of the folded normal distribution. One then proceeds using these scores as though they were the original data in a Pitman-type test. The fact that the original quantiles are not symmetrically distributed about their median and the impact of allocating negative signs detract from the intuitive appeal of the transformation. Our experience with a few examples is that the method generally gives results similar to a Wilcoxon signed-rank test.

3.4.3 Other Scores

Scores other than normal scores are sometimes appropriate. Some of these are used in more sophisticated problems. They are needed when data that are essentially non-normal occur or when there are complications due to the censoring of data. An example is given in Section 9.2. In analysing survival data, Savage or exponential scores, introduced in Section 9.3, are often appropriate.

Hypothesis tests about the median for a sample with censored data were considered in Examples 2.4 and 2.5. The use of the sign test is only possible if censorship occurs at a value above the median specified in the null hypothesis. Even in this situation, censorship may cause difficulties in the corresponding method for forming confidence intervals. There is now a vast literature on the use of both parametric and nonparametric methods for the analysis of censored data.

Some basic elementary methods for analysing censored data are discussed in Chapter 9. A more detailed treatment, especially in the context of survival data, is given in Desu and Raghavarao (2004).

SUMMARY POINTS:

- For the Pitman or raw data permutation test, a suitable statistic (Section 3.4.1) is the sum S_+ of the positive data deviations from a hypothesized mean or median θ assuming the sample is from a symmetric distribution.
- Appropriate computer software is required for significance tests. The procedure lacks robustness and often gives similar results to a *t*-test even when the latter is not appropriate. It is conditional upon each particular data set and is not widely used in practice.

3.5 Comparing Tests and Robustness

Some tests are more sensitive to departures from assumptions than others, as illustrated by the following examples.

Example 3.12

The percentage water content of agricultural land often varies only slightly over an area such as a large field apart from a few local patches of exceptionally high or low percentages associated with especially poor or good drainage. The potential range of percentage water content is exemplified in a set of readings for nearly 400 soil quadrats on 22 June 1992 found in Gumpertz, Graham and Ristiano (1997). We use two small subsets (with values slightly rounded) from this large data set to illustrate some points about single-sample nonparametric analysis. The use made of these data here has no direct relevance to the paper from which the samples were taken. The percentage water contents for the selected quadrats, arranged in ascending order of magnitude, are:

Sample A	5.5	6.0	6.5	7.6	7.6	7.7	8.0	8.2	9.1	15.1
Sample B	5.6	6.1	6.3	6.3	6.5	6.6	7.0	7.5	7.9	8.0
	8.0	8.1	8.1	8.2	8.4	8.5	8.7	9.4	14.3	26.0

For each sample, compare and comment upon the performance of the *t*-test, Wilcoxon signed-rank test, sign test and Pitman test for H_0: $\theta=9$ against the alternative H_1: $\theta\neq9$. For the last three, compare both the exact and asymptotic test performances.

Details of the computation are omitted. The results shown below were obtained using the `t.test` function from R's stats package and `pitman`, `wilcoxon.signedrank` and `sgn.test` functions from the ANSM5 package.

The relevant exact and asymptotic *P*-values are:

Sample	**A**	**A**	**B**	**B**
Test	**Exact**	**Asymptotic**	**Exact**	**Asymptotic**
t-test	0.331	–	0.823	–
Pitman	0.352	0.305	0.916	0.817
Wilcoxon	0.102	0.092	0.015	0.017
Sign	0.109	0.058/0.114*	0.003	0.002/0.004*

*Continuity correction applied.

It can be seen that using the continuity correction for the sign test has an appreciable effect in these examples. For instance, with Sample B, the continuity correction increases the two-tail asymptotic *P* to 0.004. This is not closer to the exact value, whereas for Sample A, the continuity correction does better, increasing *P* to 0.114, which is closer to the exact value. This may reflect the greater importance of the continuity correction if the sample size is not large. It indicates a need for caution about using asymptotic values even for moderate sample sizes. It is in the tails (i.e., for low values of *P*) that asymptotic values sometimes show the greatest relative departure from exact probabilities.

When there are ties, a continuity correction is not usually used for the Wilcoxon test as it has little theoretical justification, although in practice it sometimes tends to bring results closer to those for the exact permutation test. The *t*-test $P=0.331$ and 0.823 are close to the values for the Pitman test.

The data in both soil quadrat samples are slightly skewed to the right, especially in Sample B. Both the Pitman and Wilcoxon tests indicate lower efficiency relative to the sign test in the larger sample, but the effect of skewness is less for the Wilcoxon test than it is for the Pitman test. This is because the transformation to ranks does not give the more extreme values of 9.4, 14.3 and 26.0 the influence they have in the Pitman test. Indeed, they would have had the same rankings had they had the values 12.6, 12.7 and 12.8.

In view of the conclusion that the rankings would not have been altered if the three largest observations had been replaced by 12.6, 12.7 and 12.8 when there would be only a small indication of skewness, it is interesting to see how the Pitman test would perform if these values were used in it. Intuition suggests that the *P*-values should then be nearer those given by the Wilcoxon test (which, of course, are unaltered by the change). Indeed, with this change, the exact *P*-value for the Pitman test reduces dramatically to $P=0.123$ for the smaller sample and to $P=0.036$ for the larger sample, both much closer to the results for the Wilcoxon test. This should cause no surprise because there is now less indication of skewness, and so, a key validity requirement for both tests is nearer to being satisfied.

The similarity of results for the *t*-test and the Pitman test for the original data is consistent with the theoretical result that the distributional characteristics of each test are similar. The poor performance of these tests where there is a breakdown in assumptions reflects a shared lack of robustness. The Wilcoxon test, however, shows, for reasons indicated above, greater robustness against moderate skewness.

The wide availability of software to carry out asymptotic tests makes them appealing. Three notes of caution are:

- Large sample sizes are needed to ensure consistently good approximations, especially in the tails, so beware of asymptotic tests when $n<20$.
- A breakdown of an assumption such as a requirement for symmetry may make asymptotic results of little value (as is also true for exact tests).
- Practice varies regarding the use of continuity corrections. It is usually easy to check whether these are used either from documentation included with the software or from the given output, which often includes the test statistic value together with its mean and standard deviation.

Asymptotic confidence intervals are obtainable using the form of (3.2) relevant to the Wilcoxon test without ties, or for the sign test, using (3.1). For the Wilcoxon test, the asymptotic result makes it possible to obtain an approximate value of S for significance at a suitable level. The Walsh differences may then be used to get the approximate confidence interval. This works quite well if there are no ties, and it gives a starting point for trial-and-error refinement when there are many ties. For the sign test, the number of positive or negative signs that determine the end points of the interval can be estimated. An example shows how this works.

Example 3.13

For Sample B in Example 3.12, use the asymptotic formula for the Wilcoxon test and that for the sign test to establish approximate 95% confidence intervals for the population median.

The key to establishing end points of 95% confidence intervals is that if the statistic used for computing P-values for some hypothetical values of the median θ is such that $|Z| = 1.96$, then that θ is the end point of the approximate confidence interval.

First, consider the Wilcoxon test. Direct computation or use of statistical software for the asymptotic test establishes that the statistic S has a mean of 105 and a standard deviation of 26.78. Thus, if for some hypothesized mean, a calculated S is such that:

$$(S-105)/26.78 = \pm 1.96$$

it follows that such S values are appropriate values for calculating the end points of the confidence interval using Walsh averages. Solutions are $S=52.82$ and 157.18, rounded to 53 and 157 for practical purposes. In fact, only one of these values needs to be calculated in view of the symmetry in counts of positive and negative ranks; the smaller value 53 indicates that the 53 largest and 53 smallest Walsh averages should be rejected. We can avoid some tedious arithmetic by using the command `wilcoxon.signedrank(ch3$sampleB, do.asymp = TRUE)` from the ANSM5 package in R. It gives an approximate 95% confidence interval of (7.15, 8.45).

For the sign test, it follows immediately from (3.1) that when $n=20$ the mean value of the test statistic (the number of positive signs or of negative signs) is 10 and the standard deviation is $\sqrt{20}/2 \approx 2.236$, whence the lower limit is the value S such that $(S-10)/2.236=-1.96$, i.e., $S=5.62$. Rounding to the nearest integer suggests that H_0 should be rejected if there are six or fewer positive or negative signs. For these data, this gives the approximate interval (7, 8.2).

Hence, the asymptotic results give approximate 95% confidence intervals based on the Wilcoxon statistic as (7.15, 8.45) and based on the sign test statistic as (7, 8.2).

The 95% confidence interval based on the t-test statistic is (6.7, 10.9); this is considerably longer than those obtained in this example and reflects the inappropriateness of the t-test procedure when data are clearly skewed. Situations do arise where the Wilcoxon-based intervals may be similar to the t-test-based intervals in the presence of skewness. In practice, how well the Wilcoxon method performs in estimation problems depends upon the number of outlying observations in the sample because these influence only the more extreme Walsh averages. If there are only a few outliers when n is large, these are less influential than the same number of outliers for smaller n.

Given programs for exact tests, it is possible to determine the exact coverage of the intervals and also to adjust the limits by a trial-and-error method to give better coverage. In this example, the Wilcoxon interval (7.15, 8.45) is associated with a lower-tail $P=0.0235$ and an upper-tail $P=0.030$. Although one is below and the other is above 0.025, the interval is reasonable for most practical purposes. For the sign test interval with these end points, some observations coincide with the limits, so it is useful to perform an exact test for P-values just above and below these end points. Using the ANSM5 package in R and the commands `sgn.test(sampleII, 6.95, "greater")` and `sgn.test(sampleII, 8.25, "less")` indicates that the interval (6.95, 8.25) has only an 88.46% coverage (obtained by having 5.766% in each tail), but the interval (7.05, 8.15) has an even lower 73.7% coverage. Again, there is an indication here that one needs to be cautious about asymptotic results. The use of a continuity correction in determining the asymptotic limits may help.

Although programs giving exact P-values, or good Monte Carlo approximations thereto, are needed to refine asymptotic limits, limits based on asymptotic theory are usually reasonable for moderately large samples and can be computed using any program that deals adequately with the relevant asymptotic nonparametric tests.

Example 3.13 shows that an approximate 95% confidence interval for the population median using the Wilcoxon signed-rank procedure is (7.15, 8.45), while that based on the sign test is displaced slightly to the left and is slightly shorter being (7, 8.2). However, the confidence levels are not quite identical. The interval (6.7, 10.9) based on the normal theory t-test is appreciably longer, reflecting the inappropriateness of the t-test approach when applied to obviously skewed data. However, one commonly meets situations where the sign test interval is longer than that given by the Wilcoxon test, which in turn is longer than that based on the t-test.

In such examples, although there might be some evidence of departure from normality, if the departure is not severe and there is not much evidence of skewness, these results are in general agreement with the finding based on Pitman efficiency that when a sample comes from a normal distribution the efficiency of the t-test is greater than that of any other test. When a sample is from a normal distribution, the Pitman efficiency of the Wilcoxon test can be shown to be $3/\pi$ (approximately 95.5%), while that of the sign test is $2/\pi$ (approximately 63.7%).

Two kinds of departure from normality that are common are one where the sample may come from a distribution, such as the exponential or gamma distribution, which is known to be asymmetric, or one where the sample comes from a mixture of distributions. Mixtures may arise in several ways. For example, data might be results of chemical analyses carried out in two different laboratories, one of these giving less precise results than the other and perhaps also incorporating a consistent error (called *bias*) due to an equipment fault. Mixtures may also occur in accurate data, for example, because a questionnaire item is given a high weight, as in Example 3.2. Another way skewed or long-tail distributions may occur is by the intrusion of rogue observations, often called *outliers*. If the reason for why they are rogues is known, there may be a good case for rejection, but often, it is unclear as to whether suspect observations are really rogues (e.g., incorrectly recorded readings or observations on units that are not members of the population of interest) or if the sampling is from a population where a small proportion of outlying values is the norm. While many parametric tests like the *t*-test are sensitive to certain departures from assumptions, the Wilcoxon test may be less so because it makes fewer assumptions, and the sign test is even less sensitive because it in turn makes even fewer assumptions. These latter tests exhibit *robustness*, which is discussed more fully in Chapter 14. Test and estimation procedures are robust if they are little influenced by serious departures from assumptions.

Example 3.13 shows that the Wilcoxon test may be more robust than the *t*-test if there is a marked departure from symmetry. In such circumstances, the sign test, not unexpectedly, may do even better since it makes no symmetry assumption.

Example 3.14

Suppose that the data in Example 3.5, i.e.:

$$-2 \quad 4 \quad 8 \quad 25 \quad -5 \quad 16 \quad 3 \quad 1 \quad 12 \quad 17 \quad 20 \quad 9$$

are amended by omitting the observation 25 and replacing it (i) by 35 and (ii) by 65. These changes induce obvious upper-tail skewness. Table 3.8 gives the nominal 95% and 99% confidence intervals based on the normal theory *t*, the Wilcoxon signed-rank statistic and the sign test in each case.

For the moderately skewed set (i), there is a mixed picture: the *t*-test does best at the 95% level and the Wilcoxon test does best at the 99% level, but the other methods do nearly as well in each case. For the more markedly skewed set (ii), the *t* limits are highly unsatisfactory, and those based on the Wilcoxon test are moderately satisfactory at the 95% level but not satisfactory at the 99% level. However, the binomial-based sign test limits are the same as those for set (i) and are not influenced by the extreme observation at 65. For set (ii), they are the same width as the Wilcoxon limits at the 95% level although the exact coverage

TABLE 3.8

95% and 99% Confidence Intervals for Data Sets (i) and (ii) in Example 3.14 Based on the Normal Theory (*t*), the Wilcoxon Signed Ranks (*W*) and the Binomial Sign Test (*B*)

		95%	99%
	t	(2.79, 16.88)*	(−0.12, 19.79)
Data Set (i)	*W*	(2.5, 17.0)	(0.5, 20.0)*
	B	(1.0, 17.0)	(−2.0, 20.0)
	t	(0.70, 23.97)	(−4.10, 28.77)
Data Set (ii)	*W*	(2.5, 18.5)*	(0.5, 34.5)
	B	(1.0, 17.0)*	(−2.0, 20.0)*

For each set, the shortest interval at each confidence level is indicated by an asterisk.

(96.14%) is slightly greater than the exact Wilcoxon coverage (95.5%). This is more a reflection of the discontinuities in possible available levels than anything else. At the 99% level, the sign test limits are the clear winner. This is not surprising since no symmetry assumption is needed.

Are such extreme values as that in data sets (i) and (ii) realistic from a medical point of view? A change in heart rate of 35 beats per minute might just be possible, but the value of 65 in set (ii) is a data error. It is good statistical practice to check data carefully to eliminate errors, but one often has to analyse data where there is no indication that an extreme value, or apparent outlier, is not a correct observation. This is where one may do better with the simple sign test than with a test requiring more stringent assumptions. However, it is generally inappropriate to use the sign test when there is little or no evidence of skewness.

SUMMARY POINTS:

- Some tests are more sensitive to departures from assumptions than others.
- A breakdown of an assumption such as a requirement for symmetry may make asymptotic results of little value.
- Test and estimation procedures are described as being robust if they are little influenced by serious departures from assumptions.
- Tests that require fewer assumptions are generally more robust.

3.6 Exercises

3.1 Suppose that in Example 3.1 one of the scores of 104 should have been recorded as 140. Repeat the hypothesis testing for the corrected data and comment on your findings.

3.2 Explain why the 97.84% confidence interval for the first sample derived in Example 3.2 is unchanged when the recording error in the data is corrected.

3.3 A parking attendant notes the time cars have been illegally parked after their metered time has expired. For 30 such cars, the time recorded in minutes is:

10	26	24	21	18	145	16	8	9	17	5	7	15	7	32
15	9	31	52	42	23	17	4	13	11	3	19	6	19	22

Investigate a population median overstay time of 20 minutes against a two-tailed alternative using the sign test in both exact and asymptotic forms.

3.4 Using the data in Table 3.2, for the distribution of the Wilcoxon S when $n=7$, construct a bar chart like that in Figure 3.2 showing the probability function for S. Discuss the similarity, or lack of similarity, to a normal distribution probability density function.

3.5 In Example 3.5, it was mentioned that for a variety of reasons one should be cautious about extending inferences about heartbeat rates for female students to the population at large. What might some of these reasons be?

3.6 The numbers of children in a sample of 12 primary schools (Exercise 2.4) were:

126 142 156 228 245 246 370 419 433 454 478 503

Use the Wilcoxon signed-rank test to test the hypothesis that the mean number of children in the schools for which the education authority has responsibility is 400. Obtain a 95% confidence interval for the mean number of children based on the Wilcoxon test and compare it with the interval obtained using a t-test under an assumption of normality.

3.7 Kimura and Chikuni (1987) gives data for lengths of Greenland turbot of various ages sampled from commercial catches in the Bering Sea as aged and measured by the Northwest and Alaska Fisheries Center. For 12-year-old turbots, the numbers of each length were:

Length (cm)	64	65	66	67	68	69	70	71	72	73	75	77	78	83
No. of fish	1	2	1	1	4	3	4	5	3	3	1	6	1	1

Use the Wilcoxon signed-rank test to test the hypothesis that the median length of 12-year-old turbots is 73.5 cm.

3.8 Example 3.9 gave the ages in months of 12 inpatients selected at random from a children's hospital:

6 18 24 26 37 40 42 47 69 69 87 130

Form a table of Walsh averages and use it to obtain 95% and 99% confidence intervals.

3.9 Example 3.10 shows that for the observations 4, 4, 8, 8, 8, 8 and 8, the permutation distribution of the Wilcoxon signed-rank statistic for testing a median with the hypothesis H_0: $\theta=6$ is equivalent to that for the sign test of the same hypothesis. Would this equivalence hold if the null hypothesis was changed to H_0: $\theta=7$?

3.10 Before treatment with a new drug, 11 people with sleep problems had a median sleeping time of 2 hours per night. A drug is administered, and it is known for good scientific reasons that if it has an effect, it will increase sleeping time, but some doctors doubt if it will have any effect. Assess whether any doubts are justified if the hours per night slept by these individuals after taking the drug are:

2.4 1.8 2.7 2.4 2.9 0.2 3.7 5.1 8.3 2.1 3.1

3.11 Based on the observations given in Exercise 3.7, would you agree with someone who asserted that, on this evidence, the median length of a 12-year-old Greenland turbot was almost certainly between 69 and 72 cm?

3.12 The developers of a new diet claim in a television interview that adults who need to slim will be able to lose 5 kg in 2 months. On this basis, 16 overweight adults followed the diet for 2 months with weight losses in kilograms as follows:

4 6 3 1 2 5 4 0 3 6 3 1 7 2 5 6

Assuming that the developers are referring to median weight loss and the sample is effectively random, do the data support this assertion in the population of dieters? Test this without an assumption of symmetry. What would be a more appropriate test with an assumption of symmetry? Carry out this latter test.

3.13 A pathologist counts the numbers of diseased plants in randomly selected areas each 1 metre square on a large field. For 40 such areas, the numbers of diseased plants are:

21	18	42	29	81	12	94	117	88	210
44	39	11	83	42	94	2	11	33	91
141	48	12	50	61	35	111	73	5	44
6	11	35	91	147	83	91	48	22	17

Use histograms and boxplots to decide whether there is evidence of skewness or outliers. Use nonparametric tests to find whether it is reasonable to assume the median number of diseased plants per square metre might be 50 (i) without assuming population symmetry and (ii) assuming population symmetry. From these test results and the graphical evidence, is the latter assumption reasonable?

3.14 Knapp (1982) gives the percentage of births on each day of the year averaged over 28 years for Monroe County, New York. Ignoring leap years (which make little difference), the median percentage of births per day is 0.2746. Not surprisingly, this is close to the expected percentage on the assumption that births are equally likely to be on any day, that is, $100/365 \approx 0.274$. Given below is the average percentage for each day in the month of September. If births are equally likely to be on any day of the year, this should resemble a random sample from a population with a median of 0.2746. Do these data confirm this?

0.277	0.277	0.295	0.286	0.271	0.265	0.274	0.274	0.278	0.290
0.295	0.276	0.273	0.289	0.308	0.301	0.302	0.293	0.296	0.288
0.305	0.283	0.309	0.299	0.287	0.309	0.294	0.288	0.298	0.289

3.15 A standard method for assembling a routine component in a manufacturing process is known to take a trained operative a median time of 14.2 minutes to complete. A new method of assembling that same component is proposed. Twenty operatives are given practice sessions using the new method, after which each claims he or she fully understands the process. They are then each timed for assembling one new component. The times in minutes they take are:

14.2	11.3	12.7	19.2	13.5	14.4	11.8	15.1	12.3	11.7
13.2	13.4	14.0	14.1	13.7	11.9	11.8	10.7	11.3	12.2

Form a boxplot for the data. Follow this with any tests you consider appropriate to indicate whether the median assembly time in mass production is likely to differ from 14.2 minutes. Also, obtain nominal 95% confidence intervals for the median time associated with any test you use. Compare in each case the intervals based on exact procedures with the related asymptotic approximate intervals.

3.16 A group of 36 adults who see themselves as overweight start a diet in order to lose weight. Each is weighed at the start of the diet period, and after 3 weeks, any gain or loss in weight is recorded. A negative sign indicates a loss. Form a boxplot of the gain or loss in weight and comment on any features that may influence the choice of a test to determine whether the diet has led to a change in median weight. Use what you consider the most appropriate method to test whether the diet has affected median weight. The weight changes (in kg) are:

−1.2	1.4	0.2	−0.7	−6.4	−2.7	−8.6	−1.7	−2.2
0.1	−0.4	−4.2	−1.6	1.2	−1.3	−2.4	3.1	−0.2
−4.5	−6.3	−1.7	0.0	0.2	−3.7	1.1	−2.3	−0.1
−7.3	0.2	−1.4	−0.9	−2.0	0.0	1.1	−0.3	−1.1

Obtain an appropriate nominal 95% confidence interval for the median change in weight.

3.17 Example 3.1 considered the seizure severity score data:

Sample I	1	11	35	41	50	62	104	104	151	161
Sample II	2	10	27	35	38	40	47	50	58	60
	68	70	79	103	108	151	157	158	162	166

Compare and comment upon the performance of the *t*-test, Wilcoxon signed-rank test, sign test and Pitman test for H_0: $\theta=110$ against the alternative H_1: $\theta \neq 110$. For the last three, compare both the exact and asymptotic test performances.

4

Other Single-Sample Inferences

4.1 Other Data Characteristics

Inferences about location based on a group of selected observations are an important part of single-sample data analysis. However, one might then wish to test whether it is reasonable to make the further inference that the sample is one from some specific distribution, or at least from some specific family of distributions.

It might also be necessary to test whether it is reasonable to assume that the observations constitute a truly random sample. If the order of the observations is important, testing for the presence of a trend might be reasonable. Inferences may be needed for data that consist of counts rather than measurements. Some of these situations are considered in this chapter.

4.2 Empirical Distribution Function

In nonparametric inference, an important concept based on order statistics is the *sample* or *empirical* distribution function. For a random sample of size n from a population having cumulative distribution function $F(x)$, the sample, or empirical, distribution function is defined as

$$S_n(x) = \frac{\text{number of sample values} \leq x}{n}$$

For any x, the value of $S_n(x)$ is expressible in terms of the order statistics for the sample: $x(1) < x(2) < \ldots < x(n)$.

The function $S_n(x)$ is a step function with a step of size $1/n$ at $x = x_{(i)}$, $i = 1, 2, \ldots, n$. This is illustrated in Figure 4.1 for six observations 1.7, 2.5, 3.6, 5.1, 7.4, and 8.3.

Figure 4.1 shows that

$$\begin{aligned} &S_n(x) = 0 \text{ if } x < x_{(1)}, \\ &S_n(x) = i/n \text{ if } x_{(i)} \leq x < x_{(i+1)}, i = 1, 2, \ldots, n-1, \\ &S_n(x) = 1 \text{ if } x \geq x_{(n)}. \end{aligned} \tag{4.1}$$

DOI: 10.1201/9780429326172-4

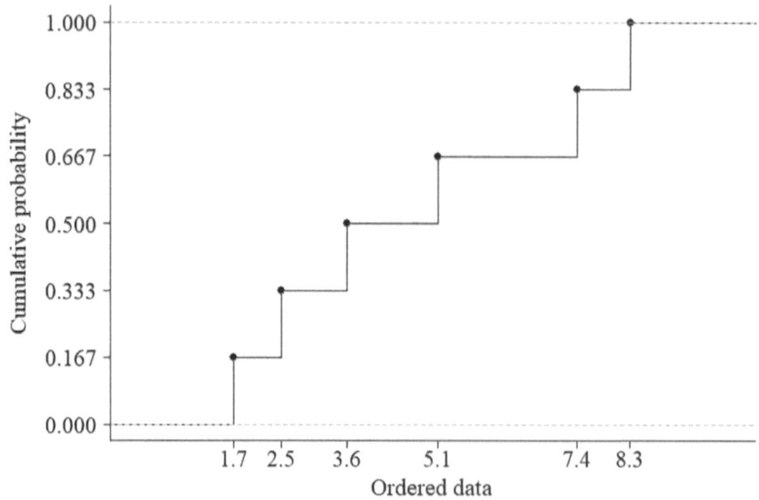

FIGURE 4.1
Sample distribution function for six observations.

The sample cumulative distribution function $S_n(x)$ is important because it is closely related to the population cumulative distribution function $F(x)$. This is reflected in the following properties:

- The mean value of $S_n(x)$ is $E[S_n(x)] = F(x)$.
- The variance of $S_n(x)$ is $\text{Var}[S_n(x)] = F(x)[1 - F(x)]/n$.
- $S_n(x)$ is a consistent estimator of $F(x)$ for any fixed x.

The term *consistent estimator* implies that $S_n(x)$ converges in probability to $F(x)$ as n tends to infinity. Gibbons and Chakraborti (2020, Section 2.3) gives a proof of these properties.

The assumption that $F(x)$ is continuous rules out, in theory, the possibility of tied observations. In practice, tied values in sample data are not uncommon. This may be due to rounding in the recording of data or to the population distribution being not strictly continuous. Later chapters describe the practical implications of tied data in specific techniques.

4.3 Matching Samples to Distributions

It is often of interest to investigate whether the observations that have been sampled are consistent with them having been drawn from some specified continuous distribution. The term consistent has a specific meaning in this context; as the sample size increases indefinitely, successive sample estimates become increasingly closer to the population value of the parameter. Kolmogorov (1933, 1941) devised a test for this purpose.

Given observations $x_1, x_2, \ldots, x_n$ the objective is to determine whether the data are consistent with their being a sample from some completely specified distribution. For example, this might be a uniform distribution over (0, 1) or over (20, 30), or a normal distribution with mean 20 and standard deviation 2.7. Kolmogorov's test is distribution-free because the procedure does not depend upon the distribution under study, providing it is completely specified under H_0. Section 4.3.3 addresses modifications of the test to answer questions like "Could the data have been drawn from a normal distribution with an unspecified mean and variance?"

4.3.1 Kolmogorov's Test

The fact that the sample, or empirical, cumulative distribution function $S_n(x)$ is a consistent estimator of the population cumulative distribution function provides the motivation for the Kolmogorov test.

The continuous uniform distribution (sometimes called the rectangular distribution) is a simple continuous distribution that arises when a random variable has the same probability of taking a value in any small interval of length δ_x lying entirely in a fixed and specified finite interval (a, b).

Suppose pieces of thread each 6 cm long are clamped at each end, and a force is applied until they break. If each thread breaks at the weakest point and this is equally likely to be anywhere along its length, then the breaking points will be uniformly distributed over the interval (0, 6), where the distance to the break is measured in centimetres from the left-hand clamp.

The probability density, or frequency, function has the form

$$f(x) = 0,\ x \le 0;\ f(x) = 1/6,\ 0 < x \le 6;\ f(x) = 0,\ x > 6.$$

The rectangular form of $f(x)$ over (0, 6) explains the name rectangular distribution.

If the thread always breaks, the probability is 1 that it breaks in the interval (0, 6), so the total area under the density curve must be 1. Only the density function above satisfies both this condition and that of equal probability of a break in any small segment of length δ_x, no matter where that segment lies within that interval. The function is graphed in Figure 4.2 and is essentially a line running from 0 to 6 at height 1/6 above the x-axis. The area under this line represents the total probability of 1 associated with the complete distribution. The black rectangle with corners labelled P, Q, R, and S between x=3.1 and

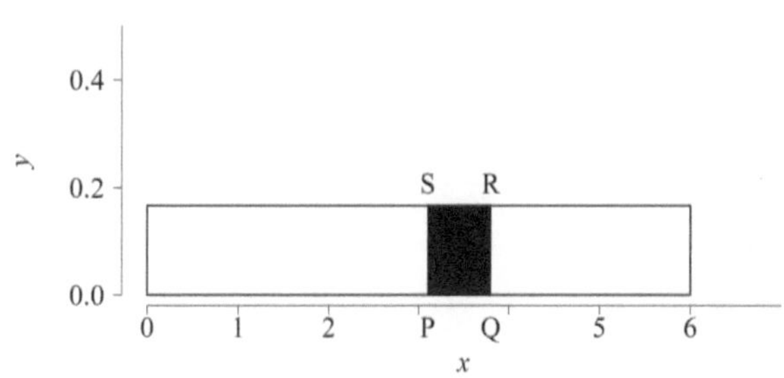

FIGURE 4.2
Probability density function of a continuous uniform distribution over (0, 6).

$x=3.8$ represents the probability that the random variable X, the distance to the break, takes a value between 3.1 and 3.8. Since the distance between P and Q is $3.8-3.1=0.7$ and the distance between P and S is 1/6, this area is $0.7/6 \approx 0.1167$. The probability of a break occurring in any segment of length 0.7 lying entirely in the interval (0, 6) is also 0.1167.

The Kolmogorov test uses not the probability density function but the cumulative distribution function, a function that gives $\Pr(X \leq x)$ for all x in (0, 6).

The cumulative distribution function is written $F(x)$ and for any x between 0 and 6, $F(x)=x/6$. It has the value 0 at $x=0$ and 1 at $x=6$. It is graphed in Figure 4.3. These notions generalize to a uniform distribution over any interval (a, b). The cumulative distribution function becomes $F(x)=(x-a)/(b-a)$, $a<x\leq b$, specifying a straight-line rising from zero at $x=a$ to 1 at $x=b$. Clearly $F(a)=0$ and $F(b)=1$. This is illustrated in Figure 4.4. Another special case of the rectangular distribution, where $a=0$ and $b=10$, is illustrated in Figure 2.4.

For any continuous distribution, the cumulative distribution function is a curve starting at zero for some x and increasingly moving from left to right until it attains the value 1. It never decreases as x increases and is said to be *monotonic non-decreasing* (or *monotonic increasing*). Figure 4.5 shows the cumulative distribution function for the standard normal distribution.

Recognizing that $S(x)$ should not depart violently from $F(x)$ for a sample from a distribution with that specific cumulative distribution function is basic to Kolmogorov's test. The test statistic, commonly referred to as D_n where n is the sample size, is the maximum difference in magnitude between $F(x)$ and $S(x)$.

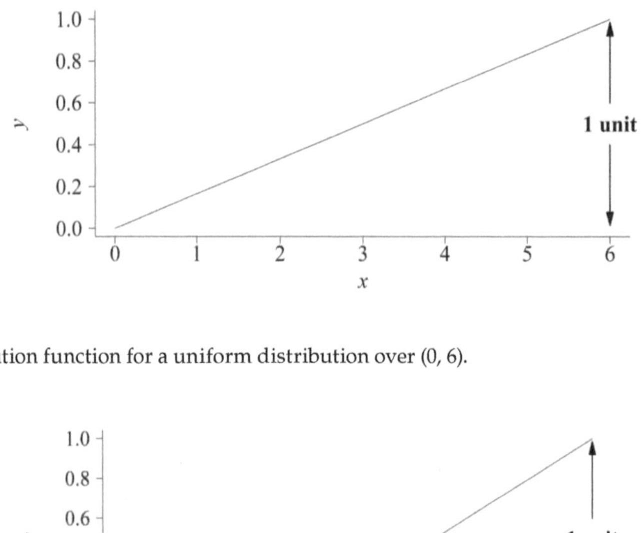

FIGURE 4.3
Cumulative distribution function for a uniform distribution over (0, 6).

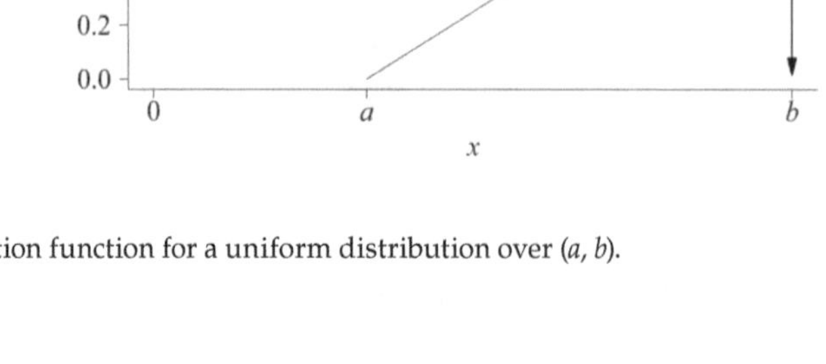

FIGURE 4.4
Cumulative distribution function for a uniform distribution over (a, b).

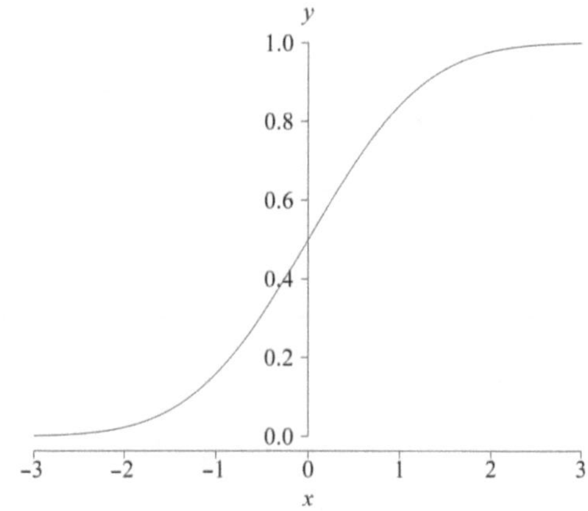

FIGURE 4.5
Cumulative distribution function for the standard normal distribution.

Example 4.1

The distances from one end at which each of the 20 threads 6 cm long breaks when subjected to strain are given below. Evaluate and graph $S(x)$. For convenience, the distances are given in ascending order.

0.6	0.8	1.1	1.2	1.4	1.7	1.8	1.9	2.2	2.4
2.5	2.9	3.1	3.4	3.4	3.9	4.4	4.9	5.2	5.9

From (4.1), it is clear that $S(x)$ increases by 1/20 at each unique x value corresponding to a break, or by $r/20$ if r break distances coincide.

When $x=0$, $S(x)=0$. It remains at this value until $x=0.6$, the first break. Then $S(0.6)=1/20$, keeping this value until $x=0.8$ when it jumps to 2/20, and so on until $x=3.4$. At $x=3.4$, there are 2 breaks so $S(x)$ increases by 2/20. Its value at each step is given in Table 4.1. Figure 4.6 shows the form of $S(x)$ for these data.

If a sample comes from a population with cumulative distribution function $F(x)$, the step function $S(x)$ should not depart markedly from $F(x)$. If breaks are uniformly distributed over (0, 6) one expects, for example, about half of these to be in the interval (0, 3), so that $S(3)$ should not have a value very different from 0.5, and so on. For comparison, Figure 4.6 also shows the cumulative distribution function for a uniform distribution over (0, 6) reproduced from Figure 4.3. Intuition may suggest this is not a good fit.

Example 4.2

Given the 20 breaking points and the corresponding $S(x)$ values in Table 4.1, is it reasonable to suppose breaking points are uniformly distributed over (0, 6)?

The goodness of fit is assessed using software that can determine the exact P-value associated with the maximum difference in magnitude between $F(x)$ and $S(x)$. Alternatively, tables can be used to find an approximate P-value.

Table 4.2 shows values of $F(x)$ and $S(x)$ at each break point given in Table 4.1. It also gives at each of these points the difference $F(x_i)-S(x_i)$. In view of the nature of $S(x)$, the maximum difference may not occur at a step point. Inspecting Figure 4.6, it is clear that instead

TABLE 4.1

Values of $S(x)$ at Step Points

x	0.6	0.8	1.1	1.2	1.4	1.7	1.8	1.9	2.2	2.4
$S(x)$	0.05	0.10	0.15	0.20	0.25	0.30	0.35	0.40	0.45	0.50
x	2.5	2.9	3.1	3.4*	3.9	4.4	4.9	5.2	5.9	
$S(x)$	0.55	0.6	0.65	0.75	0.80	0.85	0.90	0.95	1.00	

*indicates a repeated value.

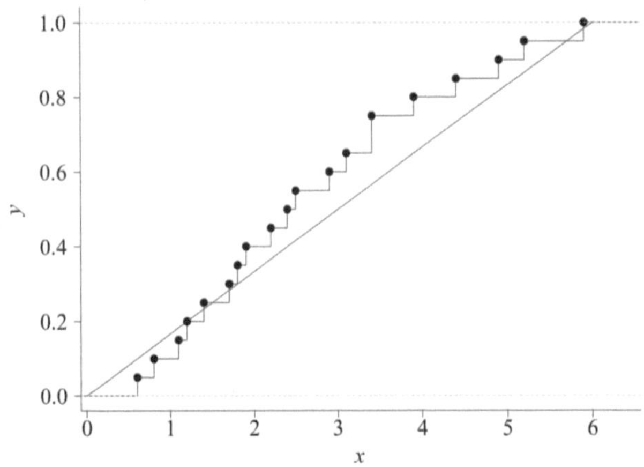

FIGURE 4.6
Sample cumulative distribution functions for the thread breaks (step function) and the uniform distribution over (0, 6) (straight line).

TABLE 4.2

Comparison of $F(x)$ and $S(x)$ for Thread Breaks

x_i	$F(x_i)$	$S(x_i)$	$F(x_i) - S(x_i)$	$F(x_i) - S(x_{i-1})$
0.6	0.10	0.05	0.05	0.10
0.8	0.13	0.10	0.03	0.08
1.1	0.18	0.15	0.03	0.08
1.2	0.20	0.20	0.00	0.05
1.4	0.23	0.25	–0.02	0.03
1.7	0.28	0.30	–0.02	0.03
1.8	0.30	0.35	–0.05	0.00
1.9	0.32	0.40	–0.08	–0.03
2.2	0.37	0.45	–0.08	–0.03
2.4	0.40	0.50	–0.10	–0.05
2.5	0.42	0.55	–0.13	–0.08
2.9	0.48	0.60	–0.12	–0.07
3.1	0.52	0.65	–0.13	–0.08
3.4	0.57	0.75	–0.18	–0.08
3.9	0.65	0.80	–0.15	–0.10
4.4	0.73	0.85	–0.12	–0.07
4.9	0.82	0.90	–0.08	–0.03
5.2	0.87	0.95	–0.08	–0.03
5.9	0.98	1.00	–0.02	0.03

a greater difference may occur immediately before such a step. There $S(x)$ has the value attained at the previous step, so a maximum may occur among the $F(x_i)–S(x_{i-1})$. These differences are also recorded in Table 4.2. Figure 4.6 shows that $S(x)$ lies above $F(x)$ for much of the interval (0, 6). The entry of greatest magnitude in the last two columns of Table 4.2 is –0.18 when $x_i=3.4$.

Gibbons and Chakraborti (2020, Table F) gives values for the minimum difference D_n required for significance at nominal 5% and 1% levels for various n, indicating for this example a critical value of 0.294 to obtain significance at the 5% level in a two-tailed test.

Output from the ANSM5 package in R can be obtained using `kstest.ANSM (ch4$breaks, "punif", min = 0, max = 6, alternative = "two.sided")`. For these data, R gives an exact $P=0.458$.

A detailed account of the computation of exact P-values can be found in Gibbons and Chakraborti (2020, Section 4.3). There is therefore no evidence against the null hypothesis.

A two-tailed test is appropriate if the alternative hypothesis specifies that the observations may come from any other unspecified distribution. Sometimes the alternative is that the cumulative distribution function must lie either wholly at or above, or wholly at or below, that specified in the null hypothesis. In the thread problem, it might be known prior to the testing that the equipment places strains on the thread that, if not uniformly distributed, will be greater at the left than towards the right. This would increase the tendency for breaks to occur close to the left end of the string. Then the cumulative distribution function would always lie above the straight-line representing $F(x)$ for the uniform distribution. In that case, a maximum difference with $S(x)$ greater than the uniform distribution $F(x)$ would, if significant, favour the alternative hypothesis, making a one-tailed test appropriate.

In commenting on Figure 4.6, it was remarked that the step function did not appear to match the population cumulative distribution function very well. The Kolmogorov test does not indicate a meaningful departure, a point taken up again in Section 4.3.2.

The Kolmogorov test appears to waste information because it only uses the difference of greatest magnitude. Tests do exist that allow for all differences, but these tend to have few, if any, advantages. This is not particularly as surprising as intuition may suggest, for the value of $S(x)$ at any stage depends on how many observations are less than the current x. Therefore, at each stage, a comparison is made on the basis of the accumulated evidence. For a discrete distribution, the Kolmogorov test tends to give too few significant results and, in this sense, is conservative. Preferred tests for discrete distributions are considered in Section 12.4.

4.3.2 Comparison of Distributions: Confidence Regions

If in Figure 4.6 step functions had been drawn everywhere at distances 0.294 units above or below that representing $S(x)$, with adjustments at each end to stop them falling below zero or above one, this would, since 0.294 is the critical difference for significance at the 5% level, give a 95% *confidence region* for $F(x)$ in the sense that any $F(x)$ lying entirely between these two new step functions would be an acceptable $F(x)$ when testing at a 5% significance level in a two-tailed test. This interval is not very useful in practice.

A common practical situation is one where it is reasonable to assume that the data may come from one of a (usually small) number of completely specified distributions. The Kolmogorov test might then be used for each. In each test, the relevant $S(x_i)$ will be the same for a given data set. As a result of these tests, some of the hypotheses may be rejected, yet more than one may be acceptable. It is then useful to get at least an overall eye comparison of how well the match of $S(x)$ is to each acceptable $F(x)$. Ross (1990, p. 85) gives a useful procedure for such eye comparisons.

Basically, this consists of plots of the maximum of the two deviations given for the last two columns of each row in a table like Table 4.2 against the $S(x_i)$ for that row. Plots may easily be produced for several different hypothesized population distributions on the same graph.

Example 4.3

Compare the fit of the thread break data in Example 4.2 with (i) a uniform distribution over (0, 6) and (ii) the distribution with cumulative distribution function over (0, 6) given by

$$F(x) = x/5,\ 0 < x \leq 3;\ F(x) = 0.2 + 4x/30,\ 3 < x \leq 6. \tag{4.2}$$

All information needed for the Kolmogorov test procedure for the distribution (i) was obtained in Example 4.2. Similar information is required for the alternative distribution in (ii). A graphical comparison is made using the format suggested by Ross.

For (i) the values of $S(x)$ of interest are those in column 3 of Table 4.2. In each case, the corresponding deviation is the value of the greatest magnitude in columns 4 and 5. For (ii) the values of $S(x)$ are again those in Table 4.2. However, fresh values of $F(x)$ for each x_i must be calculated using (4.1). For example, when $x_i = 1.2$, (4.2) gives $F(1.2) = 1.2/5 = 0.24$. Using this and the values of $S(1.1)$ and $S(1.2)$ given in Table 4.2 shows that $F(1.2) - S(1.2) = 0.24 - 0.20 = 0.04$ and $F(1.2) - S(1.1) = 0.24 - 0.15 = 0.09$. Thus, corresponding to $x = 1.2$, the maximum deviation is 0.09 just before the step in $S(x)$ at $x = 1.2$. Table 4.3 gives for each x_i the corresponding $S(x_i)$ and maximum deviations for the population distributions specified in (i) and (ii). These deviations are plotted against $S(x)$ in Figure 4.7.

TABLE 4.3

Values of x_i, $S(x_i)$, $d_i = \max[F(x_i) - S(x_i), F(x_i) - S(x_{i-1})]$ for Population Distributions Specified in (i) and (ii) of Example 4.3

x_i	$S(x_i)$	d_i(i)	d_i(ii)
0.6	0.05	0.10	0.12
0.8	0.10	0.08	0.11
1.1	0.15	0.08	0.12
1.2	0.20	0.05	0.09
1.4	0.25	0.03	0.08
1.7	0.30	0.03	0.09
1.8	0.35	−0.05	0.06
1.9	0.40	−0.08	0.03
2.2	0.45	−0.08	0.04
2.4	0.50	−0.10	0.03
2.5	0.55	−0.13	−0.05
2.9	0.60	−0.12	0.03
3.1	0.65	−0.13	−0.04
3.4	0.75	−0.18	−0.10
3.9	0.80	−0.15	−0.08
4.4	0.85	−0.12	−0.06
4.9	0.90	−0.08	−0.05
5.2	0.95	−0.08	−0.06
5.9	1.00	0.03	0.04

FIGURE 4.7
Plot of deviations $F(x) - S(x)$ against $S(x)$ for the two $F(x)$ functions specified in Example 4.3.

Although it is not essential in each case, consecutive points have been connected by straight-line segments. The parallel broken lines at $y = \pm 0.294$ represent the critical value for significance at the 5% level. The null hypothesis is rejected at a formal 5% significance level only if a deviation occurs outside these lines. There is no strong evidence against either of the population hypotheses specified in (i) and (ii).

The deviations, except for the early observations, are rather smaller for (ii), suggesting (but only marginally) that this may be a slightly better fit. The hypothesis (ii) implies a uniform but higher probability of breakages between 0 and 3 cm relative to a uniform distribution, but a lower probability between 3 and 6 cm. If other hypotheses are of interest, relevant deviations for these may be entered in Figure 4.7.

4.3.3 Tests for Normality

The Kolmogorov test is valid for any completely specified continuous distribution. If it is specified in the null hypothesis only that the sample has a normal distribution and one estimates the parameters μ, σ^2 from the data, the Kolmogorov statistic may still be used, but the associated P-values differ from those for the fully specified model.

For this important practical situation, a test proposed by Lilliefors (1967) is useful. Observations are assumed to be from a random sample from some unspecified continuous distribution and tested to see whether it is reasonable to suppose this is a member of the normal family. To this extent, the test is not distribution-free, but it is nonparametric in the sense that parameter values are not specified.

The basic idea can be extended to test compatibility with other families of distributions, such as a gamma distribution with unspecified parameters, but separate tables of critical values or appropriate software to estimate P-values are needed for each family of distributions. Conover (1999, Section 6.2) describes some difficulties in obtaining critical values. One may only get approximate values of the statistic corresponding to precise P-values such as $P = 0.05$ or 0.01. Gibbons and Chakraborti (2020, Table O) gives approximate critical values for sample sizes up to 12, then selected sizes up to 100, and asymptotic approximations for larger samples.

Another test for normality is given by Shapiro and Wilk (1965). The observed data are standardized to have a zero mean and unit standard deviation. The test is based on the correlation between the ordered standardized observed data and the expected values of the order statistics of a standard normal variable. If the data are normal, the correlation should be close to 1. The test requires special tables or appropriate software for implementation. An account of how the test works, without detail of the theory, is given by Conover (1999, Section 6.2), who also provides extensive tables that are required if relevant computer software is not available. This test has good power against a range of alternatives, but the rationale is less easy to describe by intuitive arguments. The exact distribution of the test statistic appears to be unknown (Leslie, Stephens and Fotopoulos, 1986), and approximations are very non-normally distributed for larger sample sizes.

Tests for normality may be important in deciding whether to apply nonparametric methods to a particular problem. Here Lilliefors' test is described, and results are quoted for analysis of the same data using the Shapiro–Wilk test.

Example 4.4

Data are given in Appendix 1 for ages at the death of four clans in the Badenscallie burial ground in Wester Ross, Scotland. They have already been introduced in Section 2.5. From all 117 ages recorded, a random sample of 30 was taken, and the ages for that sample were, in ascending order:

11	13	14	22	29	30	41	41	52	55	56	59	65	65	66
74	74	75	77	81	82	82	82	82	83	85	85	87	87	88

Is it reasonable to suppose the death ages are normally distributed?

In Lilliefors' test, the Kolmogorov statistic is used to compare the standard normal cumulative distribution function $\Phi(z)$ with that of the standardized sample $S(z)$ based on the transformation

$$z_i = \frac{x_i - m}{s}$$

where m is the sample mean and s is the usual estimate of population standard deviation, i.e.,

$$s = \sqrt{\frac{\sum_i x_i^2 - \left(\sum_i x_i\right)^2 \Big/ n}{n-1}}$$

Computation shows m=61.43, s=25.04. Successive z_i are calculated: e.g., for x_1=11, z_1= (11−61.43)/25.04=−2.014. This is treated as a sample value from a standard normal distribution, and tables (e.g., Neave, 1981, pp. 18–19) show $\Phi(-2.014)$=0.022. Table 4.4 is set up in an analogous manner to Table 4.2. Here $\Phi(z)$, $S(z)$ represent the standard normal and sample cumulative distribution functions, respectively. It is not essential, but the x values are included, an asterisk against a value implying it occurs more than once. The largest difference is 0.192, occurring in the final column when x=74. Since there is no analytic formula for exact P-values, only asymptotic and Monte Carlo approximations are available for Lilliefors' test. Published critical values vary somewhat between sources. For instance, Table A14 in Conover (1999) gives the 1% critical value when n=30 for a one-tailed test as 0.183, whereas in Table O of Gibbons and Chakraborti (2020), the same critical value is

TABLE 4.4

The Lilliefors Normality Test: Badenscallie Ages at Death

x_i	z_i	$\Phi(z_i)$	$S(z_i)$	$\Phi(z_i) - S(z_i)$	$\Phi(z_i) - S(z_{i-1})$
11	−2.014	0.022	0.033	−0.011	0.022
13	−1.934	0.026	0.067	−0.044	−0.007
14	−1.894	0.029	0.100	−0.071	−0.038
22	−1.575	0.058	0.133	−0.075	−0.042
29	−1.295	0.098	0.167	−0.069	−0.035
30	−1.255	0.105	0.200	−0.095	−0.062
41*	−0.816	0.207	0.267	−0.060	−0.007
52	−0.377	0.353	0.300	0.053	0.086
55	−0.257	0.399	0.333	0.066	0.099
56	−0.217	0.414	0.367	0.047	0.081
59	−0.097	0.461	0.400	0.061	0.094
65*	0.142	0.556	0.467	0.089	0.156
66	0.183	0.572	0.500	0.072	0.105
74*	0.502	0.692	0.567	0.125	0.192
75	0.542	0.706	0.600	0.106	0.139
77	0.622	0.733	0.633	0.100	0.133
81	0.781	0.782	0.667	0.115	0.149
82*	0.821	0.794	0.800	−0.006	0.127
83	0.861	0.805	0.833	−0.028	−0.005
85*	0.942	0.827	0.900	−0.073	−0.006
87*	1.021	0.846	0.967	−0.121	−0.054
88	1.061	0.856	1.000	−0.144	−0.111

An * implies a repeated value.

stated to be 0.186. Whichever table is used here, the value of the Kolmogorov statistic provides strong evidence that the sample is not from a normal distribution. Using the ANSM5 package in R and the command `lilliefors(ch4$ages)` gives a Monte Carlo $P=0.006$, which is consistent with the published critical values above for $P<0.01$.

For the Shapiro–Wilk test, using the ANSM5 package in R and the command `shapirotest.ANSM(ch4$ages)` gives an asymptotic *P*-value of 0.00117. Shapiro, Wilk and Chen (1968) carried out studies that show this test has good power against a wide range of departures from normality.

A glance at the sample data suggests that a few males died young and that the distribution is skewed with a large number of deaths occurring after age 80. Figure 4.8 shows these data on a histogram with a class interval of width 10. Readers familiar with histograms for samples from a normal distribution will not be surprised that the hypothesis of normality is rejected. Indeed, life-span data, whether for man, animal or the time a machine functions without a breakdown, will often by their nature have a non-normal distribution.

4.3.4 Relevant Inference

There are situations where inferences about means or medians are by nature irrelevant to sensible deductions. Tests such as Kolmogorov's or Lilliefors' test may then be more useful.

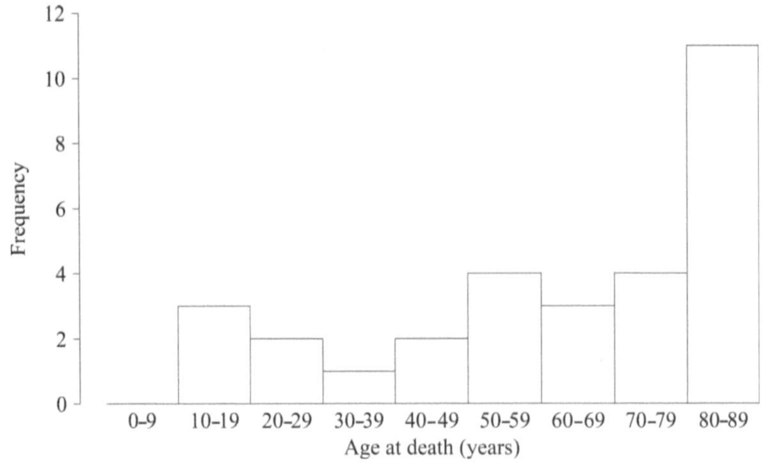

FIGURE 4.8
Histogram of death ages at Badenscallie.

Example 4.5

Figure 2.2 shows a histogram with class interval 10 for ages at death for the 59 members of the McAlpha clan recorded in Appendix 1.

The mean and median are, respectively, 61.8 and 74. The age range with most deaths is 80–89. Striking features in Figure 2.2 are the skewness of the data to the left and the bimodal nature with a small peak (secondary mode) of infant deaths in the range 0–9 followed by lower death rates in 10-year intervals until age 60. The death rate then rises sharply to the primary mode at the 80–89 range.

One might go through formal motions of attaching confidence intervals to the mean or median using an approach based on the *t*-test (obviously irrelevant because of non-normality – Lilliefors' test indicating $P<0.001$ using the ANSM5 package in R and the command `lilliefors(app1$McAlpha)`). The Wilcoxon test is not valid because of asymmetry, while the sign test is valid but not very interesting in light of the bimodality. Such approaches are at worst wrong, and at best pointless. In this example, the histogram is probably the most useful aid in interpreting the data.

What is the relevance of any formal inference procedure in a case like this? For which population does this represent a random sample? It is not a random sample of all males buried at Badenscallie, for clans and families have differing life-span characteristics.

Perhaps the pattern of death ages is a reasonable approximation to that for the McAlpha clan in Wester Ross, Sutherland and the Western Isles, where the clan is well established. The main use for these data would appear to be for comparisons with patterns for other clans.

Difficulties in the choice of analytic tools are not confined to nonparametric methods.

Analyses making various assumptions often draw attention to interpretation problems.

Software allows several tests to be applied to the same data in a single run. This freedom should not be taken too far; remember the ethical warning in the final paragraph of Section 1.7.1. If different choices are logical under differing assumptions, and for, say, location tests these lead to inconsistent results, further analysis (e.g., applying Lilliefors' test) or more simply drawing a histogram, or using some other appropriate exploratory data analysis tools may suggest the most appropriate choice. Looking at data critically is a habit well worth cultivating.

For the McAlpha data, it is clear that a sign-test-type confidence interval for the median is less inappropriate than many others. However, this begs the question as to whether the median is a useful summary statistic for bimodal data.

SUMMARY POINTS:

- The Kolmogorov test (Section 4.3.1) indicates whether data are an acceptable fit to a completely specified continuous distribution.
- Several distributions may give an adequate fit. In this case, graphical methods (Section 4.3.2) are useful to get an eye comparison of the relative goodness of fit.
- Lilliefors' test for normality with unspecified mean and variance (Section 4.3.3) uses the Kolmogorov statistic, but separate tables or relevant software are needed to test for significance or to estimate *P*-values.
- The Shapiro–Wilk test (Section 4.3.3) is theoretically more complicated than Lilliefors' test but is often more powerful.
- Consideration should be given to the relevance of the inference procedures performed to the population under study (Section 4.3.4).

4.4 Inferences for Dichotomous Data

Section 4.1 pointed out that sometimes data consist of counts. The simplest situation is when there are counts of numbers in each of two categories (binary data). These are sometimes described as *dichotomous data*. Sometimes dichotomous data arise as secondary data derived from some other type of raw data. This was the situation in Example 3.1, where the sign test was applied to data from a discrete quantitative distribution by counting the numbers of observations above or below some hypothesized median. Under the null hypothesis, the independence condition ensured that the count of observations above (or below) the hypothesized median had a binomial distribution with $p=0.5$. In this section, these ideas are extended to other situations where appropriate binomial distributions can be associated with dichotomous data.

4.4.1 Binomially Distributed Data

In Example 2.6, binomial distribution properties were used to show how to assess the strength of evidence against a hypothesis that a sample came from a B(10, 0.75) distribution, i.e., to test the hypothesis H_0: $p=0.75$ against H_1: $p>0.75$. Interest was in the parameter p.

In Example 2.4, the sign test was applied to hypotheses about the median of measured survival times (the primary data), basing the test not on these primary data but on the derived or secondary data of numbers of observations above or below a hypothesized median (12 years). For the sign test, the parameter $p=0.5$ is relevant to H_0, and although the test was based on that parameter, the primary interest was not in the parameter itself but in the median of an underlying continuous distribution of survival times.

In this section, other examples are given of inferences about a binomial parameter p. It is convenient to refer to each of the n observations as the outcome of a *trial*. The key assumptions needed for a count to have a binomial distribution are:

- The n trials are mutually independent.
- Only one of two possible outcomes A and B is observed at each trial.
- At each trial, there is a fixed probability p associated with outcome A and a probability $q=1-p$ associated with outcome B.

Outcomes A and B are conventionally referred to as *success* and *failure* (see Section 2.3.4), but these names are not always literally appropriate, so the labels of *event* A and *event* B are sometimes used. The requirement that p be constant is often violated in a strict sense. For example, in a study of diabetes among first-generation immigrants from South Asian countries to Ontario, Canada (Banerjee and Shah, 2018), the overall probability of having a diagnosis of diabetes was found to be 0.159. However, by country of origin, the probability of a diagnosis of diabetes ranged from 0.066 for those originating from Nepal to 0.228 for those from Sri Lanka. Despite these differences, for a representative random sample taken from this population of South Asian immigrants, it would be reasonable to expect approximately 16% of those in the sample to have been diagnosed with diabetes. For non-immigrant South Asians in Canada, there is evidence to suggest that the proportion with a diagnosis of diabetes may be lower. An appropriate statistical test could therefore be based on the binomial distribution corresponding to a sample of n with the null hypothesis H_0: $p=0.159$ against the alternative H_1: $p<0.159$. However, in a comment on Example 4.10, a complication that may invalidate the test is described.

Tests based on the binomial distribution also arise in quality control problems. Here interest may be in the proportion of items in large batches having a specific attribute – often a defect. A widely used procedure is to take a random sample of n items from a batch of N items and for a buyer to accept the whole batch if the number of defective items, r, in that sample, does not exceed some small fixed number s. Otherwise, the batch is rejected. The consequences of such a procedure can be studied using binomial distribution theory. The results are only approximate because the usual sampling method of selecting items without replacing each before later items are selected (sampling without replacement) does not lead to a binomial distribution for r. However, if N is very much larger than n, as it often is in practice, the binomial approximation is satisfactory. The scheme outlined below is usually embellished in practice, but the many possible elaborations are not considered here.

Example 4.6

A contract for supplying rechargeable batteries specifies that not more than 10% should need recharging before 100 hours of use. To test compliance with this requirement, a potential purchaser takes a random sample of 20 from a large consignment and finds that 3 need recharging after being used for less than 100 hours. Assess the strength of evidence that the batch is below specification.

The test is distribution-free in the sense that no assumption is made about the distribution of times before a recharge is needed. However, it is assumed that there is a fixed probability p, say, that any one battery will need recharging in less than 100 hours. In addition, it is assumed that the failure time for each battery is independent of that of any other battery. The null hypothesis is H_0: $p\leq 0.10$, and the alternative is H_1: $p>0.10$.

Tables for the binomial distribution, or standard statistical software, indicate that when $p=0.10$, the probability that a sample of 20 contains 3 or more batteries that fail to meet specifications is $P=0.323$.

Since the probability that 3 or more batteries will need recharging in less than 100 hours when $p=0.10$ is nearly one-third, there is little evidence for preferring the alternative that the failure rate in the large consignment exceeds 10%.

This example does little more than highlight the serious limitations of hypothesis testing. Three failures in 20 items represent a 15% failure rate in the sample, so the data would clearly support a null hypothesis that specified $p=0.15$ and also many higher values. Confidence intervals for p will be informative. The next example shows that a two-sided confidence interval with at least 95% coverage for the true p is the interval (0.032, 0.379). This implies that using the conventional $P \leq 0.05$ as sufficient evidence against that hypothesis, one would only reject a hypothesized value of p falling outside this wide interval. The sample evidence here suggests only that the batch percentage failing to live up to specification is likely to be between about 3.2% and 37.9%.

Output from the ANSM5 package in R can be obtained using `binom(3, 20)`.

Having obtained a confidence interval, two further questions of interest are

- What is the power of a test for H_0: $p \leq 0.10$ against various alternatives?
- How can larger samples be used to make a test more precise (in the sense of shortening the confidence interval or increasing the power for specific alternatives)?

These questions are addressed in Examples 4.8 and 4.9.

4.4.2 Confidence Intervals for Binomial Proportions

For a binomial distribution, if an event A occurs in r out of n trials, the sample estimate of the binomial probability, p, is $\hat{p} = r/n$. The notation $X=r$ where $r=0, 1, 2, \ldots, n$ is used to indicate that the event A occurs in exactly r among n trials.

To find a two-sided confidence interval for p with at least 95% coverage, the upper limit is the value p_u such that if the binomial parameters are n, p_u, then the value of $\Pr(X \leq r)$ is as close as possible to, but does not exceed 0.025. Well-known binomial distribution theory indicates that p_u must satisfy the equation

$$\sum_{i=1}^{r} \binom{n}{r} (p_u)^i (1-p_u)^{n-i} = 0.025 \tag{4.3}$$

A similar approach provides an equation for a lower limit, p_l. These equations may be modified to give any required level by replacing 0.025 with another appropriate value, e.g., 0.005 for a 99% interval.

The equation (4.3) generally has no simple analytic solution for p_u. The classic way of overcoming this difficulty was recourse to tables or charts obtained by numerical methods, where these limits were presented for several confidence levels (usually 90%, 95% and 99%) for a range of values of n and r. Table A4 in Conover (1999) is one such table. Many computer packages including Stata and SPSS compute exact confidence limits at any level chosen by the user.

For reasonably large n if neither exact tables nor relevant computer programs are available, an asymptotic approximation may be used. If there are $X=r$ occurrences of event A in n trials for any p such that $0<p<1$, then

$$Z = \frac{p - \hat{p}}{\sqrt{\hat{p}(1-\hat{p})/n}}$$

has, for large n, approximately a standard normal distribution. The approximation is good for $n>20$ providing p is close to 0.50, but appreciably larger values of n are needed if p is close to 0 or 1. An approximate 95% interval for p can be obtained by following the usual normal distribution confidence interval procedure and finding the upper and lower limits as the values of p that satisfy the equation

$$Z = \frac{p - \hat{p}}{\sqrt{\hat{p}(1-\hat{p})/n}} = \pm 1.96 \tag{4.4}$$

To obtain limits for other confidence levels, 1.96 is replaced by the appropriate value, e.g., 1.64 if a 90% interval is required.

For $r=0$, the event does not occur at all, and the estimate of the binomial probability is zero. An upper limit for p can be readily obtained using a tail probability of 0.05 as the lower tail is not meaningful and (4.3) reduces to

$$(1 - p_u)^n = 0.05$$

Correspondingly, if $r=n$ a lower limit can be obtained by solving

$$(p_l)^n = 0.05$$

Sackett et al. (1991) tabulates p_l and p_u when $r=0$ and $r=n$ for a selection of sample sizes between 1 and 300.

Example 4.7

For the data in Example 4.6 confirm that an exact 95% confidence interval for p is (0.032, 0.379). Determine also 90% and 99% intervals. Obtain an asymptotic approximation to the 95% interval.

Assuming that the number of batteries in a sample of 20 that fail to last 100 hours before needing a recharge has a binomial distribution, the exact 95% confidence limits are given by (4.3) and the corresponding equation for the lower limit. For the 90% and 99% limits, the 0.025 on the right-hand side is replaced by 0.05 and 0.005, respectively. Because the equations have no analytical solution, the required limits are obtained in practice from tables or by using appropriate software. The relevant asymptotic limits are given by (4.4).

Using the ANSM5 package in R and the commands `binom(3, 20, CI.width = 0.90)`, `binom(3, 20, CI.width = 0.95)` and `binom(3, 20, CI.width = 0.99)` gives the following exact intervals for this problem where $n=20$ and $r=3$.

90% (0.042, 0.344)

95% (0.032, 0.379)

99% (0.018, 0.450)

For the asymptotic limits (4.4) is used with $n=20$ and $\hat{p}=3/20=0.15$. The solutions give the 95% interval (–0.0065, 0.306). Most computer software will give these limits as an alternative to exact limits, but not all include a warning that these may be unreliable for small samples. Output from the ANSM5 package in R gives this interval using `binom(3, 20, do.asymp = TRUE)`.

Evidence from this sample of 20 indicates only that it seems likely that the overall proportion in a large batch lies somewhere between about 3%–4% and about 35%–40%.

As remarked already, the asymptotic approximation requires n to be considerably greater than 20 for small p. The asymptotic 95% lower limit of –0.0065 is statistical nonsense, being outside the permissible range $0 \le p \le 1$. Tables of exact values in Conover cover values of $n \le 30$, but with small values of p the asymptotic result may still be unreliable even when $n > 30$, meaning that unless one has access to a table for larger values of n one needs statistical software for an exact interval. For the case $n=30$ and $r=2$ an exact 95% interval for p is (0.008, 0.221). The normal approximation gives the nonsense interval (–0.023, 0.156). However, if $n=30$ and $r=13$ the exact 95% interval is (0.255, 0.626), and the asymptotic approximation is reasonable (0.256, 0.611).

Output from the ANSM5 package in R can be obtained using `binom(2, 30, do.asymp = TRUE)` and `binom(13, 30, do.asymp = TRUE)`.

In this example, a two-sided confidence interval was considered, but Example 4.6 concerned a one-tailed hypothesis test. The lower limit of the one-sided 95% confidence interval for the test considered there corresponds to the lower limit for the two-sided 90% interval, i.e., the 95% one-sided interval is in this example (0.042, 1).

4.4.3 Power and Sample Size

In Examples 4.6 and 4.7, there is considerable doubt about the population value of p. The purchaser in these examples would ideally like all accepted batches to contain no more than 10% that require recharging before 100 hours of use. However, taking marketing realities such as production cost into account, the purchaser may be prepared to accept a few batches where as many as, say, 20% fail to live up to this requirement but be reluctant to accept batches where more than 20% fail. By the same token, the supplier does not want the purchaser to reject batches where less than 10% will fail.

If the null hypothesis H_0: $p \le 0.10$ is tested against the alternative H_1: $p > 0.10$ and the supplier and purchaser agree the evidence is strong enough to reject H_0 if the associated P-value is $P \le 0.05$ this means there is a probability not greater than 0.05 that the purchaser will reject a batch that would be acceptable (i.e., one that in reality contains 10% or less that are defective). Be careful to distinguish between the binomial probability (lower case p) and the test P-value (capital P). In terms of acceptance sampling, the probability $P=0.05$ is called the *producer's risk* because in agreeing to accept these conditions, it is the maximum long-run probability of an acceptable batch being rejected. In conventional hypothesis testing terms, the producer's risk is the probability of an error of the first kind. If the consumer is prepared to accept a few batches with more than 10, but less than 20% defective, the consumer and producer might agree that they should devise a test procedure that ensures a certain preassigned high probability, say $P^*=0.90$, that one does not accept a batch with 20% or more defective. This means there is only a probability $1-P^*=0.10$ that a batch with 20% or more defective will slip through. This last probability is called the *consumer's risk* because in agreeing to accept these terms, it is the maximum probability that an unacceptable batch will be purchased. It is the probability of an error of the second kind, and P^* is the power. A practical problem is to determine the sample size n such that there is both a preassigned small probability P that batches for which $p \leq 0.10$ will be rejected and also a preassigned high probability P^* that batches for which $p \geq 0.20$ will also be rejected.

Here the concept of power (Section 1.5) is useful. Power depends on the value of P chosen to indicate significance, i.e., as sufficient evidence to make the null hypothesis unacceptable. It also depends upon the sample size n and upon the value of the binomial probability p specified for an alternative hypothesis.

Two relevant questions are as follows:

- Given a sample of size n what is the power of a test of H_0: $p=0.1$ against H_1: $p=0.2$ if H_0 if to be rejected for any P less than some fixed preassigned value in a one-tailed test?
- What sample size n will ensure that a test of H_0: $p=0.1$ against H_1: $p=0.2$, where H_0 is rejected for any P less than some fixed value (e.g., $P=0.05$), has a given power P^* for that alternative?

The rationale for these choices of H_0, H_1 is that if the true p is less than 0.1 a good batch is less likely to be rejected than would be the case if $p=0.1$ and that if $p>0.2$ the power of the test will exceed that when $p=0.2$. These are desirable characteristics for both producer and consumer.

Suppose that r occurrences of event A are observed in a sample of size n, and the null hypothesis H_0: $p=p_0$ is to be tested against the alternative H_1: $p>p_0$. If it is agreed that there is sufficient evidence to prefer H_1 for all P-values less than or equal to some fixed value α, then the critical region of size α consists of all values of $r \geq r_0$ where r_0 is the smallest value of r such that the binomial distribution $B(n, p_0)$ probabilities for all $r \geq r_0$ have a sum not exceeding α. In practice, r_0 would be either read from tables or now more commonly obtained using computer software.

Having determined r_0 it is in principle straightforward to find the power of this test against a single-valued alternative H_1: $p=p_1$ because this is simply $\Pr(r \geq r_0)$ when H_1 is true. This probability may be read from tables or obtained using appropriate computer software.

For large n asymptotic approximations may be used. These are based on the fact that under H_0 the distribution of

$$Z = \frac{\hat{p} - p_0}{\sqrt{p_0(1-p_0)/n}} \tag{4.5}$$

is approximately standard normal. If, for example, $\alpha=0.05$ then for an upper-tail test Z is set equal to 1.64 in (4.5). This equation is solved for $\hat{p}$ given n and p_0. To obtain an asymptotic expression for the power when H_1: $p=p_1$ is true, p_0 is replaced by p_1 in (4.4), and the solution $\hat{p}$ obtained in the previous calculation is inserted in this amended expression for Z. The resulting Z, say $Z=z_1$ is then calculated. The asymptotic approximation to the power is $\Pr(Z \geq z_1)$.

Example 4.8

For a sample of 20 batteries from a large batch, the null hypothesis H_0: $p=0.1$ is tested against H_1: $p=0.2$, where p is the population proportion of batteries requiring recharging in less than 100 hours. The alternative H_1 will be preferred if $P \leq 0.05$.

What is the minimum number of faulty batteries in the sample for which H_1 is preferred? Obtain both exact and asymptotic expressions for the power of this test.

Let X be the number of occurrences, r, of the event A (the battery needs recharging in less than 100 hours). Then under H_0, X has a binomial B(20, 0.1) distribution. The alternative H_1 is preferred if the sample gives r_0 or more occurrences of A where r_0 is the least r such that $\Pr(X \geq r_0) \leq 0.05$ when H_0 is true. Once the value r_0 is known, the power is obtained by computing for that r_0 the corresponding probability under H_1. Asymptotic results for large n use the normal approximations given above for these probabilities.

Using the command `pbinom(4, 20, 0.1)` in R indicates that when $n=20$, $p=0.10$, $\Pr(X \leq 4) = 0.9568$, whence $\Pr(X \geq 5) = 1 - 0.9568 = 0.0432$. Thus H_1 is preferred if, and only if, $r \geq 5$. To determine the power, it is necessary to find $\Pr(X \geq 5)$ for a B(20, 0.2) distribution. The R command `pbinom(4, 20, 0.2, lower.tail = FALSE)` indicates that for $p=0.2$ $\Pr(X \geq 5) = 1 - \Pr(X \leq 4) = 1 - 0.6296 = 0.3704$.

In the asymptotic approach under H_0, the use of equation (4.5) gives

$$Z = (\hat{p} - 0.1)/\sqrt{0.1 \times 0.9/20} = 1.64$$

whence $\hat{p} = 0.2100$. Thus, the relevant power is given by

$$\Pr[Z \geq (0.2100 - 0.2)/\sqrt{0.2 \times 0.8/20}] = \Pr\left(Z \geq 0.1118\right)$$

Using the R command `pnorm(0.1118, lower.tail = FALSE)`, this probability is 0.4555, which is considerably adrift of the exact value of 0.3704.

The power of this test against an alternative $p=0.20$ is only equal to 0.370 if the criterion $P \leq 0.05$ (or more precisely because of discontinuities $P \leq 0.043$) is chosen for not accepting $p=0.10$.

In the discussion preceding this example, it was suggested that a practical aim might be to reject a batch with probability (power) 0.90 if $p \geq 0.20$. The power 0.370 is far short of this requirement. The unsatisfactory nature of asymptotic power calculations with the n and p used here is less acute with larger samples.

Section 1.5 indicates that reducing the P-value necessary for evidence against H_0 generally results in reduced power against a particular alternative. In this example, if the threshold for P is lowered to 0.02 the power of the one-sided test against the alternative $p=0.2$ is reduced from 0.370 to 0.196. However, due to discontinuities in possible P-values, the exact P is not 0.02 but is 0.011.

Software can give exact P-values for one- or two-tailed tests for any n, p and r. This effectively means that the values of $\Pr(X \geq r)$, $0 \leq r \leq n$ can be obtained for any B(n, p) distribution. This is useful for developing tests such as those for attribute sampling considered in this section that have preset producer's risk (i.e., P-values for rejection of H_0) and consumer's risk (i.e., power against a specified H_1 that gives the maximum p the consumer considers tolerable).

Example 4.9

Determine the exact power of the test H_0: $p=0.10$ against H_1: $p=0.20$ when $P \leq 0.05$ for the sample sizes $n=50$, 100, 150, 200, 300. Also, determine the minimum n to ensure a power of at least 0.9 for this test.

The procedure is described fully for the case $n=100$. Results are provided for the other sample sizes, but readers should confirm these using available software. When $n=100$ and $p=0.10$, it is intuitively reasonable to expect a *P*-value not exceeding 0.05 if *r*, the number of batteries needing recharging before 100 hours, is close to 15. Using the ANSM5 package in R and the command `binom(15, 100, 0.1, alternative = "greater")` for $r=15$, with suitable amendments for $r=16$ and $r=17$ gives the following values

r	15	16	17
P	0.073	0.040	0.021

establishing $r=16$ as the critical value. The corresponding power is given by $\Pr(X \geq 16)$ when $p=0.20$. Using the command `pbinom(15, 100, 0.2, lower.tail = FALSE)` in R indicates that this probability is 0.871. Repeating the above process for the remaining sample sizes of interest with the largest exact *P*-values meeting the requirement $P \leq 0.05$, the findings can be summarized as:

n	50	100	150	200	300
r	10	16	22	28	40
P	0.025	0.040	0.044	0.043	0.038
Power	0.556	0.871	0.963	0.989	0.999

In terms of the minimum sample size to achieve a power of 0.9, this lies between 100 (associated power < 0.9) and 150 (associated power > 0.9). One may fine-tune this using exact binomial probabilities with careful adjustment of *n* and *r*. When $n=100$, the critical region is defined by $r \geq 16$. For $n=105$ and $r=16$, the power is 0.914 with $P=0.058$, so the *P*-value is a little in excess of 0.05. This excess can be corrected by increasing *r* to 17 while keeping *n* at 105. Now $P=0.032$ but with a reduced power of 0.865 showing that *n* needs to be increased. For $n=108$, the corresponding power is 0.893; with $n=109$, the power is 0.901 and $P=0.043$. These indicate that a suitable sample size is 109 with a power of 0.901.

Although one should be aware of the effects of discontinuities in *P*-values in assessing evidence for abandoning a hypothesis and in calculating power and sample sizes to achieve certain aims, the importance of such effects should not be overemphasized. In this context, it is also worth remembering that mathematical models of real-world situations are generally approximations. Hopefully, these will reflect key features of reality, but there will always be some discrepancies. For instance, as already pointed out, when *n* items are taken from a large batch of *N* items by sampling without replacement, the binomial distribution is only an approximation, albeit a good one.

4.4.4 A Caution

In Section 4.4.1, an example was given of a province within which 15.9% of first-generation immigrants from South Asia had a diagnosis of diabetes. However, variations between the countries of origin meant that each individual did not have the same probability of having diabetes. It was pointed out that if a random sample were taken from a large population, it is not unreasonable to expect this to reflect the population proportion of around 16%, because statistical theory backs an intuitive hunch that a random sample should reflect fairly closely the characteristics of a population. On this basis, first-generation South Asians could be compared with other ethnic groups in the same province using a simple test based on the binomial assumption, although, as described in Example 4.10, the analysis may still not be valid.

Example 4.10

Suppose that 7% of adults of white ethnicity in an urban population have a diagnosis of diabetes, and it is thought that this prevalence could be lower in an adjoining rural region. A random sample of adults living in this region is obtained. As part of the study, a research nurse interviews each participant. Out of the 54 adults involved, only one reports a diagnosis of diabetes. Is there sufficient evidence of a lower prevalence of diabetes in this rural region relative to the urban population?

For the purpose of this example, a one-tailed test of H_0: $p=0.07$ against H_1: $p<0.07$ is justified on the grounds that those living in rural areas may have a healthier diet and take more exercise, two factors that reduce the risk of being diagnosed with diabetes. If the direction of any effect were unknown, then a two-tailed test of H_0: $p=0.07$ against H_1: $p\neq 0.07$ would be appropriate.

Using the ANSM5 package in R and the command `binom(1, 54, 0.07, alternative = "less")` gives an exact *P*-value of 0.101. The exact two-sided nominal 95% confidence interval for p of (0.0005, 0.0989) can be obtained using the ANSM5 package in R and the command `binom(1, 54)`, showing that evidence of a reduction is not strong. If it were important to detect even a small reduction in the "success" probability to, say, $p=0.05$, a larger experiment would be needed to give reasonable power.

In Section 4.4.1, it was pointed out that difficulties might arise if particular groups of individuals within the same study, such as members of the farming community, seasonal agricultural workers, and commuters working in a nearby urban area in this example, have markedly different probabilities of having a diagnosis of diabetes. The need for a random, or effectively random sample, to overcome this problem cannot be overstressed.

Another type of problem may arise in, for instance, the testing of a new drug. For the new drug, there may be a quite different set of factors influencing the likelihood of an individual showing side effects compared with the original drug. For instance, with an established antibiotic for the treatment of pneumonia, people with asthma may be more likely than asthma-free individuals to exhibit side effects, whereas for a new antibiotic, the reverse may be the case. In these circumstances, what could be an overall beneficial effect can sometimes be masked or distorted in the sample. The solution lies either in further experimentation or at least in using more detailed observations and tests. For example, one may require an analysis that takes into account other factors that might influence outcomes for individuals, e.g., factors such as weight, age and gender of patients. The result is a more complex analytic problem, and indeed most real-life statistical problems are more complicated than the simple ones discussed in this section.

SUMMARY POINTS:

- Data consisting of the number of occurrences of one from two possible outcomes in a series of trials often give rise to a binomial distribution, and inferences are often required about the parameter p associated with that outcome. These can be obtained by performing hypothesis tests (Section 4.4.1) and calculating confidence intervals (Section 4.4.2).

- When p is close to 0 or 1, asymptotic results may be reliable only for very large samples (Section 4.4.2).
- Studies of power and the sample sizes needed to obtain a given power against a single specified alternative are relatively straightforward when the binomial distribution is relevant to primary data (Section 4.4.3).
- An overall beneficial effect can sometimes be masked by differences between subgroups within a sample (Section 4.4.4).

4.5 Tests Related to the Sign Test

4.5.1 An Alternative Approach to the Sign Test

The sign test is described in detail in Section 3.2. An alternative, but equivalent, approach more in the spirit of the work in Section 4.4 is sometimes used and this is illustrated here by an example.

Example 4.11

In Example 3.1, Sample II from the Epilepsy Risk Awareness study participants (20 individuals, none from the previous group) had seizure severity scores of:

2	10	27	35	38	40	47	50	58	60
68	70	79	103	108	151	157	158	162	166

If θ is the population median, test the hypothesis H_0: $\theta=110$ and H_1: $\theta\neq 110$ using the sign test. Obtain a 95% confidence interval for θ using the methods developed in Section 4.4.

The wide availability of programs giving exact P-values provides a useful facility for testing and estimation in a sign-test situation. For illustrative purposes, the one-sample binomial test calculator in Stata is used. Here $n=20$ and the number of observations below the hypothesized median is $r=15$. The sign test calls for the evaluation of P for a two-tailed test with $p=0.5$, which gives $P=0.041$. A nominal 95% confidence interval for p is (0.509, 0.913).

Output from the ANSM5 package in R can be obtained using `binom(15, 20, 0.5)`.

Allowing for rounding, this P-value agrees with that obtained in Example 3.1 and indeed here nothing essentially different was performed from what was done there. To obtain a confidence interval for θ the value of r is decreased by steps of 1 until a 95% confidence interval for p that just includes the value $p=0.5$ is obtained. This straightforward stepwise process indicates that when $r=15$ the confidence interval is (0.509, 0.913), while for $r=14$ it is (0.457, 0.881). Thus, when $r=14$ the value $p=0.5$ would be regarded as acceptable with $P>0.05$, but not when $r=15$. Symmetry of the binomial distribution when $p=0.5$ implies acceptance when $r=6$ but not when $r=5$. It follows that an appropriate 95% confidence interval for the median is from the sixth to the fifteenth largest observations, i.e., the open interval (40, 108), agreeing with the result in Example 3.2.

The procedures in Examples 3.1 and 3.2 and in this example are basically equivalent. Which is preferred is largely determined by the availability of appropriate tables or computational software.

4.5.2 Quantile Tests

The sign test is relevant to population medians. Other quantiles of a population can also be of interest. For example, a quantile such that three-quarters of the population values lie below it and one-quarter above, known as the upper (or third) quartile, or those such that r tenths of the observations lie below it and $(10-r)$ tenths of the observations lie above it where $r=1, 2, \ldots, 9$, known as the rth deciles. Many tests and estimation procedures involving quantiles proceed along similar lines to those about medians by simply replacing $p=0.5$ with a value appropriate to the relevant quantile. For continuous distributions, quantiles are distinct, but for discrete distributions, conventions given in most standard textbooks are needed to give distinct quantiles. The usual convention for the median of a set consisting of an even number of observations has already been met. These are arranged in ascending order, denoting the number of observations by $2m$ and the rth ordered observation by $x_{(r)}$ the median is defined as $(x_{(m)}+x_{(m+1)})/2$. Note that the fifth decile is the same as the median.

Example 4.12

In a European country, three-quarters of students taking a public examination in mathematics obtained a mark of 40 or more (i.e., the first population quartile is 40). One college entered 32 candidates for this examination of whom 13 scored less than 40. The principal explains the college's apparently poor performance by claiming that in a random sample of 32 candidates, it is quite likely that 13 would score less than the lower quartile mark, even though 8 out of 32 is the expected proportion. Is this reasoning justified?

Denoting the first quartile by Q_1, an appropriate test for the principal's assertion considers the null hypothesis as H_0: $Q_1=40$ against the alternative H_1: $Q_1 \neq 40$.

A minus sign is associated with a mark below 40 and a plus sign with the others; thus, for the sample of 32, there are 13 minuses (and 19 pluses). If the first quartile is 40 then the probability is $p=0.25$ that each candidate in a random sample has a mark below 40 (and thus is scored as a minus). The distribution of minuses is therefore binomial B(32, 0.25). The 95% confidence interval for p given $r=13$ minuses (event A) when $n=32$ is (0.24, 0.59). This includes $p=0.25$. The associated exact P-value is equal to 0.06291.

Output from the ANSM5 package in R can be obtained using `binom(13, 32, 0.25)`.

Observing 13 minus signs would not lead to rejection of the null hypothesis at a conventional 5% significance level. Nevertheless, there is some evidence against this hypothesis - enough to worry many with parental responsibility. Some may be reluctant to give the principal's claim the benefit of the doubt without further evidence.

Remember that having little or no evidence against a null hypothesis does not prove it true. It is only a statement that the evidence to date is not sufficient for rejection. This may simply be because the sample is too small. An error of the second kind can be made by nonrejection. Indeed, since the above confidence interval includes $p=0.5$, the hypothesis that the median is 40 would not be rejected.

A two-tailed test has been used. For a college with responsibility for students with learning difficulties, this might not be appropriate, and the use of a one-tailed test could be justified.

The principal's claim was made on the basis of a random sample. Students from a single college are in no sense a random sample from all examination candidates. This test only establishes that results for this particular college are not too strongly out of line with national results in the sense that they just might arise if one took a random sample of 32 candidates from all entrants.

Tests for a third quartile are symmetric with those for a first quartile if plus and minus signs are interchanged.

4.5.3 The Cox–Stuart Test for Trend

For observations ordered in time, it is often of interest to determine whether a trend is present. Cox and Stuart (1955) proposed a simple test for a monotonic trend, i.e., an increasing or decreasing trend. A monotonic trend need not be linear; it need only express a tendency for observations to increase or decrease subject to some local or random irregularities.

Consider a set of independent observations $x_1, x_2, \ldots, x_n$ ordered in time. If the number of observations is even with $m=n/2$, the differences $x_{m+1}-x_1, x_{m+2}-x_2, \ldots, x_{2m}-x_m$, are calculated. For an odd number of observations, $2m+1$, the middle value x_{m+1} is omitted and the differences $x_{m+2}-x_1, x_{m+3}-x_2$, etc. are calculated.

With an increasing trend, most of these differences are positive, whereas a decreasing trend produces mainly negative differences. If there is no trend, and observations differ only by random fluctuations about some median, these differences (in view of the independence assumption) are equally likely to be positive or negative. This implies that under the null hypothesis of no trend, the plus (or minus) signs have a B(m, 0.5) distribution and a sign test can be applied.

Example 4.13

Jeandron et al. (2015) studied the association between water supply and cholera incidence in Uvira, Democratic Republic of the Congo. Study measurements included the daily precipitation rates in mm/h. The figures below are for the first day of alternate months in 2012 and 2013 (February, April, June, August, October and December):

0 1.841 0.202 0.891 1.423 0 1.345 49.8 0 0 1.561 4.376

A decline in precipitation reduces the supply of water to the population, which in turn for this part of the world increases the incidence of cholera. If concerns are raised regarding trends in precipitation, do these data provide evidence of a monotonic trend in either direction?

Adjacent figures represent a 2-month interval, so observations will be uninfluenced by short-term correlations, which can be high. The Cox–Stuart test may therefore be applied to test for trend. A two-tailed test is appropriate as any trend may be an increasing or decreasing one.

The time period is divided into halves, and differences are calculated in the manner described above, so that February 2012 is paired with February 2013, etc. This process is simplified by presenting the figures as:

2012:	0	1.841	0.202	0.891	1.423	0
2013:	1.345	49.8	0	0	1.561	4.376

Relevant differences are 0–1.345, 1.841–49.8, 0.202–0, 0.891–0, 1.423–1.561, 0–4.376. Assigning pluses and minuses to these differences gives −, −, +, +, −, −, or 2 plus and 4 minus signs from 6 observations. The associated P-value for a two-tailed test is equal to 0.6875. Hence, there is no evidence of a monotonic trend.

Output from the ANSM5 package in R can be obtained using `cox.stuart (ch4$precipitation)`.

The fact that the first two differences are negative, the next two are positive, and the final two differences are again negative hints that the pattern of plus and minus signs might not be random. The sample is far too small to investigate this but were such a pattern to be observed for a longer series it might suggest switching between periods of increasing and decreasing trend. In Section 5.4, a runs test is described that is appropriate for testing whether fluctuations within larger samples are random.

In many parts of the world, rainfall follows a periodic pattern across the year. Time series observations can allow for this feature by taking differences between corresponding dates in adjacent years, as in this example. If the time interval between pairs of observations is such that differences cannot be aligned with the periodic cycle (e.g., if the observations are taken over 14 months and the cycle is an annual one), monotonic trends can be missed. Conover (1999, Examples 4 and 5, pp. 172–175) shows how, in certain circumstances, the Cox–Stuart test may be adapted to detect periodic trends by reordering the data.

SUMMARY POINTS:

- Inferences about quantiles of data from continuous or other measurement distributions may be based on the value of p associated with that quantile (Section 4.5.2).
- The Cox–Stuart test for trend applies the sign test to data ordered in time (Section 4.5.3).
- With the Cox–Stuart test, it is easy to miss patterns within a periodic cycle if the time span of the observations is not a multiple of the period length (Section 4.5.3).

4.6 Single-Sample Runs Tests for Randomness

4.6.1 Clustering, Alternation and Randomness

Many statistical inferences are valid only when the data are a sample or samples of independent (random) observations. Methods for obtaining a random sample – especially in simulation or Monte Carlo studies – often depend upon sophisticated mechanisms that purport to be equivalent to repeatedly tossing a fair coin or to repeatedly selecting one of the 10 digits between 0 and 9 inclusive, each having a probability 0.1 of selection.

One characteristic of the coin tossing situation is that in the long run, there should be approximately equal numbers of heads and tails and a characteristic of random digit selection is that in a long string of such digits, each digit should occur approximately the same number of times. In the coin tossing situation, a sign test would be appropriate to see whether this frequency requirement is being met. In Section 12.4, a method is given that may be applied to testing whether there is evidence for digits not occurring with equal probability.

In an ordered sequence, randomness implies more than compliance with frequency criteria. For example, if the outcomes, in order, of a computer process that purports to simulate 20 tosses of a coin were

$$\text{H H H H H T T T T T T T T T T H H H H H} \tag{4.6}$$

it would be reasonable to suspect that the process did not achieve its aim. The sequence illustrates the clustering of outcomes of the same type. It would also be surprising if the ordered outcomes were

$$\text{H T H T H T H T H T H T H T H T H T H T} \tag{4.7}$$

as this demonstrates alternation between outcomes of differing types. Randomness might seem plausible with the outcomes

$$\text{H H T H T T T H T H H T H T H H H T T H} \tag{4.8}$$

A characteristic that reflects understandable reservations about (4.6) and (4.7) is the number of runs, where a run is a sequence of one or more heads or tails. In (4.6), there are three runs – a run of 5 heads, then 10 tails, then 5 heads. In (4.7), there are 20 runs, each consisting of a single head or a single tail. Intuitively, (4.6) and (4.7) appear to have, respectively, too few, and too many, runs for a truly random sequence. The sequence (4.8) has an intermediate number of runs, namely 13.

Nonrandom sequences are commonly encountered and include, for clustering, the positions of healthy and diseased plants in a row under cultivation, and for alternation, the seating of individuals in a café, where strangers would prefer not to sit next to each other (Swed and Eisenhart, 1943).

4.6.2 A Runs Test for Two Categories

Both numbers of runs and lengths of runs are relevant to tests for randomness. Whitworth (1886, 1897) developed the distribution theory for runs [see Smeeton (2021) for details]. Bradley (1968, Chapters 11, 12) gives a detailed treatment of runs tests, and a concise account can be found in Gibbons and Chakraborti (2020, Chapter 3). The simplest form of test is based on the number of runs, r, in a sequence of N ordered observations of which m are of one kind (e.g., H) and $n=N-m$ are of another kind (e.g., T). Evidence against the hypothesis that the outcomes are independent or random is strong if too few or too many runs are observed. Computation of the probability of observing any given number of runs under the hypothesis of randomness is a subtle application of combinatorial mathematics, and only the outcome is quoted here.

The random variable R specifies the number of runs. The cases r odd and r even are considered separately. For r odd, if s is such that $r=2s+1$, this can be written as

$$\Pr(R = 2s + 1) = \frac{\binom{m-1}{s-1}\binom{n-1}{s} + \binom{m-1}{s}\binom{n-1}{s-1}}{\binom{N}{m}}$$

If r is even, and $r=2s$

$$\Pr(R = 2s) = 2 \times \frac{\binom{m-1}{s-1}\binom{n-1}{s-1}}{\binom{N}{m}}$$

The test for randomness is based on the relevant tail probabilities associated with small and large numbers of runs. Tables for $n, m \leq 20$ can be found in Swed and Eisenhart (1943), and tables based on these are given in Siegel and Castellan (1988, Table G). Sprent (1998, Example 6.16) shows how to compute the distribution of R for a small sample for given m, n. An asymptotic test is also available and is based on known results for the mean and variance of R. These are shown after some routine algebra (see, e.g., Gibbons and Chakraborti, 2020, Section 3.2) to be

$$\mathrm{E}(R) = 1 + \frac{2nm}{N}$$

and

$$\mathrm{Var}(R) = \frac{2nm(2nm - N)}{N^2\,(N-1)}$$

Asymptotically

$$Z = \frac{R - \mathrm{E}(R)}{\sqrt{\mathrm{Var}(R)}}$$

has a standard normal distribution. The approximation is improved by the usual numerator continuity correction, i.e., adding 0.5 if $R<\mathrm{E}(R)$ and subtracting 0.5 if $R>\mathrm{E}(R)$.

In practice, a two-tailed test is most often relevant, with a significant result implying nonrandomness, but one-tailed tests are meaningful in the sense that few runs imply clustering of like values while many runs imply an alternating pattern.

Example 4.14

Apply a runs test for randomness to each of the sequences of heads and tails in (4.6), (4.7) and (4.8).

In each of (4.6) and (4.7) $N=20$ and $m=n=10$, while $r=3$ for the former and $r=20$ for the latter. In (4.8) $N=20$, $m=11$, $n=9$ and $r=13$. Small or large numbers of runs indicate nonrandomness.

Tables for $N=20$ and various values of m, n such as those referenced above indicate that in both these cases a nominal $P=0.05$ is associated with $r \leq 6$ or $r \geq 16$. It is more satisfactory to compute exact P-values for the observed R.

For these examples using the ANSM5 package in R and the commands `runs.2cat(ch4$tosses1, do.asymp = TRUE)`,`runs.2cat(ch4$tosses2, do.asymp = TRUE)` and `runs.2cat(ch4$tosses3, do.asymp = TRUE)` give the following exact and asymptotic two-tail P-values:

	Exact	Asymptotic
Data set (4.6)	< 0.001	0.001
Data set (4.7)	< 0.001	< 0.001
Data set (4.8)	0.362	0.457

There is strong evidence of nonrandomness in data sets (4.6) and (4.7) but no evidence in set (4.8). The low value of the statistic R for (4.6) reflects clustering, while the high value in (4.7) suggests alternation. These are clear data characteristics in the respective cases.

The above test is not so restrictive as it might seem. For example, in sequences of supposedly random digits between 0 and 9, the test can be applied to the numbers of runs above and below the median value of 4.5. It might also be applied in this case to runs of odd and even digits.

Clustering and randomness may occur within the same sequence. Such patterns can be found in the daily weather records for the British Isles, where a series of weather fronts typically associated with alternation between wet and dry days can be followed by anti-cyclonic weather characterized by a long spell of dry weather.

Note that the runs test is likely to miss this type of deviation. For instance, the sequence

$$\text{H H H T T T T H H H T H T H T H T H T H} \quad (4.9)$$

shows a switching from clustering to alternation. Since the numbers of outcomes, heads, tails and runs are the same as those in (4.8), the test will give the same P-values as for (4.8), and the nonrandomness will remain undetected.

4.6.3 A Runs Test for Three or More Categories

If there are three or more categories and clustering or alternation is suspected, the relevant tail probabilities for the numbers of runs can be derived (Mood, 1940), although the algebra involved is not straightforward. Barton and David (1957) extends Mood's work and provides tables for the distribution of the number of runs for samples of up to 12 observations, both for 3 and 4 different outcomes. Time-ordered residuals from regression analyses can be tested for randomness by applying multiple runs distributions (Shaughnessy, 1981), although the tables of critical values in this paper are mainly for groups of similar size and so are of limited application. Schuster and Gu (1997) developed algorithms for calculating exact distributions for the number of runs for multiple outcomes using the software system Mathematica. Smeeton and Cox (2003, 2006) describe a method for estimating the distribution of the number of runs by simulating permutations of the sequence of categories using Stata. The ANSM5 package in R provides for the calculation of both asymptotic and simulation-based output. The comprehensive text by Balakrishnan and Koutras (2002) explores the theoretical aspects of a wide range of run problems.

An asymptotic test based on the normal distribution makes use of the mean and variance of the number of runs (Schuster and Gu, 1997). Let the random variable R specify the number of runs. Suppose that the total number of observations is N, with k different outcomes and n_i observations of the ith kind, $i = 1, 2, \ldots, k$. Writing $p_i = n_i/N$ for the proportion of observations of type i gives

$$\mathrm{E}(R) = N\left(1 - \sum_{i=1}^{k} p_i^2\right) + 1$$

and

$$\mathrm{Var}(R) = N\left[\sum_{i=1}^{k}\left(p_i^2 - 2p_i^3\right) + \left(\sum_{i=1}^{k} p_i^2\right)^2\right]$$

Barton and David (1957) suggests that whatever the composition of the subsets in terms of number the normal approximation is adequate for $N>12$.

Example 4.15

Smeeton and Cox (2003) gives data on the method of delivery for 17 consecutive births in a South London hospital. There are four methods of delivery, and there is interest in whether the method of delivery is clustered in time. The sequence of births is:

A A A A B A C C A A A A A D D A A

The codes are A for normal delivery, B for forceps, C for elective Caesarean and D for emergency Caesarean. The null hypothesis of randomness H_0 is used to test for the methods of delivery.

Using the notation above, $N=17$, $k=4$, $n_1=12$, $n_2=1$, $n_3=2$, $n_4=2$ and $r=7$. Since clustering is of interest, a one-tailed alternative hypothesis is appropriate. Clustering is indicated by a small number of runs, providing evidence against H_0.

The distribution of the number of runs is obtained from 100,000 simulations. The estimate for the *P*-value corresponding to $R \le 7$ is compared with the asymptotic normal approach, with and without a numerator continuity correction of 0.5, for $N=17$, and the numbers in the four categories of delivery as given. Here $\mathrm{E}(R)=9.0$, $\mathrm{Var}(R)=1.6886$. The results for the *P*-value are:

Simulation	Asymptotic Normal	Asymptotic Normal (Continuity Correction)
0.0971	0.0619	0.1242

Output from the ANSM5 package in R can be obtained using `runs.ncat(ch4$births, alternative = "less", do.mc = TRUE)` with the added argument `cont.corr = FALSE` to obtain the asymptotic normal result without the use of the continuity correction.

These results show that there is only weak evidence of clustering for the method of delivery in this sample of births.

The relatively weak evidence against H_0 is due to the small sample size. A longer sequence from the same data set produced clear evidence of clustering. This was due to elective Caesareans being almost always performed during normal working hours, whereas other types of birth can occur at any time between 0000 and 2359.

The *P*-values from the asymptotic approach are not close to those produced by simulation, suggesting that this method should only be applied to samples that are at least moderately large. Runs tests lack power to detect clustering partly because interchanging just two observations can increase the number of runs by 2. However, there are few alternatives if, as in this example, the categories cannot be ordered.

SUMMARY POINTS:

- Clustering and alternation in a series of dichotomous outcomes (e.g., heads or tails) can be investigated using a runs test (Section 4.6.1).
- In a sequence of N such ordered outcomes, the hypothesis of randomness or independence is rejected if there are too few or too many runs (clustering and alternation, respectively) (Section 4.6.1).
- Simple modifications allow testing for runs of observations above and below the median. These may provide evidence for detecting certain types of departures from independence (Section 4.6.2).
- Clustering and randomness may occur within the same sequence (Section 4.6.2).
- Runs tests can also be applied to data having several categories, usually to investigate possible clustering of outcomes of the same kind, indicated by a lower-than-expected number of runs (Section 4.6.3).

4.7 Angular Data

In some investigations, measurements are made on directions, e.g., wind direction at an airport at noon on successive days, the bearings at which released pigeons disappear over the horizon, or the successive stopping positions of a roulette wheel.

In other studies, the data, although not measured directly as angles or directions, can be represented in this way. For instance, the time of day at which babies are born in a large hospital, or the days during the year in which new cases of leukaemia are diagnosed in a certain region, are effectively angular measurements. The names *circular* or *directional* measurements are also used, the former because observations of this type are appropriately represented as points on the circumference of a circle. Such data are sometimes analysed as though they are distributed along a line without taking the directional aspect into account. This is not always a sensible course of action. Using a straight-line approach, observations of 1° and 359° would lead to a mean and median of 180°. The values then appear to be very different. If, however, the two points are plotted on the circumference of a circle, the two directions are similar, and a more appropriate average would be obtained from the direction bisecting the smaller arc between the two directions, in other words, 0°. If the data consisted of two observations of 31 December and 2 January, the "straight-line" approach leads to an average around 1 July, whereas a directional analysis leads to the more reasonable average of 1 January. Measures of location (e.g., the median) and spread (e.g., the range) may be adapted for use with angular data.

The book by Mardia (1972) on directional data analysis is written from a mathematical perspective. Fisher (1993) provides an introduction based on real-life examples from many fields. Another useful text on the theory and practice of directional data is Mardia and Jupp (2000).

One often wants to know whether data support a hypothesis that a sample is from a population uniformly distributed on the circumference of a circle or whether they indicate clustering (Mardia, 1972). The *Hodges–Ajne test* (Hodges, 1955; Ajne, 1968) and the *range test*, the latter developed from work on the circular range by Fisher (1929), address this situation.

Another question of interest is whether points are symmetrical around a particular angular direction such as the North–South line. The sign test and the Wilcoxon signed-rank test are easily adapted for this situation.

4.7.1 The Hodges–Ajne Test

This test is used to investigate whether a sample of n observations on a circle could arise from a uniformly distributed population. The alternative is that in the population observations are more concentrated within a particular arc of the circumference; outliers may nevertheless occur well away from this arc.

To carry out this test, a straight line is drawn through the centre of the circle; this divides the observations into two groups. The line is rotated about the centre to a position at which there is a minimum possible number of points, m, on one side of this line. If the points were regularly spaced they would lie at the vertices of a regular polygon and then either $m=n/2$ if n is even or $m=(n-1)/2$ if n is odd. For observations around the circle from a uniform distribution, there will be some variation in the angles between adjacent points, and the value of m will generally be lower. The lowest values of m occur when there is a clustering of points on one side of the line. Under the assumption of uniformity, the probability that m is no more than a value t is shown by Mardia (1972) to be

$$\Pr(m \le t) = \frac{\binom{n}{t}(n-2t)}{2^{n-1}}, \quad t < n/3 \tag{4.10}$$

Mardia (1972) gives a table of critical values for this test.

Example 4.16

A midwife recorded the times of birth for 12 consecutive home deliveries. She was interested in whether births tended to occur at particular times of the day. The times (rearranged in order throughout the day) were 0100, 0300, 0420, 0500, 0540, 0620, 0640, 0700, 0940, 1100, 1200, and 1720.

Since, on a 24-hour circular clock, 1 hour corresponds to $360/24=15°$, the successive angles on the circle (assuming midnight corresponds to 0°) are 15°, 45°, 65°, 75°, 85°, 95°, 100°, 105°, 145°, 165°, 180°, and 260°.

The null hypothesis H_0 is that the times of birth have a uniform distribution around the circle. To do this, a line is drawn through the centre of the circular plot of these data and rotated until the number of points on one side of the line takes the minimum value, m. A small value for m provides evidence against the assumption of uniformity made in H_0.

A circular plot of the data (Figure 4.9) shows that $m=1$ (e.g., when the line runs from 10° to 190°). Only a value of zero for m represents an even greater degree of clustering. The appropriate P-value (the probability that m is equal to zero or one, assuming that H_0 is true) given by (4.10) is 0.059. Therefore, the evidence against H_0 when $P=0.059$ casts some doubt on the assertion that births are equally likely at any time throughout the day.

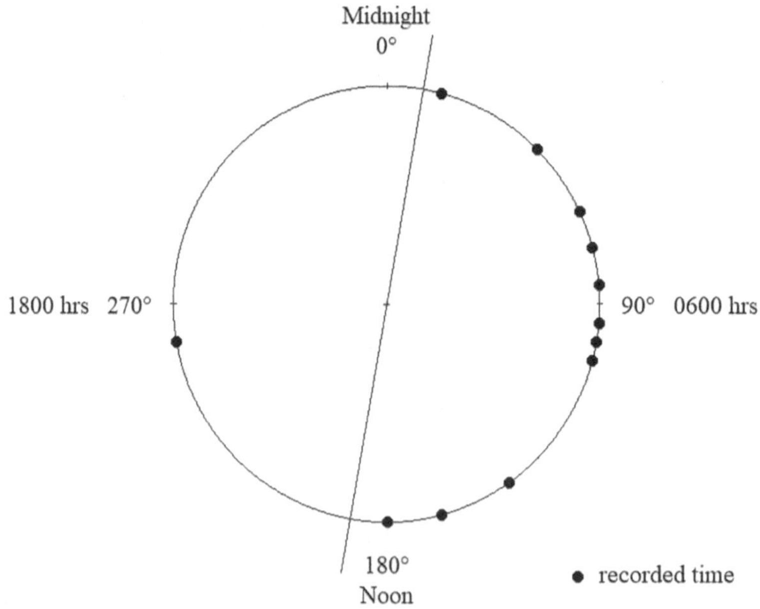

FIGURE 4.9
Circular plot of delivery time data in Example 4.15 on a 24-hour clock.

Output from the ANSM5 package in R can be obtained using `hodges.ajne(ch4$times.as.degrees)`.

Inspection of the data shows that most of the babies are born in the morning, particularly between 0300 and 0700. If further evidence for this pattern could be obtained, the information might be useful in planning maternity services and in preparing mothers for the births of their babies.

This test is especially useful where some observations may be clustered together but a few may be more distant. For example, injuries from fireworks in the UK are concentrated around 5 November and deaths from drowning are more common during the summer months, but such events are not confined exclusively to these periods.

4.7.2 The Range Test

Like the Hodges–Ajne test, the range test is also based on the null hypothesis that a sample of n observations on the circle could be from a uniformly distributed population, but now the alternative is that in the population all observations come from a particular arc. This test is therefore not appropriate in the presence of outliers.

To perform this test, the smallest arc that contains all of the points is found.

The length of this arc, w, is the circular range. If the points are spaced at regular intervals they will lie at the vertices of a regular polygon, and the circular range will then take the maximum possible value. This is $360(n-1)/n°$. The circular range will be small if all points occur close together. Under the assumption of uniformity, the probability that the circular range w is no more than r radians is shown by Mardia (1972) to be:

$$\Pr(w \le r) = \sum_{k=1}^{v} (-1)^{k-1} \binom{n}{k} \left[1 - \frac{k(2\pi - r)}{2\pi}\right]^{n-1} \tag{4.11}$$

where ν is the largest value of k such that $1-k\{(2\pi-r)/2\pi\}$ exceeds zero. Mardia (1972) gives a table of critical values for the range test.

Example 4.17

Smeeton and Wilkinson (1988) gives data for a female psychiatric patient who repeatedly attempted to commit suicide. There was evidence to suggest that these attempts occurred during one particular part of the year. Records showed that attempts had occurred on 2 June 1980, 3 June 1980, 8 June 1980, 18 June 1980, 4 July 1980, 5 June 1981, 6 June 1981 and 31 July 1981.

The successive angles on the circle (assuming 0° is the start of the year) are

151°, 152°, 157°, 167°, 182°, 154°, 155°, 209°.

The null hypothesis under test, H_0, is that the dates of the suicide attempts have a uniform distribution.

The circular range w is the length of the smallest arc that contains all of the dates. If the suicide attempts are uniformly distributed around the circle then the circular range will be relatively large. By contrast, if the dates form a tight cluster the circular range will be small, providing evidence against H_0.

In this example, the circular range is the difference between the largest and smallest angle (209°−151°=58°). This is illustrated on the circular plot in Figure 4.10. Converting 58° to radians, $r=1.0123$ radians. The appropriate P-value given by (4.11) is $P<0.0001$ providing overwhelming evidence against H_0. In view of the strength of evidence against H_0 when $P<0.0001$, it is reasonable to accept that for this patient clustering of suicide attempts occurs in June and July.

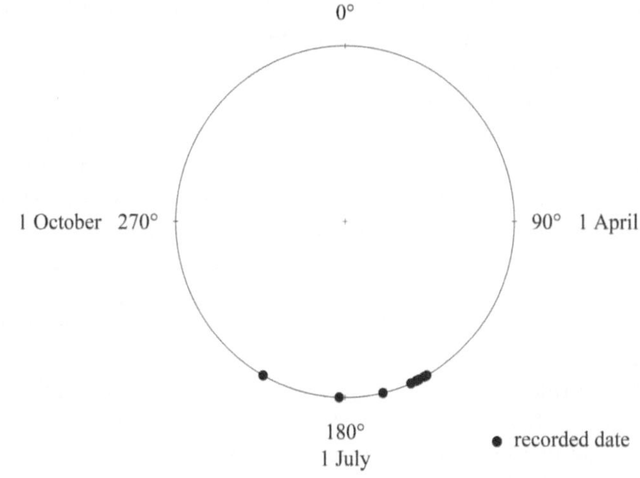

FIGURE 4.10
Recorded dates of eight suicide attempts on an annual clock where 0° represents 1 January.

Output from the ANSM5 package in R can be obtained using `rng.test(ch4$dates.as.degrees)`.

If the points within a cluster lie on both sides of 0° the difference between the largest and smallest values does not give the circular range. It is then most easily obtained from the circular plot. Unlike the Hodges–Ajne test, the range test is only useful when all observations lie in one cluster, since the circular range is highly susceptible to outliers. The occurrence of a further attempt on, say, 5 February, illustrates this.

The detection of a single cluster of points containing all the observations is particularly relevant in psychiatry. Extra support can be targeted at the patient during the appropriate period. For the patient in this example, the data suggest that each year the anniversary of a distressing event that happened around the beginning of June could trigger a series of suicide attempts that diminish in frequency over the subsequent weeks.

The circular range increases with sample size so that over subsequent years suicide attempts by this individual may be recorded outside of June and July, even if the underlying model remains the same. In fact, follow-up data for this patient (to September 1987) subsequently became available. In 1982, three attempts were recorded between 10 April and 10 June. In 1983, four attempts were recorded between 22 August and 14 December. Following this, no further attempts were recorded. Taking all of the evidence into account the most likely explanation is that any anniversary effect gradually dampened out.

4.7.3 Median or Symmetry Tests for Angular Data

In many sports (archery, cricket, target shooting, netball, bowls, etc.), the objective – at least at some phase of the activity and for some participants – is to hit a small target by dispatching an appropriate missile (e.g., throwing, kicking or pitching a ball, firing an arrow or bullet). In practice, even the most skilled exponents will tend sometimes to go left of the target and sometimes right of the target (or perhaps also above or below target).

It is often of interest to see whether there is a tendency to go more often to the left than to the right or vice versa. If the total number of times that there is a deviation to the right or to the left in n attempts is known, a sign test might be used to see if there is evidence that the median direction is or is not that of the target. If there were evidence of a tendency to deviate more often to one side than the other the players, or their coaches, may want to take corrective action. In some cases, information may also be available on the angular deviations from the target line for each attempt made by a player. Those angular deviations may be used with appropriate signs (say plus to the right and minus to the left) and after ranking absolute values use a Wilcoxon signed-rank test to carry out what is usually called in this context a *test of symmetry*. In doing so, an assumption is made that the directional distribution of the attempts is symmetric. The question of interest then is whether it is symmetric about the target direction (zero deviation).

Of course, for the serious player symmetry of shots around the target is not enough; the deviations themselves must be small. Also, in sports involving a two-dimensional target, such as archery or darts, the falling of the missile above or below the target has also to be addressed in order to ensure an accurate hit.

These ideas are also relevant in scientific and other applications and extend to testing whether angular data indicate a symmetric distribution about some specified direction. An example of the use of the Wilcoxon test this way in a geological context is given by Fisher (1993, Example 4.17). Where these tests are appropriate, they follow the usual procedure for the Wilcoxon test and the sign test given in Chapter 3.

SUMMARY POINTS:

- Angular or directional data are often best represented by points on the circumference of a circle.
- Care is needed with concepts such as the mean and median for angular data.
- The Hodges–Ajne test assesses whether population observations are more concentrated within a particular arc of the circumference (Section 4.7.1).
- The range test is not appropriate in the presence of outliers (Section 4.7.2).
- The median test assesses whether the observations are symmetric about a target direction (Section 4.7.3).

4.8 Exercises

4.1 The data sets in Appendix 1 include the ages of death for 21 members of the McGamma clan. Perform an appropriate test to determine whether it is reasonable to assume age at death is normally distributed for that clan.

4.2 For 16 individuals attending an outpatient clinic in a hospital, the times in minutes between checking in with the receptionist and the start of the consultation are:

10 42 29 11 63 145 11 8 23 17 5 20 15 36 32 15

Use the Kolmogorov test to determine whether it is reasonable to assume that the waiting times are a sample from the exponential distribution with a mean of 20. The cumulative distribution function is $F(x)=1-e^{-x/20}$, $0\leq x<\infty$.

4.3 A football club records the number of visiting team supporters at home matches. A random sample of 14 matches gives the numbers of individuals as:

1,175 1,183 1,327 1,581 1,592 1,624 1,777
1,924 2,483 2,642 2,713 3,419 5,350 7,615

Are these observations likely to have come from a normal distribution? Test using Lilliefors' test and the Shapiro–Wilk test.

4.4 Test whether the football match attendance data in Exercise 4.3 may have come from (i) an exponential distribution with a mean of 2,600 and (ii) a uniform distribution over the interval (1,100, 7,700). Comment on your findings.

4.5 A supermarket indicates in contracts given to suppliers of oranges that not more than 1% of the fruit in any consignment should show visible signs of damage. Realism forces it to agree that it will use a sampling scheme with the producer's (i.e., the supplier's) risk set at $P=0.10$ for rejection of samples with less than 1% blemished and the consumer's (i.e., the supermarket's) risk set at $1-P^{*}=0.05$ for accepting batches with 3% or more blemished. The sampling scheme used is to select a number n from a large batch of N. For each of the sample sizes $n=100$, 200,

300, 400, 500 determine the maximum number of damaged fruits, r, that would indicate the condition for the producer's risk is being met. In each case, find the corresponding producer and consumer risks. What is the smallest value of n and for what r will the above conditions for producer's and consumer's risks be met?

4.6 A commentator on a major golf championship asserted that on a good day, 10% of top-class players could be expected to complete a round with a score of 68 or less. On the fourth day of the championship, weather conditions were poor and the commentator remarked before play started that one might expect the weather to increase scores by four strokes per round, implying that 10% of players might be expected to return scores of 72 or less. In the event 26 of the 77 players competing returned scores of 72 or less. Regarding the players as a sample of 77 top-class players and assuming the commentator's assertion about scores on a good day is correct, do these fourth-day results suggest the commentator's assertion about scores in the poor weather conditions prevailing was (i) perhaps correct, (ii) optimistic, or (iii) pessimistic?

4.7 In a pilot opinion poll 18 voters were selected at random and asked if they were highly satisfied (based on a five-point scale) with the performance of the British Prime Minister. Six (one-third) said Yes and twelve (two-thirds) said No. Is this sufficient evidence to reject the hypothesis that 50% of the electorate report high satisfaction with the Prime Minister's performance?

The study was repeated using a larger sample of 225 voters. By coincidence 75 (one-third) answered Yes and 150 (two-thirds) answered No. Is the same conclusion drawn about the hypothesis that 50% of the electorate report high satisfaction with the Prime Minister's performance? Comment on your findings.

4.8 A citizens' advice centre regards the assistance provided as timely if a client receives an appropriate consultation within the three working days following the client's initial contact. The number of cases for which this standard is not met in 17 consecutive months is:

11 6 14 13 18 14 11 22 19 19 25 24 38 19 25 31 19

Is there evidence of a monotonic trend?

4.9 The UK Meteorological Office monthly weather summaries give the following annual rainfalls in mm for 15 stations in the UK during 1978. The stations are listed in order of increasing latitude. Is there evidence of a monotonic trend in rainfall from South to North?

Margate, 443; Kew, 598; Cheltenham, 738; Cambridge, 556; Birmingham, 729; Cromer, 646; York, 654; Carlisle, 739; Newcastle, 742; Edinburgh, 699; Callander, 1596; Dundee, 867; Aberdeen, 877; Nairn, 642; Baltasound, 1,142.

4.10 A psychologist is testing 16 applicants for a job one at a time. Each has to perform a series of tests, and the psychologist awards an overall point score to each applicant. A high score indicates a good performance. As each applicant may discuss the tests with later applicants before the latter are tested, it is suggested that those tested later may have an unfair advantage. Do the applicants' scores (in order of testing) given below support this assertion?

62 69 55 71 64 68 72 75 49 74 81 83 77 79 89 42

Use an appropriate runs test. Do you consider the Cox–Stuart test (Section 4.5.3) may also be appropriate? Give reasons for your decision.

4.11 In a study of the association between water supply and cholera incidence in Uvira, Democratic Republic of the Congo (Jeandron et al., 2015), the daily precipitation rates in mm/h between 1 April 2012 and 16 April 2012 were:

1.841	0.423	1.048	4.742	36.567	2.546	2.826	3.546
3.450	0.0	0.768	1.128	1.467	2.275	24.585	0.154

Use an appropriate form (or forms) of the runs test to examine nonrandomness.

4.12 For the study in Exercise 4.11, the volumes of water supplied through the piped water system (m^3) on the corresponding days in April 2012 were:

4,744	4,893	4,804	4,059	2,714	2,855	4,214	3,933
3,993	2,940	3,122	4,208	4,254	4,255	1,566	4,206

Assess these observations for nonrandomness as in Exercise 4.11. Compare your findings for precipitation with those for the volume of water supplied.

4.13 A circular ring road is constructed around a town. A map of the local area shows the town hall is situated at the centre of the circle. The positions of road accidents along the ring road are measured as bearings from the town hall. During a period of road works to the east of the town, the accidents that occur have bearings:

10°, 35°, 82°, 87°, 94°, 108°, 125°

Is there any evidence that the accidents are linked to the road works?

4.14 For a children's television quiz show, a circular board with a stationary vertical pointer attached to the centre is divided into quadrants. The board is spun and contestants are asked questions on sport, popular music, current affairs and science according to the quadrant indicated when the board comes to rest. Angles are marked around the edge of the board. For ten consecutive questions, the angles at rest are:

15°, 46°, 114°, 137°, 165°, 183°, 195°, 215°, 271°, 328°

Do you think that on average all types of questions are equally likely?

4.15 Arrows are fired at a target that is on a bearing of 145° from the archer. Angles of fire for 10 arrows are:

139°, 141°, 146°, 148°, 150°, 152°, 153°, 155°, 158°, 160°

There is concern that the arrows are tending to fall to one side of the target, so one wants to test the hypothesis H_0: $\theta = 145$ against the alternative H_1: $\theta \neq 145$. Use the sign test and the Wilcoxon test to do this. Would you question the validity of using the latter test for these data?

4.16 Twenty batteries are selected at random from a large batch and the time t to complete discharge is recorded for each. Find approximate 95% confidence limits for the lower quartile expressed in terms of the appropriate order statistics, $t_{(i)}$.

4.17 A football team obtains the following results for a series of 12 matches:

$$W\ W\ W\ D\ D\ W\ W\ L\ L\ L\ L\ W$$

where W, D and L indicate that they won, drew or lost, respectively. Use the normal approximation for the distribution of the number of runs to investigate evidence of clustering. If you have suitable software also use a simulation approach. The exact one-tail probability is $P=0.0261$. Comment on the accuracy of your estimate(s) for P.

5

Methods for Paired Samples

5.1 Comparisons in Pairs

Single-sample methods illustrate basic ideas but have limited applications. Practical problems usually involve two or more samples that may or may not be independent. This chapter addresses paired samples, a dependent sample situation where many, but not all, problems have a single-sample equivalent.

5.1.1 Studies Using Pairing

Many factors must be taken into account in the way paired samples are selected and used. In this chapter, the important ones are illustrated using medical examples. Most of these factors are important in this field, although many of them are also relevant in other applications.

To compare two stimuli, it is sometimes appropriate and advantageous, to apply both stimuli to each of, say, n individuals. The main reason for this is that responses to any stimulus often vary markedly between individuals. However, some important issues must be addressed before proceeding this way.

A dentist may wish to compare two mouthwashes for the treatment of ulcerative gingivitis in the mouth. He or she could use the number and size of the mouth ulcers in a patient to assess the severity of the problem. The patients could also indicate their perceived level of pain due to these ulcers.

One way of comparing the two mouthwashes, known as a crossover trial, is to divide the patients into two groups. Each patient in the first group is treated with one mouthwash (A) for a certain period, an assessment of the patient's response is made, and then the patient is transferred to the other mouthwash (B). Following the second course of treatment, the patient is again assessed. Each patient in the second group is treated and assessed with the mouthwashes in the reverse order. At the end of this study, the two treatments are compared based on the findings from the patient assessments.

An advantage of this design is that patients act as their own control. In an ideal world, the only difference between the two time periods would be the type of the mouthwash used, which would make a comparison straightforward. Unfortunately, there may be a carry-over effect of the first treatment into the second period if the second mouthwash is administered before the effects of the first have completely worn off. Sometimes, an interval or wash out period is allowed between the two treatments in an attempt to reduce this effect. Also, a response to either mouthwash may be affected by the prior or subsequent administration of the other (interaction is the technical statistical term). In an extreme case, administration of one mouthwash after the other might make a patient's condition worse,

DOI: 10.1201/9780429326172-5

whereas if either were given separately, both may be beneficial. Note that attention should be paid to the patient's diet as particular components might influence the occurrence or healing of mouth ulcers.

The statistical analysis of crossover trials is not considered at this stage; the procedure is described for a simple experiment in Section 8.5. Jones and Kenward (2014) gives a more general account of such designs.

Studies involving ordered application of alternative treatments can only be performed with chronic conditions where the first treatment received is unlikely to lead to complete recovery or death. Ulcerative gingivitis is such a condition (asthma and eczema are others) in which for many patients, improvement under treatment is swift but the condition deteriorates once the treatment is withdrawn. A practical difficulty is that assessment of ulceration by a dentist is notoriously subjective, as is patient assessment of pain.

To allow for subjectivity, a double blind procedure should be employed in which neither the patient nor the clinician making the assessments knows the order of presentation (which should be random). In a dental context, mouthwashes can be made to have identical appearance and taste. The washes are identified by codes available only to research workers or administrators who have no direct contact with the patient. If tablets are used, they should be of the same size, colour, and taste. Double blinding can also be used to allow for the placebo effect. This is a psychological response shown by some patients to the knowledge that treatment is being received, even though the solution or tablet (known as a placebo) does not contain an active ingredient. These aspects of a study should be carefully explained in the patient information sheet if applicable.

If it is undesirable for ethical, or practical, reasons to give patients both treatments, pairs of individuals might be chosen so that the members of each pair are as alike as possible in all relevant characteristics. This process is known as *matching*. For example, a woman aged 30 years might be matched with a similar age women of the same ethnic group, each showing a similar relevant morbidity status. In each pair, one of the women chosen at random receives the first treatment and the other the second. In this case, the assessment differences between members of each pair are of interest.

As mentioned in Section 1.7.2, patients should be free to withdraw from a study at any stage. This means that some patients may not be taking their allocated treatment by the close of the study or may have withdrawn from the study altogether. It is recommended that the results of studies that compare two or more groups of patients be analysed on an *intention to treat* basis, in which the groups are compared as they were originally chosen. An attractive alternative might be to analyse the patients by their actual treatment at the end of the study. This may introduce bias; however, as those who are unwilling, or unable, to continue in the study are likely to have different characteristics (e.g., greater health problems) compared with the other patients. *Intention to treat* analysis also gives a fairer assessment of the impact of these strategies in the real world. These comments are relevant to planning an experiment, no matter whether a parametric or nonparametric method of analysis is used.

Returning to the study of mouth ulcer treatment, a patient's assessment of pain may involve no more than indicating which mouthwash, if either, gave the greater relief (e.g., first mouthwash gave most relief, second mouthwash gave most relief, both equally good, or both ineffective). One may score that mouthwash A gave more relief as a *plus* and mouthwash B gave more relief as a *minus*, showing no difference as *zero*. Individual patient scores – *plus, minus,* or *zero* – provide the observations for a sign test of H_0: *drugs are equally effective* against an appropriate one- or two-tail alternative.

TABLE 5.1

Mean Response Time (ms) to Digital Information

Subject	1	2	3	4	5	6
LVF (1)	564	521	495	564	560	481
RVF (2)	557	505	465	562	544	448
(1)−(2)	7	16	30	2	16	33
Subject	**7**	**8**	**9**	**10**	**11**	**12**
LVF (1)	545	478	580	484	539	467
RVF (2)	531	458	560	485	520	445
(1)−(2)	14	20	20	−1	19	22

5.1.2 Further Examples

Four specific examples are as follows:

1. Geffen, Bradshaw, and Nettleton (1973) wanted to know whether certain numbers presented in random order were perceived more rapidly in the right visual field (RVF) or left visual field (LVF), or whether there was no consistent difference, it being a matter of chance whether an individual responded more quickly in one field or in the other field. For each of 12 subjects, the mean response times to digital information in each field were measured. Response times varied much more between individuals than they did between fields for any one individual. The data and the differences (i.e., LVF–RVF) for each individual are shown in Table 5.1.

   This table shows quicker response times in the RVF for all subjects expect subject 10. No difference exceeds 33 ms; whereas in either field, differences between some individuals exceed 100 ms. For example, it is 580−467=113 ms between subjects 9 and 12 in the LVF. Without matched pairs, these differences might swamp the smaller, but relatively consistent differences between fields for individuals – an aspect explored further in Exercise 6.2.
2. A course organizer might compare two teaching methods, such as lectures and online learning, by pairing students, so that, if possible, each member of a pair is of the same gender and has the same previous knowledge of the subject. For each pair, one member is allocated to lecture classes and the other to the online material. At the end of the course, the students take the same tests, and the results are interpreted in terms of pairwise differences.
3. Using double blind marking, one can compare consistency between two examiners. The two examiners mark the same series of essays without disclosing their assessments to one another. The differences between the marks awarded by each examiner for each essay are then compared to see whether one examiner consistently awards a higher mark, or whether differences have some other pattern, or whether they appear to be purely random.
4. To compare two animal diets using pairs of twin lambs, the diets are fed to each twin, and their growth is measured over a period: attention is focused on growth differences between twins within each pair. Because of genetic similarity, each of a pair of twin lambs fed on identical diets tends to grow at a similar rate. When fed different diets, any consistent differences in growth may be attributed to the effects of diets.

In summary, the aim of pairing is to make conditions, other than the treatment or factor under investigation, as alike as possible within each pair. The differences within pairs provide a measure of any treatment effect that takes the form of a shift in the distribution.

5.1.3 Single-Sample Analysis of Matched Pairs

Differences between paired observations provide a single sample that can be analysed by methods developed in Chapters 3 and 4. Consideration should be given, however, to the assumptions about observations on each member of the pair, and what precautions in experimental procedure are needed to validate analyses. These points are best brought out by examples.

Example 5.1

Using the LVF and RVF data in Table 5.1, assess the strength of the evidence for a consistent response difference between the two fields for individuals. Obtain a 95% confidence interval for that difference.

The observation on subject i in the RVF is denoted by x_i and that in the LVF by y_i. The differences $d_i=y_i-x_i$, $i=1, 2, \ldots, n$ are analysed. The differences are independent since each refers to a different individual. Under H_0: *the median of the differences is zero*, the differences are equally likely to be positive or negative, and a sign test is justified. If the distribution of the differences under H_0 is assumed to be symmetric, a Wilcoxon signed-rank (or a normal scores) test can be used. There are several different response patterns in the two visual fields that could result in a symmetric distribution of the differences under H_0.

In particular, if the response times are assumed to be identically and independently (but not necessarily symmetrically) distributed in the two fields for any individual (but this distribution need not be the same for every individual), the difference for each individual will be symmetric about zero. This follows because if X and Y have the same independent distributions, then, $X-Y$ and $Y-X$ will both have the same distribution and, therefore, must be symmetrically distributed about zero.

A symmetric distribution of differences is also obtained if the LVF and RVF responses for any individual have different symmetric distributions, as long as each has the same median or mean if the latter exists. It is well to be aware of such subtleties, but a Wilcoxon signed-rank test is clearly justified when it is assumed that distributions for LVF and RVF for any one individual are identical under the null hypothesis. The alternative hypothesis of interest is usually that there is a shift in location only, as indicated by a shift in one field or the other in the median of otherwise identical distributions. Inspection of the differences may clearly indicate whether there is serious asymmetry that suggests this condition may not hold. If it is also assumed that the differences are approximately normally distributed, a t-test is appropriate. Whether normality is a reasonable assumption for the differences can be tested using the Lilliefors' test or the Shapiro–Wilk test, but these tests may have low power for small samples, unless the departure from normality is very marked.

The observations in this example can be treated as a single sample composed of the differences d_i. Therefore, they can be investigated by the methods described in detail in Chapter 3, and the analyses are only sketched for this study. The location parameter or measure of treatment effect (assuming this to be only a location shift) is denoted by θ, and the null hypothesis is H_0: $\theta=0$, so the ordered signed deviations are simply the ordered differences obtained from Table 5.1, i.e.,

$$-1 \quad 2 \quad 7 \quad 14 \quad 16 \quad 16 \quad 19 \quad 20 \quad 20 \quad 22 \quad 30 \quad 33$$

and the corresponding signed ranks are as follows:

$$-1 \quad 2 \quad 3 \quad 4 \quad 5.5 \quad 5.5 \quad 7 \quad 8.5 \quad 8.5 \quad 10 \quad 11 \quad 12.$$

Using the sign test for zero median difference, there are 1 minus and 11 plus signs.

Appropriate tables or software indicate $P=0.0063$, providing very strong evidence that responses differ between fields. The nominal 95% confidence interval based on the sign test is (7, 22) since H_0 is rejected at a nominal $P=0.05$ level only for two or fewer, or ten or more, minus signs. Using the Wilcoxon signed-rank test, $S_-=1$, corresponding to a two-tailed test of $P<0.001$. In fact, for this example, $P\approx 0.00098$.

For a confidence interval with coverage of at least 95%, tables indicate that the 14 greatest and least Walsh averages are required.

The interval turns out to be (9.5, 23.5). The Hodges–Lehmann point estimator of the median (i.e., the median of the Walsh averages) is 17.25.

If normality is assumed, a two-tailed t-test gives $P<0.001$, and a 95% confidence interval (10.1, 22.9) symmetrical about the mean of 16.5. Normal theory gives the shortest confidence interval. The fact that the interval is only slightly displaced for the sign test implies reasonable symmetry. With a Lilliefors' test statistic of 0.1525, well below the value required for significance, the test gives a Monte Carlo P-value of 0.610. For the Shapiro–Wilk test, $P=0.729$, so there is no evidence of non-normality.

Output from the ANSM5 package in R can be obtained using the following commands: `sgn.test(ch5$LVF - ch5$RVF, 0)`, `wilcoxon.signedrank(ch5$LVF - ch5$RVF, 0)`, `lilliefors(ch5$LVF - ch5$RVF)` and `shapirotest.ANSM(ch5$LVF - ch5$RVF)` with the standard stats package in R providing the command `t.test(ch5$LVF - ch5$RVF)`.

In summary, the sign, Wilcoxon signed-rank, and t-tests all point to strong evidence against H_0, so that it is reasonable to conclude that response times in the RVF are faster. The consistency of the confidence intervals given by these approaches suggests that the mean difference is between about 10 and 22 ms.

It is important to remember that the set of differences compared are independent (a necessary condition for validity of the location tests) because each difference is calculated for a different individual, under the proviso that participants are unrelated by family associations, etc.

If the response rate in the LVF had been measured before that in the RVF for all subjects, a difficulty in interpretation would arise. The result might then be explainable, at least in part, by a learning process; people responding more quickly in the RVF because they were learning to react more rapidly. There could be a mixture of a learning effect and an inherent faster response in the RVF. This difficulty is avoided if the field to be tested first — left or right — is decided in a random manner for each individual. That was done in this experiment. This should balance out, and largely annul, any learning factor. Another approach is to achieve balance by selecting six subjects (preferably at random) to be tested first in the LVF. The remaining six subjects are tested first in the RVF. Such balanced designs provide a basis for separating a learning effect from an inherent difference between field responses, although a somewhat larger experiment would be needed to do this by appropriate parametric or nonparametric methods.

A modification to the above procedure allows testing of the hypothesis that a location difference has a pre-specified value θ. As in Chapters 3 and 4, deviations from that value are considered. This effectively shifts the origin to the hypothetical median, so that for

TABLE 5.2

Marks (out of 90) in Two Arithmetic Tests

Pupil	A	B	C	D	E	F	G	H	I	J	K
First test	45	61	33	29	21	47	53	32	37	25	81
Second test	53	67	47	34	31	49	62	51	48	29	86
(Second – First)	8	6	14	5	10	2	9	19	11	4	5

the revised data (the deviations), the null hypothesis is that the location parameter for the population under study takes the value zero.

Example 5.2

A total of 11 children are given an arithmetic test; after additional practice using online material specifically designed to develop these skills, they are given a further test of equal difficulty. Their marks in each test (out of 90) and the individual differences are given in Table 5.2. Do these support a claim that the average improvement due to the online practice is 10 marks?

The question essentially is whether, if these children are assumed to be a random sample from some hypothetical population (perhaps children of the same age trained in the same educational system or studying the same syllabus), it is reasonable to suppose the mean mark difference is 10.

The mark differences in the last line of Table 5.2 are considered in terms of deviations from 10, i.e., (*Second* – *First*) – 10. These deviations are as follows:

$$-2 \quad -4 \quad 4 \quad -5 \quad 0 \quad -8 \quad -1 \quad 9 \quad 1 \quad -6 \quad -5$$

The differences arranged in order of magnitude with appropriate signs are as follows:

$$0 \quad -1 \quad 1 \quad -2 \quad -4 \quad 4 \quad -5 \quad -5 \quad -6 \quad -8 \quad 9$$

and the ranks without signs are as follows:

$$1 \quad 2.5 \quad 2.5 \quad 4 \quad 5.5 \quad 5.5 \quad 7.5 \quad 7.5 \quad 9 \quad 10 \quad 11$$

Using the rules given in Section 3.3.4 for mid-ranks with ties, together with the convention mentioned at the end of that section for replacing rank 1 associated with the zero difference by 0, the signed ranks are as follows:

$$0 \quad -2.5 \quad 2.5 \quad -4 \quad -5.5 \quad 5.5 \quad -7.5 \quad -7.5 \quad -9 \quad -10 \quad 11$$

In this example, the statistic needed is the sum of the positive ranks $S_+ = 0 + 2.5 + 5.5 + 11 = 19$. Using the Wilcoxon signed-rank test, an exact two-tailed test $P = 0.248$ is obtained, demonstrating that there is no strong evidence against H_0. While the sample is too small to justify an asymptotic procedure, the asymptotic two-tailed $P = 0.229$, comparing reasonably well with the exact P-value.

Output from the ANSM5 package in R can be obtained using `wilcoxon.signed-rank(ch5$arithmetic, 10, do.asymp = TRUE)`.

How does one decide if two arithmetic tests are equally difficult? Might not the improved marks in the second test imply it was easier? The statistician should seek the assurance of the educationalist conducting the test that reasonable precautions have been taken to ensure equal difficulty. Sometimes standard tests are used in these situations. Such tests have been tried on large groups of students with results that show convincingly that they are for all practical purposes of equal difficulty.

Confidence intervals for the mean or median difference may be obtained in the usual way. The nominal 95% confidence interval for these data is (5, 12), in close agreement with the normal theory *t*-test interval (5.14, 11.76).

It may not be realistic to test simply for a location shift. Sometimes pupils who perform well initially benefit little from extra tuition; in this example, Pupil K cannot possibly improve by more than 9 marks. Likewise, pupils with strengths in other areas, such as music or sport, may find the concepts of arithmetic difficult to grasp. In such a situation, those in the mid-ability range improve the most. A statistician may find evidence of this simply by looking at the data, and there are, in fact, tests for such tendencies. Deciding what should be tested or estimated is often a topic for fruitful discussion between statistician and experimenter.

Example 5.3

The data below show the differences (systolic blood pressure after exercise – systolic blood pressure before exercise) measured in mm Hg for a random sample of 24 males aged 18 years measured by a first-year medical student for an undergraduate project. Obtain 95% confidence limits for the population difference based on (i) the sign test, (ii) the Wilcoxon signed-rank test, and (iii) the *t*-test and comment on the appropriateness of each. For convenience the differences have been arranged in ascending order as follows:

–5 –5 0 2 10 15 15 15 18 20 20 20 20
22 30 30 34 40 40 40 41 47 80 85

The standard methods developed in Chapter 3 and in this chapter are used. Computational details are omitted since examples of similar calculations have already been given. The nominal 95% confidence intervals quoted below were obtained using relevant statistical software:

Sign test	(15, 40)
Wilcoxon signed-rank test	(17.5, 35)
t-test	(16.87, 35.96)

The output can be obtained from the ANSM5 package in R using `sgn.test(ch5$bp)` and `wilcoxon.signedrank(ch5$bp)` and from the standard stats package using

`sgn.test(ch5$bp)`.

Not surprisingly, the sign test interval is the longest. There is slight evidence of skewness, but none of the intervals is seriously displaced relative to the others. The Wilcoxon interval being shorter than that based on normal theory may be attributed to a slightly heavier upper tail (especially the values 80 and 85) than one would expect with a sample from a normal distribution. Lilliefors' test for normality gives a test statistic value of 0.161 and a Monte Carlo of $P = 0.102$. For the Shapiro–Wilk test, there is stronger evidence against normality, the statistic 0.901 corresponding to $P = 0.023$.

The output from the ANSM5 package in R can be obtained using `lilliefors(ch5$bp)` and `shapirotest.ANSM(ch5$bp)`.

Obtaining a sample of more than 24 participants might be an unrealistic expectation for one medical student but it would be necessary for any detailed studies of the effects of exercise on blood pressure (and interest would be in the distribution of the differences and not just in the mean or median difference).

The variation in the size of the differences is unsurprising with blood pressure measurements taken by inexperienced first-year medical students. Readings are usually recorded to the nearest mm Hg, but in practice, they may only be recorded to the nearest 5 mm Hg by those who are unfamiliar with the technique. The exercises would probably be performed with a range of enthusiasm depending on the degree of physical fitness and level of motivation of each participant. A few extremely high differences and a few small negative changes are, therefore, understandable.

The initial data records should be checked for accuracy particularly if handwritten. Had a recording error in a single blood pressure reading led to a calculated difference −103 rather than the actual value of −5 the following 95% confidence intervals would have been obtained:

Sign test	(15, 40)
Wilcoxon signed-rank test	(16, 32.5)
t-test	(7.83, 36.83)

The output can be obtained from the ANSM5 package in R using `sgn.test(ch5$bp.incorrect)` and `wilcoxon.signedrank(ch5$bp.incorrect)` and from the standard stats package using `t.test(ch5$bp.incorrect)`.

The reader who has difficulty in explaining the differences between these results and those recorded above in terms of the effect of an outlier should refer to Section 3.4.2. In particular, it is worth noting that the Wilcoxon interval is little changed because the introduction of a value of −103 has a tendency to make the sample distribution, if anything, a little more symmetric than the original sample. However, the tails become rather longer than one would expect for a normal distribution, and this has the effect of elongating the *t*-based interval. For these amended data, the Lilliefors' test gives a Monte Carlo $P=0.0093$ and a Shapiro–Wilk test gives $P<0.001$, both hinting strongly at non-normality implicit in the tail values.

Practical experience in handling data highlights the crucial importance of detecting data errors. Many computer programs include output to help detect outliers; e.g., a printout of maximum and minimum values often (but not always) highlights a glaring data error.

SUMMARY POINTS:

- Observations from a simple crossover trial can be analysed as paired samples (Section 5.1.1).
- Appropriate matching of observations can highlight small but consistent differences between two groups that might otherwise be missed (Section 5.1.2).
- Differences between paired observations provide a single sample that can be analysed by the methods described in Chapters 3 and 4 (Section 5.1.3).

5.2 A Less Obvious Use of the Sign Test

The way the data are presented in Example 5.4 does not make it obvious that a sign test is relevant.

Example 5.4

There has long been disagreement among members of a mountaineering club about whether which of two rock climbs is more difficult. Hoping to settle the issue, one member checks the club log book. This records for any climb by a member whether it is successfully completed. The log shows that 108 members have attempted both climbs with the outcomes summarized in Table 5.3. Is there evidence that one climb is more difficult?

A climber succeeding in both climbs, or failing in both, provides no information about relative difficulty. Such cases are ties so far as comparing difficulty is concerned. If additional information was available, for instance, about each climber's personal assessment of the difficulty, the situation would be different. As it is, the only meaningful comparators of difficulty are numbers who succeed at one climb but fail at the other. Table 5.3 shows that nine climbers succeed at the first climb but fail at the second. This can be regarded as a *plus* for the first climb. Also, 14 climbers fail on the first climb but succeed at the second; this can be thought of as a *minus* for the first climb.

This is a sign test situation. If the climbs are of equal difficulty, a *plus* or *minus* is equally likely for each climber. Thus, under H_0: *climbs are equally difficult*, the number of "plus" signs has a $B(n, 0.5)$ distribution where n is the total number of plus and minus signs.

There are 9 pluses and 14 minuses, so $n=23$ and an exact test $P=0.4049$. Alternatively, the normal approximation to the binomial distribution indicates, using (3.1), that $Z=(9.5-11.5)/0.5\sqrt{23}\approx-0.83$ giving a two-tailed test $P=0.4065$, so there is no evidence that one climb is easier. Hence, the null hypothesis that the climbs are equally difficult is retained.

Output from the ANSM5 package in R can be obtained using `sgn.test(c(rep(0, 14), rep(1, 9)), 0.5, do.asymp = TRUE)`.

Most of the pairs (73+12=85 from 108) provide no information on the relative difficulty of the climbs, and this may seem wasted data, but such data give an indication of how big or small any difference might be. In some situations, many observations are needed because a small difference is being sought that may not be distinguishable when the only relevant criteria are *success/failure* or *failure/success* categories. In the context of these data, it is likely that most of those who failed at both categories were less experienced or less enthusiastic than those who achieved one success.

TABLE 5.3

Outcome of Two Rock Climbs

		First Climb	
		Succeeded	Failed
Second Climb	Succeeded	73	14
	Failed	9	12

Those who succeeded at both categories were likely to be the more experienced or the more enthusiastic. It would be interesting to know whether club members generally attempted the climbs in a particular order as a learning effect might then be involved.

The continuity correction of replacing 9 by 9.5 in the numerator of the asymptotic approximation has a marked effect in this example. If it is omitted, the two-tailed test P is reduced from 0.4065 to 0.2971. However, in this case, the conclusion would not be altered by this appreciable numerical change.

Many packages include programs for this simple test that may be used to obtain the relevant P-value for any sign test. The test is called *McNemar's test*, having been proposed by McNemar (1947). Conover (1999, Section 3.5) presents it more formally, but it is effectively a paired-sample sign test.

SUMMARY POINTS:

- McNemar's test is relevant to paired observations to assess changes in attitude or for assessments of relative difficulty.
- It is equivalent to the sign test (Section 3.2).

5.3 Power and Sample Size

Section 4.4.3 explored the power of some tests for binomial probabilities of the form H_0: $p=p_0$ against single-valued alternatives, such as H_1: $p=p_1$ where the choice of p_0, p_1 was determined by their relevance to producer's and consumer's risks associated with a sampling scheme. Exact power calculations were relatively straightforward, and they were used to find sample sizes for meeting specified producer and consumer risks.

Consideration is now given to relationships between power and sample size for single-sample location tests relevant both to basic single-sample situations and to differences between matched pairs. Exact results are only available in a limited number of cases, and in this context, the power for a specified alternative is in general not distribution free even when the test itself is. This restricts the overall value of exact power computations. Nevertheless, providing care is taken in interpretation, computing approximate power for given sample sizes before beginning an experiment helps one make efficient use of resources.

First, some key features and limitations are highlighted using simple examples for the sign test, where only relatively straightforward computations are needed. Then, brief consideration is given to the Wilcoxon signed-rank test where power and sample size determinations introduce greater difficulties in both theory and practice. A clear text on calculus, e.g., Spiegel and Wrede (2010), may be helpful in working through this section.

5.3.1 Power and Sample Size for the Sign Test

Examples 5.5–5.7 cover three situations where a sign test is used for a median θ of the form H_0: $\theta=\theta_0$ against H_1: $\theta>\theta_0$ where a specific alternative H_1: $\theta=\theta_0+1$ reflects a minimal departure from H_0 of practical importance. Without loss of generality (see, for instance, Example 5.2), set $\theta_0=0$. Attention is confined to a one-tailed test, but modifications for a two-tailed test are straightforward if one uses the principle of doubling one-tail *P*-values for a two-tail equivalent. If the power of a test for the alternative H_1: $\theta=1$ is computed, then if the true shift exceeds one unit, the power will, in general, be greater.

From these examples, it will emerge that for three different population distributions all having the same variance, the power of the equivalent tests is different for each, confirming that power is not a distribution-free property. This should cause no surprise in the light of the well-known result that the Pitman efficiency of tests depends on the underlying population distribution.

Example 5.5

A city council funds a summer holiday club for children in order to increase the average time spent by young people in physical activity (PA). Participant-reported information of hours spent per week on PA is collected prior to and on completion of the intervention. Council members and the club team believe that the increase in hours per week is likely to take a normal distribution with a standard deviation of 2, but the size of the mean change is disputed. The finance committee believes that the intervention is ineffective. However, the club team claim an average increase of 1 hour. The holiday club will be financed in future years only if the team's assertion is plausible. The records of 10 participants are assessed using the sign test, observations scored according to an increase or decrease in the time spent. Assuming that all participants experienced a change in time spent in PA, and taking a significance level of 0.05, how likely is it that the team's assertion will be accepted? The team think that basing a decision on just 10 children is unfair. What sample size would ensure that their assertion, if correct, will be accepted with a probability of 0.9?

Expressed more formally, for a sample of 10 from a normal distribution with unknown mean θ and standard deviation 2, what is the power of the sign test for H_0: $\theta=0$ against H_1: $\theta=1$ when the probability of a type I error is $\alpha=0.05$? What sample size would ensure a test with power 0.9? There are no values of 0.

Under H_0, the probability of a positive value (plus) is 0.5, the number of pluses, r, in a sample of size 10 has a binomial B(10, 0.5) distribution, and the number of plus signs can be taken as a test statistic. For the alternative H_1, the number of pluses will also have a binomial distribution, but with a different value of p given by the probability p_1 of obtaining a positive value from a normal distribution with mean 1 and standard deviation 2. It follows that $Z=(X-1)/2$ has a standard normal distribution, whence p_1 is given by $\Pr(X>0)=\Pr(Z>-0.5)=0.6915$. The R command `pnorm(-0.5, lower.tail = FALSE)` gives this probability.

The procedure is equivalent to testing H_0: $p=0.5$ against H_1: $p=p_1$ in the quality control process of Example 4.9, with 0.5 as the producer's risk and p_1 as the consumer's risk. A value r is required for which the *P*-value α is close to 0.05 with sample size $n=10$ and $p=0.5$. Once r has been determined, the power can be calculated for $p=p_1=0.6915$.

Using the ANSM5 package in R and the command `binom(r, 10, 0.5, alternative = "greater")` gives the following values for $n=10$, $p=0.5$:

r	7	8	9
P	0.172	0.055	0.011

with corresponding power values for $p=0.6915$ given by `binom(r, 10, 0.6915, alternative = "greater")`:

r	7	8	9
Power	0.6268	0.3604	0.1365

With α set at 0.05, the corresponding power is 0.1365. However, discontinuities are influential here. Had $\alpha=0.05$ been replaced by $\alpha=0.011$, the power would still be 0.1365, whereas setting $\alpha=0.055$ gives a power of 0.3604. It is unlikely that even if true, the team's claim will be upheld, so they are correct in believing that this test is unfair.

To find the minimum sample size for a power of 0.9, a range of larger sample sizes and associated values of r are examined. Findings including sample sizes of 20, 30, 50, and 100 can be tabulated as follows:

n	10	20	30	50	100
r	9	15	20	32	59
P	0.011	0.021	0.049	0.032	0.044
Power	0.137	0.384	0.695	0.827	0.988

In terms of the minimum sample size required to achieve a power of 0.9, this lies between 50 (associated power<0.9) and 100 (associated power>0.9). One may fine tune using exact binomial probabilities by gradually increasing both n and r. When $n=58$, taking $r=36$ gives a P-value of 0.043 with a power of 0.903. The assessment should involve at least 58 children.

The club team could rightly claim that the sign test is hardly the best tool for this assessment. The Pitman efficiency of the sign test relative to the t-test for samples from a normal population is 0.64. This implies that for large n, the relative efficiency is less than two-thirds, suggesting that a sample of about two-thirds the size should have the same power if a t-test is applied. Many statistical packages allow power calculations for the t-test when it is optimal, and in this case, these indicate that for a $N(\theta, 4)$ distribution, the necessary sample size with power 0.90 with the chosen values of θ in H_0 and H_1 is $n=36$, broadly in line with that suggested by the Pitman efficiency.

The computed $p_1=0.6915$ was based on the strong assumption that the sample was from an $N(\theta, 4)$ distribution. Knowing the population variance without knowledge of the value of the mean is a situation that seldom arises. Should this occur (or if normality could only be assumed with the variance unknown), one should prefer the normal theory-based inference to that using the sign test. However, the results obtained in this example are illuminating for comparisons with the situations covered in Examples 5.6 and 5.7 where a non-normal population, again with variance 4, is assumed.

Example 5.6

Suppose the changes in PA follow a double-exponential distribution with a standard deviation of 2 and unknown mean. Given a sample of 10 children, find the power of the sign test for H_0: $\theta=0$ against H_1: $\theta=1$ if the probability of a type I error is $\alpha=0.05$. What sample size would ensure a power at least 0.9?

The double-exponential distribution, also known as the Laplace distribution, with mean θ and standard deviation 2 (variance 4), has probability density function as follows:

$$f(x)=\frac{1}{2\sqrt{2}}\exp[-|x-\theta|/\sqrt{2}] \tag{5.1}$$

The distribution is symmetric about the mean, θ. Probabilities associated with the tails are greater than those for the normal distribution with the same mean and variance, and this is a classic example of a long-tail symmetric distribution.

Under H_0, the number of positive sample values (pluses) again has a B(10, 0.5) distribution. If the number of plus signs is used as the test statistic, these will also have a binomial distribution under H_1, but the parameter p is now given by $\Pr(X>0)=p_1$, say, where X has the double-exponential distribution (5.1) with $\theta=1$. Thus, the original test is equivalent to testing H_0: $p=0.5$ against H_1: $p=p_1$, and the power study for the given sample size and for determining the sample size required for any specified power proceed as described in Section 4.4.3.

This probability can be obtained using the function `plaplace` from the R package VGM for the upper tail area of a double-exponential distribution. Note that the square root of the standard deviation (in this example, $\sqrt{2} \approx 1.4142$) is required. `plaplace(0 1, sqrt(2), lower.tail = FALSE)` shows $p_1=0.7534$. Alternatively, p_1 can be calculated by setting $\theta=1$ in (5.1) and then integrating over the interval $(0, \infty)$ (see Exercise 5.13). For testing H_0: $p=0.5$ against H_1: $p=0.7534$ when $\alpha=0.05$, using the R function `binom` from the ANSM5 package gives the exact power 0.2518, which again does the club team a disservice.

Discontinuities in possible P-values are the same as those in Example 5.5 and replacing $\alpha=0.05$ by $\alpha=0.011$ does not alter the power, whereas replacing it by $\alpha=0.055$ gives for the double-exponential distribution a power of 0.5358. Fine tuning using exact results gives a minimum required sample size of 30 corresponding to an exact $P=0.0494$ and an exact power 0.9023.

The sample requirement of 30 children is approximately half the size (58) needed for the same power when using the sign test with a normal distribution with the same variance. This is in line with the Pitman efficiency of the sign test relative to the t-test, which is equal to 2 when sampling from a double-exponential distribution. Note, however, that the t-test itself has lower efficiency for a sample from a double-exponential distribution than it has for a normal distribution with the same mean and variance.

The reason the sign test is appreciably more powerful for testing the same basic hypotheses about θ in this test than it was for an equivalent test in the previous example is that although the same values of θ are specified in both examples, and that for H_0 translates to $p=0.5$ in each case, that for H_1 transforms to $p=0.6915$ and $p=0.7534$, respectively, so that for the second example, there is a larger departure from the value of p under H_0. That this produces a more powerful test is in line with the notion of a good test being one where its power increases as the difference between the value specified in H_1 increases relative to a fixed value specified in H_0.

Examples 5.5 and 5.6 featured samples from symmetric distributions. A skew distribution with a similar one-unit shift to the right is now considered. The case in this example is unusual in practice but it provides a direct comparison with Examples 5.5 and 5.6 because again the sample is from a distribution with a variance of 4.

Example 5.7

The changes in PA follow an exponential distribution as follows:

$$f(x)=\frac{1}{2}\exp\left[-\frac{1}{2}(x+1.3862-\theta)\right],\quad x\geq\theta-1.3862 \tag{5.2}$$

which has median θ and standard deviation 2. For a sample of 10 children, find the power of the sign test for H_0: $\theta=0$ against H_1: $\theta=1$ if the probability of a type I error is $\alpha=0.05$. What sample size would ensure a power at least 0.9?

The distribution specified by (5.2) is skew with a long right tail. As in the preceding example under H_0, the number of positive sample values (pluses) has a B(10, 0.5) distribution. Using the number of plus signs as the test statistic, this also has a binomial distribution under H_1 but now $p_1=\Pr(X>0)$. Thus, the original test is equivalent to testing H_0: $p=0.5$ against H_1: $p=\Pr(X>0)$ and the power study for the given sample size, and for determining the sample size required for any specified power, again proceed as described in Section 4.4.3.

This probability can be obtained using the function `pexp` from the standard R stats package for the upper tail area of an exponential distribution (`pexp(0.3862, rate = 1 / 2, lower.tail = FALSE)`), which shows $p_1=0.8244$. Alternatively, integration of the probability density function over $(0, \infty)$ shows that under H_1, $\Pr(X>0)=0.8244$ (Exercise 5.14). For testing H_0: $p=0.5$ against H_1: $p=0.8244$ when $\alpha=0.05$, using the ANSM5 package in R and the command `binom(9, 10, 0.8244, alternative = "greater")` gives an exact power of 0.4539. Discontinuities in possible P-values described in Example 5.5 are again relevant. If $\alpha=0.05$ is replaced by $\alpha=0.011$, the test still has a power of 0.4539, whereas if it is replaced by $\alpha=0.055$, the power becomes 0.7499. Fine tuning using exact results from the `binom` function from the ANSM5 package in R shows that only $n=18$ children are needed for the minimum required sample size corresponding to an exact $P=0.0481$ and giving an exact power equal to 0.9194.

The sign test is particularly useful for skew distributions and in this example in which there is a long right tail, moderate power is obtained for quite small samples for the alternatives considered. The combination of skewness and a long tail reduces the power of a t-test appreciably due to the breakdown of normality assumptions.

As indicated, the above example is unrealistic from the practical viewpoint because it is unusual to meet exponential distributions where the interest is in simple shifts of the complete distribution that affect the mean or median alone. More often, the issue relates to whether a sample is from one or other of a set of exponential distributions where all positive values of X are possible but where the scale parameter λ is unknown. Changes in λ lead to changes not only in the mean and median but also in the variance and higher moments. This is more complicated than a simple location shift. Nevertheless, the sign test is still valid for studying such changes, provided the test is based on allocation of signs to observations depending on whether they are above or below the median specified in H_0. Exercise 5.15 is intended for readers interested in exploring the power and sample size relationship for this case.

Finding an adequate sample size to give a good chance of detecting median shifts of interest is a useful exercise but it has practical limitations. Examples 5.5–5.7 confirm that even if a test is distribution free, the power computations associated with specific hypotheses about a population mean or median are no longer distribution free. Furthermore, exact power calculations can only be made easy if all observations are assumed to come from a population with a specific distribution. In contrast, an advantage of the sign test is its applicability to situations where each observation may come from a different distribution, providing only that all these distributions have a common median.

There is also a certain irony that a nonparametric or distribution-free test can be chosen because of uncertainty about the distribution from which the sample is taken, yet computation of a sample size to guarantee a desired power requires knowledge of the distribution. Nevertheless, reasonable approximations are often obtainable if a few rational assumptions about symmetry, length of tails, or other prominent characteristics can be made.

A further limitation to the usefulness of power and sample size calculations applies to both parametric and nonparametric inference. Using a test with good power for rejection of H_0 for differences of importance does not imply that if H_0 is rejected, then the difference is important. A confidence interval is more informative about this. To illustrate this point, the situations in Examples 5.5–5.7 are considered. In each situation, a zero median, θ, is assumed under H_0, and only a value $\theta \geq 1$ represents a departure of interest. If H_0 is rejected and a 95% confidence interval for θ is calculated and this turns out to be (0.2, 1.3), there is considerable doubt about whether $\theta \geq 1$. The departure from H_0 may still be unimportant. It is important to remember the distinction between statistical significance and practical importance that was highlighted in Section 1.6.2.

Despite their limitations, approximate power computations are a useful first step in many experimental situations. They may indicate that a proposed experiment is too small to be likely to detect interesting departures from H_0. More rarely, they may suggest that resources are being squandered when a smaller experiment would give all the information required.

Studies on power and sample size provide target sizes for experiments likely to be of value, or, if resources are limited, indicate whether using all of these will be enough to achieve an experimenter's goals. Even though the power of a simple test like the sign test is not distribution free, power studies are still useful when broad assumptions about the type of population distribution can be made, e.g., is it symmetric and long tailed, or perhaps more like a uniform distribution? Unless it is very small, a sample will often give some hint as to whether it comes from a distribution with one or more such characteristics. If there is evidence that the distribution is fairly symmetric but has longer tails than those associated with the normal distribution, power studies that are optimal for the double-exponential distribution may provide good approximations.

Example 5.7 shows that for a particular skew distribution with one long tail, the sign test performs well compared with the *t*-test. This is broadly true for many skew distributions with a long tail. Very often, a quick eye inspection of sample values aided perhaps by some exploratory data analysis tool like a box and whisker plot (Section 2.5) will indicate that a sample appears to have come from, say, a fairly symmetric long-tailed distribution not unlike the double exponential. Its variance might be estimated by the sample variance. For a null hypothesis about the median, θ, H_0: $\theta = \theta_0$ and a specific alternative H_1: $\theta = \theta_1$, one might compute the value $p_1 = \Pr(X > \theta_0)$ assuming that the distribution of X is really a double-exponential distribution with variance equal to the sample variance. This effectively assumes that the correct p_1 is likely to differ little from that for the double-exponential distribution. Suppose, for the alternative hypothesis of interest $p_1 = 0.82$ for a double-exponential distribution, then one might be conservative and work out the power for this situation, and also that for a slightly lower p_1, say, $p_1 = 0.78$. It is likely that the true power will lie somewhere near the values given by these approximations.

Noether (1987) gives a good asymptotic approximation for power of a sign test. If the probability of a type I error is denoted by α and that of a type II error by β, so that the power is $1-\beta$ (Section 1.5), then the sample size required to obtain that power with a sign test is given by:

$$n \approx \frac{\left(z_\alpha + z_\beta\right)^2}{4\left(p_1 - 0.5\right)^2} \tag{5.3}$$

where p_1 has the meaning assigned to it throughout this section and z_α is the value of a standard normal variable that must be exceeded corresponding to a *P*-value α when H_0 holds. For example, in a one-tailed test with $\alpha=0.05$, $z_\alpha=1.645$. A corresponding meaning is attached to z_β.

Example 5.8

Apply the Noether approximate formula to compute the sample size n for the test considered in Example 5.6 to give a power of at least 0.9 when $\alpha=0.05$ in a one-tailed test.

The appropriate value of p_1 is shown in Example 5.6 to be $p_1=0.7534$ and it is easily verified that $z_\alpha=1.645$ and $z_\beta=1.282$. Substitution of these values in the Noether formula gives $n\approx 33.4$. Conservative rounding up suggests that a sample size of 34 is appropriate.

Output from the ANSM5 package in R can be obtained using `noether(p1 = 0.7534, alpha = 0.05, power = 0.9)`.

This asymptotic result is close to the size $n=30$ found in Example 5.6. Bearing in mind that such calculations are frequently used when there is uncertainty about the precise population distribution, calculations using the Noether formula will often be adequate in practice.

For the problem in Example 5.5, the Noether formula gives $n=58.4$, close to the exact value $n=58$, and in Example 5.7, it gives $n=20.3$ (exact value $n=18$), suggesting that the approximation is reasonable even when observations are from a highly skewed distribution.

5.3.2 Power and Sample Size for the Wilcoxon Signed-Rank Test

This topic is considered in less detail than for the sign test. There are additional difficulties in performing exact tests. Approximate results are often all that can be used in practice.

A simple property of the sign test under the alternatives H_1 considered in Section 5.3.1 is that no matter from what distribution the sample comes, the test statistic, the number of plus signs, has a binomial distribution. For this statistic, for any given value of θ in H_1, all that differs between samples from different distributions is the value of the binomial parameter p_i. This simplicity is a feature of the statistic used, the number of plus signs.

Although the corresponding statistic for the Wilcoxon signed-rank test, the sum of the positive signed ranks, has a relatively simple symmetric distribution under H_0, it generally has a rather intractable distribution under any alternative hypothesis for all but a few simple population distributions. Exact power calculations have only been made for small sample sizes and for a limited number of population distributions, such as the normal and a few *t*-distributions with small numbers of degrees of freedom, see, for instance, Klotz (1963) and Arnold (1965). Even approximations are limited in their usefulness. An approximate formula for power for a given sample size against alternatives close to that in H_0 is discussed in detail by Lehmann (2006, Section 4.2). Hollander, Wolfe, and Chicken (2014, Section 3.1) gives an example of its use. The result is applicable only if all observations are from the same distribution. In addition, one needs to know the value of the population frequency function at the median or mean specified in the null hypothesis (which, without loss of generality, may be set at $\theta=0$) and also the value at this median of the frequency function for the sum of two independent variables having this same distribution. Except for a few distributions, such as the normal, where the sum also has a normal distribution, computation of the latter requires a good understanding of distribution theory and calculus.

The reason the distribution of the sum of two independent observations comes into the calculation is closely allied to the relevance of the Walsh averages in test and estimation procedures associated with signed ranks. This sum for any two sample values has the same sign as the corresponding Walsh average (which is simply that sum divided by 2), and under H_0: $\theta=0$, the sign is equally likely to be positive or negative, i.e., $\Pr(x_i+x_j>0)=0.5$; whereas, under H_1: $\theta=\theta_1$, if θ_1 is positive, then $\Pr(x_i+x_j>0)=p_1$, where $p_1>0.5$. The value of p_1 depends on the population distribution and is not always easy to calculate.

Although the signed-rank statistic no longer has a binomial distribution, the approximation due to Noether given in (5.3) may still be used to estimate the sample size having a given power. However, for a given θ_1, this may be difficult to calculate except for some simple distributions. It may also be sensitive to an incorrect choice of population distribution, again illustrating the difficulty due to uncertainty about the population distribution when making power calculations for nonparametric methods.

Example 5.9

Using an approximation based on (5.3), estimate the sample size needed to guarantee a power of 0.80 using a Wilcoxon signed-rank test when sampling from a normal $N(\theta, 4)$ distribution for a one-tailed test for H_0: $\theta=0$ against H_1: $\theta=1$ with probability of a type I error not exceeding $\alpha=0.05$.

First, $p_1=\Pr(X_1+X_2>0)$ is calculated where X_1 and X_2 are independently N(1, 4). This together with appropriate values of z_α, z_β, is substituted in (5.3) to estimate n.

Using normal distribution theory, $U=X_1+X_2$ is distributed N(2, 8) under H_1. Thus $Z=(U-2)/(2\sqrt{2})$ has a standard normal distribution.

It follows that $\Pr(U>0)$ implies $p_1=\Pr(Z>-1/\sqrt{2})=\Pr(Z>-0.7071)=0.7602$.

From previous examples, $z_\alpha=1.645$ when $\alpha=0.05$. The power, $1-\beta$, is 0.8 so $\beta=0.2$ and hence $z_\beta=0.842$. Substituting these values in (5.3), gives a sample size estimate of $n\approx 23$.

Output from the ANSM5 package in R can be obtained using `noether(p1 = 0.8244, alpha = 0.05, power = 0.9)`.

Some packages provide programs to determine the sample size to give required power when the optimum normal theory *t*-test is used in these circumstances and indicate a sample size $n=25$. This suggests that the asymptotic test size of 23 for the Wilcoxon signed-rank test may be underestimated since the Pitman efficiency of the Wilcoxon signed-rank test relative to the optimal test in this case is $3/\pi$, which is slightly less than 1. However, the result is of the right order of magnitude, and Pitman efficiency is only a rough guide for small samples.

SUMMARY POINTS:

- Power and sample size calculations for the single sample (Section 5.3.1) or matched pair differences (Section 5.3.2) are in general not distribution free.
- Reasonably good approximations depend on assumptions about the population distribution.
- Some power computations are relatively easy for the sign test because the statistic still has a binomial distribution under the alternative hypothesis (Section 5.3.1).
- The distribution of the signed-rank statistic often proves intractable under alternative hypotheses (Section 5.3.2).
- Asymptotic approximations may be useful in determining approximate sample sizes needed to attain a given power.

5.4 Runs Test for Paired Groups

In Section 4.5.3, it was remarked that the series of plus and minus signs produced in the Cox-Stuart trend test could be investigated further using a runs test. For instance, in Example 4.13, the daily precipitation rate measurements (mm/h) for Uvira can be presented as follows:

	February	April	June	August	October	December
2012:	0	1.841	0.202	0.891	1.423	0
2013:	1.345	49.8	0	0	1.561	4.376

with the signs of the differences between corresponding months for 2012 relative to 2013 being –,–,+,+,–,–. This represents a run of 2 pluses followed by a run of 2 minuses and then a run of 2 pluses or 3 runs from 6 observations. The question of interest in this example addresses whether the plus and minus signs occur at random, as with the outcomes of tossing a fair coin. The observations for the years 2012 and 2013 constitute two matched groups, the pairings being February, April, June, August, October, and December. Clustering would raise the possibility of periods of persistent excess (or shortfall) of precipitation for 2013 relative to 2012, of importance in the study of medium-term changes in rainfall patterns.

Since the Cox-Stuart test is used to investigate monotonic rather than linear trends, time gaps between adjacent observations do not need to be identical. It follows, in particular, that the formulation above can also be used if the two halves of the observations are separated by a relatively long period of time.

Example 5.10

The supporting information for Jeandron et al. (2015) gives the volumes of water supplied through the piped water system (m^3) in Uvira on the first day of each month between May and April in 2009/2010 and 2013/2014 as follows:

	May	June	July	August	September	October	November	December	January	February	March	April
09/10	4,282	5,233	4,306	3,408	4,177	2,986	3,872	3,739	2,095	3,735	3,028	2,691
13/14	4,443	2,847	3,291	3,699	3,071	3,801	3,815	3,527	3,939	4,104	4,307	3,939

Can the signs of the differences for corresponding dates in the two 12-month periods be regarded as a random sequence?

The null hypothesis is that sequence of signs is random. The difference in volumes for 2009/2010–2013/2014 have signs: –,+,+,–,+,–,+,+,–,–,–,–. This constitutes 7 runs with 5 pluses and 7 minuses from a total of 12 observations. The one-tail *P*-value of 0.6515 shows no evidence against the null hypothesis.

The output from the ANSM5 package in R can be obtained using `runs.2cat(sign (ch5$yr0910- ch5$yr1314), alternative = "less")`.

Despite the lack of evidence against randomness, one should note the run of 4 minuses inclusive between January and April. The remaining runs have either length 1 or 2. Therefore, this sequence could be an example of a mixture of alternation and clustering, which can be overlooked in testing for randomness (Section 4.6).

SUMMARY POINTS:

- In the runs test for paired groups, time points within the groups are matched on a one-to-one basis.
- Time gaps between observations do not need to be identical.
- The paired groups can be separated by a relatively long period of time if the time points within the groups can be matched.
- Clustering is indicated by a small number of runs of pluses/minuses. Alternation is suggested by a relatively high number of runs.

5.5 Exercises

5.1 The blood pressures (mm Hg) of 11 patients are measured before and after administration of a drug. The differences in systolic blood pressure (*pressure before–pressure after*) for each patient are as follows:

7 5 12 –3 –5 2 14 18 19 21 –1

Use the Wilcoxon signed-rank test to investigate whether the sample (assumed random) provides evidence that the drug causes a significant change in blood pressure. Is the use of this test justified? Compare your findings with the performance of the sign test.

5.2 Samples of cream from each of 10 dairies (A–J) are each divided into two portions. One portion from each is sent to Laboratory I and the other to Laboratory II, for bacterial counts. The counts (thousands bacteria ml^{-1}) are as follows:

Dairy	A	B	C	D	E	F	G	H	I	J
Lab I	11.7	12.1	13.3	15.1	15.9	15.3	11.9	16.2	15.1	13.6
Lab II	10.9	11.9	13.4	15.4	14.8	14.8	12.3	15.0	14.2	13.1

Use the Wilcoxon signed-rank test to assess the evidence for any consistent difference between laboratories for subsamples from the same dairy. Obtain also nominal 95% and 99% confidence intervals for the difference and compare these differences with the intervals obtained when normality is assumed.

5.3 A hormone is added to one of otherwise identical diets given to each of 40 pairs of twin lambs. Growth differences over a 3-week period (a positive difference indicating greater growth for the hormone diet) are recorded for each pair, and signed ranks are allocated to the 40 differences. The lower rank sum, corresponding to the negative differences, was $S_1=238$. There was only one rank tie. Use the Wilcoxon signed-rank test to investigate the evidence that the hormone may be affecting the growth rate in an increasing or decreasing manner.

5.4 A psychologist interviews both father and mother of each of 17 unrelated children with learning difficulties, asking each individually a series of questions designed to test how well they understand the problems that their child is likely to face in adult life. The psychologist records whether the father (*F*) or mother (*M*) shows the better understanding of these potential problems. For the 17 families, the findings are as follows:

F M M F F F F F F F M F F F M F F

Is the psychologist justified in concluding that fathers tend to show better understanding? Justify your choice of test.

5.5 For each of nine matched pairs of students, one student is allocated to a series of lectures and the other to appropriate online learning material. At the end of the course, the students take the same exam. The marks achieved (out of 100) are as follows:

Pair	1	2	3	4	5	6	7	8	9
Online	50	56	51	46	88	79	81	95	73
Lectures	25	58	65	38	91	32	31	13	49

Identify possible methods of analysis. Investigate these results by what you consider the most appropriate parametric or nonparametric methods to determine whether they provide acceptable evidence that online material leads to better examination results. Give reasons for your choice of methods.

5.6 A total of 100 general practitioners attend a health promotion workshop. At the start of the workshop, they are asked to indicate whether they are in favour of routinely asking patients about alcohol consumption. They are then shown a video on the health and social problems caused by the excessive consumption of alcoholic drinks. The video is followed by discussion in small groups. After the video and discussion, they are asked the original question again. Do the results given below indicate a significant change in attitudes subsequent to the video and group discussion?

		Before Video and Discussion	
		In Favour	*Against*
After Video and Discussion	*In Favour*	41	27
	Against	16	58

5.7 A canned-soup manufacturer is experimenting with a new-formula tomato soup. A tasting panel of 70 each taste samples of the current product and the new one (without being told which is which). Of the 70, 32 prefer the new-formula product, 25 the current product, and the remain cannot distinguish between the two. Is there enough evidence to reject the hypothesis that consumer preference is equally divided? Do the data support a claim that as many as 75% of those who have a preference may prefer the new formula?

5.8 To produce high-quality steel, one of two hardening agents A B may be added to the molten metal. Hardness of steel varies from batch to batch; so to test the two agents, 40 batches are each subdivided into two portions. For each batch, agent A is added to one portion and agent B to the other. To compare hardness, sharpened specimens for each pair are used to make scratches on each other; that making the deeper scratch on the other is the harder specimen. For the 40 pairs, that with agent B is adjudged harder in 24 cases and that with agent A in 16. Is this sufficient evidence to reject the hypothesis of equal hardness?

5.9 For a subsample of 10 pairs from the steel batches in Exercise 5.8, a more expensive test is used to produce a hardness index. This index is greater for harder steel. The indices recorded are as follows:

Batch Number	**1**	**2**	**3**	**4**	**5**	**6**	**7**	**8**	**9**	**10**
Additive A	22	26	29	22	31	34	31	20	33	34
Additive B	27	25	31	27	29	41	32	27	32	34

Use an appropriate test to determine whether these data support the conclusion reached in Exercise 5.8.

5.10 On the day of the third round of the Open Golf Championship in 1987 before play started, a television commentator said that conditions were such that the average scores of players were likely to be at least three higher than those for the second round. For a random sample of 10 of the 77 players participating in both rounds, the scores are as follows:

Player	**A**	**B**	**C**	**D**	**E**	**F**	**G**	**H**	**I**	**J**
Round 2	73	73	74	66	71	73	68	72	73	72
Round 3	72	79	79	77	83	78	70	78	78	77

Do these data support the commentator's claim? Consider carefully whether a one- or two-tailed test is appropriate.

5.11 Scott, Smith, and Jones (1977) gives a table of estimates of the percentages of UK electors predicted to vote for the Conservative Party by two opinion polling organizations, A and B, in each month in the years 1965–1970. For a random sample of 15 months during that period, the paired percentages are as follows:

A 43.5 51.2 46.8 55.5 45.5 42.0 36.0 49.8 42.5 50.8 36.6 47.6 41.9 48.4 53.5

B 45.5 44.5 45.0 54.5 49.5 43.5 41.0 53.0 48.0 52.5 41.0 47.5 42.5 45.0 52.5

Use the Wilcoxon signed-rank test to investigate whether these results indicate a significant tendency for one of the organizations to return higher percentages than the other. Obtain an appropriate 95% confidence interval for any mean or median difference between predictions during the period covered.

5.12 Pearson and Sprent (1968) give data for hearing loss (in decibels below prescribed norms) at various frequencies. The data below show these losses for 10 individuals aged between 46 and 54 at frequencies of 0.125 and 0.25 kHz, respectively. A negative loss indicates hearing above the norm. Is there an indication of a different loss at the two frequencies?

Participant	A	B	C	D	E	F	G	H	I	J
0.125 kHz	2.5	−7.5	11.25	7.5	10.0	5.0	7.5	2.5	5.0	8.75
0.25 kHz	2.5	−5.0	6.25	6.25	7.5	3.75	1.25	0.0	2.5	5.0

5.13 Confirm the value given in Example 5.6 for $\Pr(X>0|\theta=1)$ for the distribution given in (5.1). What is the corresponding probability conditional upon $\theta=2$?

5.14 Confirm the value given in Example 5.7 for $\Pr(X>0|\theta=1)$ for the distribution given in (5.2). What is the corresponding probability conditional upon $\theta=1.2$?

5.15 Determine the sample size needed to have power at least 0.80 for the sign test that the median θ is H_0: $\theta=1$ against the alternative H_1: $\theta=2$ with the one-tail $P=0.05$ if the observations are known to be a random sample from an exponential distribution with frequency function $f(x)=\lambda e^{-\lambda x}$, $x\geq 0$. (Hint: You will need to find values of λ that give medians corresponding to those specified in the null and alternative hypotheses.)

5.16 A dentist working in the Accident and Emergency centre of a dental hospital investigates whether boys are more likely than girls to present with accidental damage to their teeth. Starting with the null hypothesis that boys and girls arrive in equal proportions, use the Noether approximation to estimate the sample size required for a study using a one-tailed sign test if the alternative hypothesis is that the proportion of boys is 0.8. Assume that a power of 0.9 is required when a significance level α = 0.05 is chosen. A colleague suggests that girls might in fact outnumber boys presenting, making a two-tailed test more appropriate. What is the impact of this on the sample size requirement ?

6

Methods for Two Independent Samples

6.1 Location Tests and Estimates

6.1.1 Extension from Single Samples

The distribution-free tests and estimation procedures introduced in Chapters 3 and 4 were applied to single samples. Chapter 5 considered two samples for which observations can be matched in a one-to-one manner and treated as a single sample of differences. This chapter considers two independent random samples. That is, samples in which members of the first sample, or group, are independent of those in the second. The methods covered here as with those in earlier chapters also make the assumption that members within each group are independent of each other.

In the study of citrus fruits and scurvy (Example 2.3), the 12 sailors can be viewed as two independent samples. One pair received citrus fruits as part of their diet, whereas the other 10 sailors did not. In performing a test of the null hypothesis that citrus fruit consumption has no effect on the progress of scurvy, a statistic was used to provide a number that reflects the degree of divergence from what might be expected were the null hypothesis to be true. Sailors were ranked in terms of health following 2 weeks of intervention from the least sick (rank 1) to the most sick (rank 12). The statistic chosen was the sum of the two ranks for the sailors who received citrus fruits, denoted by S. In Lind's study, the sailors who received citrus fruits had the best outcomes and were ranked as 1 and 2 giving $S=3$. This gives strong evidence against the null hypothesis and indicates that the consumption of citrus fruits is beneficial for scurvy. Given that there were 12 participants altogether, the maximum value that S could have taken is 23, with the sailors who received citrus fruits having the worst outcomes, again providing evidence against the null hypothesis. Had the two sailors been given intermediate rankings, S would have taken a value around, say, 10 to 16, indicating little evidence against the null hypothesis.

This procedure can be generalized to two independent groups of any size, where the sum of the ranks of one of the groups is used as the test statistic. This method of dealing with the data is known as the Wilcoxon–Mann–Whitney approach.

6.1.2 Wilcoxon–Mann–Whitney Test

For two independent samples, the analogue of the one-sample Wilcoxon signed-rank test is what has become known as the *Wilcoxon rank-sum test*. Although this method is generally accredited to Wilcoxon (1945), its origins stem from a paper published by Deuchler (1914). Readers should note that there is an error in Deuchler's algebra (Kruskal, 1957).

Mann and Whitney (1947) independently developed an equivalent test. It is convenient, though not universal practice, to refer to the tests jointly as the *Wilcoxon–Mann–Whitney*

DOI: 10.1201/9780429326172-6

test, or for brevity, as the *WMW test*. The Wilcoxon formulation is more widely encountered as it is intuitively simpler. One should be aware that in the scientific literature, the Wilcoxon formulation is commonly referred to as the *Mann–Whitney U test*.

In the Wilcoxon formulation of the test, the two samples are combined, and the data are ranked overall. The original samples are then separated, with each rank being attached to the corresponding observation. The usual null hypothesis is that the two samples are from identical populations. A common alternative hypothesis is that the population distributions differ only in the mean or median. If both samples come from the same population (which may be of any continuous form and need not be symmetric), a mix of low, medium and high ranks is expected in each sample. Under the alternative hypothesis, one expects lower ranks to dominate in one population and higher ranks in the other. Such a shift in location is often referred to as an *additive* treatment effect, i.e., there is a constant difference between the two treatments.

Conventionally, the numbers of observations in the two samples are denoted by m and n, respectively, the observations being labelled as $x_1, x_2, \ldots, x_m$ and $y_1, y_2, \ldots, y_n$. For convenience and without loss of generality, the second sample is assumed to be at least as large as the first.

The sum of the ranks in the first sample, S_m, can be used to determine the strength of evidence against H_0. The sum S_n for the second sample is an equivalent alternative statistic.

Values of the rank sum close to the minimum possible or the maximum possible provide strong evidence against the null hypothesis. Intermediate values of S_m indicate that there is little evidence against H_0.

The test may also be used when samples are from two distributions with identical cumulative distribution functions under H_0, but under the alternative hypothesis, one cumulative distribution curve lies beneath the other apart from some points where the curves touch. In this situation, under H_1 low or high ranks should dominate in one sample, in contrast to a fairly even distribution of ranks under H_0. Such H_1 are sometimes referred to as *dominance* alternatives. *P*-values may be determined from the permutation distribution of rank sums under the null hypothesis.

Example 6.1

Students attending an after-school mathematics club are interested in finding out whether using the calculator facility on a mobile telephone enables calculations to be performed more quickly compared to a standard calculator. The 21 students attending on the afternoon of the experiment divide themselves randomly into groups A (using a mobile telephone) and B (using a standard calculator), with 10 and 11 students in the respective groups. The students perform a list of calculations devised by a member of staff who supervises the time keeping. Students stay in the room until all have finished. The total times in minutes for each member of each group to complete the calculations were

Group A	23	18	17	25	22	19	31	26	29	33	
Group B	21	28	32	30	41	24	35	34	27	39	36

Do the data indicate that the use of a mobile telephone is superior (i.e., leads to more rapid computations)?

A two-tailed test is appropriate. The sum of ranks associated with the smaller sample, Group A, in a joint ranking of all data is required. Alternatively, the rank sum for Group B could be used. The *P*-value associated with this sum allows assessment of the strength of evidence against H_0: *the population medians are identical*, whereas the alternative is that one sample is from a population with a greater median. It is more useful to think in terms of medians rather than means because there is no need to assume symmetry. The test will also be valid for a dominance alternative, i.e., for a tendency for the computations to be done more quickly with one of the calculation tools, although that time difference may vary appreciably between students.

The sample sizes are $m=10$ and $n=11$. Most software programs for the Wilcoxon test will calculate the ranks automatically, but to obtain them manually, it helps to arrange data in ascending order within each sample. For the calculation data, the groups are:

Group A	17	18	19	22	23	25	26	29	31	33	
Group B	21	24	27	28	30	32	34	35	36	39	41

Following the ranking of the combined observations these become:

Group A	1	2	3	5	6	8	9	12	14	16	
Group B	4	7	10	11	13	15	17	18	19	20	21

For Group A, the rank sum of $S_m=1+2+3+5+6+8+9+12+14+16=76$. The rank sum for Group B is $S_n=155$.

If software to give exact P-values is not available, some relevant values for S_m are given in many tables. For this example, using the Applied Nonparametric Statistical Methods, fifth edition (ANSM5) package in R and the command `wilcoxon.mann.whitney(ch6$GroupA, ch6$GroupB)` indicates that for a two-tailed test the relevant exact $P=0.0159$. The output also confirms the value of S_m given above. Hence, the low P-value provides fairly strong evidence against H_0. On average, students appear to perform calculations more quickly with their mobile telephones.

Tables for the Wilcoxon–Mann–Whitney statistic are given by Gibbons and Chakraborti (2020, Table J) and by Neave (1981, p. 30). Conover (1999, Table A7) and others give various quantiles for S_m, S_n. Actual, rather than nominal, significance levels may be obtained from computer programs giving the exact permutation distribution.

It would be unreasonable for the teacher to recommend that students use their mobile telephones during mathematics lessons. One reason for this is that attendees at the mathematics club are probably not representative of all the students, as they may have more interest in mathematics or possibly need additional assistance with their course work. In addition, the use of mobile telephones is likely to encourage social networking during lessons, which could act as a distraction.

Suppose that only 40 minutes can be allocated to the experiment due to the transport available to the students for returning home from school. The fifth pupil in Group B is then unable to complete the calculations, and this observation is censored. In the ranking of the data, however, since this observation is the only one with a time exceeding 40 minutes, it still has a rank of 21. However, a t-test cannot validly be applied to the observations since an exact value is required for each observation.

6.1.3 The Mann–Whitney Formulation

A statistic U, which is a function of the rank sum S, can be calculated for either group to determine the strength of the evidence against the null hypothesis. For the first of the two samples, this statistic is given by $U_m=S_m-m(m+1)/2$. The equivalent statistic for the (perhaps larger) sample is $U_n=S_n-n(n+1)/2$. It is only necessary to compute one of S_m or S_n, for the sum of all the ranks from 1 to $m+n$ is $(m+n)(m+n+1)/2=S_m+S_n$. Consideration of the relations $U_m=S_m-m(m+1)/2$ and $U_n=S_n-n(n+1)/2$ shows that each has a minimum value of zero, and that

$$U_m = mn - U_n, \tag{6.1}$$

so that again only one of U_m, U_n needs to be computed. Either may be used in a test, although U_m is generally given in tables.

In the Mann–Whitney approach, either U_m or U_n is calculated directly. To obtain either, a count is made of the number of observations in one sample exceeding each member of the other sample. Ranks are not needed, and the procedure eases calculation if no computer program is available for the test. It also forms the basis for determining a confidence interval for a location difference.

Example 6.2

Calculate a test statistic for the data in Example 6.1 using the Mann–Whitney approach.

First, inspect the observations in each sample; they need not be ordered, but counting is easier when they are. It is visually easier to count the number of times each observation in Group A is exceeded by an observation in Group B. This gives U_n, from which U_m can be determined using equation (6.1).

The data from Example 6.1, arranged for convenience in ascending order in each group, are:

Group A	17	18	19	22	23	25	26	29	31	33	
Group B	21	24	27	28	30	32	34	35	36	39	41

The first observation, 17, in Group A is exceeded by all 11 observations in Group B.

So also are the observations 18, 19. The observations 22 and 23 are exceeded by 10 observations in Group B. This procedure is continued until the final observation in Group A is reached. Its value (33) is exceeded by 5 observations in Group B. The counts determined by this process are then added to give

$$U_n = 11+11+11+10+10+9+9+7+6+5 = 89.$$

Using (6.1) gives $U_m = 110 - 89 = 21$. Since $U_m = S_m - m(m+1)/2$, it follows that

$S_m = U_m + m(m+1)/2 = 21 + 10(10+1)/2 = 21 + 55 = 76$, consistent with the value of S_m found in Example 6.1.

Again using the ANSM5 package in R and the command `wilcoxon.mann.whitney(ch6$GroupA, ch6$GroupB)` gives exact tail probabilities corresponding to the value of U (in this case $P = 0.0159$), showing the same finding as in Example 6.1. Note that this equivalence of the Wilcoxon and Mann–Whitney formulations is true in general.

As with the statistic S, some general statistical packages compute U but leave the user to consult tables or else give an asymptotic result based on the normal distribution (see Section 6.1.5). Many tables only give values of the statistic that correspond to nominal conventional significance levels rather than exact P-values, but most explain what is given and how to use the tables. For instance, Neave (1981) indicates that if either U_m or U_n do not exceed 26, the two-tailed P-value for Example 6.2 is less than 0.05.

6.1.4 Wilcoxon–Mann–Whitney Confidence Intervals

The Mann-Whitney statistic compares all possible differences when observations in the first sample are linked with observations in the second. Denote these differences as $d_{ij} = x_i - y_j$. In computing U_m, a score of 1 is allocated if d_{ij} is positive and 0 if d_{ij} is negative. To calculate U_n, these scores are reversed, i.e., score 1 if d_{ij} is negative and 0 if d_{ij} is positive. To calculate a confidence interval for the difference in location, the actual values of some or all

d_{ij} are required. If c is the value of the Mann–Whitney statistic U that would be regarded as just providing sufficient evidence to indicate significance at the 100α% significance level, the 100(1 − α)% confidence limits are given by the c+1 smallest and c+1 largest d_{ij}.

The reasoning here is not unlike that used in establishing a confidence interval using Walsh averages in the one-sample situation based on the Wilcoxon signed-rank approach.

Example 6.3

Obtain nominal 95% confidence limits for a population median shift based on the WMW method using the data in Example 6.1.

If computer software that calculates a confidence interval directly is not available, tables such as those in Neave (1981) indicate that for sample sizes of 10, 11, the critical value is U_m=26 for significance at a nominal 5% level. Denoting the sample values by $x_1, x_2, \ldots, x_{10}$ and $y_1, y_2, \ldots, y_{11}$ the 27 largest and smallest $d_{ij}=x_i-y_j$ are required. It is not essential to compute all d_{ij}, but for completeness, these are given in Table 6.1. These need not be computed manually if suitable software is available, either to determine the confidence limits directly or even if only to compute the required differences.

It is easier to compute differences manually if the sample values are ordered. Those for the first sample are written in the top row, and those for the second sample are in the first (left) column. The entries in the body of Table 6.1 are the differences between the data entries at the top of that column and at the left of that row, e.g., the first entry, –4, is 17 − 21 = −4, the fifteenth entry moving along rows, –1, is 23–24. Note that the difference between each entry in any pair of rows is a constant equal to the difference between the corresponding entries in the left-hand data column; there is an analogous constant difference between entries in pairs of columns. The largest entries appear in the top right of the table, and entries decrease and eventually change sign, moving towards the bottom left.

A count shows 21 positive and 89 negative values whence U_n=89 and U_m=21 as found in Example 6.2. The critical value is U_m=26 for significance at least at a 5% level, so the lower limit for a 95% confidence interval for the median difference $\theta_1 - \theta_2$ is obtained by eliminating the 26 largest negative differences; the next largest negative difference is the required lower limit. Using Table 6.1 gives this lower limit as –13. Similarly, the elimination of the 26 largest differences gives the upper limit as –2. Hence, a nominal 95% confidence interval for the difference $\theta_1 - \theta_2$ is (–13, –2), or equivalently, for $\theta_2 - \theta_1$ the interval is (2, 13).

TABLE 6.1

Paired Differences for Times Taken to Complete Calculations

	17	*18*	*19*	*22*	*23*	*25*	*26*	*29*	*31*	*33*
21	–4	–3	–2	1	2	4	5	8	10	12
24	–7	–6	–5	–2	–1	1	2	5	7	9
27	–10	–9	–8	–5	–4	–2	–1	2	4	6
28	–11	–10	–9	–6	–5	–3	–2	1	3	5
30	–13	–12	–11	–8	–7	–5	–4	–1	1	3
32	–15	–14	–13	–10	–9	–7	–6	–3	–1	1
34	–17	–16	–15	–12	–11	–9	–8	–5	–3	–1
35	–18	–17	–16	–13	–12	–10	–9	–6	–4	–2
36	–19	–18	–17	–14	–13	–11	–10	–7	–5	–3
39	–22	–21	–20	–17	–16	–14	–13	–10	–8	–6
41	–24	–23	–22	–19	–18	–16	–15	–12	–10	–8

If normality is assumed, the 95% confidence interval based on the relevant t-distribution is (−12.6, −1.9). The close agreement between this and the Wilcoxon interval is reassuring because there is little to indicate serious nonnormality or a difference between population variances in these data apart from a small indication that the Group B data might have a slightly greater spread.

Confidence intervals at other levels can be obtained. For example, a 99% interval is obtained by rejecting the 18 extreme differences, and using Table 6.1, it is seen to be (−16, 1).

Output from the ANSM5 package in R can be obtained using `wilcoxon.mann.whitney(ch6$GroupA, ch6$GroupB)` and `wilcoxon.mann.whitney(ch6$GroupA, ch6$GroupB, CI.width = 0.99)`.

The computational process is reminiscent of that with Walsh averages given in Section 3.3.2.

To explore the exact confidence level, further results of hypothesis testing situations at or near the end points of the interval may be examined. This is achieved in the case of a 95% confidence interval by an appropriate addition or subtraction to all observations in one sample and then obtaining exact P-values for the corresponding test of zero median difference. While the ANSM5 package in R computes exact intervals, some software packages may use asymptotic approximations based on the asymptotic theory given for hypothesis testing in Section 6.1.5. If both samples are moderate or large in size, asymptotic approximations are usually quite good. They may be unreliable if the samples differ greatly in size, especially if one of the samples is small.

The appropriate point estimator of the median difference based on the WMW procedure used above, known as the *Hodges–Lehmann estimator*, is the median of the differences in Table 6.1. It can be verified that in the above example this is −7.5.

Note that in general, a Hodges–Lehmann median difference is *not* the same parameter as a difference between medians. Methods for estimating a median difference, such as the Hodges–Lehmann approach, should be viewed as distinct from those that estimate the difference between two medians (Newson, 2017).

6.1.5 Asymptotic Approximation

For the WMW test, it can be shown that the test statistic S_m has mean $m(m+n+1)/2$ and variance $mn(m+n+1)/12$, whence it follows that for large-sample sizes

$$Z = \frac{S_m - m(m+n+1)/2}{\sqrt{mn(m+n+1)/12}} \tag{6.2}$$

has approximately a standard normal distribution. For large m, n a continuity correction may have little effect, but if one is used, the appropriate correction is to add 0.5 to S_m if $Z<0$ and to subtract 0.5 if $Z>0$.

For the Mann–Whitney formulation, if the statistic U_m is used (6.2), it is replaced by

$$Z = \frac{U_m - mn/2}{\sqrt{mn(m+n+1)/12}} \tag{6.3}$$

Modifications to the denominators in (6.2) and (6.3) are required for tied data. These are discussed in the next section.

6.1.6 Ties

Mid-ranks are used for ties as for the Wilcoxon signed-rank test (Section 3.3.4). If software to compute exact *P*-values is not available and there are only a few ties, basing significance tests on the appropriate critical values for the no-tie case is unlikely to be seriously misleading. If m, n are both reasonably large (say, 15 or greater), the normal approximation developed below may be used with reasonable confidence. Adjustment is essential only when there are a moderate to large number of ties. Example 6.4 illustrates a situation where ties dominate.

In the Mann–Whitney formulation, if an observation in the second sample equals an observation in the first sample, it is scored as 1/2 in counting the number of observations in the second sample exceeding that observation in the first.

For small to medium-sized samples, ties present no difficulty if a computer program is available to generate exact probabilities for the WMW test based on the appropriate permutation distribution.

Example 6.4

The experiment described in Examples 6.1 and 6.2 is performed at a different school. The observations recorded include some ties in the times taken by different participants. For convenience, the data are given in ascending order, but this is not essential, especially if suitable software is used.

Group A	16	18	19	22	22	25	28	28	28	31	33		
Group B	22	23	25	27	27	28	30	32	33	35	36	38	38

Do the data indicate that one of the calculating aids is superior (i.e., leads to more rapid computations)?

The WMW test with mid-ranks for ties is used. The group sizes are $m=11$ and $n=13$. Computer programs usually assign ranks or mid-ranks automatically, but doing this manually is straightforward since the data are ordered. This procedure assigns the ranks as:

Group A	1	2	3	5	5	8.5	13.5	13.5	13.5	17	19.5		
Group B	5	7	8.5	10.5	10.5	13.5	16	18	19.5	21	22	23.5	23.5

Unless S_m is needed specifically and no software is available to compute it, allocation of ranks can be avoided. It is easier to obtain U_n by simply counting for each first sample value the number of observations that exceed it in the second sample (scoring one half for between-sample ties). These are summed. For example, for each value 22 in Group A, there is one tied value, scored as 1/2, and 12 values exceeding 22 in Group B, giving a contribution of 12.5 to U_n. Proceeding in this way, starting with the value 16, gives

$$U_n = 13 + 13 + 13 + 12.5 + 12.5 + 10.5 + 7.5 + 7.5 + 7.5 + 6 + 4.5 = 107.5$$

From (6.1), $U_m = 11 \times 13 - 107.5 = 35.5$. For a two-tailed test, the exact $P=0.0359$. There is some evidence against H_0 and the results suggest that, as in the original experiment, computations are completed more rapidly using a mobile telephone.

Output from the ANSM5 package in R can be obtained using `wilcoxon.mann.whitney(ch6$groupA.sch2, ch6$groupB.sch2)`.

Tables such as those in Neave (1981) indicate that values of $U_m \leq 37$ imply $P < 0.05$ for a two-tailed test. This one example suggests that a few ties do not seriously upset conclusions based on no-tie critical values for moderate sample sizes.

The situation using conventional no-tie tables is less satisfactory when there are many ties, or for ties in unbalanced samples. For example, a sample of 3 consisting of the values 1, 2, 2, and a sample of 13 with values 1, 1, 4, 5, 5, 5, 7, 8, 9, 9, 9, 9, 10 leads to $U_m = 5$. In a no-tie situation, the two-tailed test gives $\Pr(U_m \leq 5) = 0.0571$, whereas the exact permutation test allowing for ties gives $\Pr(U_m \leq 5) = 0.0464$ as the true P-value for this specific tie pattern. Although the strength of evidence against the null hypothesis is not markedly different, adhering to a rigid 5% significance level would lead to differing conclusions.

As seen in Example 3.8, if ties result from rounding and greater accuracy allows ties to be broken, the way the ties break may alter appreciably conclusions regarding significance.

Software to determine exact permutation probabilities associated with the test statistics U or S is particularly valuable when there are ties. The ANSM5 package in R allows this. Some general programs take ties into account by using mid-ranks but provide only an asymptotic test (which may be unreliable for small $m+n$ or when one of m, n, is small). Alternatively, one may resort to tables appropriate to the no-tie situation, but nominally significant levels can no longer be guaranteed.

A common tie situation is one where precise measurements are not available and only grouped data can be obtained. For example, instead of the complete sample values in Example 6.4 only numbers of participants taking between 10 and 19 minutes, 20 and 29 minutes, 30 and 39 minutes are given. It is still possible to calculate the U or S statistics making allowances for ties, but it may now be misleading to use tabulated critical values for these statistics.

Example 6.5

Suppose that for the experiment described in Example 6.4 only the numbers taking 10–19, 20–29, 30–39 minutes were recorded as given below. Perform the test in Example 6.4 using this reduced information.

No. of Minutes	10–19	20–29	30–39
Group A	3	6	2
Group B	0	6	7

A WMW test based on mid-ranks is carried out using, if available, a program giving exact permutation probabilities.

If no suitable program is available, calculating the required value of U_n is reasonably straightforward but requires care. For each of the three ties in the category 10–19 for Group A, there are $6+7=13$ greater values in Group B. Thus, between them, these three ties contribute a total of $3 \times 13 = 39$ to U_n. Similarly, each of the six Group A ties in 20–29 corresponds to six ties and seven greater values in Group B. Thus each contributes a score of $(1/2 \times 6) + 7 = 10$, so that together they contribute a score of $6 \times 10 = 60$. Finally, by a similar argument, the two Group A ties in 30–39 each contribute 3.5, whence $U_n = 39 + 60 + 7 = 106$ and now from (6.1) $U_m = 143 - 106 = 37$. In Example 6.4, it was noted that $U_m = 37$ is the minimum value that implies $P \leq 0.05$ in a two-tailed test if there are no ties. The exact two-tail permutation probability for $U_m = 37$ for these tied data is $P = 0.0442$. Hence, there is reasonably strong evidence against the hypothesis of equal medians.

Output from the ANSM5 package in R can be obtained using `wilcoxon.mann.whitney(groupA.sch2.grp, groupB.sch2.grp)`.

In this specific example, the use of conventional no-tie tables is not seriously misleading; this is not always so. In particular, heavy tying often leads to major discontinuities in possible *P*-values since the tying may reduce appreciably the number of possible values that U_m may take.

By pooling the data into groups, less information is available, so it is not surprising that comparing these findings with those from Example 6.4, there is a reduction in the strength of evidence against H_0 that is reflected by the slightly higher *P*-value.

In the presence of ties, the denominator in formulae (6.2) and (6.3) needs modification, giving:

$$\frac{mn(m+n+1)}{12}\left[1-\frac{\sum_{k=1}^{K}\left(t_k^3-t_k\right)}{(m+n)\left((m+n)^2-1\right)}\right] \tag{6.4}$$

where t_k is the number of ties for the kth rank, and K is the total number of unique ranks with ties (e.g., for the ranks 1 2.5 2.5 4 6 6 6 8, K=2, and t_1=2, t_2=3).

For large m, n

$$Z=\frac{T-E(T)}{\sqrt{\mathrm{Var}(T)}} \tag{6.5}$$

has approximately a standard normal distribution.

The approximation is improved by a continuity correction replacing T by T+0.5 if T is less than the value expected under the null hypothesis and by T – 0.5 otherwise.

Example 6.6

Ages at death for members of two clans (see Appendix) in the Badenscallie burial ground (arranged in ascending order) are:

McGamma	13	13	22	26	33	33	59	72	72	72	77	
	78	78	80	81	82	85	85	85	86	88		
McBeta	0	19	22	30	31	37	55	56	66	66	67	67
	68	71	73	75	75	78	79	82	83	83	88	96

Use the WMW large-sample approximation to test H_0: *the population distributions are identical* against H_1: *the population distribution medians differ.*

With sample sizes of 21 and 24, respectively, a large-sample approximation should suffice. To adjust for ties, mid-ranks are required. The appropriate mid-ranks are:

McGamma	2.5	2.5	5.5	7	10.5	10.5	15	23	23	23	28	
	30	30	33	34	35.5	40	40	40	42	43.5		
McBeta	1	4	5.5	8	9	12	13	14	16.5	16.5	18.5	18.5
	20	21	25	26.5	26.5	30	32	35.5	37.5	37.5	43.5	45

One may use either S or U, but once the ranks are recorded, it is easier to use S. For McGamma, addition gives $S_m = 518.5$. The mean, E(T), is $21 \times (21 + 24 + 1)/2 = 483$ (the same as that for the no-tie situation). Using (6.4) to compute the variance, it can be shown that the denominator in (6.5) is 43.92, while giving $Z = (518.5 - 483)/43.92 \approx 0.808$. In addition, if ties had been ignored and (6.2) used, the only change would be that 43.92 in the denominator would have been replaced by 43.95, so in this case the effect of ties is trivial. This is often the situation unless ties dominate. Since $P = \Pr(|Z| > 0.808) = 0.419$, there is no plausible evidence against the hypothesis of identical population medians. Use of the continuity correction gives $Z = (518.5 - 483 - 0.5)/43.92 \approx 0.797$, $P = \Pr(|Z| > 0.797) = 0.426$, and the same conclusion is reached.

Output from the ANSM5 package in R can be obtained using `wilcoxon.mann.whitney(appl$McGamma, appl$McBeta, do.asymp = TRUE)`.

For both m, n greater than 20, one expects asymptotic results to agree closely with those for the exact WMW permutation test. With even the modest sample sizes here of 21 and 24, the algorithms necessary to deal with the very large number of possible combinations are beyond the scope of the ANSM5 package in R. The `wilcox.test` function in the standard stats package in R (version 4.2.3) (upon which the `wilcoxon.mann.whitney` function in the ANSM5 package is based for exact calculations in the absence of ties and for asymptotic calculations), refuses to compute an exact P-value when ties exist in the data. However, in order to judge the accuracy of the asymptotic results, one can remove ties using the `jitter` function in R (which adds/subtracts a small amount to/from each value), compute the exact P-value for the jittered values and repeat a number of times, taking the mean of the exact P-values as an approximation to the exact P-value.

Output from the ANSM5 package in R can be obtained in this way, using 10,000 jittered versions of the data, using `mean(sapply(1:10000, function(x){wilcoxon.mann.whitney(jitter(appl$McGamma), jitter(appl$McBeta))$pval.exact}))`.

The above command gives $P = 0.430$, indicating that the asymptotic P-value is quite good. Using software that can give an exact P-value in the presence of ties for samples of the size involved here, we get $P = 0.426$, which is identical to the asymptotic P-value when the continuity correction is used, at least to this three-decimal place level of accuracy.

It might be argued that rather than H_1 specifying a median shift, a hypothesis of the dominance of one distribution over the other may be more appropriate on genetic grounds. Inspection of the data gives a slight hint of a greater spread for the McBeta data, indicating a situation analogous to the classic Behrens–Fisher problem of testing for equality of means for two normal distributions with different variances. This more general problem has been considered by Pothoff (1963), Fligner and Rust (1982) and others.

With efficient algorithms, computation of the exact permutation distribution tail probability in this case takes only seconds on most PCs, but as pointed out above, the asymptotic results are nearly as good for samples of this size, and there may seem little point in calculating exact probabilities when an asymptotic result suffices. Of particular interest when using an asymptotic result is whether a reasonable fit in the tails is obtained. For the case $m = 21$, $n = 24$, if $S_m = 641$ the exact test one-tail probability is 0.0001 and the asymptotic probability is 0.0002; if $S_m = 557$ the corresponding probabilities are 0.0317 and 0.0311.

6.1.7 Pitman Two-Sample Test

Pitman (1937a) developed a raw data permutation test for two independent samples along the lines of the (Fisher) Pitman test described in Section 3.4.1. His prime aim was to show, at least for samples that are not very small, that it leads to similar inferences to those based on a two-sample *t*-test that requires a normality assumption for strict validity. In effect, he verified an earlier assertion by Fisher that using the *t*-test or other normal theory tests gave a good approximation to an exact randomization test even when the normality assumption was no longer valid. This justified continued use of the then easier-to-compute *t*-test as the basis for inferences. The Pitman two-sample test is even more computationally challenging than the (Fisher) Pitman test and in practice, it has rarely been used.

SUMMARY POINTS:

- The Wilcoxon rank-sum test is used to investigate differences in location or domination of one distribution over the other by ranking the data (Section 6.1.2).
- Mann and Whitney gave an alternative formulation (Section 6.1.3).
- The method is often referred to as the Wilcoxon–Mann–Whitney test.
- Confidence intervals can be calculated using a computational process similar to the use of Walsh averages (Section 6.1.4).
- Asymptotic approximations based on the normal distribution are available for large samples (Section 6.1.5).
- Ties can be adjusted for using mid-ranks (Section 6.1.6).
- Pitman developed a two-sample permutation test similar in nature to the (Fisher) Pitman test for a single sample (Section 6.1.7).

6.2 The Median Test

6.2.1 A Warning

The median test comes with a health warning. The price one pays for its simplicity is that its power is usually appreciably lower than other more appropriate alternatives for all but large samples of 100 or more observations. This is true even when its Pitman efficiency is higher than that of its competitors. This is a situation where the asymptotic or large-sample property of the Pitman efficiency is not reflected in smaller samples, especially if the sample sizes m, n differ considerably.

Freidlin and Gastwirth (2000) make an extensive study of the power of the median test relative to commonly used alternatives. These include the WMW test and tests for the median shift in otherwise identical distributions based on scores described in Section 6.3 and others described elsewhere. If the test is used for a pure median shift (assuming this is the only distributional difference between the sampled populations), the test is also not robust against breakdowns in assumptions such as equality of variance or other measures of spread. Freidlin and Gastworth rightly recommend that the median test *is not appropriate for routine use*. Nevertheless, because of its simplicity, it is useful for illustrating features such as the importance of clarifying assumptions and the basic ideas behind power studies. A thorough account of the theory behind the median test, and a wider but closely related class of tests, is given in Gibbons and Chakraborti (2020, Section 6.4).

6.2.2 The Basic Test

Consider samples of m, n observations that can be assumed to come from distributions that differ, if at all, only by a shift in location. That is, if $F(x)$ and $G(y)$ are the population cumulative distribution functions, then $G(y)=F(x-\delta)$ where $\delta=\theta_x-\theta_y$. The usual null hypothesis is H_0: $\delta=0$ implying identical distributions. If the population medians are θ_x, θ_y an equivalent form of H_0 is H_0: $\theta_x=\theta_y$.

The test is usually against one- or two-sided alternatives that the medians differ. Under H_0 the sample number of observations above the common median has a $B(m+n, 0.5)$ distribution and the number of observations, T_m, above the median in the first sample has a $B(m, 0.5)$ distribution.

The *median test* was proposed by Brown and Mood (1948). It does not test whether that common median has a particular value, but as is also the situation in the two-sample t-test or the WMW test, only whether it is reasonable to suppose both populations have the same unknown median. If both populations have the same median, then the combined sample median, M, say, provides a point estimate of that median.

The median test may be based on the statistic T_m defined above. As already indicated, if the samples are from populations with the same median, the distributions of the numbers in each above M will be approximately $B(m, 0.5)$ and $B(n, 0.5)$, where the relevant probabilities are now conditional upon the numbers above M in each sample adding to the number above M in the combined sample. When $m+n$ is even and no sample value equals M, in the combined sample the numbers of values above and below M will each be $(m+n)/2$. If $m+n$ is odd, at least one sample value equals M. In all cases where some values equal M, it is reasonable to omit these and proceed with a reduced sample. Unless many values equal M, this is usually satisfactory.

Conditioning on the total numbers above or below M in the combined sample leads to permutation tests that have much in common with some procedures developed more generally in Chapter 12. In view of the generally poor performance of this test, the examples given are mainly used to illustrate principles.

Example 6.7

Following a lecture on an aspect of algebra, students were given a problem based on the material presented and asked to press a buzzer when finished. The times taken in minutes by the 7 male and 21 female students working individually are given below. Do these observations indicate a difference between males and females in the median time taken to solve the problem?

Males	15	16	20	24	27	31	34				
Females	8	9	11	11	14	15	15.8	16.2	16.6	17	18
	19	20	21	22	23	24	25	26	27	28	

To apply the median test, the samples are pooled, and the median time, M, for the combined sample is determined. Ranking the 28 observations and taking the mean of the 14th and 15th observations gives a value of 19.5 minutes. None of the observations is equal to the median, so there are 14 values greater and 14 less than M. For the males, there are 5 values greater than M. For the females, there are 9 such observations. If the unknown population medians are θ_1 and θ_2 respectively, an appropriate test is H_0: $\theta_1=\theta_2$ against H_1: $\theta_1 \neq \theta_2$.

Since, for males, there are only 2 values below M and 5 above, there is a suggestion that males may tend to take longer than females. That is, there is evidence against H_0 being true.

To explain the rationale behind the test, it is convenient to set out the information on numbers above and below M in each sample in a table:

	Above M	**Below M**	**Total**
Males	5	2	7
Females	9	12	21
Total	14	14	28

An important feature of this table is that the row and column totals and the grand total are fixed. There are two rows and two columns of data, so such a table is called a two-by-two or 2×2 *contingency table.* The four data positions are often referred to as cells. That at the intersection of the ith row and jth column is called cell (i, j). For example, in the table above, cell (2, 1) contains the entry 9.

The row totals are determined by the sample sizes m, n and the column totals by the rule that M is the combined sample median. The test is based on the permutation distribution of all possible cell entries in a table like that above consistent with the marginal totals. Since the marginal totals are fixed, knowledge of one cell value means that all the other cell values can be determined by subtraction. It is convenient to focus on the possible values in cell (1, 1) (i.e., the number of values above M in sample 1). Suppose that a value of 3 is considered. To have the correct marginal totals, by subtraction the entries in the table are:

	Above M	**Below M**	**Total**
Males	**3**	4	**7**
Females	11	10	**21**
Total	**14**	**14**	**28**

Such an outcome is more favourable to H_0 than the one observed in this example, because the proportions of observations greater than M for each gender are closer to equality. Indeed, the only outcomes as or less favourable to H_0 than that observed are those where cell (1, 1) takes the values 0, 1, 2, 6 or 7.

These allocations, together with the one observed in this example, form an appropriate tail critical region for assessing the evidence for H_1 against H_0. It is then necessary to calculate the probability associated with this tail when H_0 is true. Assume temporarily that if H_0 is true, then M is the value of the common median for the two populations. This may not be true, but it is intuitively a best estimate of the common median to be obtained from these data. As seen below, this assumption is not critical to the line of reasoning. As in the sign test, under H_0 the probability of obtaining r observations above the median in a sample of 7 is the binomial probability

$$p_r = \binom{7}{r}\left(\frac{1}{2}\right)^7$$

That of getting s above the median in a sample of 21 is

$$p_s = \binom{21}{s}\left(\frac{1}{2}\right)^{21}$$

These are independent in the sense that they refer to different samples. They are, however, conditional upon the fixed column total, i.e., $r+s=14$. Turning to the column totals, in the combined sample of 28, the probability of observing 14 out of 28 observations above M is

$$p_c = \binom{28}{c}\left(\frac{1}{2}\right)^{28}$$

with $c=14$. The probability of observing, respectively, 5 and 9 observations above the median in Samples 1 and 2, conditional upon observing 14 above the median in the combined sample is now required. From the definition of conditional probability, this is $P^*=p_r p_s/p_c$. Thus, for the data in this example,

$$P^* = \frac{\binom{7}{5}\left(\frac{1}{2}\right)^7 \times \binom{21}{9}\left(\frac{1}{2}\right)^{21}}{\binom{28}{14}\left(\frac{1}{2}\right)^{28}}$$

A generalization of this result to samples of any size is considered in Exercise 6.9.

The binomial probability $p=1/2$ cancels out in the expression for P^*. This will happen if the binomial analogues of p_r, p_s, p_c are written down for any p; i.e., P^* is independent of p. It suffices to assume only that M is the common value of some unspecified quantile. Strictly, this requires an assumption of identical population distributions, but since for the choice of M the unspecified quantile is likely to be close to the median, the test is unlikely to mislead if one assumes only that the distributions differ little in the neighbourhood of the median. The relevant tail probability is the sum of the P^* for the configurations that are as or more extreme than that observed. That is, when the numbers above the median in the first sample (males) are 0, 1 2, 5, 6 or 7.

For the evaluation of P^*, it is commonly written in an equivalent form. For a general 2×2 contingency table with cell entries

$$\begin{matrix} a & b \\ c & d \end{matrix}$$

$$P^* = \frac{(a+b)!(c+d)!(a+c)!(b+d)!}{(a+b+c+d)!a!b!c!d!} \tag{6.6}$$

where $a!=a\times(a-1)\times(a-2)\times\ldots\times3\times2\times1$. Conventionally, $a!$ is referred to as *factorial a*. Also, by definition, $0!=1$.

The factorials in the numerator of (6.6) are those for the marginal totals. Those in the denominator are for the grand total and the individual cell entries. For the given data $P^*=[7!\times21!\times14!\times14!]/[28!\times5!\times2!\times9!\times12!]=0.1539$. For the value 6 in cell (1, 1), a similar computation gives $P^*=0.0355$, while for the value 7 in cell (1, 1) $P^*=0.0029$. A moment's reflection shows that for values 2, 1 or 0 in cell (1, 1), these probabilities are repeated. Thus the relevant P-value for the two-tail alternative to H_0 is $P=2\times(0.1539+0.0355+0.0029)=0.3846$.

Output from the ANSM5 package in R can be obtained using `med.test(ch6$males, ch6$females)`.

TABLE 6.2

Exact Permutation Distribution for the Median Test in Example 6.7

Cell (1, 1) Entry	Probability
0	0.0029
1	0.0355
2	0.1539
3	0.3077
4	0.3077
5	0.1539
6	0.0355
7	0.0029

The high *P*-value in this example reflects the poor power performance noted by Freidlin and Gastwirth, especially when the size of one of the samples is small. The corresponding values for the *t*-test and the WMW test are respectively $P=0.056$ and $P=0.123$.

There are marked discontinuities in the permutation distribution, despite there being 28 observations. The fact that only 7 of these are in one sample means that the test statistic can take only one of eight possible values.

This test is usually referred to as the Fisher (exact) test, a test discussed more fully in Section 12.2.1. The ANSM5 package in R generates the complete permutation distribution over all possible cell (1, 1) values as given in Table 6.2. This confirms the symmetry of the distribution and explains the marked probability discontinuities between successive possible *P*-values. In Chapters 3 and 4, it was seen that for most statistics discontinuity decreased rapidly with increasing sample size. Because there are 28 observations here the level of discontinuity may surprise, but as explained above, this is because one sample is small relative to the other.

Markedly different sample sizes (in Example 6.7 one was three times the other) often occur in clinical trials where there may be only a few patients with a rare illness. These patients may then be compared with a much larger *control* group of individuals without the illness.

6.2.3 Confidence Intervals

Given full measurement data, a confidence interval for the difference between population medians based on the median test permutation distribution may be established. Writing the true difference $\delta=\theta_2-\theta_1$, a nominal 95% confidence interval for δ is (d_1, d_2), where the end points d_1, d_2 are chosen so that if each observation in sample 2 (or sample 1) is changed by more than these amounts, the hypothesis that the medians are equal for Population 1 and the amended Population 2 is rejected. This is relevant because increasing or decreasing all sample (or population) values by a constant k increases or decreases the median by k. In view of the general poor power performance of the median test, the details are omitted, outlining the process only in an example. Several statistical software packages give approximate confidence intervals for the median difference.

Example 6.8

For the data in Example 6.7 determine a nominal 90% confidence interval for the difference in population medians. The procedure is based on the Fisher exact test developed in Example 6.7.

From Table 6.2, it is evident that if the numbers above the median [i.e., those in cell (1, 1) of the 2×2 tables developed in Example 6.7] were 0 or 7 a two-tailed test of the hypothesis that the medians are equal for samples of the given sizes (7 and 21) would be rejected with an exact P-value of $0.0029\times2=0.0058$. Similarly, with 1 or 6 the corresponding P-value is equal to $(0.0029+0.0355)\times2=0.0768$. Having established this, the next step is to determine what constant adjustment is needed to each observation in Sample 2 (the females) to avoid rejecting H_0 in a median test based on Sample 1 and the adjusted Sample 2, but that value being such that H_0 would be rejected for any greater adjustment. These adjustments provide nominal 90% (exact 92.32%) confidence limits.

Inspecting the original data, since H_0 is rejected if only one Sample 1 observation exceeds the median, this implies that the joint sample median must not exceed 31 (the second highest value for the males). Since the number above the median in both samples must not exceed 14, this implies that for the adjusted Sample 2 (females) if H_0 is not to be rejected it must have no more than 12 values at or above the adjusted median. If the adjusted median is to be no more than 31 this means the adjustment must ensure there are fewer than 9 values below 31.

Rejection will occur if any number greater than the difference between 31 and the smallest of the 13 largest observations in Sample 2 is added to all values. Counting forwards from the lowest observation shows that this value is 16.6. Therefore, $31 - 16.6=14.4$ is added to all second sample values. This implies that if $\theta_2-\theta_1$ is less than –14.4 H_0 would be rejected. Similarly, since H_0 is rejected if fewer than 2 values in Sample 1 are above the combined sample median, H_0 should be rejected if $\theta_2-\theta_1$ is greater than 4. A nominal 90% confidence interval for the difference $\theta_2-\theta_1$ is therefore the closed interval [–14.4, 4]. Note that the actual confidence level of this interval is 92.32% since the exact $P\approx0.077$.

Output from the ANSM5 package in R can be obtained using `med.test(ch6$males, ch6$females, CI.width = 0.90)`.

As usual, at the ends of this interval there are ties at the combined sample medians for the first sample and the adjusted second sample. Checking that H_0 would be accepted at the end points amounts to performing a test with allowance for ties at the joint median. In this example, the end points are included in the interval.

Assuming normality, the t-based 95% confidence interval for the mean difference is (–11.04, 0.14). This is shorter than that based on median test theory and is necessarily symmetric about the sample mean difference. The WMW-based interval is (–12, 1), in close agreement with the interval based on the t-test.

Given a program for the Fisher exact test for a 2×2 contingency table, one can determine for any given values of m, n the number above the median in the first sample, which just gives significance, and, because there is only one degree of freedom, all other entries in the 2×2 table follow automatically. It is then, as indicated in the above example, relatively simple to determine confidence limits by appropriate additions or subtractions from all second-sample observations.

6.2.4 The Control Median Test

A modification of the median test proposed by Mathisen (1943) is called the control median test. Given two independent samples of size m, n, one is designated the *control sample*. There is no loss of generality if this is labelled Sample I. In applications, this often consists of observations for an existing standard procedure. If the Sample I median is M_1, the test statistic, V, is the number of Sample II observations that are less than M_1. If m is even and there are no ties and if no Sample I value is equal to M_1, then if m is set equal to $2r$ and there are j Sample II values less than M_1, and no Sample II value equals M_1 there will be $n-j$ Sample II values greater than M_1. It follows that in the combined sample there will be $r+j$ values less than M_1 and $r+n-j$ values greater than M_1.

Due to the constraints on the marginal totals that follow from the modification, the probabilities P^* in the median test are multiplied by a factor $(m+1)/(m+n+1)$ (Gart, 1963). Hence, under the null hypothesis that both samples are from populations with the same common median, this adjustment gives

$$\Pr(V=j)=\frac{\binom{r+j}{j}\times\binom{r+n-j}{n-j}}{\binom{m+n+1}{n}} \tag{6.7}$$

Small values of j relative to $n-j$ suggest that the Population II median may be greater than the Population I (or control) median. Large values of j relative to $n-j$ suggest the opposite. In a one-tailed test in the first case, the corresponding P-value may be computed using (6.7) and summing over all j less than or equal to that observed. In the second case, there is an equivalent summation over the observed and larger j.

If m is odd then we can, without loss of generality, act as if the size of the first sample were $m+1$ rather than m, effectively allowing the median value (which will be one of the observed values) to contribute to both r values below the median and r values above the median.

Example 6.9

Apply a control median test to the data below regarding the Sample I as a control group to test the null hypothesis against the alternative that the Sample II is from a population with a greater median.

Sample I	49	58	75	110	112	132	151	276	281	362
Sample II	50	58	96	139	152	159	189	225	239	242
	257	262	292	294	300	301	306	329	342	346
	349	354	359	360	365	378	381	388		

The median M_1 of the control group is first obtained. Then it is necessary to ascertain how many Sample II values exceed M_1. The relevant one-tail P-value is found by calculating the probabilities associated with this or a greater number, j, above the median using (6.7).

The control (first sample) median is $M_1=(112+132)/2=122$. Removing the 3 smallest observations from Sample II shows that there are 25 values in that sample exceeding 122. The probability of this scenario occurring is calculated. Next, the probabilities for the more extreme situations are found, these being when 26, 27 or all 28 Sample II values exceed 122. From the 2×2 contingency tables corresponding to each of these cases, the associated probabilities can be calculated using (6.7). The relevant 2×2 tables are

5	5	5	5	5	5	5	5
25	3	26	2	27	1	28	0

with associated probabilities

$$\Pr(V = 25) = 0.004761, \ \Pr(V = 26) = 0.002129,$$

$$\Pr(V = 27) = 0.000721, \Pr(V = 28) = 0.000142.$$

It follows that the required one-tail probability is $P=0.004761+0.002129+0.000721+0.000142=0.007753$. Hence, there is strong evidence that the median time for the Sample II population is greater than that for the control sample population.

Output from the ANSM5 package in R can be obtained using `control.median(ch6$sampleI, ch6$sampleII, alternative = "greater")`.

Taking the median of the second sample as the control median generally results in *P*-values that are different, but usually not alarmingly so. This may be regarded as a weakness of the test. It does, however, have Pitman efficiency 1 relative to the ordinary median test.

Asymptotic normal approximations to the control median test are given by Gibbons and Chakraborti (2020, Section 6.5), but these also give different results dependent on the choice of the control sample.

The concept is most useful when one sample may be logically chosen as a control because it corresponds to an existing treatment, or perhaps to an untreated sample or a set of patients not suffering from the condition under investigation.

As mentioned earlier, a program for the Fisher exact test can be used to obtain the relevant *P*-value for the median test. This is not so for the control median. This is because in the case of the median test, all marginal totals in the 2×2 tables are fixed, whereas in the control median test, the fixed elements are entries in row 1 and the row totals. It is evident by inspection of the four tables given above that here the column totals are not fixed. One therefore requires a program dedicated to the control median test or one that will compute probabilities equivalent to those in (6.7) for all relevant j.

SUMMARY POINTS:

- The median test is not recommended if assumptions allow the use of generally more powerful tests such as the WMW test (Section 6.2.1).
- It is a test for location based on numbers in each of the two samples above and below the combined sample median (Section 6.2.2).
- The median test is a form of the Fisher exact test (Section 6.2.2).
- Asymptotic approximations for the median test are not required if software for the Fisher exact test is available (Section 6.2.2).
- Confidence intervals can be calculated when full measurement data are available (Section 6.2.3).
- A modified test is the control median test (Section 6.2.4).

6.3 Normal Scores

As in the one-sample case, transforming ranks to give scores with something like the characteristics of samples from a normal distribution is intuitively appealing. In Section 3.4.2, a brief reference was given to van der Waerden and expected normal scores. For the two independent samples scenario corresponding to the WMW test van der Waerden scores, replace the rank r by the $r/(m+n+1)$th quantile of the standard normal distribution ($r=1, 2, \ldots, m+n$).

Expected normal scores may also be used, but there is usually little difference between results using these and van der Waerden scores, so only the latter are considered here. Symmetry implies that the mean of the $m+n$ van der Waerden scores is zero if there are no ties. The mean of all van der Waerden scores is not exactly zero with ties if quantiles based on mid-ranks are used. For a few ties, the effect is negligible, as adjacent quantile scores are replaced by scores close to their mean. For ties, an alternative to the mid-rank quantile (which is the mean of the tied ranks) is to take the mean of the quantiles corresponding to each of the ranks that are tied. In practice, the difference is usually slight, except for large numbers of ties or for ties in certain positions.

If one has an exact permutation test program that covers the Pitman two-sample test (Pitman, 1937a) or allows arbitrary choice of scores, that program may be used to compute exact P-values for a test using van der Waerden scores by substituting these scores for the original data. The ANSM5 package in R provides a `normal.scores.test` function that uses van der Waerden scores, forming these directly from the original data with the convention that for ties the mean of the quantiles corresponding to the ranks that are tied (the second of the options mentioned above) is used.

Example 6.10

Consider the data for computation times given in Example 6.1, viz.,

Group A	23	18	17	25	22	19	31	26	29	33	
Group B	21	28	32	30	41	24	35	34	27	39	36

Use van der Waerden scores to assess whether the data indicate that one calculating tool is superior (i.e., leads to more rapid computations).

The data are ranked and van der Waerden scores are assigned to each. Using these scores in a permutation test program allowing arbitrary scores, an exact P-value may be computed.

It is easiest to use any available software that generates the scores automatically. However, they may also be obtained using tables of the standard normal distribution. For example, for these data $m=10$, $n=11$, so the van der Waerden score for the observation ranked 3 (which here is the observation 19 in Group A) is the 3/22nd quantile of the standard normal distribution, i.e., the value of x such that $\Phi(x)=3/22\approx 0.1364$. From tables of the standard normal distribution, this is found to be $x=-1.097$. Using the ANSM5 package in R and the command `normal.scores.test(ch6$groupA, ch6$groupB)` gives a two-tail $P=0.0117$ for this example, showing strong evidence against the null hypothesis of equal medians.

The P-value is less than that obtained using the WMW procedure and almost identical to that for a t-test. This is not surprising. The data suggest only slight departures from normality and are of a type where one would not have serious reservations about using a t-test. In these circumstances, although one would expect based on Pitman efficiency that the Wilcoxon test may be not quite as powerful as the t-test, it would be reasonable to expect the normal scores test to perform much like the t-test as it moulds the data to a form with the characteristics of a sample from a normal distribution. Indeed, when sampling from normal populations that differ only in mean, the van der Waerden scores test has a Pitman efficiency of 1.0 relative to the t-test.

If software for computing exact P-values with van der Waerden scores is not available, an asymptotic approximation valid for large samples may be based on (6.4) and (6.5), where the scores used are the van der Waerden scores. In practice, this is reasonable for samples of moderate size because of the normalizing influence of the transformation that enhances the rate of convergence towards the asymptotic approximation.

Normal scores are usually not appropriate for the analysis of survival data or for certain other situations where long-tail distributions (often with censored observations as an added complication) are involved. Alternative scoring methods are available. Some of these are discussed in Chapter 9.

SUMMARY POINTS:

- Among tests based on a transformation of ranks, using van der Waerden scores (Section 6.3), is straightforward.
- Results are often similar to those given by the WMW test.

6.4 Tests for Equality of Variance

Historically, tests about differences between medians or means have dominated inference for two-sample problems, but there is increasing interest in differences in variation, or spread, which may or may not be associated with differences in means or medians. In the field of quality control, for example, this reflects consumer demand that all products of a particular kind should be consistently reliable as well as performing well on average.

Care is needed in testing these characteristics when an assumption of normality is dropped. For two independent samples $x_1, x_2, \ldots, x_m$ and $y_1, y_2, \ldots, y_n$ from normal distributions, the appropriate statistic to test the equality of population variances, irrespective of whether the means are equal, is

$$F = \frac{\sum (x_i - \bar{x})^2 / (m-1)}{\sum (y_i - \bar{y})^2 / (n-1)},$$

which has an F distribution with $m-1$, $n-1$ degrees of freedom under the hypothesis that the population variances are equal, irrespective of whether the means are equal.

The normal distribution family is the best-known example of what are called location–scale parameter families. These are families of distributions of random variables X, Y

having location parameters θ_1, θ_2 and scale parameters ϕ_1, ϕ_2 such that $U=(X-\theta_1)/\phi_1$ and $V=(Y-\theta_2)/\phi_2$ are identically distributed.

For the normal distribution in conventional notation $\theta_i=\mu_i$ and $\phi_i=\sigma_i$, $(i=1, 2)$ and U, V each have a standard normal distribution.

Interest may be in spread differences between members of families of distributions that are not of the location–scale parameter form, as seen below. Also, even with location–scale parameter families, many of the tests for differences in dispersion assume either that $\theta_1=\theta_2$, or that the value of the difference $\delta_1=\theta_2-\theta_1$ is known.

The F-statistic relevant to normal distributions is almost unique. It provides a test for differences between the population spreads or dispersions without needing to know the values of the population location parameters.

So far in this chapter the tests considered have been basically designed for shifts in means or medians, but it has been pointed out that several of the tests are also relevant to tests of dominance, where the alternative hypothesis is of the form H_1: $G(x)\le F(x)$ or else H_1: $G(x)\ge F(x)$ with strict inequality for at least some x.

This is important in the context of dispersion because, for many distributions, even when they belong to the same family, it is impossible to have a shift in mean or median without altering other distribution characteristics such as variance. A simple example is the exponential distribution with parameter λ. The probability density function is $f(x)=\lambda e^{-\lambda x}$, $x\ge 0$, and $E(X)=1/\lambda$, $Var(X)=1/\lambda^2$ and $med(X)=\ln(2/\lambda)$. Thus, a change in λ changes the mean, median and variance, reflecting a dominance situation. The parameter λ is often referred to as a scale parameter.

Where both location and spread are influenced by parameter changes, there are difficulties in developing completely distribution-free tests for differences in variance or for more general measures of spread. As already indicated, tests for differences in variance require an assumption that the means or medians are equal or that the difference between them is known. These points are discussed more fully by Hollander, Wolfe and Chicken (2014, Sections 5.1 and 5.2) and by Gibbons and Chakraborti (2020, Chapter 9). Most tests for a difference in spread either implicitly or explicitly assume samples are from populations with identical medians, or that the difference between medians is known, and that an adjustment has been made to align the samples by addition or subtraction of that difference for the data in one of the samples.

If the sample values, or if a test, indicate equality of medians is untenable, it is often suggested that one should adjust sample values to align the sample medians. This has intuitive appeal and is often a sensible thing to do, but examples exist to show that it may sometimes be counterproductive and, in general, it will tend to give rather too many small P-values.

The focus here is mainly on tests that assume population means or medians are identical or can be made so by an addition to or subtraction from one set of sample values, but in practice, such assumptions are often at best approximations.

6.4.1 The Siegel–Tukey Test

This test proposed by Siegel and Tukey (1960) is easy to carry out, but like many nonparametric tests for spread or dispersion, it is not very powerful. The Pitman efficiency relative to the F-test, when samples are from normal distributions, is only 0.61.

The basic idea behind the test is that if two samples come from populations differing only in variance, the sample from the population with greater variance will be more spread out. If there is a known location difference as well, the samples are first aligned by subtracting the median difference from all values in one sample. In practice, this difference is seldom known. If an unknown location difference is assumed, a common practice, as suggested above, is to align the populations by shifting the median (or mean) of one sample to coincide with that of the other sample. This requires an appropriate addition to, or subtraction of, a constant for all observations in one sample. The sample variances are unaltered by this change, but, as indicated above, the procedure is not optimal unless the difference between sample medians happens to equal that between the population location parameters in a location–scale model situation. However, the change usually works reasonably for the Siegel–Tukey test if there is no strong sample evidence of differences in skewness.

If the combined samples are arranged in order and rank 1 is allocated to the smallest observation, rank 2 to the largest, rank 3 to the next largest, ranks 4 and 5 to the next two smallest, ranks 6 and 7 to the next two largest, and so on, the sum of the ranks for the sample from the population having the greater variance should be smaller than if there were no difference in variance.

Example 6.11

Davis and Lawrance (1989) give (as part of a larger data set collected for different purposes) the time in hours for two different types of tire failure under similar test conditions. Failure type A is rubber chunking on the shoulder. Failure type B is cracking of the side wall. Do the following data suggest the population variances may differ?

Type A	177	200	227	230	232	268	272	297				
Type B	47	105	126	142	158	172	197	220	225	230	262	270

The medians of the two samples are aligned, and then the ranks are allocated as described above. The lesser sum of ranks associated with one sample is calculated and tested for significance as in the WMW test. The null hypothesis is that the variances are identical. Without further information, a two-tailed test is appropriate.

The first sample median is $(230+232)/2=231$ and the second sample median is 184.5. Alignment for location is obtained by subtracting $231-184.5=46.5$ from all first sample values and working with the adjusted sample values below for type A, those for type B being unaltered.

Type A	130.5	153.5	180.5	183.5	185.5	221.5	225.5	250.5
Type B	47	105	126	142	158	172		
	197	220	225	230	262	270		

Assigning ranks using the scheme outlined above is easier if all sample values are arranged in ascending order before allocation. The result of this operation is given in Table 6.3, where type A (adjusted) values are indicated in bold. The sum of the ranks for type A faults is $S_m=106$, giving $U_m=S_m-m(m+1)/2=70$ and $U_n=mn-U_m=26$. Tables for the WMW test give $U=22$ as the critical value for significance at a nominal 5% level in a two-tailed test for samples of 8 and 12 observations. The ANSM5 package in R provides a program for an exact Siegel–Tukey test. This indicates that the exact two-tail P-value corresponding to $S_m=106$ is $P=0.0979$. Hence there is only weak evidence against the hypothesis of equal variance, not sufficient to reject H_0 at a nominal 5% significance level. For these data, the WMW test two-tail $P=0.0268$.

TABLE 6.3

Allocation of Ranks for Siegel–Tukey Test

Value	47	105	126	**130.5**	142	**153.5**	158	172	**180.5**	**183.5**
Rank	1	4	5	**8**	9	**12**	13	16	**17**	**20**
Value	**185.5**	197	220	**221.5**	225	**225.5**	230	**250.5**	262	270
Rank	**19**	18	15	**14**	11	**10**	7	**6**	3	2

Type A (Adjusted) in Bold.

Alignment was obtained by equating the sample medians. Since variance is based on squared deviations from the mean, equating means has intuitive appeal. In practice, which alternatives are chosen usually makes little difference. The median is easier to calculate – hardly a justification with modern computing methods – but it may also be more robust if there is some skewness. Lehmann (2006) recommends an alignment based on the Hodges–Lehmann estimator of the median difference.

6.4.2 Ansari–Bradley Type Tests

Several almost equivalent tests based on the dispersion of ranks in each sample about the combined sample mean rank were formulated at about the same time. If there are $N=m+n$ observations and these are ranked 1–N in ascending order, as in the WMW test procedure, the mean rank is $(N+1)/2$. The deviation of rank i from this mean is $d_i=i-(N+1)/2$. The sum of these deviations over all N ranks is zero. If the spread in one sample is greater than that in the other, low and high ranks tend to dominate in that sample. Since the d_i corresponding to low and high ranks have opposite signs, the magnitudes of the d_i in each sample are relevant, and these are reflected by taking either absolute values of the d_i or the squares of the d_i as scores. The sum of these for one sample is an appropriate statistic in a Pitman-type permutation test for a difference in spread under the assumption that the population medians are identical.

Tests that are in essence equivalent and use the sum of the absolute values $|d_i|$ were proposed with minor variations by Freund and Ansari (1957), David and Barton (1958) and Ansari and Bradley (1960). Gibbons and Chakraborti (2020) refers to them collectively as the Freund–Ansari–Bradley–David–Barton tests. While one may use any program for the Pitman test with these scores as data, the ANSM5 package in R has a function that calculates scores given the original data and uses the name Ansari–Bradley test. Mid-ranks are widely used for ties. This does not alter the mean rank but affects some of the scores.

Example 6.12

Apply the Ansari–Bradley test to the data in Example 6.11.

The test assumes the medians of both populations are equal. The data are aligned as in Example 6.11 by subtracting 46.5 from all first sample values. The adjusted sample values are used for the calculations. The combined sample data are ranked and the $|d_i|$ are summed over one sample for use in a Pitman-type permutation test or in a dedicated Ansari–Bradley test program.

The original data after adjustment with the ranks and the scores $|d_i|$ are given in Table 6.4. The type A data are shown in bold. The sum of the scores $|d_i|$ for type A is $S=6.5+4.5+1.5+0.5+0.5+3.5+5.5+7.5=30$.

TABLE 6.4

Ranks and Scores for Ansari–Bradley Test in Example 6.11

Value	47	105	126	**130.5**	142	**153.5**	158	172	**180.5**	**183.5**
Rank	1	2	3	**4**	5	**6**	7	8	**9**	**10**
Score	9.5	8.5	7.5	**6.5**	5.5	**4.5**	3.5	2.5	**1.5**	**0.5**
Value	**185.5**	197	220	**221.5**	225	**225.5**	230	**250.5**	262	270
Rank	**11**	12	13	**14**	15	**16**	17	**18**	19	20
Score	**0.5**	1.5	2.5	**3.5**	4.5	**5.5**	6.5	**7.5**	8.5	9.5

Type A (adjusted) in bold.

Using either a program for a Pitman test with these scores or a dedicated Ansari–Bradley program gives a two-tail $P=0.1437$ (slightly greater than that for the Siegel–Tukey analysis). Hence, there is little evidence of a difference between variances.

Output from the ANSM5 package in R can be obtained using `ansari.bradley(ch6$typeA, ch6$typeB)`.

The effect of lining up the samples on the basis of their medians implies that the test is not exact in the sense that there is a tendency to give too many small *P*-values. Simulation studies have confirmed this tendency, although the exact *P*-values are approached for very large samples.

The alternative to using the $|d_i|$ as scores is to use the squared d_i (Mood, 1954). The Mood test is otherwise analogous to the Ansari–Bradley test. The ANSM5 package in R provides a function for this test, and for the data considered in Examples 6.9 and 6.10, the exact test indicates $P=0.1293$ using the command `mood(ch6$typeA, ch6$typeB)`. Asymptotic approximations valid for large samples are available for the Ansari–Bradley test and for the Mood test.

6.4.3 The Conover Squared-Rank Test for Variance

If the means of X, Y are, respectively, μ_x, μ_y, then equality of variance implies

$\mathrm{E}[(X-\mu_x)^2]=\mathrm{E}[(Y-\mu_y)^2]$, where $\mathrm{E}[X]$ is the expectation of X.

Conover (1980) proposed a test for equality of variance based on the joint squared ranks of squared deviations from the means, i.e., from $(x_i-\mu_x)^2$, $(y_i-\mu_y)^2$. The population means are seldom known, so again it is assumed reasonable to replace them by their sample estimates m_x, m_y. The deviations do not need to be squared to obtain the required rankings because the same order is achieved by ranking the absolute deviations $|x_i - m_x|$, $|y_i - m_y|$. These deviations are ranked, and the squares of the ranks are used as scores. The test statistic T is the sum of the scores for one of the samples. For large samples, Z given by (6.5) has a standard normal distribution. Conover (1999, Table A9) gives quantiles of T for a range of sample sizes in a no-tie situation.

Programs generating the permutation distribution for arbitrary scores may also be used if exact tail probabilities are required and the ANSM5 package in R provides a specific function for the test.

Example 6.13

For the data in Example 6.11, use the squared-rank test for equality of variance.

Deviations of each observation from its sample mean are required. The absolute deviations are then ranked, the ranks squared, and the statistic T is calculated.

TABLE 6.5

Absolute Deviations (Adev) and Squared Ranks (Sqr), Example 6.12

Adev	**6**	7.5	**7.875**	**11**	17	22	**30**	**34**	37.5	**37.875**
Sqr	**1**	4	**9**	**16**	25	36	**49**	**64**	81	**100**
Adev	41	46	51	54	**59**	**61**	75	83	91	133
Sqr	121	144	169	196	**225**	**256**	289	324	361	400

Type A in bold.

Denoting type A sample values by x_i and type B sample values by y_i gives $m_x=237.875$ and $m_y=179.5$. The absolute deviation for each sample value is computed to the nearest integer unless greater accuracy is required to split ties, e.g., for the type A observation 227, since no ties are involved, the deviation used is $227-238=-11$, giving an absolute deviation of 11. Table 6.5 gives the ordered absolute deviations for the combined samples together with squared ranks. Type A deviations are in bold. The sum of the bold squared ranks is $T=720$. From (6.4) with squared ranks as scores $\text{Var}(T)=71{,}979.6$. Also $\text{E}(T)=1{,}148$, whence from (6.5), $Z=(720-1{,}148)/268.29=-1.595$, indicating an asymptotic $P=0.1107$. However, using the ANSM5 package in R and the command `conover(ch6$typeA, ch6$typeB)` gives an exact two-tail $P=0.1267$, showing that the evidence against H_0 is not strong.

Whereas the Siegel–Tukey test with median alignment would establish significance at the formal 5% level in a one-tailed test if that were appropriate, the Conover squared-rank test just fails to do so. While the test appears in this case to be less powerful than the Siegel–Tukey test, it is reasonably robust. The results are broadly in line with those for the Ansari–Bradley test, which in fact uses a scoring system not very different from that for the Siegel–Tukey test.

Rather than using squared ranks, the ranks of absolute deviations could have been analysed, performing what is essentially a WMW test. This leads to a less powerful test than the squared-rank test (Conover, 1999, Section 5.3).

The Conover scores may be used in any permutation test program for a Pitman-type test if a dedicated program is not available.

6.4.4 The Moses Test for Extreme Reactions

Tests so far discussed have considered spread or dispersion as a *variance* type of characteristic. Consider two samples, A and B, such that Sample A has a smaller range of values than Sample B. One might then wish to investigate whether extreme values (e.g., for reaction or response times) are more likely to occur in the population from which Sample B was drawn.

Moses (1952) proposed a test of extreme reactions. The medians of two associated populations are assumed to be equal. The null hypothesis is that extreme values are equally likely to occur in both populations. The alternative is that extreme values are more likely in the population from which the sample with the larger range was drawn. The two samples are combined, ranked overall, and the observations are put into rank order. Each observation is labelled either A or B according to the sample from which it originally came.

If both samples come from the same population, then each is likely to contain some high, some low and some intermediate values. For instance, if there are seven observations in Sample A and eight in Sample B, this might be indicated by the ordering:

B B A A B A B A A B A B B B A

Under the alternative hypothesis, values for Sample A will be concentrated in one part of the series, for example:

B B B A A A A A B A A B B B B

Once the values have been ordered, the ranks of the lowest and highest scores from Sample A are noted. The number of cases contained within these values, including the extremes, is defined as the span of the A scores. In the first example above, the span is relatively large (13) whereas in the second case, it is quite small (8). The smaller the value of the span, the stronger is the evidence against the null hypothesis. The Moses test determines the probability of observing a span of no more than that recorded, assuming that the null hypothesis is true.

If samples A and B have m and n observations, respectively, then the span of A must be between m and $m+n$ inclusive. Suppose that the observed span is $m+k$. Under the null hypothesis, the probability that the span S is no more than this is given by

$$\Pr(S \le m+k) = \frac{\sum_{i=0}^{k} \binom{i+m-2}{i} \binom{n+1-i}{n-i}}{\binom{m+n}{m}} \tag{6.8}$$

Probabilities for the upper tail of the scan length distribution can be calculated in a similar manner. Exact P-values for the one-tailed Moses test can be found using the ANSM5 package in R, although tail probabilities are tedious to compute using tabulated values of binomial coefficients.

Example 6.14

Nineteen women who had received results from cervical cancer screening completed a questionnaire regarding their attitude towards receiving detailed information about cervical cancer, with a high score indicating a great degree of interest.

Those women who had received a negative screening result (Group I) were compared with those with a positive result (Group II). The questionnaire scores (out of 25) were as follows:

Group I:	8	9	10	12	13	15	18	20	21	
Group II:	1	3	5	6	11	14	19	22	23	25

Investigate the null hypothesis that extreme values are equally likely to occur in the positive test and negative test populations of women.

The two groups are combined and the observations are ranked. The ranks are written in ascending order. The span of Group I is then obtained by recording the number of cases, including the extremes, contained within the lowest and highest ranks relating to this group. A one-tailed test is appropriate, as there are grounds for believing that the scores for patients with a positive test result might be more variable (Miller, 1995).

The group sizes are m=9, n=10. The ranks for the combined observations are obtained. These are shown in Table 6.6. Ranks relating to Group I are given in bold. The lowest rank for a Group I observation is 5, and the highest is 16, giving a span of 12. The minimum possible value for the span is m=9, thus k=3. Using (6.8) gives P=0.0149. Using the ANSM5 package in R and the command `moses.extreme.reactions(ch6$groupI, ch6$groupII)` confirms this. There is therefore considerable evidence against the null hypothesis. One may reasonably conclude that extreme questionnaire scores occur more frequently for patients with a positive screening result.

TABLE 6.6

Allocation of Ranks for Moses Test, Example 6.14

Value	1	3	5	6	**8**	9	**10**	11	**12**	**13**
Rank	1	4	5	8	**9**	**12**	**13**	16	**17**	**20**
Value	14	**15**	**18**	19	**20**	**21**	22	23	25	
Rank	11	**12**	**13**	14	**15**	**16**	17	18	19	

Group I in bold.

The span of the Group I observations is often referred to as the range, and it is a measure highly susceptible to outliers. One way of taking this into account is to decide before the investigation to remove (or trim) one or more Group I observations from the upper and lower extremes. The span is then based on the values that remain. For the above data, R gives a *P*-value of 0.036 when one observation is trimmed from each end. There is little reason to use a trimmed span in this example, as there are no obvious outliers. Trimming when there are outliers in other contexts is discussed briefly in Section 14.4.

If ties occur between two or more members of the same group, the tied values are adjacent in the ordering, and the span is not affected. In addition, if ties occur between members of the two different groups, the span will generally not be affected. In the absence of trimming, the only exception is where a tie involves either the highest or the lowest observation in a group; the value for the span is then unclear. In R, the span is calculated by assuming that the value which is highest/lowest is more extreme than its tied counterpart. This can overestimate the value of the span and hence the *P*-value. In the above example, if the highest value in Group I is changed to 22, the resulting tie leads to a *P*-value of 0.04.

The findings in this example are in line with the belief of many psychologists that individuals who receive a positive diagnosis for a serious illness tend to be either *blunters* whose reaction to the bad news is denial or *monitors* who wish to be informed about every detail regarding their illness (Miller, 1995). Individuals who do not have to face up to such bad news are assumed to have an intermediate degree of interest in that illness. Thus, the blunters would have low scores on the cervical cancer questionnaire, with the monitors having relatively high scores. This could explain why Group B has more extreme values than Group A.

Since $n = m + 1$, there is a considerable simplification in the numerator of (6.8) for this example.

6.4.5 Some Related Procedures

Klotz (1962) proposed using van der Waerden scores in a test analogous to the Mood test, while Capon (1961) suggested a similar test using expected normal scores. Sukhatme (1957) proposed a test having analogies with the Mann–Whitney formulation of the WMW test, but the scoring is more complicated and there are practical limitations to its use. It is described by Gibbons and Chakraborti (2020, Section 9.7), who point out that it can be adapted to construct confidence intervals for a scale parameter.

Gastwirth (1965) proposed statistics for detecting location and scale differences based on simple scores assigned to top and bottom nonoverlapping fractions of the combined sample ordered data, with zero scores given to any data with intermediate ranks. The motivation for his approach was that tests for location based on the transformation of ranks, such as those using van der Waerden scores, that have high Pitman efficiency relative to optimal normal theory tests when the latter are appropriate give greater weights to scores associated with extreme ranks. Gastwirth was especially interested in tests that were more efficient for dispersion than the tests discussed here. A general account of his approach is given in Sprent (1998, Section 6.6).

SUMMARY POINTS:

- For spread or dispersion, the Siegel–Tukey test (Section 6.4.1) or Ansari–Bradley type tests (Section 6.4.2) often lack power and are not robust.
- The Conover squared-rank test (Section 6.4.3) is more robust than the Siegel–Tukey and Ansari–Bradley tests.
- The Moses test for extreme reactions (Section 6.4.4) is relatively simple, but being based on a range of ranks, it is susceptible to outliers.
- Many dispersion tests require for strict validity that for both populations the medians or means are the same or that the difference, if any, between them is known.

6.5 Tests for a Common Distribution

Section 4.2 described tests of the hypothesis that a single sample was drawn from some specified distribution. The analogous issue for two independent samples is whether it is reasonable to suppose each comes from the same unspecified distribution. The *Smirnov* test (Smirnov, 1939, 1948) has similarities to the Kolmogorov test developed in Section 4.2.1.

6.5.1 Smirnov Test for a Common Distribution

The null hypothesis H_0: *two samples come from the same distribution* may be tested against the alternative H_1: *the distributions have different cumulative distribution functions (cdfs).* The nature of any differences is not specified further. The distributions may have the same mean but different variances; one may be skew, the other symmetric, etc. The sample cdfs are compared; the test statistic is the difference of greatest magnitude between the two functions.

Example 6.15

The following data are rates of saliva production (ml/min) for 7 female and 21 male adults, arranged for convenience in ascending order. Is it reasonable to conclude that the samples are from populations with the same distribution?

Females	0.45	0.60	0.80	0.85	0.95	1.00	1.75				
Males	0.40	0.50	0.55	0.65	0.70	0.75	0.90	1.05	1.15	1.25	1.30
	1.35	1.45	1.50	1.85	1.90	2.30	2.55	2.70	2.85	3.85	

TABLE 6.7

Calculation of the Smirnov Test Statistic

Female (x_i)	Male (y_i)	$S_1(x)$	$S_2(x)$	$S_1(x)-S_2(y)$
	0.40	0.000	0.048	–0.048
0.45		0.143	0.048	0.095
	0.50	0.143	0.095	0.048
	0.55	0.143	0.143	0.000
0.60		0.286	0.143	0.143
	0.65	0.286	0.190	0.096
	0.70	0.286	0.238	0.048
	0.75	0.286	0.286	0.000
0.80		0.429	0.286	0.143
0.85		0.571	0.286	0.285
	0.90	0.571	0.333	0.238
0.95		0.714	0.333	0.381
1.00		0.857	0.333	0.524
	1.05	0.857	0.381	0.476
	1.15	0.857	0.429	0.428
	1.25	0.857	0.476	0.381
	1.30	0.857	0.523	0.334
	1.35	0.857	0.571	0.286
	1.45	0.857	0.619	0.238
	1.50	0.857	0.667	0.190
1.75		1.000	0.667	0.333
	1.85	1.000	0.714	0.286
	1.90	1.000	0.762	0.238
	2.30	1.000	0.809	0.191
	2.55	1.000	0.857	0.143
	2.70	1.000	0.904	0.096
	2.85	1.000	0.952	0.048
	3.85	1.000	1.000	0.000

The sample cumulative distribution functions $S_1(x)$, $S_2(y)$ are computed at each sample value. The difference $S_1(x)-S_2(y)$ is also computed and recorded. For samples of size m, n, respectively, $S_1(x)$, $S_2(y)$ are step functions with steps $1/m$, $1/n$, respectively, at each sample value (with multiple steps if there are ties).

Table 6.7 gives in successive columns the sample values and corresponding values of $S_1(x)$, $S_2(y)$ and $S_1(x)-S_2(y)$ at each sample point. The difference of greatest magnitude (final column) is 0.524. The difference is positive, suggesting that at the saliva production rate at which it occurs, namely 1 ml/min, the cdf for females is greater than that for males. The preponderance of positive values for the difference $S_1(x)-S_2(y)$ suggests that the corresponding population cdf for females increases more rapidly than that for males. This reflects the greater spread of the male data, suggesting both a shift in location and an increasing spread in the form of an elongated right tail. If there were prior reasons to believe males salivate more rapidly, a one-tailed test might be justified. However, a two-tailed test is appropriate for unspecified general alternatives. An exact test is preferable if a suitable program is available. One is provided in the ANSM5 package in R, and in this example using the commands `kstest.ANSM(ch6$salivaF, ch6$salivaM, alternative = "greater")` and `kstest.ANSM(ch6$salivaF, ch6$salivaM)` gives an exact one-tail $P=0.0465$ and an exact two-tail $P=0.0930$. There is a slight indication that the cdf for females may be consistently above that for males for at least many identical values of x, y.

Sometimes a test for location or variance establishes a difference when a Smirnov test indicates no overall distributional difference; Kolmogorov–Smirnov type tests are often less powerful than tests for specific characteristic differences such as those for location.

Like the Kolmogorov test, the Smirnov test may appear not to be making full use of the data because it uses only the maximum difference. As pointed out in Section 4.3.1, the statistic uses cumulative information. Section 6.5.2 explains that power may be increased by considering all differences.

Tables of nominal significance levels may be used if no program for an exact test is available. Asymptotic approximations are available for large samples. Many published tables for the Smirnov test give not critical values but quantiles that must be exceeded for significance at a given level. Neave (1981, p. 31) gives values for the equivalent $mn[\max|S_1(x)-S_2(y)|]$ for significance for a wide range of sample sizes.

Several standard packages give programs for the Smirnov test (often referred to as the Kolmogorov–Smirnov test), but care should be taken to see whether each test uses only an asymptotic approximation or a more appropriate approach for small samples.

Chandra, Singpurwalla and Stephens (1981) developed a test of this type for the Weibull distribution, often met with survival data. Modifications of the Smirnov test for discrete data are discussed by Eplett (1982), while Saunders and Laud (1980) devised a multivariate Kolmogorov goodness-of-fit test.

6.5.2 The Cramér–von Mises Test for Identical Populations

The differences $S_1(x)-S_2(y)$ at each sample value are not independent. This makes it difficult to work out the distribution for statistics that take account of all differences. However, for one such test, the *Cramér–von Mises test* (Cramér, 1928; von Mises, 1931), an approximate theory gives simple significance tests that, except for very small samples, are virtually independent of sample size, and these are useful if one only wants broad measures of evidence for or against identity rather than more precise *P*-values.

The test statistic is a function of the sum of squares of the differences $S_1(x)-S_2(y)$ at all sample points. Denoting this sum of squares by S^2_d the test statistic is $T=mnS^2_d/(m+n)^2$. In a two-tailed test for significance at the 5% level, T must exceed 0.461; for significance at the 1% level, T must exceed 0.743.

Example 6.16

Perform the Cramér–von Mises test for the data in Example 6.15.

The differences in the last column of Table 6.7 are squared and added to form the statistic T.

From the last column of Table 6.7, the sum of squares of the differences (using figures before rounding) is 1.7823, whence $T=7\times 21\times 1.7823/(28\times 28)=0.334$. Since $T<0.461$, it is not possible to conclude that the population cumulative distribution functions differ for at least some x.

Output from the ANSM5 package in R can be obtained using `cramer.von.mises(ch6$salivaF, ch6$salivaM)`.

The Cramér–von Mises test is often more powerful than the Smirnov test and is easy to use because of the approximation. Accurate tables exist for some values of m, n. The only labour additional to that for the Smirnov test is the calculation of the sums of squares of differences. In this example, the result is not inconsistent with that of the two-tailed Smirnov test.

6.5.3 The Wald–Wolfowitz Runs Test

Wald and Wolfowitz (1940) proposed a test based on runs for a population difference between two independent samples. Observations in the first sample are coded as *zero*, and those in the second sample are coded as *one*. They are then combined, ranked and placed in order of size. Each observation is replaced by its code to form runs of zeros and ones. The null hypothesis H_0 is that the zeros and ones are randomly ordered. These runs are analysed for clustering or alternation by the method described in Section 4.6.

Example 6.17

Consider the computation time data given in Example 6.1, viz.,

Group A	23	18	17	25	22	19	31	26	29	33	
Group B	21	28	32	30	41	24	35	34	27	39	36

Use the Wald–Wolfowitz runs test to assess whether the distributions of computation times differ between Group A and Group B.

There are $N=21$ observations in all, with $m=10$ in Group A and $n=11$ in Group B.

The observations in Group A are coded as 0, and Group B observations as 1. The data are combined and ranked overall, with each observation labelled by its code as shown in Table 6.8. Inspection of this table indicates $r=12$ runs leading to an exact one-sided $P=0.680$, indicating no evidence of clustering.

Output from the ANSM5 package in R can be obtained using the following series of commands.

```
code <- c(rep(0, length(ch6$groupA)), rep(1, length(ch6$groupB)))
ranked_code <- code[order(c(ch6$groupA, ch6$groupB))]
runs.2cat(ranked_code, alternative = "less")
```

TABLE 6.8

Ranking and Coding of the Computation Times for the Combined Groups

Time	17	18	19	21	22	23	24	25	26	27	28
Rank	1	2	3	4	5	6	7	8	9	10	11
Code	0	0	0	1	0	0	1	0	0	1	1
Time	29	30	31	32	33	34	35	36	39	41	
Rank	12	13	14	15	16	17	18	19	20	21	
Code	0	1	0	1	0	1	1	1	1	1	

The Wald–Wolfowitz test has zero asymptotic efficiency for testing either location or dispersion relative to normal theory alternatives (Mood, 1954). This finding is therefore unsurprising despite an exact two-sided $P=0.0159$ for the Wilcoxon test. However, as the Wilcoxon test is specifically designed to look at the possibility of differences in location, whereas the Wald–Wolfowitz test is designed to pick up unspecified differences, it would be fairer to compare performance with that of the Smirnov test or the Cramér–von Mises test.

Difficulties arise in assigning runs if there are tied values in different samples. Gibbons and Chakraborti (2020) recommends a conservative approach of breaking all ties in all possible ways, performing the test on each possible resolution of ties, and taking the result that is least likely to reject the null hypothesis.

6.5.4 The Two-Sample Runs Test on a Circle

Suppose that two independent samples of angular data, A and B, have been obtained. In this test, the points are plotted on a circle and labelled A and B according to the sample from which they came. The number of runs around the perimeter of the circle is noted. In proceeding around the circle once, the final run and initial run must be from different groups. The number of runs has therefore to be even (if the extreme points are from the same group when arranged on a line, the number of runs is reduced by one when the points are arranged on a circle).

Where there is a tendency for points from the same group to cluster, the number of runs is relatively small; for alternation, the number of runs is relatively large (see Section 4.6).

If R and R_C represent the number of runs when the data are arranged on a line and on a circle, respectively, then using the notation of Section 4.6, the lower tail probabilities for the number of runs on the circle, $2s$, are given by

$$\Pr(R_c \le 2s) = \Pr(R \le 2s+1)$$

with upper tail probabilities calculated in a similar manner.

Example 6.18

The midwife in Example 4.16 also recorded the gender of each of the 12 home deliveries. She was interested in whether or not the times of births for boys and girls occurred in clusters. The data were as follows (M=male, F=female)

Time	0100	0300	0420	0500	0540	0620
Sex	M	F	F	F	M	M
Time	0640	0700	0940	1100	1200	1720
Sex	F	F	F	F	M	M

The null hypothesis H_0 is that for the populations of male and female births, the times of delivery are randomly ordered with respect to gender. Since the midwife is interested only in clustering and not in alternation, a one-tailed test is appropriate. A small value for the number of runs, $2s$, indicates possible clustering.

The data are plotted on a circle. The numbers of boys and girls are $m=5$ and $n=7$, respectively. The count of runs starts from the beginning of a particular run as viewed on the circle. Runs are therefore counted from the female birth that occurred at 0300 rather than the male birth at 0100. The number of runs, $2s$, is four. As reasoned above, for these two groups, five runs could not occur with a circle. The relevant *P*-value is therefore $\Pr(R \leq 5)$, where R is the number of runs for such data arranged along a line. This *P*-value is 0.197, implying that there is no strong evidence against the null hypothesis. On the basis of these data, the times of delivery for boys and girls do not seem to cluster with respect to gender.

Output from the ANSM5 package in R can be obtained using `runs.2cat(ch6$sex, alternative = "less")`.

Recall that if the number of runs for the data on a line is odd, the number of runs is one fewer when the data are plotted on a circle. This implies that low *P*-values are only obtained from small studies in extreme situations. For instance, unless both m and n are at least eight, there will only be strong evidence against H_0 (formally $P<0.05$) if the number of runs is two.

SUMMARY POINTS:

- For comparison of continuous distributions, the Smirnov test (Section 6.5.1) has analogies with the one-sample Kolmogorov test (Section 4.3.1).
- The Cramér–von Mises test (Section 6.5.2) is easy to use as conventional significance levels are almost independent of sample size except for very small samples.
- The Wald–Wolfowitz runs test operates by replacing each observation by a label that indicates its group (Section 6.5.3).
- A two-sample runs test can be applied to angular data. The number of runs on a circle is always even (Section 6.5.4).

6.6 Comparing the Degree of Clustering in Two Independent Groups

It may be of interest to compare the level of clustering of categories in two sequences. For instance, suppose that two teachers are to be appraised on their verbal interaction with students during an observed lesson using classroom interaction analysis (Flanders, 1970). A recording is made of the class and the type of interaction occurring during the lesson at equally spaced points in time. The most basic model has two categories: "teacher addresses class" (T) and "student addresses class" (S). High verbal interaction, which is favoured by many educationalists, occurs when the teacher and students are confident in having a two-way discussion with each other during the lesson. With high interaction, one might expect short interchanges that could be represented as a random pattern, such as:

T T S T S S S T S T T S T S T T T S S T

If a teacher's style is directive (as in more traditional methods of instruction), clusters of teacher input will be observed, for example:

T T T T T S T T T T T T T T S S T T T T T

In comparing the styles of the two teachers, the difference between the number of runs in the observed lessons is important. Assume that the two groups are independent and that under the null hypothesis, the categories occur in a random manner. Denoting the run distributions as R_1 and R_2, from Chapter 4, each distribution is asymptotically normal. The distribution of $R_1 - R_2$ is therefore asymptotically normal with mean and variance obtained from the relationships $E(R_1 - R_2)=E(R_1) - E(R_2)$ and $Var(R_1 - R_2)=Var(R_1)+Var(R_2)$, respectively.

In practice, the assumption that the sequence for each group is random is highly restrictive, and dependency on previous observations is generally taken into account in the null hypothesis. Clustering level comparison then amounts to testing the equality of transition probability matrices for two stochastic processes. Darwin (1959) developed an asymptotic method for Markov chains with multiple categories based on a chi-squared distribution and applied it to teacher-student interaction data obtained from two classrooms. Teacher training is the main context in which further developments of this approach have been made.

SUMMARY POINTS:

- If two sequences are to be compared on their level of clustering, the assumption under the null hypothesis that categories in the two sequences occur in a random manner is highly restrictive.
- An approach that takes clustering into account under the null hypothesis is to test for the equality of the transition matrices of the sequences, treating the observations as coming from a Markov chain.

6.7 Power and Sample Size

In the light of the discussion in Section 5.3 for single samples, it is unsurprising that power and sample size results for the median test and the WMW test are not distribution-free. As one might expect, the situation for the WMW test is more complicated than that for the median test because the latter is an extension from the sign test and therefore essentially only involves binomial distributions, whereas the WMW test involves the distribution of ranks, which may be complicated under H_1.

6.7.1 Power and Sample Size for the Median Test

The hypotheses only involve possible differences between unspecified population medians θ_1 and θ_2. If this difference is denoted by $\delta = \theta_2 - \theta_1$, the null hypothesis is H_0: $\delta = 0$ and the median test is equivalent to testing whether the samples support a hypothesis that the probability that a sample value is above the common median is $p = 0.5$ for each sample. If the combined sample median is used as the estimate of the common median, then under H_0 the probability is only approximately $p = 0.5$ for reasons indicated in Section 6.2. Under an alternative hypothesis H_1 specifying some value $\delta \neq 0$ the probabilities p_1, p_2 for sample values to lie above the combined sample median will be unequal, but the estimation of p_1, p_2 and the interpretation of the relationship between the combined sample median and the two population medians depends on both the nature of those populations and the relative sample sizes.

The combined sample median is only easy to interpret if samples are from populations that differ, if at all, only by a median shift. The interpretation is simplest when $m = n$. Then, if H_1 specifies a fixed difference δ between the medians, i.e., H_1: $\theta_2 - \theta_1 = \delta$, the combined sample median is an intuitively reasonable estimator of a constant midway between θ_2 and θ_1, i.e., an estimator of $\theta_1 + 0.5\delta = \theta_2 - 0.5\delta$. Figure 6.1 shows probability density functions for two arbitrary distributions that differ only in their medians θ_1, θ_2. The dotted vertical line from point P to point M meets the x-axis at the point M where $\theta_1 + 0.5\delta = \theta_2 - 0.5\delta$.

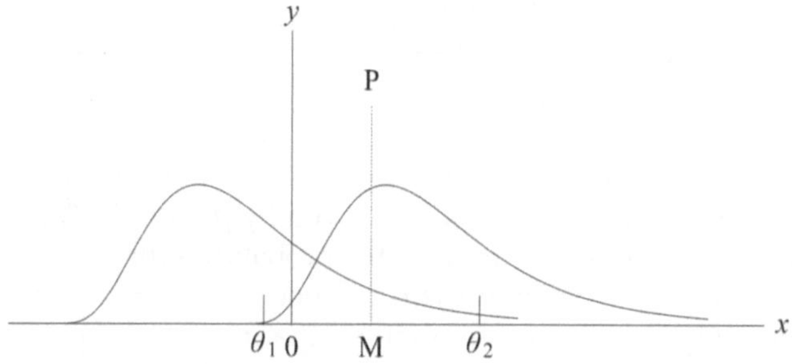

FIGURE 6.1
Probability density functions for two distributions that differ only by a shift in median or mean. The point M lies midway between the population medians θ_1, θ_2 and is the median of a population consisting of an equal mixture of the two populations. The combined sample median provides an estimate of this point if both samples are of equal size $m=n$.

If it is assumed that under H_0 the common value of the median is θ_1 and that the common probability density function $f(x)$ is completely defined, the approximate power of the median test may be calculated under these restrictive assumptions by computing a probability p_1 that under H_1 a sample value from the specified distribution with median θ_1 takes a value greater than $\theta_1+0.5\delta$. Likewise, p_2 is computed for a similar distribution with the median shifted to $\theta_1+\delta$. Programs are now fairly readily available for power calculations in this situation when there are assigned median shifts for completely specified distributions. The method is illustrated in Examples 6.19 and 6.20. However, this approach is unrealistic in many practical situations where the median test is likely to be used, for these are conditions where a WMW test may in general be more powerful.

In more complicated situations where a median shift is the hypothesized difference, an approach based on order statistics may be used. However, here again, some knowledge of the precise form of the distribution is required. The method is outlined by Gibbons and Chakraborti (2020, Section 6.4), who also give an asymptotic approximation available for unequal sample sizes.

An alternative approach to obtaining the sample size required to achieve a given power that is probably a reasonable approximation if $m=n$ is to use any available program to work out the sample size for a corresponding t-test with the required power and to adjust on the basis of the Pitman efficiency relative to the t-test where this is known. This is discussed briefly in the comments on Examples 6.19 and 6.20. This approach is only appropriate for m, n reasonably large in the light of the findings of Freidlin and Gastwirth (2000) referred to in Section 6.2.1.

Example 6.19

For samples of size 20 from each of two normal distributions with standard deviation 2 and unknown medians θ_1, θ_2, find the power of the median test of the null hypothesis H_0: $\theta_2=\theta_1$ against H_1: $\theta_2=\theta_1+1$ when the probability of a type I error is $\alpha=0.05$. What equal sample sizes would be needed to ensure a power of at least 0.80?

Under H_0 the rows in the 2×2 table for the Fisher exact test in Section 6.2 may be considered as the number of positive and negative outcomes in independent samples each of size 20 when sampling from binomial distributions for which the p are equal with approximately the value $p=0.5$. Under H_1 the probability that any value in the sample from the first population has a value greater than $\theta_1+0.5$ is given by $p_1=\Pr(X>\theta_1+0.5)$, where X has a $N(\theta_1, 4)$ distribution. This is approximately the probability that a sample value from this distribution takes a value greater than the combined sample median.

Similarly, under H_1 the probability that any value in the sample from the second population has a value greater than $\theta_1+0.5$ is given by $p_2=\Pr(X>\theta_1+0.5)$, where X has a $N(\theta_1+1, 4)$ distribution. This is approximately the probability that a sample value from this distribution takes a value greater than the combined sample median. As a consequence of the symmetry assumption, it is easily verified that $p_2=1-p_1$. Approximate power and sample size determinations now reduce to considering those for the problem of testing H_0: $p_1=p_2=0.5$ against H_1: p_1, p_2 *take the values computed* when H_1: $\theta_2=\theta_1+1$ holds. Computer software exists for the latter test, giving either exact or asymptotic approximations.

For the first population, $Z=(X-\theta_1)/2$ has a standard normal distribution, since X is $N(\theta_1, 4)$. Thus

$$p_1 = \Pr(X > \theta_1 + 0.5) = \Pr\left(Z > 0.25\right) = 0.4013.$$

Since $p_2=1-p_1$ this implies $p_2=0.5987$, which may be verified directly. Using the `power.fisher.test` function in the statmod package in R for power calculations for the Fisher exact test for comparing two binomials provides a Monte Carlo estimate of power. For $n=m=20$, for the test considered here, the program gives an exact power of around 0.22 using the command `power.fisher.test(0.4013, 0.5987, 20, 20, alpha=0.05, nsim=1000, alternative = "less")`.

Larger samples are required to achieve a power of 0.80. If software is not available to estimate the sample size, a trial and error approach is needed. For $m=n=100$, the `power.fisher.test` function gives the power at around 0.85. For $n=m=90$, the required power of 0.80 is achieved.

Since the samples are from normal distributions differing only in mean, the optimal test is a *t*-test and for the situation in this example, the relevant computations (available in many statistical packages) indicate $m=n=51$ are the sample sizes needed to attain a power of 0.8. This is broadly in line with the expected size based on the Pitman efficiency of 0.64 for the median test when the *t*-test is optimal. Making the assumption that the Pitman efficiency is reasonable for the test under consideration would suggest the sample size $51/0.64=80$ approximately. However, it should be kept in mind that this is only a limiting result concerning the power for small departures independent of the choice of α, β and that appreciable variations from this estimate are possible for finite samples, specific alternatives, and particular choices of α, β. Nevertheless, using a sample size based on the relevant Pitman efficiency will usually give an indication of the size needed to give reasonable power in a practical context, providing samples are not small or unbalanced.

The restriction to $m=n$ is not essential to the binomial comparison problem, but it is needed to induce the property that the sample median is a reasonable estimate of $\theta_1+0.5\delta$. This in turn depends on the assumption that samples come from identical distributions apart from possible differences between medians. Relaxation of these assumptions changes the function of θ_1 and θ_2 that is estimated by M.

In the light of common uncertainty about the exact nature of the populations being sampled, asymptotic power calculations for comparing two binomials available in many software packages will often suffice for obtaining sample sizes needed to give substantial power since these sample sizes are often fairly large, and an approximate sample size is often all that is required.

Example 6.20

For samples of size 20 from each of two double exponential distributions with standard deviation 2 and unknown medians θ_1, θ_2, find the power of the median test of H_0: $\theta_2 = \theta_1$ against H_1: $\theta_2 = \theta_1 + 1$ when the probability of a type I error is $\alpha = 0.05$. What equal sample sizes would be needed to ensure a power of at least 0.80?

As in Example 5.6, a double exponential distribution with mean θ and standard deviation 2 has a probability density function

$$f(x) = \frac{1}{2\sqrt{2}} \exp[-|x-\theta| / \sqrt{2}] \tag{6.9}$$

Using arguments similar to those in Example 6.19 under H_0 the rows in the 2×2 table for the Fisher exact test in Section 6.2 may be considered as the number of positive and negative outcomes in independent samples each of size 20 when sampling from binomial distributions for which the p are equal with approximately the value $p = 0.5$. Under H_1 the probability that any value in the sample from the first population has a value greater than $\theta_1 + 0.5$ is given by $p_1 = \Pr(X > \theta_1 + 0.5)$ where X has a double exponential distribution given by (6.9) with mean θ_1 and this is approximately the probability that a sample value from this distribution takes a value greater than the combined sample median. Similarly, under H_1 the probability that any value in the sample from the second population has a value greater than $\theta_1 + 0.5$ is given by $p_2 = \Pr(X > \theta_1 + 0.5)$ where X has a double exponential distribution with mean $\theta_1 + 1$. This is approximately the probability that a sample value from this distribution takes a value greater than the combined sample median. Because of symmetry, it can be verified that, as in the previous example, $p_2 = 1 - p_1$. Approximate power and sample size determinations now reduce to considering those for the problem of testing H_0: $p_1 = p_2 = 0.5$ against H_1: p_1, p_2 *take the values computed when* $\theta_2 = \theta_1 + 1$.

For the first population sampled p_1 under H_1 is computed by integrating (6.9) over the interval $(\theta + 0.5, \infty)$ where $\theta = \theta_1$ (see Exercise 6.18). This gives $p_1 = 0.3511$. Since $p_2 = 1 - p_1$ this implies $p_2 = 0.6489$. This may be verified directly by integrating (6.9) over the interval $(\theta + 0.5, \infty)$, where now $\theta = \theta_1 + 1$. Using the statmod package in R and the command `power.fisher.test(0.3511, 0.6489, 20, 20, alpha = 0.05, nsim = 1000, alternative = "less")` gives a Monte Carlo power of around 0.45. Larger samples will be needed to achieve a power of 0.80. If no software is available to estimate the sample size directly a trial and error approach is needed. For $m = n = 40$, the `power.fisher.test` function gives a power of around 0.79. For $n = m = 41$ the required power of 0.80 is achieved.

The t-test is no longer optimal so a direct comparison is inappropriate. However, in view of its nonoptimality, the t-test would require a larger sample size to achieve the same power as it would have for the corresponding normal test, where, as indicated in the remarks on Example 6.19 sample sizes $m = n = 51$ are required to attain a power 0.8, indicating, in line with the Pitman efficiency of 2.0 relative to the t-test, that the median test is preferable here.

The median test has higher Pitman efficiency than the WMW test for double exponential distributions, that differ if at all, only in medians. However, several authors have indicated that for small or unbalanced samples (i.e., samples of very different sizes) the median test may not have as high a power as the WMW test for some alternatives specified in H_1.

6.7.2 Power and Sample Size for the WMW Test

Added complexities of power and sample size computations for the Wilcoxon signed-rank test relative to the sign test carry over to the WMW test relative to the median test. Only an approximation due to Noether (1987) is considered here. This method is similar to that given for the signed-rank test in (5.3) for determining the equal sample sizes to ensure a given power when testing for a median shift only in otherwise identical distributions (which need not be symmetric).

The analogous formula to (5.3) is

$$n \approx \frac{(z_\alpha + z_\beta)^2}{6(p_1 - 0.5)^2} \tag{6.10}$$

where $p_1=\Pr[(Y-X)>0]$. The samples are from populations of random variables X, Y that differ only in medians, and $n=m$ is the common sample size. The other terms have the same meaning as in (5.3). The distribution of $Y-X$ depends on the nature of the distributions of X and Y and is in general obtained by standard distribution theory for change of variables. In simple situations like that, when X, Y are both normally distributed, p_1 is easily computed as seen in the following example.

Example 6.21

Samples of equal size n are to be taken from two normal distributions each with a standard deviation of 2 and unknown medians θ_1, θ_2. Use (6.10) to estimate n so that the power of the WMW test of H_0: $\theta_2=\theta_1$ against H_1: $\theta_2=\theta_1+1$ is at least 0.80 when the probability of a type I error is $\alpha=0.05$.

Relevant values of z_α, z_β are obtained from tables or using appropriate software. Normal distribution theory indicates that $Y-X$ is in this case distributed N(1, 8) whence p_1 can be obtained. Relevant values are substituted in (6.10).

It can be shown (Exercise 6.19) that $z_\alpha=1.645$, $z_\beta=0.842$ and $p_1=0.6381$, whence substitution in (6.10) gives $n\approx 54.05$. Rounding up the result suggests samples of size 55 are needed to attain the required power.

These results for the normal distribution are consistent with the known Pitman efficiency of $3/\pi\approx 0.95$ relative to the t-test in this case because, as noted in Example 6.20, the t-test sample sizes needed for this power are $n=51$, giving a relative efficiency for the WMW test of $51/55\approx 0.93$, in close agreement with the Pitman efficiency $3/\pi$.

Hollander, Wolfe and Chicken (2014, Section 4.1) generalizes (6.10) to unequal sample sizes and discusses some further results for asymptotic approximate power calculations.

SUMMARY POINTS:

- Studies of power and sample size are reasonably straightforward only for the median test using samples of the same size from distributions that differ only in mean or median.
- This situation often occurs where the median test is not the most appropriate (Section 6.7.1).
- Reasonable asymptotic approximations are available for the WMW test (Section 6.7.2).

6.8 Exercises

6.1 An alloy is composed of zinc, copper and tin. It may be made at one of two temperatures H (higher) or L (lower). To determine whether hardness is affected by temperature, a sample is taken from each of 9 batches at L and 7 at H. To arrange them in ascending order of hardness, all specimens are scraped against one another to see which makes a deeper scratch (a deeper scratch indicates a softer specimen). On this basis, the specimens are ranked 1 (softest) to 16 (hardest) with the results given below. Use the Wilcoxon-Mann-Whitney test to assess whether hardness is unaffected by temperature. State any assumptions needed for the validity of this test.

Temp	H	L	H	H	H	L	H	L	L	H	H	L	L	L	L	L
Rank	1	2	3	4	5	6	7	8	9	10	11	12	13	14	15	16

6.2 Suppose that the data for response times in the left visual field and the right visual field in Table 5.1 are available but the information that they are paired is omitted. Construct a table that presents the information appropriately. Use the Wilcoxon-Mann-Whitney test, adjusting for tied observations with mid-ranks, to investigate whether it would be reasonable to conclude that responses in the two fields differed. Does your conclusion agree with that found in Example 5.1? Explain your findings.

6.3 In dentistry, the DMF score is the total of the number of Decayed, Missing and Filled teeth. The following data are DMF scores for 34 male (M) and 54 female (F) first-year dental students.

M	8	6	4	2	10	5	6	6	19	4	10	4	10	12	7	2	5
	1	8	2	0	7	6	4	4	11	2	16	8	7	8	4	0	2
F	4	7	13	4	8	8	4	14	5	6	4	12	9	9	9	8	12
	4	8	8	4	11	6	15	9	8	14	9	8	9	7	12	11	7
	4	10	7	8	8	7	9	10	16	14	15	10	4	6	3	9	3
	10	3	8														

Use an asymptotic Wilcoxon-Mann-Whitney test to determine whether the DMF score differs significantly between males and females. Do tied ranks have much influence on the appropriate test statistic?

6.4 Hill and Padmanabhan (1984) give body weights (g) of diabetic and normal mice. Is there evidence of a significant difference in mean body weight? Obtain the Hodges–Lehmann estimate of the difference together with a 95% confidence interval. Compare this interval with that based on the t-distribution.

Diabetic	42	44	38	52	48	46	34	44	38				
Normal	34	43	35	33	34	26	30	31	27	28	27	30	37
	38	32	32	36	32	32	38	42	36	44	33	38	

6.5 A firm uses standard insulation in its oven production. To test its effectiveness, random samples are taken from the production line, and the ovens selected are heated to 400°C, noting the time taken to cool to 350°C after switching off. For a sample of 8 ovens, the times in minutes are:

15.7 14.8 14.2 16.1 15.3 13.9 17.2 14.9

A cheaper insulation is then explored, and using this on a sample of 9 ovens, the times taken for the same temperature drop are:

13.7 14.1 14.7 15.4 15.6 14.4 12.9 15.1 14.0

Can it be asserted that there is no real evidence of a different rate of heat loss? Obtain a 95% confidence limit for the difference in median heat loss (a) with and (b) without a normality assumption. Comment critically on any differences between your conclusions.

6.6 The journal *Biometrics* published data on the number of completed months between receipt of a manuscript for publication and the first reply to the authors for each of the years 1979 and 1983 (Armitage, 1984). The data are summarized below. Is there evidence of a difference in average waiting times between 1979 and 1983?

Completed Months	0	1	2	3	4	5	6
1979	26	28	34	48	21	22	34
1983	28	27	42	44	17	6	16

6.7 Brooks et al. (2014) reported for a group of young people the number of days per week on which they spent a total of 60 minutes or more in physical activity. Is there evidence of a difference between boys and girls?

Number of Days	0–2	3–6	7
Boys	292	1,152	555
Girls	421	1,381	307

Given the additional information that the group were a mixture of 11-year olds, 13-year olds and 15-year olds, and that they were recruited from school classes, comment on your findings.

6.8 A psychologist wants to know whether men or women are more upset by delays in being admitted to the hospital for routine surgery. She devises an anxiety index measured on patients 1 week before scheduled admission and records it for 17 men and 23 women. These are ranked 1–40 on a scale of increasing anxiety. The sum of the ranks for the 17 men is 428. Is there evidence that anxiety is gender-dependent? If so, which gender appears to be more anxious?

6.9 Show that for independent samples of size m, n with r observations above the combined sample median in the first sample and no sample observations are equal to the combined sample median, then the expression for P^* given in Example 6.7 generalizes to

$$P^* = \frac{\binom{m}{r} \times \binom{n}{k-r}}{\binom{m+n}{k}},$$

where $k=(m+n)/2$.

6.10 Two samples of sizes m and n, respectively, where $m+n=20$ have only one tie, the observations involved being those that would otherwise have ranks 2 and 3. What would be the differences between the van der Waerden scores if the scores for these tied values are based on the mid-rank 2.5 and the van der Waerden scores if these are based on the mean of the van der Waerden scores corresponding to rank 2 and rank 3? Carry out a similar comparison if there were ties for ranks 10, 11 and 12.

6.11 Perform a Siegel–Tukey test on the data in Example 6.11 after shifting one set of sample values to align the sample means.

6.12 Using the data in Example 6.11, carry out a test analogous to the Conover squared-rank test but using absolute ranks instead of squared ranks.

6.13 A researcher notes the total time (in seconds) needed to perform a series of simple manual tasks for each of eight children with learning difficulties and seven children without learning difficulties. The times are:

Without Difficulties	204	218	197	183	227	233	191	
With Difficulties	243	228	261	202	343	242	220	239

Discuss whether one-tailed or two-tailed tests would be more appropriate for this study. Use a Smirnov test to find whether the researcher is justified in asserting these samples are likely to be from different populations.

6.14 For the samples in Exercise 6.13 investigate differences in dispersion, indicating why this might give further insight.

6.15 Carter and Hubert (1985) give data for percentage variation in blood sugar over 1-hour periods for each of two groups of nine rabbits given different dose levels of a drug. Is there evidence of a response difference between levels?

Dose I	0.21	−16.20	−10.10	−8.67	−11.13
	1.96	−10.19	−15.87	−12.81	
Dose II	1.59	2.66	−6.27	−2.32	−10.87
	7.23	−3.76	3.02	15.01	

6.16 A television series examines various aspects of life in a particular European country. One episode looks at travel and another focuses on the political system. The two episodes are compared on the clarity of presentation, taking as an indicator the number of words in a random sample of sentences. The observations are:

Travel	16	15	12	20	24	16	27	13	27	26
Politics	40	9	8	26	30	15	53	36	59	20

Perform tests to determine whether there is evidence that in these episodes (i) sentence lengths show a difference in location; (ii) the variances of sentence lengths differ between themes; (iii) the distributions of sentence lengths differ in an unspecified way; (iv) the sentence lengths for either episode are not normally distributed.

6.17 Records from a maternity hospital show that during a particular year, identical and nonidentical twins were born on the following days:

Identical	Jan 3	Feb 27	Apr 14	May 31	Aug 6
	Oct 2	Dec 11			
Nonidentical	Jan 30	Feb 28	Mar 5	Apr 13	Jun 2
	Jun 29	Jul 8	Jul 17	Aug 4	Sep 17
	Oct 9	Oct 26	Nov 17		Dec 7

Is there evidence of a possible clustering of births of identical and nonidentical twins?

6.18 Confirm the value of p_1 under H_1 for the double exponential distribution introduced in Example 6.19.

6.19 Verify the correctness of the required sample sizes given in Example 6.20 for the WMW test in the circumstances given there.

7

Basic Tests for Three or More Samples

7.1 Extension to Multiple Samples

Having looked at nonparametric methods for a single sample, two paired samples and two independent samples, this chapter considers situations in which three or more groups are analysed. For example, so far, Lind's investigation into the treatment of scurvy at sea has been addressed as a comparison of two sailors who were provided with citrus fruits and ten sailors who did not receive this intervention. In fact, Lind's study design (Lind, 1753) involved pairs of sailors, each pair being assigned to one of six interventions as shown in Table 7.1.

Although a meaningful analysis of these six groups might be difficult due to the small sample size, it is apparent that by merging five of the six groups into one category, potentially useful information will be discarded. Methods designed for multiple groups should therefore be used where possible. As in earlier chapters, the techniques described here require the assumption that members within each group are independent of each other. Many of these are based on ranking the observations.

It should be borne in mind that parametric one- or two-sample analyses of continuous data using tests and estimation procedures based on normality are modified for three or more samples. Emphasis shifts from the *t*-distribution and related tests and estimation procedures to the analysis of variance (ANOVA) and tests based on the *F*-distribution. Spiegel and Stephens (2017) give a useful introduction to parametric ANOVA based on normally distributed populations.

This chapter concerns largely, but not exclusively, nonparametric analogues of situations arising in the one-way classification ANOVA and the simplest two-way classifications. Multiway grouping, especially more general two-way classifications including factorial structures, and some more advanced related topics are covered in Chapter 8.

TABLE 7.1

Dietary Supplements in Lind's Study

Pair	Supplement
1	Quart [1.1 L] of cider a day
2	25 gutts [drops] of elixir vitriol 3 times a day
3	Two spoonfuls [30 mL] of vinegar three times a day
4	Course of seawater – half a pint [280 mL] per day
5	Two oranges and a lemon a day
6	Bigness of a nutmeg

DOI: 10.1201/9780429326172-7

7.2 Location Tests for Independent Samples

7.2.1 The Kruskal–Wallis Test

Kruskal and Wallis (1952) extended the Wilcoxon-Mann-Whitney (WMW) test using the Wilcoxon formulation to three or more samples. This test is commonly referred to as Kruskal–Wallis ANOVA. It corresponds to the parametric one-way ANOVA applied to normal distributions, an extension of the unpaired *t*-test. As well as being applicable to data consisting only of ranks, it is relevant as an overall test for equality of population means or medians when samples are from otherwise identical and continuous distributions. It also has reasonable power for testing for identical population cumulative distribution functions against an alternative that one or more of these cumulative distributions are distinct in the sense of dominating the others. As in the two-sample case, a shift in mean or median is often referred to as an additive treatment effect.

If all samples come from the same population (which may be of any continuous form and need not be symmetric), a mix of low, medium and high ranks is expected in each sample. However, if the populations differ in terms of their means or medians, observations having high ranks will tend to be concentrated in one or more of the samples, with the lower ranking observations concentrated in other samples. The Kruskal–Wallis test involves summing the ranks in each group and comparing these totals against what might be expected were the populations to be identical. As the test is reasonably straightforward, details of the algebraic formulation have been left as an exercise.

Suppose that k random samples have been obtained. To apply the Kruskal–Wallis test, it is necessary first to calculate the value of S_k, a quantity based on the sums of ranks within the samples. As with the WMW test, assuming no ties, the observations are combined and ranked from the smallest to the largest. Observations are separated into their original groups and for each sample the ranks are added together. Each total is then squared and divided by its sample size. The terms so created are added to give S_k.

The test statistic is then:

$$T = \frac{12S_k}{N(N+1)} - 3(N+1) \tag{7.1}$$

Consequently, under the alternative hypothesis, S_k will contain the square of at least one relatively large rank sum, leading to larger values of T. Given that this statistic is based on squared values, small values of T are difficult to interpret and one-tailed tests are used.

Computing exact permutation probabilities under H_0 by hand is impractical except for small samples. For N moderate or large, T has a chi-squared distribution with $k-1$ degrees of freedom under H_0.

Example 7.1

In purchasing a home, an indicator of affordability is the house price-to-earnings ratio. In the United Kingdom, it is typically between around 3 and 10, with wide geographical variations. For a sample of randomly selected towns in three regions of the United Kingdom, the affordability values are:

A	3.4	3.7	4.5	
B	4.0	4.3	5.2	5.9
C	6.2	7.8	10.2	

TABLE 7.2

Affordability and Ranks in Three Regions of the United Kingdom

Region A	Affordability	3.4	3.7	4.5	
	Rank	1	2	5	
Region B	Affordability	4.0	4.3	5.2	5.9
	Rank	3	4	6	7
Region C	Affordability	6.2	7.8	10.2	
	Rank	8	9	10	

Use the Kruskal–Wallis test to examine the validity of the hypothesis that affordability values for towns in these regions come from identical populations.

A suitable null hypothesis is H_0: *the populations from which the samples were drawn have identical medians* while the alternative is H_1: *the population medians are not all equal.*

The $N=10$ affordability values are ranked overall and the ranks are assigned to the appropriate groups. These are shown in Table 7.2.

The sum of ranks is $1+2+5=8$ for the towns in Region A, $3+4+6+7=20$ for Region B and $8+9+10=27$ for Region C, whence $S_k=(8)^2/3+(20)^2/4+(27)^2/3=364.33$.

Hence, (7.1) gives:

$$T=\frac{12\times 364.33}{10\times 11}-3\times 11=6.745.$$

This corresponds to an exact $P=0.010$. There is strong evidence that median affordability differs between regions.

Values of the house price-to-earnings ratio tend to be greater in Region C. In comparing regions, it is important to assess whether the property stock is similar in the towns chosen. For instance, for historical and other reasons, the proportion of terraced (cheaper) properties varies widely between towns. In addition, towns close to each other may have similar characteristics, e.g. a history of heavy manufacturing, and it might be inappropriate to analyse them as independent observations.

An alternative test statistic in this example is S_k since all other quantities in T are invariant in the exact permutation distribution (i.e. are the same for any permutation). The use of T is preferred for practical reasons (e.g. availability of tables and for asymptotic approximations).

Calculating the P-value by hand is tedious. The `kruskal.wallis` function in the ANSM5 package in R computes an exact P directly from the above data. Tables giving critical values of T for small sample sizes are given by Neave (1981, pp. 32–34), while Hollander and Wolfe (1999, Table A12) gives upper-tail P-values for large T for a range of sample sizes. These are strictly relevant only to a no-tie situation. Meyer and Seaman (2014) gives exact probabilities for three samples of equal size between 5 and 35, and four samples of equal size between 2 and 10. Good estimates of exact probabilities are also available using the Monte Carlo facility in the `kruskal.wallis` function in the ANSM5 package in R, which also gives the asymptotic chi-squared approximation, as do many other general statistical packages. With $T=6.745$, this gives $P=0.0343$, considerably different from the exact value indicating, not surprisingly, that the asymptotic result can be misleading for small samples.

Output from the ANSM5 package in R can be obtained using `kruskal.wallis(ch7$affordability, ch7$regions, do.asymp = TRUE)`.

7.2.2 The Kruskal–Wallis Test with Ties

As in the WMW test, mid-rank values are given to tied data. With ties, a more general formula for T is needed. In addition to the terms calculated above, it is necessary to calculate $S_r = \sum_{i=1}^{j} r_{ij}^2$, where some of the r_{ij} will now be mid-ranks.

As noted, the Kruskal–Wallis test is a form of ANOVA. Those familiar with ANOVA in the context of normally distributed populations may recognize S_k and S_r as uncorrected treatments and total sums of squares for ranks. An appropriate correction for the mean is subtracted from each, namely $C = N(N+1)^2/4$. The test statistic is:

$$T = \frac{(N-1)(S_k - C)}{(S_r - C)} \tag{7.2}$$

If there are no ties, this formula reduces to (7.1), since with no ties, $S_r = N(N+1)(2N+1)/6$.

Example 7.2

In a study of retired footballers and dementia (Ling et al., 2017) players were grouped according to the position played. The categories used for the analysis are centre forward, centre half and other. The observations on age at symptom onset are:

Centre Forward	55	56	66				
Centre Half	52	58	66	67			
Other	40	63	66	69	77	77	78

Use the Kruskal–Wallis test to examine the validity of the hypothesis that these may be samples from identical populations.

The null hypothesis is that the populations are identical. The alternative hypothesis considered is that the samples are from populations that are not identically located. Manual calculation of the required statistic is outlined, but this is only needed if no program is available to compute an exact P-value. Ranks are shown in Table 7.3.

Hence, $s_1 = 3+4+8 = 15$, $s_2 = 25$ and $s_3 = 65$, whence:

$S_k = 15^2/3 + 25^2/4 + 65^2/7 = 75 + 156.25 + 603.57 = 834.82$.

TABLE 7.3

Age at Symptom Onset of Dementia for Retired Football Players by Position Played

Centre Forward	Age at Symptom Onset	55	56	66				
	Rank	3	4	8				
Centre Half	Age at Symptom Onset	52	58	66	67			
	Rank	2	5	8	10			
Other	Age at Symptom Onset	40	63	66	69	77	77	78
	Rank	1	6	8	11	12.5	12.5	14

The sum of squares of allocated ranks is $S_r=1012.5$. Since $N=14$, $C=14 \times (15)^2/4=787.5$. It follows that:

$$T = \frac{13 \times (834.82 - 787.5)}{(1{,}012.5 - 787.5)} = 2.734.$$

Using the ANSM5 package in R, the command `kruskal.wallis(ch7$age, ch7$positions)` gives an exact $P=0.271$. There is no real evidence to support a difference in age at symptom onset of dementia between the playing position groups.

There is increasing evidence that repeated heavy blows to the head might increase the chance of being diagnosed with dementia (Ling et al., 2017). Football players in the centre forward and the centre half positions, being more likely to head the ball during a match, would then be at a greater risk of developing dementia at a younger age than those in other positions. There is indeed a visual impression of a concentration of lower ranks in the centre forward and the centre half groups. However, the total number of players involved in this study is small ($N=14$). A larger study would be required to identify what might be a true difference.

The asymptotic chi-squared approximation gives $P=0.255$, which is close to the exact value despite the number of observations being so small.

Monte Carlo estimates can also be provided by the `kruskal.wallis` function in the ANSM5 package in R. These may be useful for sample sizes a little larger than those in this example. The computational burden for exact P-values rapidly increases as N increases.

For the above example, most statisticians would consider a parametric ANOVA to be suitable. This leads to basically similar conclusions (Exercise 7.3).

Boos (1986) developed more comprehensive tests for k samples using linear rank statistics that test specifically for location, scale, skewness and kurtosis. Willemain (1980) tackles the interesting problem of estimating the mean of a population, given only the largest observation in each of the n samples. That scenario may arise if each of n consultants is given one dose of a new or expensive drug and asked to give this to the patient of greatest need on his or her list. A record may be taken of the change, say, in some level of a blood constituent and the aim may be to estimate the mean, or median, change that could be expected if the drug were used for all patients suffering from the relevant disease.

Shirley (1977) proposed a nonparametric test to determine the lowest dose level at which there is a detectable treatment effect compared with the level for untreated control subjects. An improved version of this test was given by Williams (1986) and House (1986) extended the idea to a randomized block design.

7.2.3 Tests Based on Transformation of Ranks

A modification of the Kruskal–Wallis test is to use van der Waerden scores in place of ranks. The rank, or mid-rank, r is replaced by the $r/(N+1)$th quantile of a standard normal distribution. Ignoring a slight discrepancy in the case of ties, the mean of these scores is zero. By evaluating quantities corresponding to S_k and S_r in Section 7.2.1 with ranks replaced by van der Waerden scores and setting $C=0$, a statistic T analogous to (7.2) can be calculated for the van der Waerden scores. Again, asymptotically, T has a chi-squared distribution with $k-1$ degrees of freedom. The asymptotic result is usually reasonable for all but very small samples. The ANSM5 package in R includes a function `kruskal.wallis.vdW` giving exact P-values for samples that are not too large and provide Monte Carlo estimates for larger samples. The exact P-value given by using the ANSM5 package in R and the command `kruskal.wallis.vdW(ch7$age, ch7$positions)` with van der Waerden scores for the data in Example 7.2 is 0.336. Expected normal scores may also be used.

7.2.4 The Jonckheere-Terpstra Test

The Kruskal–Wallis test is an omnibus test for differences in location or for differences due to dominance. If treatments represent, for example, steadily increasing doses of a stimulant, it might be appropriate to test hypotheses about means or medians, θ_i, of the form H_0: *all θ_i are equal* against

$$H_1 : \theta_1 \leq \theta_2 \leq \theta_3 \leq \ \dots \ \leq \theta_k$$

where at least one of the inequalities is strict, or against an ordered alternative,

$$H_1 : \theta_1 \geq \theta_2 \geq \theta_3 \geq \ \dots \ \geq \theta_k,$$

where again at least one inequality is strict. These are one-tailed tests.

A test for such ordered alternatives was given by Jonckheere (1954), but it had been conceived independently by Terpstra (1952). In essence, it extends the WMW test using the Mann–Whitney formulation. The samples must be ordered in the sequence specified in H_1. Exact permutation tests are more readily available for small samples than they are for the Kruskal–Wallis test. An asymptotic test is also available. This is discussed after an example. The asymptotic test is widely available in standard statistical software packages.

Example 7.3

Ling et al. (2017) give the ages at symptom onset in a group of retired football players. Characteristics of dementia (recurrent hallucination, behavioural changes, mood changes and motor impairment) were recorded as present or absent for each player. The observations grouped according to the number of features are:

	Number of Features	Age at Symptom Onset (Years)					
Sample 1	1/2	66	66	69	77	77	78
Sample 2	3	40	52	56	58	66	67
Sample 3	4	55	63				

Use the Jonckheere-Terpstra test to assess the evidence for a tendency for age at symptom onset to decrease as the number of symptoms increases.

The null hypothesis is H_0: *age at symptom onset is independent of the number of features* against H_1: *age at symptom onset decreases with the number of features.* The samples are already arranged in the order of increasing number of features implicit as the natural order under H_1.

If there are k samples, the sum, U, of all Mann–Whitney statistics, U_{rs}, relevant to the rth sample ($r=1, 2, \dots, k-1$) and any sample, s, for which $s>r$ is calculated. Thus, for the three samples above U_{12}, U_{13} and U_{23} should be calculated. For example, U_{12} is the number of sample 2 values that exceed each sample 1 value with ties being scored as 0.5 as in the Mann–Whitney statistic. Here, $U_{12}=1.5+1.5+0+0+0+0=3$. Similarly, $U_{13}=0+0=0$ and $U_{23}=2+4=6$. Adding all U_{rs} gives a total of $U=9$. The exact test using the ANSM5 package in R and the command `jonckheere.terpstra(ch7$dementia.age, ch7$features, alternative = "greater")` indicates $P=\Pr(U\le 9)=0.005$, for the one-tailed test relevant here. There is thus clear evidence that experiencing several of the characteristics of dementia is associated with a younger age at symptom onset.

These findings should be interpreted with care as in this example, only a simple count of the number of characteristics is made. In practice, these characteristics can vary in frequency and severity between individuals.

The asymptotic Jonckheere-Terpstra test is based on U as defined in Example 7.3 having a mean $\mathrm{E}(U)=\left(N^2-\sum_{i=1}^{k}n_i^2\right)/4$ and variance $\mathrm{Var}(U)=\left\{N^2(2N+3)-\sum_{i=1}^{k}\left[n_i^2(2n_i+3)\right]\right\}/72.$

For large N, and the individual n_i not too small, the distribution of

$$Z=\frac{U-\mathrm{E}(U)}{\sqrt{\mathrm{Var}(U)}}$$

is approximately standard normal.

The sample sizes in Example 7.3 are too small for an asymptotic result to inspire confidence, but for this example $Z=-2.528$ corresponds, as indicated above, to a one-tail $P=0.0057$.

The expression for Var(U) needs adjustment for ties. The adjustment is trivial for relatively few ties but Section 12.3.2 shows that the existence of many ties may have a dramatic effect. A formula for adjusting Var(U) for ties is given by Lehmann (2006) and a modified version is given by (12.11) in Section 12.3.2.

The `jonckheere.terpstra` function in the ANSM5 package in R used this modified version and the command `jonckheere.terpstra(ch7$dementia.age, ch7$features, alternative = "greater", do.asymp = TRUE)` gives $P=0.0055$.

Investigating ordered treatment differences is sometimes of interest. This includes the testing of H_0: *all θ_i are equal* against

$$H_1:\theta_1\le\theta_2\le\ldots\le\theta_{q-1}\le\theta_q\ge\theta_{q+1}\ge\ldots\ge\theta_k$$

where at least one inequality is strict. These tests are called *umbrella tests*. They were proposed by Mack and Wolfe (1981), and are discussed in some detail by Hollander, Wolfe and Chicken (2014, Section 6.3). An umbrella test would be relevant if it were believed that a treatment applied at increasing doses was beneficial (e.g. increased yield) up to a certain dose level, while at higher doses, a toxic effect made the treatment harmful (e.g. reduced yield).

7.2.5 The Median Test for Several Samples

The median test in Section 6.2 can be generalized to three or more samples. The alternative hypothesis is now one of the differences between population medians without specifying which populations differ in location, how many differences there are, or their direction. As in the two-sample case, each sample may come from any unspecified population (they need not all have the same distribution). Given k samples where the unknown population medians are $\theta_1, \theta_2, \ldots, \theta_k$, one tests H_0: $\theta_1=\theta_2=\ldots=\theta_k$ against H_1: *not all θ_i are equal.*

The procedure generalizes that in Section 6.2.1. If M is the combined sample median for all observations, the numbers of observations above and below M are counted for each sample. As in Section 6.2.1, sample values equal to M are rejected and the reduced samples are analysed. Assuming sample values that equal M have already been dropped, suppose that, among the n_i observations in sample i, there are $a_i > M$, $b_i < M$ and $i = 1, 2, \ldots, k$. The numbers above and below M are recorded in a $k \times 2$ contingency table with k rows and 2 columns. The values a_i, b_i are constrained for each sample so that $a_i + b_i = n_i$, the number in that sample. When no observations equal M, the column totals A and B both equal $N/2$. Table 7.4 illustrates this for a general contingency table with k rows and 2 columns.

The probability P^* of the above cell values when the row and column totals are fixed is given by a generalization of (6.7):

$$P^* = \frac{A!B!\prod_{i=1}^{k}(n_i!)}{N!\prod_{i=1}^{k}(a_i!)\prod_{i=1}^{k}(b_i!)} \tag{7.3}$$

where $\prod_{i=1}^{k}(x_i!)$ is the product $(x_1!) \times (x_2!) \times \ldots \times (x_k!)$. As illustrated in the following example, other $k \times 2$ contingency tables with the same marginal totals must be considered to obtain the P-value.

Example 7.4

Six dental surgeons (I, II, III, IV, V and VI) perform the removal of a third molar (or wisdom) tooth from adult patients. To investigate the assertion that some surgeons extract these teeth more quickly, the following data were obtained. These relate to 28 patients, each of whom had just one such tooth removed. The time (in minutes) taken for the six dental surgeons to operate were:

I	23	25	34	45			
II	7	11	14	15	17	24	40
III	5	8	16	20	26		
IV	12	21	31	38			
V	30	43					
VI	4	9	10	18	19	35	

The median extraction time, M, for all 28 patients is 19.5 minutes, so no sample value is equal to the median. Table 7.4 shows the number of extraction times above and below M for each surgeon. Use a median test to assess the strength of any evidence against H_0: *all six samples come from populations with the same median extraction time.*

The test is an extension of the Fisher exact test developed in Section 6.2.2 for the two-sample situation to the $k \times 2$ table having the form in Table 7.4. Details are omitted here but the formula (7.3) is obtained by extending arguments similar to those leading to (6.7). For the data in Table 7.5, a P-value is calculated based on a critical region consisting of all 6×2 contingency tables having the given marginal totals for which, under H_0, P^* does not exceed that observed. For the 2×2 tables in Example 6.7, the more extreme configurations

TABLE 7.4

A General Contingency Table for the Median Test

Above M	**Below** M	**Total**
a_1	b_1	n_1
a_2	b_2	n_2
.	.	.
.	.	.
a_k	b_k	n_k
—	—	—
A	B	N

TABLE 7.5

Numbers of Observations Above and Below the Combined Median M in Example 7.4

Dental Surgeon	**Above** M	**Below** M	**Total**
I	4	0	4
II	2	5	7
III	2	3	5
IV	3	1	4
V	2	0	2
VI	1	5	6
Total	14	14	28

were clear. For larger contingency tables, this is not the case, nor can these extreme configurations be attributed to a particular tail; i.e. the alternative hypothesis H_1: *not all medians are equal* is essentially two-tail, not unlike the situation in the Kruskal–Wallis test (Section 7.2.1), or indeed the F-test in parametric ANOVA based on assumed normal distributions.

For Table 7.5, the calculation using (7.3) reduces to

$$P^* = \frac{(14!)(14!)(4!)(7!)(5!)(4!)(2!)(6!)}{(28!)(4!)(2!)(2!)(3!)(2!)(1!)(0!)(5!)(3!)(1!)(0!)(5!)}$$

where, by definition, $0!=1$, giving $P^*=0.00013$ using only a pocket calculator.

The P^* for all other contingency tables with the same marginal totals that have the same or lower values of P^* can be found similarly. These probabilities sum to $P=0.046$ for the comparison of third molar extraction times. There is moderately strong evidence against H_0, indicating some variation in median extraction times between the dental surgeons.

Output from the ANSM5 package in R can be obtained using `med.test(ch7$time, ch7$surgeon)`.

Individual sample sizes here are all small (particularly, for dental surgeon V). The main differences in Table 7.5 seem to be that surgeon I probably has a median above M and surgeon VI (possibly also surgeon II) has a median below M. If separate sign tests of H_0: $\theta_i=M$ were to be performed for each sample, it would not be possible to demonstrate strong evidence against H_0 (even in a one-tailed test) since, for any particular surgeon, the number of patients operated on is very small. In practice, much larger samples of patients would be selected for such a study.

The results from a comparison of dental surgeons need to be interpreted carefully. Most extractions are completed in around 15–20 minutes, but complicated cases can take 1 hour or more. These can be identified in advance from a radiograph, so consultants may deal with serious cases, with less experienced staff operating on straightforward patients. Higher extraction time for surgeon I could indicate greater responsibility, rather than inexperience.

A situation in which it might be of interest to test for a common median when samples come from distributions differing in other respects (e.g. spread, skewness, etc.) arises in a comparison of alternative components that could be used in an industrial process. It may be possible to measure a characteristic such as time to failure for a number of replicates of each type of component. A preliminary step may then be to test whether it is reasonable to suppose all may have the same median time to failure. If there is clear evidence they do not, one might reject from further consideration those components that appear to have a lower median time. Further choice may depend on preference for a component with little variability in its failure time distribution, or on nonstatistical factors, such as cost.

The difficulty in carrying out an exact test without a suitable computer program makes an asymptotic test of interest. In the notation used in Table 7.4, one option makes use of the statistic:

$$T = \sum_{i=1}^{k} \frac{\left(a_i - \dfrac{n_i A}{N}\right)^2}{n_i A / N} + \sum_{i=1}^{k} \frac{\left(b_i - \dfrac{n_i B}{N}\right)^2}{n_i B / N} \tag{7.4}$$

The term n_iA/N is equal to the expected number of cases above M for the *ith* surgeon, assuming that the null hypothesis is true. Readers familiar with the chi-squared test for independence in contingency tables will recognize (7.4) as the sum of

$$\frac{(\text{observed value} \; - \text{expected value})^2}{\text{expected value}}$$

for all cells, which is discussed in Section 12.2.2. There, an alternative form easier to compute manually is given by (12.8). For a $k \times 2$, table T has asymptotically a chi-squared distribution with $k-1$ degrees of freedom. Widely available chi-squared tables give minimal critical values for significance at conventional levels and nearly every statistical software package has a program for this test.

Example 7.5

For the tooth extraction data in Example 7.4, test for equality of population medians using the asymptotic test based on equation (7.4).

All quantities in (7.4) are calculated from Table 7.5 and the resulting T is compared with the tabulated critical value. The expected numbers (n_1A/N, etc.) are calculated for each cell in Table 7.4, e.g. with surgeon I, the expected number of cases above M is $4 \times 14/28 = 2$. The expected numbers, in the order given in Table 7.5, are:

2	2
3.5	3.5
2.5	2.5
2	2
1	1
3	3

Substituting these and the observed cell values from Table 7.5 in (7.4) gives:

$$T = 2\times\left(2^2/2 + 1.5^2/3.5 + 0.5^2/2.5 + 1^2/2 + 1^2/1 + 2^2/3\right) = 11.15.$$

Tables indicate $T \geq 11.07$ is the critical value of chi-squared with $k-1=5$ degrees of freedom for significance at the 5% level. There is hence moderately strong evidence against H_0.

All samples are small, but exact and asymptotic results agree well. For the chi-squared distribution with 5 degrees of freedom, $\Pr(T \geq 11.15)=0.048$, close to the exact $P=0.046$ obtained in Example 7.4.

Output from the ANSM5 package in R can be obtained using `med.test(ch7$time, ch7$surgeon, do.asymp = TRUE)`.

The alternative hypothesis that not all medians are equal is essentially two-tailed, but the asymptotic test uses a single tail of the chi-squared distribution. This is because the squares of the discrepancies *observed − expected* are used and these are necessarily zero or positive, no matter the sign of any actual discrepancy.

In practice, there may be interest in whether there are differences in medians for some subset of the k populations. One might take pairs of samples and compute, say, 95% confidence limits for the median differences for each pair using the method given in Section 6.2.2. Since these confidence intervals will generally be determined with varying precision for different pairs (due to differences in sample size and perhaps differences in the population distributions), direct comparisons involving these intervals sometimes lead to what at first sight appear bizarre consequences. For example, if zero is included in the 95% confidence interval for the difference $\theta_2-\theta_1$ and also in the 95% confidence interval for the difference $\theta_3-\theta_1$, there is no strong evidence against the null hypotheses that $\theta_2=\theta_1$ or that $\theta_3=\theta_1$.

It is tempting to conclude that these 95% confidence intervals imply a lack of evidence against the hypothesis $\theta_3=\theta_2$. This may not necessarily be the case. For instance, if most of the confidence interval for $\theta_2-\theta_1$ is positive and most of that for $\theta_3-\theta_1$ is negative, the likely differences with respect to θ_1 are in opposite directions. Consequently, differences between possible values for θ_2 and θ_3 may be quite large, and the 95% confidence interval for $\theta_2-\theta_3$ might not include zero. Similar considerations apply for pairwise comparisons in parametric analyses and the practice is not recommended.

If assumptions needed for, say, a Kruskal–Wallis test appear to hold that test should be preferred to the median test, the latter then usually being appreciably less powerful. For the data in Example 7.4, however, one may have reservations about the Kruskal–Wallis test because the sample values suggest appreciable differences in the spread of times taken by different surgeons.

SUMMARY POINTS:

- The Kruskal–Wallis test (Section 7.2.1) is a rank analogue of the one-way classification ANOVA.
- Asymptotically, the test statistic takes a chi-squared distribution with $k-1$ degrees of freedom (Section 7.2.1).
- The test statistic can be applied to data with tied observations using a simple adjustment (Section 7.2.2).

- An alternative is to transform ranks to van der Waerden scores (Section 7.2.3).
- The Jonckheere-Terpstra test (Section 7.2.4) is appropriate if the alternative hypothesis orders the treatments.
- An asymptotic approximation is widely used, but care is needed with ties (Section 7.2.4).
- The median test (Section 7.2.5) is applicable to k samples from any population (these may differ from one another in aspects other than the median).
- The test is essentially a special case of the Fisher exact test, or asymptotically a Pearson chi-squared test applied to a $k \times 2$ contingency table.
- The median test often has low power relative to many of its competitors.

7.3 Repeated Measurements

7.3.1 The Friedman Test

Friedman (1937) extended the sign test to the case of three or more related samples. It corresponds to the simplest form of repeated ANOVA applied to normal distributions, an extension of the paired t-test. Typically, these samples arise from repeated measurements made on the same individuals at several time points. In the terminology of study design, these individuals are referred to as blocks. In the Friedman test, the scores within each block are ranked. For example, if measurements are made at five successive times, the five measurements will be ranked from 1 to 5 in order of the corresponding scores. With just two time points, the ranks can be labelled as+and –, and Friedman's test reduces to the sign test.

The null hypothesis is that the population distributions at each time point are the same so that, for instance, the population medians are equal. If the null hypothesis is true, inspection of the readings for a particular time point will reveal a mixture of ranks. If the population medians differ, then for at least one of the time points, the ranks should be mainly high or mainly low.

Suppose that there are b individuals measured at t time points and for each individual, the t values are ranked. Assume that there are no tied ranks. For each time point, the sum of the ranks is calculated to give values s_i, $i=1, 2, \ldots, t$. For example, with $b=10$ individuals measured at $t=8$ time points, the ranks for Time 1 are added together to give s_1, ranks for Time 2 are added to give s_2, etc., producing 8 rank totals calculated over the 10 individuals.

The Friedman statistic is

$$T = \frac{12\sum_{i=1}^{t} s_i^2}{bt(t+1)} - 3b(t+1) \tag{7.5}$$

If b and t are not too small, T has approximately a chi-squared distribution with $t-1$ degrees of freedom. For this no-tie situation, some tables are available for conventional critical values of T, e.g. Neave (1981, p. 34), Hollander and Wolfe (1999, Table A22), Gibbons and Chakraborti (2020, Table N). Most statistical software provides programs for exact or asymptotic tests.

TABLE 7.6

Pulse Rates for Students Before and After Exercise

	Time		
Student	**I**	**II**	**III**
A	72	120	76
B	96	120	95
C	88	132	104
D	92	120	96
E	74	101	84
F	76	96	72
G	82	112	76

TABLE 7.7

Pulse Rates Ranked Within Students

	Time		
Student	**I**	**II**	**III**
A	1	3	2
B	2	3	1
C	1	3	2
D	1	3	2
E	1	3	2
F	2	3	1
G	2	3	1
Total	10	21	11

Example 7.6

For a group of seven students, the pulse rate (per minute) was measured before exercise (I), immediately after exercise (II) and 5 minutes after exercise (III) (Table 7.6). Use the Friedman statistic to test for differences between pulse rates on the three occasions.

The null hypothesis is H_0: *population distributions of pulse rates are identical* against H_1: *at least one time point has a different distribution of pulse rates reflected by a median shift or dominance.* There are no ties within students, so (7.5) may be used.

For each student, the three pulse rate values are ranked as shown in Table 7.7. Next, for each time point, ranks are totalled across the students to give $s_1=1+2+1+1+1+2+2=10$, $s_2=21$ and $s_3=11$.

To use (7.5), the sum of the squared totals is found, $\sum_{i=1}^{3} s^2 = 10^2 + 21^2 + 11^2 = 662$. There are $b=7$ students and $t=3$ time points, so the test statistic is given by:

$$T = \frac{12 \times 662}{7 \times 3 \times (3+1)} - 3 \times 7 \times (3+1) = 10.57.$$

This gives an exact $P=0.0027$. By contrast, the chi-squared approximation with $T=10.57$ (2 degrees of freedom) gives $P=0.0051$. There is strong evidence of a difference in pulse rates between the three times.

Output from the ANSM5 package in R can be obtained using `friedman(ch7$pulse, ch7$time.period, ch7$student, do.asymp = TRUE)`.

The difference between the *P*-values given by the exact and approximate methods is a cause for concern. For such small data sets, it is unwise to rely on the chi-squared approximation.

Inspection of the raw data shows that the strength of evidence against the null hypothesis is not surprising. From a physiological point of view, the pulse rate should increase on exercise and fall off to the resting value once the exercise has been completed. Some individuals (e.g. C) appear to take longer than others to return to the resting pulse rate.

While the physiological explanation of the differences is sensible for this example, care must be taken with measurements repeated in time because the order of the times is fixed. This means that evidence against H_0 may be due to some factor other than that under study. In the present study, if all individuals recorded their initial pulse rate standing but were seated following exercise, the reduction in pulse rate between occasion II and occasion III might be greater than that observed with standing throughout.

Note that in this example, the blocks (students) are presented as rows. Blocks may also be presented as columns, so it is important to read the table carefully.

7.3.2 The Friedman Test with Ties

If there are ties among the observations for a specific individual, mid-rank values are used. A more general formula is required for *T*. In addition to the above terms in (7.5), it is necessary to calculate $S_r = \sum_{i,j} r_{ij}^2$ where r_{ij} is the rank allocated to individual *i* at time point *j*. With no ties $S_r = bt(t+1)(2t+1)/6$. It is also convenient to define $S_t = \sum_{i=1}^{t}(s_i^2)/b$.

As with the Kruskal–Wallis test, a correction factor is required. This is $C = bt(t+1)^2/4$. The Friedman statistic can be written as:

$$T = \frac{b(t-1)(S_t - C)}{S_r - C}. \tag{7.6}$$

If there are no ties, this is equivalent to (7.5). As with the no-ties case, if *b* and *t* are not too small, *T* has an approximate chi-squared distribution with $t-1$ degrees of freedom. Iman and Davenport (1980) suggests that a better approximation than (7.6) is

$$T_1 = \frac{(b-1)(S_t - C)}{S_r - S_t}. \tag{7.7}$$

which, under a null hypothesis of no treatment difference, has approximately an *F*-distribution with $t-1$ and $(b-1)(t-1)$ degrees of freedom. If the ranking is identical in all blocks, the denominator of T_1 is zero. Iman and Davenport show that in formal significance testing terms, this may be interpreted as a result significant at a $P = (1/t)^{b-1}$ significance level. T_1 is relevant regardless of ties.

Example 7.7

Pearce (1965, p. 37) quoted the results of a greenhouse experiment carried out by J. I. Sprent (unpublished). The data are the number of nodes to flower initiation (first flowering) summed over four plants in each pot for the pea variety Greenfeast subjected to $t=6$ treatments – one an untreated control, the others named growth substances.

The observations are shown in Table 7.8 with the blocks as columns rather than rows. There were $b=4$ blocks (corresponding to the individuals in Example 7.6) allowing for differences in light intensities and temperature gradients depending on proximity to greenhouse glass. The blocks were arranged to make these conditions as similar as possible for all units (pots of four plants) in any one block. Use the Friedman statistic to test for differences between treatments in the number of nodes to flower initiation.

Within each block, the six observations are replaced by their ranks. The Friedman statistic is then calculated to test H_0: *no difference between treatments* against H_1: *at least one treatment has a different location parameter from the others.* Testing is performed using (7.6) or (7.7).

Relevant ranks are given in Table 7.9, where a column is added giving rank totals s_i for each treatment. Squaring each rank and adding gives $S_r=361$.

Adding the squared row totals leads to $S_t=(6.5^2+22^2+8^2+19^2+15^2+13.5^2)/4=339.625$. Also, $C=4\times 6\times 7^2/4=294$.

Applying (7.6) gives: $T=4\times 5(339.625-294)/(361-294)=13.62$, and (7.7) gives:

$$T_1 = 3(339.625 - 294)/(361 - 339.625) = 6.40.$$

Using T_1, as recommended by Iman and Davenport, the respective degrees of freedom for the F-distribution are $t-1=5$, and $(b-1)(t-1)=15$. Critical values for significance at the 5%,

TABLE 7.8

Nodes to First Flower, Total for Four Plants

	Block			
Treatment	**I**	**II**	**III**	**IV**
Control	60	62	61	60
Gibberellic Acid	65	65	68	65
Kinetin	63	61	61	60
Indole Acetic Acid	64	67	63	61
Adenine Sulphate	62	65	62	64
Maleic Hydrazide	61	62	62	65

TABLE 7.9

Nodes to First Flower, Ranks within Blocks

	Block				
Treatment	**I**	**II**	**III**	**IV**	**Total**
Control	1	2.5	1.5	1.5	6.5
Gibberellic Acid	6	4.5	6	5.5	22
Kinetin	4	1	1.5	1.5	8
Indole Acetic Acid	5	6	5	3	19
Adenine Sulphate	3	4.5	3.5	4	15
Maleic Hydrazide	2	2.5	3.5	5.5	13.5

1% and 0.1% significance levels are respectively 2.90, 4.56 and 7.57, so the result would be regarded as significant at the 1% level. If instead, T is compared with the critical values of the chi-squared distribution with $t-1$ or 5 degrees of freedom the result is significant at the 5%, but not the 1%, level. For the chi-squared distribution with 5 degrees of freedom, $\Pr(T \geq 13.620) = 0.0182$. Comparing T with the critical values given in the tables referenced above for six treatments and four blocks significance is indicated at the 1% level if $T \geq 12.71$, so again, significance is indicated at that level, but remember that Neave's table applies strictly to the no-tie situation. For smaller samples, the `friedman` function of the ANSM5 package in R gives exact P-values and for larger samples, Monte Carlo P-values are available. The Monte Carlo P-value is $P = 0.0048$, while the asymptotic P-value based on the chi-squared distribution is $P = 0.0182$ as already stated. There is hence strong evidence of a difference between treatments on the number of nodes to flower initiation.

Using the Iman and Davenport approximation for ties, the Monte Carlo P-value remains unchanged at $P = 0.0048$, while the asymptotic P-value based on the F-distribution is $P = 0.0023$.

Output from the ANSM5 package in R can be obtained using `friedman(ch7$nodes, ch7$treatment, ch7$block, do.asymp = TRUE)` and `friedman(ch7$nodes, ch7$treatment, ch7$block, use.Iman.Davenport = TRUE, do.asymp = TRUE)`.

While for practical reasons T or T_1 is used as the test statistic, it would be reasonable to use S_t because both S_r and C remain unaltered in the permutation distribution, which involves only permuting ranks within blocks ("individuals").

It is possible to replace ranks within blocks with normal scores. Experience suggests that such rescoring has few advantages. Indeed, if there are ties, the block differences removed by ranking may be reintroduced, though usually not dramatically.

Ranking within blocks is robust against many forms of heterogeneity of variance, in that, it removes any inequalities of variance between blocks (see Exercise 7.9). Durbin (1951) gives a rank-based Friedman-type test for data in incomplete blocks.

SUMMARY POINTS:

- For related samples, the Friedman test (Section 7.3.1) is applicable to block designs where ranks are allocated within blocks.
- It is an analogue of repeated measures ANOVA (Section 7.3.1).
- Asymptotically, the test statistic takes a chi-squared distribution (Section 7.3.1).
- In the layout of the data, it is important to establish whether the blocks are rows or columns (Section 7.3.2).
- The Friedman test statistic requires a correction factor for data with ties (Section 7.3.2).
- An alternative test statistic asymptotically takes an F-distribution (Section 7.3.2).

7.4 The Cochran Test for Binary Responses

For some studies involving repeated measurements, only binary data are available (win or lose, succeed or fail, male or female) so the Friedman test cannot be applied. In this situation, the Cochran test can be used, with responses scored as 0 or 1. For instance, a group of students may supply information about whether they regularly participate in football, cricket and tennis, with the responses being coded as 1 (yes) or 0 (no).

Sport	Football	Cricket	Tennis
Student			
A	0	1	0
B	1	0	0
C	1	1	1
D	0	1	1
E	0	0	1
F	1	0	0
G	1	0	0

Cochran (1950) proposed a method applicable to such situations to test the null hypothesis H_0: *population proportions of students who participate are identical for each sport* against H_1: *for at least one sport, the population proportions differ.* In conventional statistical terms, the sports may be regarded as treatments and students as blocks. If there are t treatments in b blocks and binary (i.e. 0, 1) responses, the appropriate test statistic is:

$$Q = \frac{t(t-1)\sum_{i=1}^{t} T_i^2 - (t-1)N^2}{tN - \sum_{j=1}^{b} B_j^2} \tag{7.8}$$

where T_i is the total of 1s for treatment i, B_j is the total for block j and N is the number of 1s. The exact permutation distribution of Q is not easily obtainable, but for large samples, Q has approximately a chi-squared distribution with $t-1$ degrees of freedom.

For two treatments, the test reduces to McNemar's test, and in Section 12.5, an alternative form of the McNemar test is given, which is exactly equivalent to Q above.

Example 7.8

Five members, A, B, C, D and E, of a mountaineering club, each attempt three rock climbs at each of which they either succeed or fail. If a success is recorded as 1 and a failure as 0, the outcomes may be summarized as shown in Table 7.10. Are the climbs the same in difficulty?

The hypotheses to be tested are H_0: *all climbs are equally difficult* against H_1: *the climbs vary in difficulty.*

TABLE 7.10

Outcomes for Three Rock Climbs

Member	A	B	C	D	E
Climb 1	1	1	0	0	1
Climb 2	1	0	0	1	0
Climb 3	0	1	1	1	1

There are $t=3$ treatments (climbs) and $b=5$ blocks (members). In (7.8), $T_1=3$, $T_2=2$, $T_3=4$, $B_1=2$, $B_2=2$, $B_3=1$, $B_4=2$, $B_5=2$, $N=9$.

$$Q = \frac{3(3-1)\left(3^2+2^2+4^2\right)-(3-1)9^2}{3\times 9-\left(2^2+2^2+1^2+2^2+2^2\right)}$$

$$= 1.2.$$

Using the ANSM5 package in R and the command `cochran.q(ch7$outcome, ch7$climb, ch7$member)` provides an exact P-value=0.852. Asymptotically, Q has approximately a chi-squared distribution with 2 degrees of freedom, giving $P=0.549$. The difference between the two P-values is unsurprising given the small sample size. However, they both show that there is no evidence that the climbs vary in difficulty.

SUMMARY POINTS:

- The Cochran test (Section 7.4) is applicable to multiple samples consisting of binary data.
- The test statistic has an asymptotic chi-squared distribution.
- For two treatments, the procedure reduces to McNemar's test.

7.5 Tests for Heterogeneity of Variance

Most tests for heterogeneity of variance discussed in Section 6.4 extend to several samples. This is illustrated for the Conover squared-rank test given in Section 6.4.3. The procedure parallels that for the Kruskal–Wallis test with squared ranks of absolute deviations replacing data ranks as scores.

The absolute deviations of all sample values from their sample mean are ranked over all samples and these ranks are squared. The statistic T is identical to (7.2) except that now s_i is the sum of the squared-rank deviations for sample i and C is the square of the sum of these squared ranks divided by N.

Symbolically, denoting such a squared rank by r_{ij}^2, the quantity $C=\left[\sum_{i,j} r_{ij}^2\right]^2/N$. S_r is here the sum of squares of the squared ranks, i.e. $S_r=\sum_{i,j} r_{ij}^4$.

If all population variances are equal, then for a reasonably large sample, T has a chi-squared distribution with $t-1$ degrees of freedom. If there are no rank ties, the sum of squared ranks is $N(N+1)(2N+1)/6$. Also, for no ties, the denominator of T reduces to:

$$S_r - C = (N-1)N(N+1)(2N+1)(8N+11)/180.$$

Example 7.9

Suppose that as well as decreasing with the number of features observed, it is thought that age at symptom onset may vary more between patients diagnosed with dementia as the number of features increases. Use the squared-rank test to test for heterogeneity of population variance based on the age at symptom onset data from Ling et al. (2017) used in Example 7.3:

	Number of Features	Age at Symptom Onset (Years)					
Sample 1	1/2	66	66	69	77	77	78
Sample 2	3	40	52	56	58	66	67
Sample 3	4	55	63				

The absolute deviation $|x_{ij}-m_i|$ of each observation x_{ij} from its sample mean m_i is obtained. These deviations are then ranked over combined samples and the statistic T is calculated and compared to the relevant chi-squared critical value.

The means for samples 1, 2 and 3 respectively are $m_1=72.167$, $m_2=56.5$ and $m_3=59$. The absolute deviations together with their ranks (1 for least, 14 for greatest, with mid-ranks for ties) are as follows:

1/2 Features	Absolute Deviation	6.17	6.17	3.17	4.83	4.83	5.83
	Overall Rank	10.5	10.5	3	7.5	7.5	9
3 Features	Absolute Deviation	16.5	4.5	0.5	1.5	9.5	10.5
	Overall Rank	14	6	1	2	12	13
4 Features	Absolute Deviation	4	4				
	Overall Rank	4.5	4.5				

The test statistic required is $T = \dfrac{(N-1)(S_k - C)}{S_r - C}$ as defined in (7.2).

In principle, the calculation of T is straightforward. First, find the value of S_k. The sum of squared ranks for the group "1/ 2 features" is $s_1=10.5^2+10.5^2+3^2+7.5^2+7.5^2+9^2=110.25+110.25+9+56.25+56.25+81=423$. Similarly, $s_2=550$, $s_3=40.5$, so the sum of all squared ranks is 1,013.5. Hence $S_k=(423)^2/6+(550)^2/6+(40.5)^2/2=81{,}058.292$.

Next, calculate C. The square of the sum of all squared ranks$=1{,}013.5^2$. Dividing this quantity by the sample size ($N=14$) gives $C=73{,}370.16$. Finally, S_r, the sum of the ranks raised to the fourth power$=10.5^4+10.5^4+3^4+7.5^4+7.5^4+9^4+14^4+6^4+1^4+2^4+12^4+13^4+4.5^4+4.5^4=127{,}126.375$. The use of (7.2) gives $T=1.859$. For a chi-squared distribution with 2 degrees of freedom, this corresponds to $P=0.395$. Hence, there is no evidence of heterogeneity of variance.

One might have reservations about an asymptotic result with such small samples, although in this example, evidence for heterogeneity of variance is clearly lacking. Little better can be done except in extremely small samples where exact permutation distributions can be calculated. For the original data, there may of course be differences in location whether or not the null hypothesis H_0: *no variance difference in a squared-rank test* is retained.

Using the ANSM5 package in R and the command `conover(ch7$dementia.age, ch7$features, do.asymp = TRUE)` gives the above result and an exact $P = 0.667$.

An alternative approach would be a Kruskal–Wallis test using overall ranks of absolute deviations.

SUMMARY POINTS:

- Most two-sample tests for heterogeneity of variance discussed in Section 6.4 extend to several samples (Section 7.5).
- The procedure that extends the squared-rank test (Section 6.4.3) parallels that for the Kruskal–Wallis test with squared ranks of absolute deviations replacing data ranks as scores.

7.6 Some Miscellaneous Considerations

7.6.1 Power Sample Size and Efficiency

In light of the problems hinted at in the discussions about power calculations in the simpler one- and two-sample situations in earlier chapters, it is unsurprising that exact power calculations for most of the procedures considered in this chapter are only available for a few very restricted, and often unrealistic, situations. Exploration of the circumstances in which a specific method is likely to be beneficial and any indication of sample sizes needed to meet specific objectives may be aided by information about Pitman efficiency. Pitman efficiencies for some of the methods developed in this chapter as well as in other situations discussed by them are considered by Hollander, Wolfe and Chicken (2014, Sections 6.10 and 7.16). Bünning and Kössler (1999) give some asymptotic power results for the Jonckheere-Terpstra test and some closely related tests and report that simulation studies indicate that their asymptotic results provide good approximations even for moderate sample sizes.

7.6.2 Runs Test for Three or More Samples

In their paper on multiple runs, Barton and David (1957) describe how the method developed for data consisting of several categories can be applied to quantitative measurements, referring to this as an ANOVA runs test. This is essentially an extension of the Wald–Wolfowitz runs test to three or more independent samples. Within each sample, observations are given the same code. The observations are then ranked in order of size and replaced by their allocated codes. Runs in the codes representing these observations are investigated for clustering or alternation using the method described in Section 4.6.3.

This technique may prove useful in detecting the presence of repeated measurements that have been pooled together to create a group of seemingly independent observations.

Statistical consultants need to be aware that inexperienced researchers involved in small-scale studies are sometimes tempted to combine repeated observations simply to increase the sample size (Bland, 2015). Suppose that several doctors are each timed performing a specific task on a number of occasions. If the variation in the times recorded for each doctor is much less than that between the doctors, the application of this test may highlight clusters of similar values relating to particular doctors.

Example 7.10

A new surgical procedure is performed on five occasions by each of two consultants, two medical registrars and two clinical medical students. The time (in seconds) taken for each performance of the procedure were:

Consultant A	97	157	181	215	216
Consultant B	520	527	540	566	580
Registrar C	305	320	345	350	356
Registrar D	454	456	471	485	489
Student E	414	422	423	451	467
Student F	544	550	599	661	665

Members of the surgical team present their data to a medical statistician not as above but arranged by seniority as three groups of seemingly independent observations without mentioning that the data contain repeated measurements.

Consultants	97	157	181	215	216	520	527	540	566	580
Registrars	305	320	345	350	356	454	456	471	485	489
Students	414	422	423	451	467	544	550	599	661	665

They ask for assistance in looking for group differences in the times taken. Use the multi-sample form of the Wald–Wolfowitz test to investigate whether these observations might not be truly independent.

There are $N=30$ observations, $k=3$ groups, $n_1=n_2=n_3=10$ observations in each group.

The null hypothesis under test is H_0: *ordered times within level of seniority occur at random.* Similar values from a particular individual may result in a cluster of observations with the same seniority code. Clustering being of interest, a one-tail alternative hypothesis H_1: *ordered times within level of seniority are clustered* is appropriate. A small number of runs would indicate clustering, providing evidence against H_0.

Consultants' observations are coded as 0, those for registrars as 1 and students' time as 2. The data are combined and ranked overall, each observation being labelled by its code as shown in Table 7.11.

Inspection of Table 7.11 indicates $r=10$ runs. The group sizes are too large for the exact P-value to be obtained. The distribution of the number of runs generated by 100,000 simulations gives a Monte Carlo $P=0.00003$ for the tail probability of $R \le 10$.

For the asymptotic test (Section 4.6.3), $p_1=p_2=p_3=1/3$, $E(R)=30(1-(1/9+1/9+1/9))+1=(30\times 2/3)+1=21$. Var($R$) is straightforward since $p_1^2+p_2^2+p_3^2=1/3$ and $p_1^3+p_2^3+p_3^3=1/9$. $\text{Var}(R)=30[1/3-(2\times 1/9)+(1/3)^2]=6.6667$ gives $Z=-4.260$, one-tail $P=0.000010$. With the continuity correction, $Z=-4.067$, one-tail $P=0.000024$.

Output from the ANSM5 package in R can be obtained using `runs.ncat(team.member[order(procedure.time)], alternative = "less", do.mc = TRUE)`.

TABLE 7.11

Ranking and Coding of the Times to Complete a Surgical Procedure

Time	97	157	181	215	216	305	320	345	350	356
Rank	1	2	3	4	5	6	7	8	9	10
Code	0	0	0	0	0	1	1	1	1	1
Time	414	422	423	451	454	456	467	471	485	489
Rank	11	12	13	14	15	16	17	18	19	20
Code	2	2	2	2	1	1	2	1	1	1
Time	520	527	540	544	550	566	580	599	661	665
Rank	21	22	23	24	25	26	27	28	29	30
Code	0	0	0	2	2	0	0	2	2	2

There is very strong evidence for the clustering of values having the same seniority code. This should raise concern about the independence of these observations. Looking at the data as originally collected shows that these concerns are well-founded.

The small variation in the times recorded for specific team members, and the much greater variation between individuals, allow similar observations to be easily detected as clusters of the same code. This becomes much more difficult if individuals show greater variability and overall differences between individuals are smaller.

Although, overall, there is a tendency for consultants to be the speediest, and students, the slowest at this task, Consultant B is an exception. A relationship between seniority and operative time is not necessary for clusters to be detected, only that repeated values for specific individuals are similar.

Tied values within an individual's observations may be dealt with by the usual method of averaging ranks. Ties between observations from different individuals will cause problems in the assignment of clusters if those individuals have different codes. In such a situation, a conservative approach is to order the tied values to maximize the number of runs. Such an approach is likely to have little power with a large number of ties between codes.

For three or more categories, the algebra involved in calculating tail probabilities is complex. Barton and David (1957) give tables for small samples consisting of three and four categories. However, multiple runs distributions can be simulated using the ANSM5 package in R, as above.

7.6.3 A Quantile-Based Procedure for Minimum Dispersion

Sometimes, samples come from several populations for each of which the kth quantile is known to be the same. It may then be of interest to select the population with the smallest variance, or some other measure of spread.

For example, if several different methods of producing an item all lead to the same proportion of defectives (e.g. falling below a minimum permitted weight), one may want to select the production method with the smallest variability of weight among the items produced. Gill and Mehta (1989, 1991) give a nonparametric approach to this problem.

SUMMARY POINTS:

- Exact power calculations for most multiple sample nonparametric procedures are only available for a few very restricted, and often unrealistic, situations (Section 7.6.1).
- A multiple sample extension of the Wald–Wolfowitz runs test (Section 7.6.2) may be useful in checking for possible dependence of observations within a group.
- Finding exact *P*-values for the multiple sample runs test involves complex algebra. Simulation is a much easier alternative (Section 7.6.2).

7.7 Exercises

7.1 Consider k random samples, the ith sample ($i=1, 2, \ldots, k$) consisting of n_i observations. Writing x_{ij} for a typical element of the ith sample, where $j=1, 2, \ldots, n_i$, explain how the Kruskal–Wallis test statistic T is calculated using an algebraic approach.

7.2 Uniform editions by each of three writers of detective fiction are selected. The number of sentences per page, on randomly selected pages in a work are:

C.E. Vulliamy	13	27	26	22	26		
Ellery Queen	43	35	47	32	31	37	
Helen McCloy	33	37	33	26	44	33	54

Use the exact Kruskal–Wallis test to examine the validity of the hypothesis that these may be samples from identical populations and compare the findings with the results from the asymptotic test.

7.3 Perform an ANOVA assuming normally distributed populations on the observations in Example 7.2, comparing your result with those for the Kruskal–Wallis test.

7.4 Lubischew (1962) gives measurements of maximum head width in units of 0.01 mm for three species of *Chaetocnema*. Part of his data is given below. Use a Kruskal–Wallis test to see if there is a species difference in head widths.

Species 1	53	50	52	50	49	47	54	51	52	57	
Species 2	49	49	47	54	43	51	49	51	50	46	49
Species 3	58	51	45	53	49	51	50	51			

7.5 Hinkley (1989) gives braking distances taken by motorists to stop when travelling at various speeds. A subset of his data is:

Speed (mph)	Breaking Distances (feet)			
20	48			
25	33	59	48	56
30	60	101	67	
35	85	107		

Use the Jonckheere-Terpstra test to assess the evidence for a tendency for braking distance to increase as speed increases.

7.6 Chris Theobald supplied the following data from a study of 40 patients suffering from a form of cirrhosis of the liver. Blood platelets form in the bone marrow and are destroyed in the spleen, so it was thought that an enlarged spleen might lead to more platelets being eliminated and hence to a lower platelet count. The spleen size of each patient was found using a scan and scored from 0 to 3 on an arbitrary scale, 0 representing a normal spleen, and 3, a grossly enlarged spleen. The platelet count was determined as the number in a fixed volume of blood. Do these data indicate an association between spleen size and platelet count in the direction anticipated by the experimenter?

Size													
0	156	181	220	238	295	334	342	359	365	374	391	395	481
1	65	105	121	150	158	170	214	235	238	255	265	390	
2	33	70	87	109	114	132	150	179	184	241	323		
3	79	84	94	259									

7.7 Snee (1985) gives data on average liver weights per bird for chicks fed three levels of growth promoter (none, low, high). Blocks correspond to different birdhouses. Use a Friedman test to see if there is evidence of a growth promotor effect.

Block	1	2	3	4	5	6	7	8
None	3.93	3.78	3.88	3.93	3.84	3.75	3.98	3.84
Low Dose	3.99	3.96	3.96	4.03	4.10	4.02	4.06	3.92
High Dose	4.08	3.94	4.02	4.06	3.94	4.09	4.17	4.12

7.8 Biggins, Loynes and Walker (1987) considered various ways of combining examination marks where all candidates sat the same number of papers but different candidates selected different options from all those available. The data below are the marks awarded to 12 candidates by four different methods of combining results. Do the schemes give a consistent ranking of the candidates? Is there any evidence that any one scheme treats some candidates strikingly differently than the way they are treated by other schemes so far as rank order is concerned? Is there any evidence of a consistent difference between the marks awarded by the various schemes?

Cand.	1	2	3	4	5	6	7	8	9	10	11	12
A	54.3	30.7	36.0	55.7	36.7	52.0	54.3	46.3	40.7	43.7	46.0	48.3
B	60.6	35.1	34.1	55.1	38.0	47.8	51.5	44.8	39.8	43.2	44.9	47.6
C	59.5	33.7	34.3	55.8	37.0	49.0	51.6	45.6	40.3	43.7	45.5	48.2
D	61.6	35.7	34.0	55.1	38.3	46.9	51.3	44.8	39.7	43.2	44.8	47.5

7.9 Berry (1987) gives the following data for the number of premature ventricular contractions per hour for 12 patients with cardiac arrhythmias when each is treated with 3 drugs, A, B and C.

Patient	1	2	3	4	5	6	7	8	9	10	11	12
A	170	19	187	10	216	49	7	474	0.4	1.4	27	29
B	7	1.4	205	0.3	0.2	33	37	9	0.6	63	145	0
C	0	6	18	1	22	30	3	5	0	36	26	0

Use a Friedman test to investigate differences in response between drugs. Note the obvious heterogeneity of variance between drugs.

7.10 Carry out an ordinary randomized block ANOVA on the data in Exercise 7.9. Do you consider it to be valid? Is the Friedman analysis to be preferred? Why?

7.11 Cohen (1983) gives data for the number of births in Israel for each day in 1975. Shown below are the data for the number of births on each day in the 10th, 20th, 30th and 40th weeks of the year.

	Day						
Week	**Mon**	**Tue**	**Wed**	**Thu**	**Fri**	**Sat**	**Sun**
10	108	106	100	85	85	92	96
20	82	99	89	125	74	85	100
30	96	101	108	103	108	96	110
40	124	106	111	115	99	96	111

Perform Friedman analyses to determine whether the data indicate (i) a difference in number of births between days of the week that shows consistency over the four selected weeks and (ii) any differences between the number of births in the 10th, 20th, 30th and 40th weeks.

7.12 At the beginning of a tutorial, 12 names are read out in random order to 10 medical students. Four are names of prominent sporting personalities (Group A), four are national and international politicians (Group B) and four are scientists who regularly feature in the media (Group C). At the end of the session, students are asked to recall as many of the names as possible. The numbers recalled were:

Student	1	2	3	4	5	6	7	8	9	10
Group A	3	1	2	4	3	1	3	3	2	4
Group B	2	1	3	3	2	0	2	2	2	3
Group C	0	0	1	2	2	0	4	1	0	2

Rank the data within each block (student) and use a Friedman test to assess evidence of a difference between recall rates for the three groups.

7.13 Carry out an ANOVA on the data in Exercise 7.12. Do the conclusions agree with the Friedman test? If not, why not?

7.14 A media expert claims that the pattern of social media use by young people across the day is the same, regardless of the time of the year, weekends or holidays. Four young people who agree to participate in an investigation into the use of social media keep a diary of their social media engagement each day for 1 year. As part of the study, they note down whether or not they are using social media at 8 pm each day. Records for 10 randomly selected days are tabulated below (Yes=1, No=0). Are there differences in terms of evening use?

Day	**1**	**2**	**3**	**4**	**5**	**6**	**7**	**8**	**9**	**10**
Participant A	1	0	0	1	1	1	1	0	1	1
Participant B	1	1	1	1	0	1	1	0	0	0
Participant C	1	1.	0	1	1	1	1	1	0	1
Participant D	1	1	0	0	0	1	1	1	0	1

7.15 As well as increasing with speed, it is thought that braking distance may vary more between drivers as speed increases. Use the squared-rank test to test for heterogeneity of population variance based on the following subsample from Hinkley (1989) of initial speeds (mph) and braking distance (feet).

Speed (mph)	**Breaking Distances (feet)**			
5	2	8	8	4
10	8	7	14	
25	33	59	48	56
30	60	101	67	

7.16 Analyse the data in Example 7.1 using the multiple samples runs test. Compare your findings with the results from the Kruskal–Wallis analyses.

8

Analysis of Structured Data

8.1 Factorial Treatment Structures

The nonparametric methods considered up to this point, whether for one, two or multiple groups, have all been applied to studies in which just one factor is under investigation. For instance, in the analyses of the data from the study of retired footballers and dementia (Ling et al., 2017), one example looks at age of symptom onset by position played. The issue of interest is the potential impact of a single factor, position played, on age of onset.

Often greater insight can be obtained by considering two or more factors in the same design. For instance, the age of symptom onset can in principle be investigated by looking at both the position played and family history of dementia (yes/no) in the same analysis. There will then be six subgroups under study, these being: centre forward with family history, centre half with family history, other with family history, and centre forward with no family history, centre half with no family history, other with no family history.

Analysing the observations by position played provides an average age of onset for each of the three position groups. Differences between the groups represent the effect on the mean age of one position relative to another. In the same way, group averages can be obtained for players with a family history of dementia and those without a family history of dementia, from which the effect of having a family history of dementia can be estimated. An important question might be whether certain combinations of family history and position are associated with a particularly high risk of early onset of symptoms beyond that expected by simply adding together the individual effects of position and family history (*additive effects*). Unfortunately, the sample size for the retired footballer study (14 participants) is too small for a sensible comparison of the six subgroups. In the example below, these concepts are illustrated with a study involving somewhat larger subgroups.

Example 8.1

A study of the driving performance of younger and elderly drivers was conducted in Braunschweig, Germany (Schwarze, Ehrenpfordt and Eggert, 2014). Participating drivers (8 between 27 and 31 years, 8 between 67 and 72 years) completed a planned route through the city that involved a mixture of easy and difficult manoeuvres. On completing the route, the drivers were asked to rate the workload (effort and concentration required) for the easy and difficult manoeuvres on a scale from 0 to 10. The mean reported workload scores for each age group by the difficulty of the manoeuvre are shown in Table 8.1. Is it reasonable to conclude that the effect due to age group and difficulty combined is similar to the sum of the effects due to age and difficulty added together?

DOI: 10.1201/9780429326172-8

TABLE 8.1

Mean Workload Scores after Completing the Drive by Age Group and Difficulty of Manoeuvre

	Easy	Difficult
Younger	2.38	6.20
Elderly	3.75	3.75

For the easy manoeuvres, the effect of being younger relative to the elderly is 2.38 – 3.75 = –1.37. For the elderly drivers, the effect due to difficulty is 3.75 – 3.75 = 0; the mean workload remains constant regardless of the manoeuvre's difficulty. If the combined effect of the two factors is equal to the sum of the two parts, then the mean score for younger drivers performing difficult manoeuvres should be equal to 3.75 + (2.38 – 3.75) = 2.38. Table 8.1 shows that the observed mean is 6.20. Therefore, the mean workload score for younger drivers performing difficult manoeuvres is much greater than that expected under the assumption of additive effects. This finding is referred to as a positive *interaction* between youth and task difficulty.

This chapter concerns largely, but not exclusively, nonparametric analogues of situations arising in the one-way classification analysis of variance and the simplest two-way classifications. More general designs, including factorial structures, along with some more advanced related topics, are covered in Chapter 13.

The concept of interaction introduced above applies more widely, but first, it is considered in the context of linear models (Spiegel and Stephens, 2017), where the model may include an interaction term as well as additive effects.

The simplest case is that of two factors, each at two levels. For instance, a chemical reaction may be allowed to proceed for two different times – 2 hours and 3 hours. These are the two levels of the first factor, *time*. Suppose that the experiment is carried out at two different temperatures – 75°C and 80°C. These are the two levels of a second factor, *temperature*. The responses, or yields, are the amounts of a chemical produced by a fixed amount of input material for each of the factor combinations

- 2 hours at 75°C (level 1 time, level 1 temperature);
- 2 hours at 80°C (level 1 time, level 2 temperature);
- 3 hours at 75°C (level 2 time, level 1 temperature);
- 3 hours at 80°C (level 2 time, level 2 temperature).

The experiment is repeated, or replicated, a specified number of times. The average, or the total output, for the same number of replicates for each factor combination is calculated. Apart from random variation in the form of an additive error, which in the classic parametric model is assumed to be distributed $N(0, \sigma^2)$, the output y_{ij} for the first factor at level i and the second at level j ($i, j = 1, 2$) is specified by a linear model described in the next section. If each factor has purely additive effects, this is known as a *no-interaction* model.

To clarify these concepts, assume temporarily an unrealistic state of perfection where there is no random variation to create error. Suppose that operating for 2 hours at 75°C gives an output X and that this increases to $X+3$ if time is left unaltered but temperature is increased to 80°C. Also, suppose that output increases to $X+8$ if the temperature is left at 75°C, but time is increased to 3 hours. As explained in Example 8.1, there is *no interaction*

TABLE 8.2

Expected Yields in a No-Interaction Model

	Temperature (°C)	
Process Time (hours)	**75**	**80**
2	X	X+3
3	X+8	X+11

if the result of increasing both time from 2 to 3 hours and temperature from 75°C to 80°C is to increase yield from X to $X+3+8=X+11$. These results for expected yield are summarized in Table 8.2.

In general, for a combination of level i of the first, or row, factor with level j of the second, or column, factor the true yield (including now an additive error) may be denoted by y_{ij} and the corresponding expected yield (output apart from error) by x_{ij}. Using the notation in Table 8.2:

$$x_{11} = X,\ x_{12} = X + 3,\ x_{21} = X + 8,\ x_{22} = X + 3 + 8 = X + 11,$$

and $x_{11}+x_{22}=2X+11=x_{12}+x_{21}$, i.e., the diagonal or cross-sums in Table 8.2 are equal. This is a fundamental characteristic of an additive (or no interaction) model for an experiment with two factors each at two levels.

In many real-life situations, the effect of increasing both factors is either to boost or to diminish the effect of changing only one factor. This would happen if, for example, x_{22} in Table 8.2 took the value $X+17$ or $X+2$ instead of $X+11$, the other x_{ij} being unaltered. In Example 8.1, changing both factors boosts the score expected from additive effects. The cross-sums are then no longer equal. For the 2×2 factorial model with no interaction, the requirement that the cross-sums are equal implies

$$x_{11} + x_{22} - x_{12} - x_{21} = 0. \tag{8.1}$$

The presence of random error means that this equality will not hold for the observed yields in each of r replicates, y_{ijk}, where $k=1, 2, \ldots, r$, or for the corresponding total or mean yields over all replicates for each factor combination.

The standard analysis of variance test for interaction is essentially one of assessing whether the observed means, temporarily denoted by y_{ij}, are such that $y_{11.}+y_{22.}-y_{12.}-y_{21.}$ departs from zero by an amount which is too great to be attributed to random variation. If the departure is too great, this implies an interaction.

More formally, asserting there is an interaction implies acceptance of a hypothesis

$$\mathrm{H}_1 : x_{11} + x_{22} - x_{12} - x_{21} = I, \text{ where } I \neq 0 \tag{8.2}$$

rather than H_0: $I=0$.

In the absence of additive errors, if there is no interaction, from (8.1) at *either* temperature, the difference in yield at 3 hours minus that at 2 hours is the same, i.e.,

$$x_{21} - x_{11} = x_{22} - x_{12}.$$

If there are additive errors, and no interaction, for the observed mean yields, $y_{ij\cdot}$, in general

$$y_{21.} - y_{11.} \neq y_{22.} - y_{12.},$$

but their mean

$$\left(y_{21.} - y_{11.} + y_{22.} - y_{12.}\right)/2,$$

is a sensible estimate of the effect of time on yield. This is usually referred to as the *main effect* of time. Analogous arguments can be applied to the estimation of the main effect of temperature.

If there is interaction, in general, this implies

$$x_{21} - x_{11} \neq x_{22} - x_{12},$$

and the concept of using

$$\left(y_{21.} - y_{11.} + y_{22.} - y_{12.}\right) / 2,$$

as an estimate of the main effect becomes somewhat meaningless. The concept of the main effect is inappropriate when there is interaction. It is then more profitable to examine separately the effect of any one factor at each level of the other factor. Two-way tables of means corresponding to each factor combination, together with appropriate standard errors, are valuable tools to aid the interpretation of factorial treatment data (Spiegel and Stephens, 2017).

The above arguments extend to any study design with two factors at each of two levels. In the next section, numerical examples are used to indicate the analysis and interpretation of 2 × 2 factorial structures using parametric methods, in order to discuss why nonparametric methods might be preferred, and later which might be appropriate ones. This may depend upon whether or not the linear model is retained.

SUMMARY POINTS:

- A factor is a qualitative variable that can be used to divide the observations into categories.
- One or more factors can be incorporated into a study design such that analyses are performed in terms of the subcategories. This is described as a factorial structure.
- The categories within a factorial design are known as levels.
- If the effect of two factors in combination is equal to the sum of the effects of the individual factors, the factors are described as being linear additive.
- A positive interaction is present if the sum of the combined effects exceeds the sum of the individual effects. A negative interaction is present if the sum of the combined effects is less than the individual effects.
- Random error means that in practice, the existence and size of interactions are not always obvious. It is usually necessary to perform hypothesis tests for interaction.

8.2 Balanced 2 × 2 Factorial Structures

Examples 8.2, 8.3 and 8.4 below cover situations where it is appropriate to use the standard parametric analysis of variance, and possible interpretations of the results are indicated in each case.

Example 8.2

Pots each with five plants of the same variety of pea are grown in greenhouses at each of the four combinations of two levels of growth hormone (factor A) and with two levels of undersoil heating (factor B). There are six replicates (pots) of each treatment combination, and after 9 weeks, all plants are harvested and weighed. The total weights for the five plants in each plot are recorded.

The conventional notation used here differs from that in Section 8.1. The levels of factor A are denoted by a_1, a_2 and those of factor B by b_1, b_2. The four possible combinations of levels are a_1b_1, a_1b_2, a_2b_1, a_2b_2. The weights in grams totalled for the five plants in each pot are:

a_1b_1	16.2	16.6	17.1	17.3	17.5	17.7
a_1b_2	17.2	17.4	17.5	17.9	18.1	18.3
a_2b_1	18.2	18.5	18.6	19.1	19.2	19.5
a_2b_2	18.1	18.9	19.6	19.9	20.1	20.3

The usual linear model with the individual factors providing the main effects and allowing for an interaction term is assumed, and the usual homogeneous normal error structure is specified. Most standard software programs will produce an analysis of variance table that is similar to this:

Source	Degrees of Freedom	Sum of Squares	Mean Square	*F*	*P*
Main A	1	18.727	18.727	51.66	0.000
Main B	1	2.535	2.535	6.99	0.016
Interaction	1	0.002	0.002	0.00	0.947
Error	20	7.250	0.363		
Total	23	28.513			

and a table of means

	b_1	b_2	Row Means
a_1	17.07	17.73	17.40
a_2	18.85	19.48	19.17
Column Means	17.96	18.61	18.29

The small *P*-values for the main effects of A and B indicate both are significant. There is no indication of an interaction.

Output from R can be obtained using `summary(aov(ch8$plant.weight ~ ch8$growth.hormone * ch8$undersoil.heating))` and `addmargins(tapply(ch8$plant.weight, list(ch8$growth.hormone, ch8$undersoil.heating), mean), FUN = mean)`.

In the table of means, the difference between the row means is greater than the difference between the column means. This reflects the more significant main effect of A. The small mean difference between the diagonally opposite means in the body of the table, i.e.,

$$(17.07 + 19.48)/2 - (18.85 + 17.73)/2 \approx 0.01$$

is clear confirmation of a lack of interaction. In the absence of interaction, the main effects of factors A and B are sensible measures of the additive effects of each factor.

Example 8.3

Suppose that in the experiment in Example 8.2 the data were as given there except that for the factor combination a_2b_2, the observations were all increased by four units, i.e., they were

a_2b_2 22.1 22.9 23.6 23.9 24.1 24.3

What impact does this have on the analysis of variance and the interpretation of results? Following the procedure used in Example 8.2, the analysis of variance table is:

Source	Degrees of Freedom	Sum of Squares	Mean Square	F	P
Main A	1	85.127	85.127	234.83	0.000
Main B	1	42.135	42.135	116.23	0.000
Interaction	1	0.002	0.002	0.00	0.000
Error	20	7.250	0.363		
Total	23	158.113			

and the table of means is

	b_1	b_2	Row Means
a_1	17.07	17.73	17.40
a_2	18.85	23.48	21.17
Column Means	17.96	20.61	19.29

In the presence of interaction, a study of main effects is potentially misleading. For example, the main effect of B is epitomized by the difference in the marginal column means, i.e., 20.61 − 17.96 = 2.65. However, at the lower level of A, the mean difference is only 17.73 − 17.07 = 0.66 whereas at the higher level, it is 23.48 − 18.85 = 4.63. These last two differences are often referred to as *particular effects*. They are usually of more interest to experimenters than main effects when there are interactions.

Output from R can be obtained using `summary(aov(ch8$plant.weight.2 ~ ch8$growth.hormone * ch8$undersoil.heating))` and `addmargins(tapply(ch8$plant.weight.2, list(ch8$growth.hormone, ch8$undersoil.heating), mean), FUN = mean)`.

In this example, if the experimenter were interested in which factor level produced the greatest mass of plant material, it is important to recognise that the effect of factors is not purely additive. Having both factors at the higher level increases yield above that expected with a purely additive model.

Situations where factors interact in a manner like that described in this example are common in many fields involving experimental or observational data. For example, there is strong evidence that smoking increases the risk of lung cancer, as does exposure to certain forms of asbestos dust. Further, the effects are not merely additive. The lung cancer incidence rate for smokers exposed to asbestos dust is appreciably higher than it would be if the effects were simply additive.

Interactions may sometimes mask main effects if the effect of increasing the level of one factor reverses the direction of response to the other factor. This is illustrated in Example 8.4.

Example 8.4

Suppose that the data were the same as in Example 8.2 except that for the factor combination a_2b_2 the observations were smaller with values

a_2b_2 17.9 18.3 18.4 18.6 18.7 18.8

What impact does this have on the analysis of variance and the interpretation of results? Following the procedure used in Example 8.2, the analysis of variance table is:

Source	Degrees of Freedom	Sum of Squares	Mean Square	F	P
Main A	1	9.375	9.375	43.64	0.000
Main B	1	0.107	0.107	0.50	0.489
Interaction	1	1.707	1.707	7.94	0.011
Error	20	4.297	0.215		
Total	23	15.485			

and the table of means is

	b_1	b_2	Row Means
a_1	17.07	17.73	17.40
a_2	18.85	18.45	18.65
Column Means	17.96	18.09	18.02

Here, with the presence of interaction, the main effects do not shed much light. In principle, the main effect of B is given by the difference in marginal column means, i.e., 18.09 − 17.96 = 0.13. However, at the lower level of A, the mean difference is 17.73 − 17.07 = 0.66, while at the higher level, it is 18.45 − 18.85 = −0.40. These last two differences are the *particular effects*. Here the indication is that each particular effect of B is in opposite directions at each level of A.

Output from R can be obtained using `summary(aov(ch8$plant.weight.3 ~ ch8$growth.hormone * ch8$undersoil.heating))` and `addmargins(tapply(ch8$plant.weight.3, list(ch8$growth.hormone, ch8$undersoil.heating), mean), FUN = mean)`.

The above examples cover situations where standard normal theory analysis of variance is appropriate. Standard errors for the individual means and for the marginal means provide a tool for determining confidence intervals for main effects and particular effects, but this standard aspect of the classic parametric analysis is not pursued here.

Nonparametric procedures become important when assumptions needed for parametric methods may break down. Earlier chapters have shown that the use of statistics based on ranks is then sometimes appropriate in the sense that they often give similar results to a parametric analysis performed under appropriate conditions, but at the same time overcome difficulties associated with the breakdown of assumptions needed in a parametric approach.

A simple idea discussed by Conover and Iman (1976, 1981) is to rank all observations in ascending order and to perform an analysis of variance on the ranks instead of the original data. This has been shown by many writers to produce unfortunate anomalies. On a more positive note, it provided a stimulus for new approaches to the modelling and study of interactions.

Example 8.5

The data in (i) Example 8.2, (ii) Example 8.3 and (iii) Example 8.4 are ranked in ascending order, and an analysis of variance is performed on these ranks.

For ranks based on the data in Example 8.2, the following analysis of variance is obtained:

Source	Degrees of Freedom	Sum of Squares	Mean Square	F	P
Main A	1	805.0	805.0	60.93	0.000
Main B	1	77.0	77.0	5.83	0.025
Interaction	1	2.7	2.7	0.20	0.658
Error	20	264.3	13.2		
Total	23	1,149.0			

This leads, as in Example 8.2, to the conclusion that there is no interaction.

Analysis of variance for ranks based on the data in Example 8.3 gives:

Source	Degrees of Freedom	Sum of Squares	Mean Square	F	P
Main A	1	840.17	840.17	114.11	0.000
Main B	1	155.04	155.04	5.83	0.000
Interaction	1	7.04	7.04	0.96	0.340
Error	20	147.25	7.36		
Total	23	1,149.50			

Unlike the situation in Example 8.3, evidence of an interaction has disappeared. One reason for this is that other possible yields at the higher level of each factor would result in the same rankings as those in this example.

Since the absence of interaction makes interpretation of main effects meaningful, this might seem desirable. However, as indicated in Example 8.3, the presence of interaction in the original data may be important in practice, indicating that effects are not strictly additive. The practical implication in that example is that an increased dose of growth hormone together with the higher level of undersoil heating enhances growth appreciably more than increasing only one or the other level. This may have important economic or other implications.

For ranks based on the data in Example 8.4, the following analysis of variance is obtained:

Source	Degrees of Freedom	Sum of Squares	Mean Square	F	P
Main A	1	770.7	770.7	52.70	0.000
Main B	1	4.2	4.2	0.28	0.599
Interaction	1	80.7	80.7	5.52	0.029
Error	20	292.5	14.6		
Total	23	1,149.5			

Unlike the situation arising with the data in Example 8.3, converting the observations to ranks still provides evidence for an interaction. The values for a_2b_2 are now much closer to the values in the other three groups, so there is little flexibility in terms of changing these values while keeping the same rankings. Any changes possible under this restriction will retain the interaction in the data.

Output from R can be obtained using `summary(aov(rank(ch8$plant.weight)~ ch8$growth.hormone * ch8$undersoil.heating))` for the data in Example 8.2 with `ch8$plant.weight.` replaced with `ch8$plant.weight.2` and `ch8$plant.weight.3` for the data in Examples 8.3 and 8.4.

Classical parametric methods are appropriate for each of the data sets considered in this section. Example 8.5 illustrates a possible nonparametric approach that is unsatisfactory in the sense that it gives answers incompatible with those given by the parametric method when the latter is appropriate.

Is it possible to develop nonparametric approaches that may be valuable when the core assumptions of conventional analysis of variance break down? Can methods be designed that are not unduly disturbed by a few outliers? Some answers can be provided only after a more detailed study of the concept and meaning of interactions in a wider context than that of 2×2 factorial structures.

SUMMARY POINTS:

- In a factorial design, the impact of an individual factor is called a main effect.
- The appropriate analysis of an $a \times b$ factorial design is heavily dependent on whether there are interactions. In the presence of interaction, a study of the main effects is misleading, and attention should then be focused on the interactions.
- Standard analysis of variance can be performed if the most important assumptions, e.g., the outcome variable can be assumed to come from a normal population, are plausible.
- The nonparametric approach of ranking all observations and performing an analysis of variance on the ranks is not always helpful, as it can produce anomalies.

8.3 The Nature of Interactions

The concept of interactions is often first encountered in the context of factorial treatment structures in designed experiments. A wide-ranging discussion of these and other important types of interactions and their implications is given in Cox (1984).

It is well established that in a parametric context, interactions may sometimes be reduced, or even eliminated, by a data transformation. Whether it is desirable to do so depends upon the nature of the interaction, the physical nature of the data involved, and the questions being asked.

With factorial treatment structures and continuous data, interest is often in the linear model. If there is no interaction, in the classic analysis of variance for two factors A and B at each of a, b levels, respectively, and r replications of each factor combination, the yields, or responses, y_{ijk} may be written

$$y_{ijk} = \mu + \alpha_i + \beta_j + \varepsilon_{ijk},$$

where $i=1, 2, \ldots, a$, $j=1, 2, \ldots, b$, $k=1, 2, \ldots, r$ and

$$\sum_{i=1}^{a} \alpha_i = \sum_{j=1}^{b} \beta_j = 0.$$

In the classic parametric model, the ε_{ijk} are independently and identically distributed (iid) $N(0, \sigma^2)$. Generally, σ is unknown.

The model with interaction is

$$y_{ijk} = \mu + \alpha_i + \beta_j + \gamma_{ij} + \varepsilon_{ijk}$$

with additional constraints $\sum_i \gamma_{ij} = \sum_j \gamma_{ij} = 0.$

The notation $y_{ij.}$ is used for the mean over the r replicates for yields corresponding to factor level combination (ij), $y_{i..}$ for the mean over all units with factor A at level i, $y_{...}$ for the mean of all observations, and so on. Some writers use these notations for the corresponding totals, writing the means as $\bar{y}_{ij.}$, etc. The estimates of the parameters are

$$\hat{\mu} = y_{...}, \quad \hat{\alpha}_i = y_{i..} - y_{...}, \hat{\beta}_j = y_{.j.} - y_{...}, \hat{\gamma}_{ij} = y_{ij.} - y_{i..} - y_{.j.} + y_{...}$$

If the assumption of a homoscedastic normally distributed error is relaxed and the error is merely assumed to have the same unknown distribution with zero median and the remaining features of the above linear models are retained, interest centres on finding appropriate distribution-free procedures for testing for interaction, and perhaps for estimating the parameters in the model. Here the term distribution-free is used rather than nonparametric as there is still interest in hypotheses about, or estimation of, the parameters μ, α_i, β_j, γ_{ij}. Distribution-free refers to the error structure.

Some writers call this approach either *semiparametric* or *robust parametric*. Example 8.5 showed that a rank ordering of all observations followed by a formal analysis of variance of ranks can be unsatisfactory in the context of the linear model since it may miss a very evident interaction. The transformation to ranks, although a monotonic transformation of the data, effectively destroys the linearity.

The concept of interaction is often relevant in a broader context than linear models and is used to cover virtually any departure from an additive linear model. It also includes cases where it may not even be possible to assume a linear model with additive error for the original response variable.

It is well established that even in the one-way classification model, homogeneity of error variance can often be obtained by some appropriate transformation of the data. In a broader context, a model such as a multiplicative model may sometimes be converted to a linear model (with or without interaction terms) by a logarithmic transformation.

While there are many situations – such as those considered in Examples 8.2, 8.3 and 8.4 – where retention of the linear model is sensible if there is interest in responses in the measured units, there are situations where retaining the linear model is of less interest, or where that model may not even be appropriate. A linear model is irrelevant when responses, for example, are of a qualitative nature or measured only on an ordinal scale. In these circumstances, only certain types of interaction may be of interest.

One consequence of these generalities has been the development of two distinct approaches to nonparametric methods for dealing with interactions. The first retains the *parametric* element of the linear model, e.g., the parametric structure in the two-factor situation with the form $\mu+\alpha_i+\beta_j+\gamma_{ij}$. The other approach replaces this structure with an entirely different model that is both completely nonparametric and distribution-free. In this latter situation, an analysis may then often be appropriately performed using a rank transformation. This is because, without the linear model, a new conception of interaction is relevant.

Before a decision is made about which model is appropriate in a particular case, it helps to consider different types of interaction along the lines given in Cox (1984). Interactions are categorized according to their association with different aspects of data. Cox suggested that in essence, an observation y may be explained in terms of factors of three types.

The first type of factor comprises *treatment factors* that are under the control of the experimenter (e.g., temperature, pressure, the time a reaction is allowed to proceed in a chemical process). Treatment factors may be quantitative, ordinal, nominal or binary.

The second type of factor Cox called *intrinsic factors*. These include variables such as the age or gender of animals in an experiment. The experimenter has no control over such factors in the sense that they cannot be altered, although they might be taken into account in the choice of units and in the allocation of treatments to these units. Variables outside the experimenter's control, such as meteorological conditions, may also be regarded as intrinsic factors. Again, intrinsic factors may be quantitative, ordinal, nominal or binary. In many experiments, intrinsic factors and evaluation of their effects are of interest, as is any interaction between them and treatment factors.

The third type of factor consists of *nonspecific* or not uniquely characterized factors, such as block characteristics in a randomized block experiment.

Cox pointed out that three types of *two-factor* interaction are commonly of interest, these being,

- treatment×treatment;
- treatment×intrinsic;
- treatment×nonspecific.

Some of these may not be estimable. For example, in a randomized block experiment where each treatment occurs once and only once in each block, it is not possible to test for a *blocks×treatment* interaction. In effect, this is assumed to be zero. If it is not zero, it is incorporated in and will increase the error mean square in an analysis of variance.

If, in a randomized block experiment, each treatment is replicated r times within each block, the situation is technically equivalent to a two-factor experiment with factor A (treatments) at, say, t levels and factor B (blocks) at b levels. Any program for analysis of variance of a two-factor design will provide a test for *blocks×treatment* interaction. In Cox's terminology, this is an interaction between a *treatment* and a *nonspecific* factor.

Under the usual parametric linear model assumptions, a *blocks×treatment* interaction implies that treatment and block effects are not strictly additive. In practice, this is an indication that in one or more blocks treatments are responding differently from the way they respond in the remaining blocks. With the usual normal homogeneous error assumptions, this approach is essentially a parametric one, so this is not discussed further here.

Interest is usually focused on the effect of treatment factors on responses. Intrinsic and nonspecific factors must be allowed for only in the way they may bear on the treatment factors or if they are in themselves of interest (e.g., one may be interested in whether the response to a drug differs with the intrinsic factor *gender*). If the issue of interest is the additional hours of sleep achieved by individuals with sleeping problems given two different drugs, A and B, and one obtained the following results for four patients of each gender,

Male	A	1	2	2	2
	B	4	4	4	4
Female	A	4	4	4	5
	B	0	0	0	1

it does not need a formal analysis to show that there is an interaction between the intrinsic factor *gender* and the treatment factor *drug*, for it is evident that drug B is of greater benefit than drug A for males, but that the opposite applies for females.

Cox drew attention to the fact that certain types of interaction are order dependent in the sense that they are not removed by replacing the original data with monotonic transformations. In particular, this applies to the transformation to ranks, as this is a monotonic transformation (i.e., one that does not change the order of the data). If the data given above on the increased hours of sleep induced by two drugs are replaced by their ranks in increasing order (using mid-ranks for ties), the data become

Male	A	4.5	7	7	7
	B	12	12	12	12
Female	A	12	12	12	16
	B	2	2	2	4.5

and interaction in the sense that the drugs show an opposite ordering in responses between genders is still clear without any formal analysis.

In this example, a linear response model has little justification, and certainly, the error structure for the original data could hardly be described as homoscedastic normal by any stretch of the imagination.

The simple examples considered so far in this chapter have indicated that the rank transformation may remove interactions of interest when the additive linear model is appropriate (because linearity will not in general be preserved under this monotone transformation). In fact, when order-based interactions are of interest, there may be no particular reason to consider a linear additive model. A key feature of order-based interactions is that they are preserved under any monotonic, or order preserving, transformation, which includes transformation to ranks.

The distinction between these two situations has led to two different approaches to nonparametric or distribution-free analyses involving interactions. These are discussed in Sections 8.4.1 and 8.4.2, respectively.

SUMMARY POINTS:

- An observation may be explained in terms of treatment factors, intrinsic factors and nonspecific factors.
- Three types of two-factor interaction of common interest are treatment×treatment, treatment×intrinsic and treatment×nonspecific.
- Certain types of interaction are order dependent and are not removed by replacing the original data with monotonic transformations.

8.4 Alternative Approaches to Interactions

In Section 8.4.1, a method proposed by Hettmansperger and Elmore (2002) is considered in which it is assumed that a linear model that may include an additive interaction term is appropriate. Interest is in detecting any interaction and studying its nature.

In Section 8.4.2, a different approach where a rank transformation is appropriate is considered. This uses an alternative to the additive linear model. The basic notions stem from work by Akritas and Arnold (1994).

8.4.1 Interaction in an Additive Linear Model

Hettmansperger and Elmore considered the traditional model

$$y_{ijk} = \mu + \alpha_i + \beta_j + \gamma_{ij} + \varepsilon_{ijk},$$

but assuming only that the errors are identically and independently distributed continuously with zero median. They considered the balanced case with r replicates of each factor combination, the factor A having a levels and the factor B having b levels.

Their approach is motivated by the form of the usual test statistic in the analysis of variance for testing the hypothesis H_0: $\gamma_{ij}=0$ for all (i, j). The test statistic has an F-distribution with $m=(a-1)(b-1)$ and $n=ab(r-1)$ degrees of freedom under H_0 with the usual normal error assumption.

They point out that two relevant features of the test statistic for interaction are (i) that the numerator is a sum of squares of cell mean residuals formed by removing estimates of the A-factor and B-factor and overall (mean) effects, and (ii) the denominator is an estimator of error variance. They propose as the first step obtaining residuals after fitting estimates of μ, α, β appropriate to the model

$$y_{ijk} = \mu + \alpha_i + \beta_j + \varepsilon_{ijk}.$$

That is the additive linear model without an interaction term. These estimates might be, but need not necessarily be, the least squares estimates. If there is interaction, the residuals from fitting this model will reflect the fact that nonzero γ_{ij} have effectively been combined with the error term. This will inflate the error.

Hettmansperger and Elmore base a test for interaction on the *ranks* of the residuals formed above, i.e., $r_{ijk} = \text{rank}(y_{ijk} - \hat{\mu} - \hat{\alpha}_i - \hat{\beta}_j)$. Their test statistic Q is a modified analogue of the usual analysis of variance test statistic for interaction. However, the usual F-test is no longer appropriate if the normality assumption for errors is dropped, which is likely to be the situation when a nonparametric approach might be taken.

This statistic is, in effect, the interaction sum of squares of the ranks divided by the total mean square for ranks. It can be shown to have, for large values of r, approximately a chi-squared distribution with $(a-1)(b-1)$ degrees of freedom under the hypothesis of no interaction.

This approximation is based on theory developed in Chapters 3 and 4 of Hettmansperger and McKean (2011). The statistic is somewhat analogous to that used in the Kruskal–Wallis test.

Some studies based on approximate permutation distributions given by Hettmansperger and Elmore suggest that this asymptotic result is reasonable for all but very small data sets. Although the approximation has not been investigated in detail, it is believed that it may well be a conservative test in that it uses the total mean square in the denominator as a scaling factor rather than the error mean square, as in the analysis of variance.

Example 8.6

Use the method described above to test for interaction in the data in Example 8.3 using least squares estimates for the parameters in the additive model without interaction, then analysing the ranked residuals.

If least squares estimators are used, the first step is to fit an additive model without an interaction term. The analysis of variance program in most packages provides this facility and will usually give residuals. The residuals are then ranked, and an analysis of variance program fitting a model with an interaction term to these residual ranks provides the necessary information to calculate the Hettmansperger and Elmore statistic.

If a two-way analysis of variance program is used to fit the pure additive model without interactions, the residuals are

a_1b_1	0.125	0.525	1.025	1.225	1.425	1.625
a_1b_2	−1.525	−1.325	−1.225	−0.825	−0.625	−0.425
a_2b_1	−1.642	−1.342	−1.242	−0.742	−0.642	−0.342
a_2b_2	−0.392	0.408	1.108	1.408	1.608	1.808

Ranking the residuals gives:

a_1b_1	14	16	17	19	21	23
a_1b_2	2	4	6	7	10	11
a_2b_1	1	3	5	8	9	13
a_2b_2	12	15	18	20	22	24

These ranks are subjected to an analysis of variance fitting the model, allowing for interaction. The analysis of variance for this analysis is:

Source	Degrees of Freedom	Sum of Squares	Mean Square	F	P
Main A	1	0.0	0.0	0.00	1.000
Main B	1	0.2	0.2	0.01	0.918
Interaction	1	840.2	840.2	54.26	0.000
Error	20	309.7	15.5		
Total	23	1,150.0			

Output from R can be obtained using `resids <- aov(ch8$plant.weight.2 ~ ch8$growth.hormone + ch8$undersoil.heating)$resid` and `summary(aov(rank(resids) ~ ch8$growth.hormone * ch8$undersoil.heating))`.

Not surprisingly for the main effects, here labelled A, B, both sums of squares are near zero because the residuals are being analysed after having fitted the additive constants representing these factors. The *P*-values given in the above analysis are not relevant as they are based on an analysis of ranks and not all assumptions needed for the usual variance-ratio based *F*-test apply.

The information from this analysis for determining the Hettmansperger and Elmore statistic Q is the interaction sum of squares 840.2 and the total mean square, which is 1,150/23=50. The latter may be calculated independently. For ranked data, this quantity is given by $rab(rab+1)/12$ where, in this example, $r=6$, $a=b=2$. Thus, for this example, $Q=840.2/50 \approx 16.80$. Comparing this with the chi-squared distribution having $(a-1)(b-1)=1$ degree of freedom shows that $P<0.001$. Hence there is strong evidence of an interaction.

Output from the ANSM5 package in R can be obtained using `hettmansperger.elmore(ch8$plant.weight.2, ch8$growth.hormone, ch8$undersoil.heating)`.

Such evidence has already been found in Example 8.3, and, as there need be no concern about the normality assumption made in that example, it is somewhat artificial to apply the nonparametric procedure here. However, the procedure here may prove more efficient with a more general error assumption. If the modification outlined below, in which an alternative to least squares estimators of the α_i, β_j is applied, the procedure may be more robust against outliers.

While the method above is a test for interaction, it is not directly relevant to the estimation of effects. Its importance lies in the fact that strong evidence of interaction indicates that it may be more appropriate to study particular effects than it is to study main effects. How this might be done is considered below.

The result obtained here is only asymptotic. Hettmansperger and Elmore point out technical difficulties in developing an exact permutation test since mean ranks for each factor combination are not invariant under transformation. However, they indicate that, for an example in which they considered Monte Carlo sampling of permutations, a *P*-value that closely approximated their asymptotic result was obtained.

It is possible to use alternatives to the least squares estimators of the mean and main effects parameters. One suggested by Hettmansperger and Elmore is to use *median polish estimators* as described and studied in detail by Emerson and Hoaglin (1983). When this method is applied to the original data, the residuals have row and column medians zero rather than means zero. It is equivalent to fitting by least absolute deviations. An option for using the median polish is available in the `hettmansperger.elmore` function in the ANSM5 package in R. This calculates the residuals, which may then be ranked and the ranks analysed as in Example 8.6. Because outliers have little influence on the fit using median polish, this may result in one or two outliers having very large residuals, which may in some cases make the test less sensitive to the detection of interactions, although giving a model more appropriate for the reliable data (i.e., those excluding the outliers). This point is commented on in the next example.

Example 8.7

Consider again the data in Example 8.3 with the modification of replacing the final entry of 24.3 for both factors at the higher level by an extreme outlier taking the value 16. Use both least squares and median polish to fit a no-interaction model and use the Hettmansperger and Elmore rank of residuals method to test for interaction in each case.

Proceeding as in Example 8.6, least squares estimation gives $Q=11.21$, indicating $P<0.001$. Applying median polish and then ranking the residuals using these estimators of μ, α_i, β_j leads to $Q=9.72$, indicating $P\approx0.02$. There is strong evidence of interaction.

Output from the ANSM5 package in R can be obtained using `hettmansperger.elmore(ch8$plant.weight.4, ch8$growth.hormone, ch8$undersoil.heating, median.polish = TRUE)`.

That median polish gives a greater P-value than least squares is perhaps surprising because for each procedure the residual for the outlying 16 is ranked 1 since it is the negative residual of greatest magnitude. In each case, other residuals for that same factor combination are all positive and ranked 20, 21, 22, 23 and 24. However, there is greater variability among ranks of residuals for other treatment combinations using median polish, and this tends to dilute the evidence for interaction. This may reflect the slightly lower efficiency of median-based methods when the classic normality assumptions apply.

A classic least squares analysis of these data allowing for interaction gives an interaction $P=0.062$, meaning that interaction would then be ignored using conventional significance levels.

As indicated already, in the presence of interaction it is usually inappropriate to make comparisons on the basis of main effects, but more important to look at the particular effects of one factor at each fixed level of the other. In the case of a 2×2 structure considered in examples so far in this chapter, if there are interactions, one might compare the response at each level of factor A at the lower level of factor B using the WMW test and then do a similar comparison at each level of factor A at the higher level of factor B. Similar inspections might be made for particular effects of factor B.

In the more general situation of more than two levels of one or both factors, an obvious extension would be to perform Kruskal–Wallis tests across all levels of factor A separately at each level of factor B. An example of a 3×2 experiment is given by Hettmansperger and Elmore.

Tables of means, together with appropriate standard errors, are commonly used as aids to the interpretation of data with factorial structures, as already mentioned. A useful EDA extension of these, particularly with larger data sets, is to present boxplots for all treatment combinations. Hettmansperger and Elmore include an excellent illustration of this in the data set they use for illustrative purposes.

A similar approach to testing for interaction is described briefly by Higgins (2004, Chapter 9) who describes the procedure as an *aligned-rank transformation*. He also indicates related methods for testing for main effects, but it would seem desirable to use these only if no significant interaction is detected.

8.4.2 The Akritas and Arnold Distribution-Free Response Model

As indicated already, a linear model involving additive factor effects and a possible interaction term may not be appropriate. Also, even if it is appropriate for the initial data, the linear model will in general no longer hold for overall rank transformations of the data.

Akritas and Arnold (1994) proposed a remodelling of the response variable in terms of distributions rather than parameters. The models they proposed are very general and wide ranging, covering not only factorial structures but extending to experiments in which repeated measurements on the same units are made. In their initial presentation, they also relaxed the restriction made so far in this chapter of equal replication of all factor combinations. There are parametric models and analytic procedures available for many of the situations they cover, but their approach allows an appreciable relaxation of assumptions.

The generality of their model means that in practice fairly complicated algebra is needed to cover all possible scenarios. Their original paper has spawned a vast literature, but perhaps because of the generality of most of the procedures, the basic concepts do not seem to have received much coverage in nonparametric textbooks at the elementary level.

Some of the fundamentals of the approach are discussed here. For those who wish to apply the method to specific problems, the comprehensive review by Brunner and Puri (2001) is a good starting point, perhaps after first reading at least the introduction to Akritas and Arnold (1994). Other key papers in the area include Akritas, Arnold and Brunner (1997) and Brunner and Puri (1996). All contain numerous references to related and earlier work in this field, much of the latter being of a somewhat *ad hoc* nature, assuming either interactions did not exist, or if they did, that they were of some particular kind.

Inspiration for the Akritas and Arnold model comes from the simplest one-way linear model involving t groups or treatments, which in the classic parametric set-up may be written

$$y_{ij} = \mu + \alpha_i + \varepsilon_{ij}, \quad i = 1,2,\ldots,t, \quad j = 1,2,\ldots,n_i. \tag{8.3}$$

where $\sum_i \alpha_i = 0$ to ensure uniqueness of the α_i, and the ε_{ij} are iid N(0, σ^2).

The usual nonparametric approach is to drop the normality assumption on the ε_{ij} requiring only that they have a median of zero. In the case $t=2$, the standard nonparametric procedure is then WMW. The model (8.3) is then a model for location shift. An alternative way of writing this is to specify the distribution of the y_{1j} by a cumulative distribution function $F_1(x)$ and that of the y_{2j} by $F_2(x)=F_1(x-\theta)$. If $\theta \neq 0$, the implication is that the distributions differ only in location. Section 6.1.2 noted that the WMW test is also appropriate if the alternative to equality of distribution functions is that one distribution *dominates*. That is, for all x, either $F_1(x) \geq F_2(x)$ or $F_1(x) \leq F_2(x)$, with strict inequality for at least some x.

In this latter form, the test is completely nonparametric, whereas, in the linear shift form, there is a parameter θ, even if the test is distribution-free in the sense that the cdfs are not confined to any particular family.

Akritas and Arnold generalized this idea and instead of the model (8.3), assumed only that the y_{ij} had a distribution $F_i(x)$ such that

$$F_i(x) = M(x) + A_i(x), \quad \sum_{i=1}^{t} A_i(x) = 0. \tag{8.4}$$

That is, F_i is the distribution function of all y_{ij}, $j=1, 2, \ldots, n_i$, and it is expressed as the sum of two distributions.

The null hypothesis of no treatment difference is H_0: *all* $A_i(x)=0$, or equivalently, the $F_i(x)$ are the same for all i. Let $\bar{F}_{.}(x)$ denote the average over all i of the $F_i(x)$ and set $M(x) = \bar{F}_{.}(x)$. Then, $A_i(x) = F_i(x) - \bar{F}_{.}(x)$. Under the null hypothesis, which implies all $F_i(x)$ are equal, the models in (8.3) with only the normality assumption dropped and that in (8.4) are equivalent. However, under the alternative hypothesis, the latter is more general as it does not assume that the Y_{ij} are from families differing only in location. Akritas and Arnold refer to the model using (8.3) with the normality assumption relaxed as the *robust parametric model* and that in (8.4) as the *nonparametric model.*

The nonparametric model may be extended to a two-way factorial structure, and this extension shows appreciable differences from the corresponding robust parametric model considered in the preceding sections of this chapter, i.e.,

$$y_{ijk} = \mu + \alpha_i + \beta_j + \gamma_{ij} + \varepsilon_{ijk}$$

with the usual constraints and the errors iid with zero median.

For each given i, j assume now that the independent Y_{ijk} have the distribution function F_{ij} which can be expressed in the form

$$F_{ij}(x) = M(x) + A_i(x) + B_j(x) + C_{ij}(x),$$

subject to constraints

$$\sum_i A_i(x) = \sum_j B_j(x) = \sum_i C_{ij}(x) = \sum_j C_{ij}(x) = 0$$

where, extending the notation developed for the one-way classification model, the components are expressed in terms of appropriate averages of certain of the F_{ij}. Specifically,

$$M(x) = \bar{F}_{..}(x),\ A_i(x) = \bar{F}_{i.}(x) - \bar{F}_{..}(x),$$

$$B_j(x) = \bar{F}_{.j}(x) - \bar{F}_{..}(x), C_{ij}(x) = F_{ij}(x) - \bar{F}_{i.}(x) - \bar{F}_{.j}(x) + \bar{F}_{..}(x).$$

For the nonparametric model, the hypothesis of no interaction implies that $C_{ij}(x)=0$.

Careful reflection shows that this hypothesis is unchanged by monotonic transformations of the observed y_{ij}. This includes overall ranking of the data. An immediate implication is that an appropriate test will only pick up interactions that are preserved under transformation to ranks in the sense considered by Cox (1984).

Applied to conventional factorial designs, the usual assumption that the responses y_{ijk} are independent is appropriate. It is not essential to assume independence in other situations, and indeed, when applied to repeated measurements on the same units, such an assumption is inappropriate. Unlike the linear additive model, there is no specified error distribution and thus no assumptions about the variance of the y_{ijk}. The variance need not even exist, as would be the case for a Cauchy distribution.

Hypotheses that can be tested will still hold for monotone transformations of the original data, so all the analyses may be based on an overall ranking of the data. As Brunner and Puri (2001) point out, this means the procedure may be applied not only to designed experiments giving continuous responses but also to data consisting of counts and indeed to any observational data that may be ranked. In their most general form, the relevant test statistics allow for unequal replication in factorial structures, and ties may be dealt with by using mid-ranks.

Methods based on this model are especially apposite when linear additive models are inappropriate or irrelevant. Situations of this kind are common in drug testing and other clinical trials where, for example, data may consist of numbers of patients per group of, say, 50 or 100, responding to one of several different drugs, each drug being in this sense a treatment.

TABLE 8.3

Numbers of Patients Exhibiting Nausea and Vomiting in Groups of 50

	Males					Females				
Drug I	7	9	9	5	6	6	11	9	10	10
Drug II	8	11	10	10	9	3	4	7	2	4
Drug III	7	9	16	4	8	18	32	19	14	39

If three types of chemotherapy are assigned each to several groups of 50 patients undergoing treatment for cancer and counts made of the number of patients experiencing nausea and vomiting, typical counts by gender might be those indicated in Table 8.3. Brunner and Puri (2001) give an example of count data arising in a different context.

Attention here is confined to the Akritas and Arnold model applied to two-factor experiments with the first factor A at a levels and the second factor B at b levels. The simplest case is the 2×2 structure with $a=b=2$. The derivation of the results used here may be found in a much broader context in Brunner and Puri (2001). The key features are

i. Determination of an appropriate measure of the *relative* treatment effect of each factor combination,
ii. Forming a relevant statistic to test for interactions and when these are not significant equivalent statistics to test for main effects.

Appropriate asymptotic tests then need to be formulated, or exact small sample tests if feasible. Only asymptotic tests for what Brunner and Puri describe as ANOVA-type statistics are considered here. All tests are carried out on the overall ranked data, and effects are estimated on the basis of the sample, or empirical, cdfs for these ranked data.

For a 2×2 design with r replicates when using ranks, the relative treatment effects are analogous to treatment effects in the classic model, where the parameters α_i are estimated by the mean $y_{i..}-y_{...}$, i.e., the mean of all units receiving the ith level of factor A minus the overall mean. The corresponding relative treatment effect in the Akritas and Arnold model is $r_{i..}-r_{...}$.

In the 2×2 situation, the main effects and interactions are estimated by the mean rank analogues of the data means in the classic analysis.

However, the analogues of the ANOVA test statistics for mean and interaction differ in that, unlike the linear model with an additive homoscedastically distributed error term, the assumption of constant error variance is no longer relevant.

In light of remarks on the inappropriateness of considering main effects in the presence of interaction, it is recommended that the first test be one for interaction.

The ANOVA-type test statistic for interaction proposed by Brunner and Puri takes, for equal replication in an $a\times b$ design, the form

$$F_{AB} = \frac{abr(r-1)}{(a-1)(b-1)\sum_{i=1}^{a}\sum_{j=1}^{b} d_{ij}^2}\sum_{i=1}^{a}\sum_{j=1}^{b}(r_{ij.}-r_{i..}-r_{.j.}+r_{...})^2, \tag{8.5}$$

where

$$d_{ij}^2 = \sum_{k=1}^{r}(r_{ijk} - r_{ij.})^2$$

This is numerically equal to the standard ANOVA statistic for interaction computed for overall ranks. Inspection of the individual d_{ij} in typical examples shows that these differ markedly. This has a considerable effect on the relevant degrees of freedom for the asymptotic F-test. Approximate degrees of freedom for general cases are best expressed in matrix notation. Their evaluation is not difficult with appropriate computer software for handling matrix operations. These general forms are not discussed here, but they can be expressed in simple algebraic closed formulae for the 2×2 and one or two other simple structures.

Under the hypothesis of no interaction for the 2×2 structure, asymptotically, F_{AB} given by (8.5) has an F-distribution with degrees of freedom ν_1, ν_2 given by

$$\nu_1 = \frac{\left(\sum_{i=1}^{2}\sum_{j=1}^{2} d_{ij}^2\right)^2}{\sum_{i=1}^{2}\sum_{j=1}^{2} d_{ij}^4 + \sum_{i=1}^{2}\sum_{j=1}^{2} d_{ij}^2 d_{kl}^2}, \tag{8.6}$$

where the suffix $ij \neq kl$, and

$$\nu_2 = (r-1)\frac{\left(\sum_{i=1}^{2}\sum_{j=1}^{2} d_{ij}^2\right)^2}{\sum_{i=1}^{2}\sum_{j=1}^{2} d_{ij}^4}, \tag{8.7}$$

It can be verified (see Exercise 8.4) that if all d_{ij}^2 are equal, then $\nu_1=1$ and $\nu_2=4(r-1)$, the degrees of freedom associated with the test for interaction in classic ANOVA.

The complexity of the form of the degrees of freedom may seem surprising. However, the form is similar to that proposed by Satterthwaite (1946) for dealing with the classic Behrens–Fisher problem of extending the t-test for equality of means to samples from populations with different variances.

The formula for ν_2 extends to $a \times b$ factorials by taking summation over i from 1 to a and that over j from 1 to b. Although the matrix extension is straightforward, closed expressions for ν_1 become very complicated for a, $b>2$. Given below is an example for $a=3$, $b=2$ when

$$\nu_1 = \frac{2\left(\sum_{i=1}^{3}\sum_{j=1}^{2} d_{ij}^2\right)^2}{2\sum_{i=1}^{3}\sum_{j=1}^{2} d_{ij}^4 + 4\sum_{i=1}^{3} d_{i1}^2 d_{i2}^2 + \sum_{\text{rem}} d_{ij}^2 d_{kl}^2}, \tag{8.8}$$

where the final denominator sum $\sum_{\text{rem}}$ is over all pairs with suffix $ij \neq kl$ not occurring in the previous term, i.e., not in $\sum_{i=1}^{3} d_{i1}^2 d_{i2}^2$. If all d_{ij}^2 are equal, this reduces to 2, the degrees of freedom for interaction in classic ANOVA (Exercise 8.5).

Example 8.8

Apply the Akritas and Arnold method to the data in Table 8.3 to determine if there is evidence of a significant interaction between factors A (drug) and B (gender).

The statistic for testing for interaction is given by (8.5), which has asymptotically an F-distribution with degrees of freedom given by (8.7) and (8.8).

The interaction statistic is computed either by extracting the interaction variance ratio from a standard ANOVA analysis of the ranked data or by direct substitution of the relevant values in (8.5). The latter can be done once the d_{ij}^2 have been obtained and these are needed to evaluate (8.7) and (8.8). With the more advanced models discussed by Brunner and Puri, it is best to use programs that perform matrix operations. The calculations here can be handled by R and statistics packages through the writing of macros.

For purposes of discussion, the values of d_{ij}^2 are given below. Their calculation is straightforward, with d_{11}, for example (for males receiving drug I), being formed from the ranks associated with the numbers 7, 9, 9, 5 and 6 in Table 8.3. Taking ranks across all the data in Table 8.3 (and averaging in the presence of ties), these are 21, 15, 15, 25 and 23.5. The differences between these ranks and their average (19.9) are squared and summed to give 88.2.

$$d_{11}^2 = 88.2,\ d_{12}^2 = 156.2,\ d_{21}^2 = 75.2,\ d_{22}^2 = 48.8,\ d_{31}^2 = 265.8,\ d_{32}^2 = 14.8$$

whence (8.7) and (8.8) give $\nu_1 \approx 1.9$ and $\nu_2 \approx 15.2$. The value of the test statistic for interaction is $F_{AB} = 20.14$. With 2, 15 degrees of freedom, this gives $P < 0.001$. There is strong evidence of an interaction, the nature of which can be deduced from the means of the data given in Table 8.3 or from the table of rank means given below.

The presence of interaction suggests that particular effects rather than main effects should be examined in any further analysis. That is, effects should be compared within each gender separately.

The means for the original data in Table 8.3 are

	Males	*Females*
Drug I	7.2	9.2
Drug II	9.6	4.0
Drug III	8.8	24.4

A clear indication of a likely interaction is that while Drug II shows for males the highest mean side-effect rate by a small margin, in the case of females, Drug II has a markedly lower rate than that for either alternative drug. Table 8.3 also indicates immediately that the marked variation in the numbers of females showing side effects with drug III suggests considerable variability between the five groups in this factor combination, implying that the error is heteroscedastic if a linear model is considered.

That the interaction effect is still exhibited by the rank transformation is indicated by the following table of rank means:

	Males	*Females*
Drug I	11.1	17.6
Drug II	18.6	4.2
Drug III	13.7	27.8

The transformation to ranks does not result in the homogeneity of error variance required in the classic analysis. This is evident from the marked differences between the six d_{ij}^2 displayed above. This is why the classic analysis applied to ranks would be inappropriate. Also, even with the classic model, monotone transformation to ranks may destroy linearity. The Akritas and Arnold model, where it is appropriate, compensates for these difficulties. As previously emphasized, it may not be appropriate if linearity is an important, physically meaningful concept.

If the interaction test indicates no significant interaction, then the main effects can be investigated. These follow broadly similar lines to the test for interaction described in this example for $a \times b$ factorial structures with equal replication. Details are omitted even for the simple cases, but these may be found in Brunner and Puri (2001).

In this example, it may be appropriate to apply Kruskal–Wallis or some other test to data for each gender separately in view of the strong evidence of interaction.

Brunner and Puri (2001) discuss precisely what hypotheses can be tested and how the methods extend to designs with unequal replication and to repeated measurements where the independence assumption is no longer valid and correlations between observations on the same unit at different times have to be taken into account. It is interesting to note that the WMW, the Kruskal–Wallis, Page and several other well-known tests are special cases of the Akritas and Arnold one-way model. Details are given in Brunner and Puri (2001).

SUMMARY POINTS:

- The Hettmansperger and Elmore test for interaction is based on the ranks of the residuals obtained from an analysis of variance of the original observations that take into account the main effects but not the interaction term (Section 8.4.1).
- These ranks are then subjected to an analysis of variance that allows for interaction (Section 8.4.1).
- The Hettmansperger and Elmore statistic, Q, is based on the interaction sum of squares and the total mean square (Section 8.4.1).
- For larger samples, Q can be approximated by a chi-squared distribution (Section 8.4.1).
- If the additive linear model is irrelevant and only interactions that are preserved under monotonic transformations are of interest, a model proposed by Akritas and Arnold that defines interaction in terms of distribution functions is often appropriate (Section 8.4.2).
- This takes a simple form for two-way factorial structures and extends very generally to cover cases of unequal replication as well as repeated measurement structures (Section 8.4.2).

8.5 Crossover Experiments

When several treatments are involved, experimenters often wish to give each treatment to each study participant. Response to any treatment may be influenced by treatments previously received. The situation may occur in clinical trials involving several drugs where the response to one drug may be influenced by the residual effects of a drug previously administered, something mentioned in Section 5.1.1.

A similar situation may arise if several different tools are used to perform a specific task. The performance with tool A might be more satisfactory if it is used before tool B rather than after. This could happen, for example, if tool B created considerable vibration that left the operator's hands shaky when he or she started to use tool A afterwards.

In educational or psychological experiments involving different tests, each being given to all participants, possible learning effects should be considered. Procedures tested early in the sequence may in that way influence performance in procedures tested later in the sequence.

It is sometimes possible to eliminate residual or carryover effects by allowing appreciable time lapses between treatments. Otherwise, a design should be used that allows for these effects.

One device useful in such circumstances is a *crossover design,* mentioned briefly in Section 5.1.1. Only the simplest such design is described here, in which two treatments are compared using what is called a *two-period crossover design.* The method may be extended to any finite number of treatments over more than two periods.

Crossover designs are widely applied in studies of the effects of drugs on chronic medical conditions such as asthma, where neither death nor complete cure is likely. Jones and Kenward (2014) give an account of their use, together with descriptions of both parametric and nonparametric methods of analysis. Putt and Chinchilli (2004) review a number of nonparametric approaches to the analysis of crossover trials and include references to more detailed studies.

Koch (1972) gives an example of a nonparametric analysis of one such experiment, which is considered in Example 8.9. Two treatments (A and B, say) are applied to each of an even number of patients. Half of these are selected at random to receive treatment A first. The treatment order is reversed for the remaining patients. The design enables the testing of hypotheses about

- Direct effect of treatments;
- Any residual effects of treatments;
- Period effects.

where *period effects* represent a genuine difference in responses in each period that is independent of the actual treatment applied in that period. This is only meaningful if there are no residual effects of treatments.

Example 8.9

Koch (1972) analysed the data below from an experiment at the Dental Research Center, University of North Carolina. Ten children were randomly assigned to two groups of five. Two treatments were orally consumed by each child, these being

A 100 ml grapefruit juice followed by an elixir of Pentobarbital;
B 100 ml water followed by an elixir of Pentobarbital.

The first group of five children received A in Period I and B in Period II (sequence AB). The second group received B, then A (sequence BA). Data are the amount of the elixir in a 10 ml blood sample taken 15 minutes after administration expressed in units of μg ml^{-1}. The data were

Child	1	2	3	4	5	6	7	8	9	10
Sequence	AB	AB	AB	AB	AB	BA	BA	BA	BA	BA
Period I	1.75	0.30	0.35	0.20	0.30	7.20	7.10	0.75	2.15	3.35
Period II	0.55	1.05	0.63	1.55	8.20	0.35	1.55	0.25	0.35	1.50

The large values for treatment B for children 5, 6 and 7 relative to other readings suggest heterogeneity, so a nonparametric test seems appropriate. Koch based his analysis on the WMW test.

The data are first examined for each child summed over Period I and Period II. If there is a different carryover effect (i.e., if the joint response summed over the two periods is different), it would be expected that the WMW test would indicate a significant difference between the sums for children 1–5 (sequence AB) and 6–10 (sequence BA). The sums for each child in the order given in the data are

Sequence	AB	2.30	1.35	0.98	1.75	8.50
Sequence	BA	7.55	8.65	1.00	2.50	4.85

The R commands `groupAB.sum <- (ch8$periodI + ch8$periodII)[ch8$sequence == "AB"]` and `groupBA.sum <- (ch8$periodI + ch8$periodII)[ch8$sequence == "BA"]` can be used to obtain the above sums. The ANSM5 package in R and the command `wilcoxon.mann.whitney(groupAB.sum, groupBA.sum)` then give an exact two-tail $P=0.3095$ for the WMW test. A possible alternative would be a normal scores test, using the ANSM5 package in R and the command `normal.scores.test(groupAB.sum, groupBA.sum)`, which gives $P=0.2778$, so there is no evidence of a differential residual effect. This does not mean that there is no residual effect from a previous treatment, but if there is any, then it applies equally to either sequence. When there is no evidence of differential residual effects, the next step, and the most important one, is to test for a direct effect of treatments. This is based on the *differences* Period I–Period II for all individuals. If the effects of the treatments depend on the order of administration, then the Period I–Period II differences for the first five children should exhibit a different location pattern to those same differences for the remaining children because one is measuring the effect A–B, the other B–A. Under H_0: *no treatment difference,* both sets of differences should have the same distribution.

The relevant differences are

Sequence	AB	1.20	−0.75	−0.28	−1.35	−7.90
Sequence	BA	6.85	5.55	0.50	1.80	1.85

The R commands `groupAB.diff <- (ch8$periodI - ch8$periodII)[ch8$sequence == "AB"]` and `groupBA.diff <- (ch8$periodI - ch8$periodII)[ch8$sequence == "BA"]` can be used to obtain the above differences. The ANSM5 package in R and the command `wilcoxon.mann.whitney(groupAB.diff, groupBA.`

`diff`) then give an exact two-tail $P = 0.0159$, a fairly strong indication of a real treatment effect. There is no evidence of a differential residual effect, but evidence of a location difference between A and B.

A further test proposed by Koch is only relevant if the hypothesis of no differential residual effect is accepted. This is a test for a period effect, i.e., that the responses A–B may be different for each sequence order. The A–B difference for those in the sequence AB is that given above, while for those given the sequence BA, it is the difference given above after reversal of sign. For the above data, the WMW test gives a two-tail $P = 0.309$, so there is no evidence of a period effect.

In practice, most crossover experiments used in the pharmaceutical industry or in psychological or educational tests, for example, are more complicated than the one described in Example 8.9.

SUMMARY POINTS:

- In crossover designs, each participant or unit receives each treatment.
- Crossover studies have one group of participants for each possible ordering of the treatment. For the simplest two-period crossover design, the two groups consist of those who receive Treatment A first and those who receive Treatment B first.
- The periods during which the treatments are received should be separated by a time lapse to reduce the chance of responses to the second treatment being contaminated by residual effects from the first treatment.

8.6 Specific and Multiple Comparison Tests

In parametric analysis of variance, after a preliminary overall test, interest usually shifts to specific treatment comparisons. In factorial experiments, for example, interest may centre on main effects and interactions. If interactions are indicated, then the focus is on the particular effects of each factor. In other contexts, one may be interested in pairwise comparisons for a selected subgroup of treatments or in comparing one specified treatment, often called a *control*, with each of the remaining treatments. One may also want to make comparisons between groups of treatments; for example, if three of a set of seven fertilizer treatments include a nitrate component while the remaining four do not, one may want to make inferences about differences between the overall means or medians for treatments with the nitrate fertilizers and the corresponding measure for nitrate-free fertilizers.

There are nonparametric procedures for assessing such contrasts, covering both hypothesis tests and estimation (e.g., obtaining confidence intervals for differences). A detailed treatment is beyond the scope of this elementary text. Hettmansperger and McKean (2011, Chapter 4) give an account of the basic theory illustrated with a range of examples. Many contrasts of practical interest are considered by Hollander, Wolfe and Chicken (2014, Chapters 6 and 7) and by Conover (1999, Chapter 5).

Careful thought should be given to which of many possible contrasts are relevant and to the implications of the fact that not all contrasts are independent of one another. In an experiment in which one treatment may reasonably be regarded as a control, it is often both relevant and sensible to compare the mean or median for this treatment with the corresponding measure for each remaining treatment. Such contrasts may be of interest, for example, when comparing the efficacy of three different drugs in treating a disease, and the experiment also includes a control group consisting of individuals not receiving any drug.

If an experiment is conducted where the different treatments consist of administering the same drug at a sequence of increasing dose levels, the main contrast of interest will be in how the response changes with dose level. Where each dose is given to a different group of patients, overall tests for increasing or decreasing responses with dose are provided by the Jonckheere–Terpstra or the Page test (the latter is not covered in this text). In some circumstances, umbrella tests of the type mentioned briefly at the end of Section 7.2.5 may also be appropriate. Further tests using specific contrasts may also be relevant to find, for example, if there is a threshold dose which must be exceeded before there is evidence of a real treatment effect.

For both parametric and nonparametric analyses, many more-or-less automatic procedures known as *multiple comparison tests* have been devised. They are designed to protect against making unwarranted inferences when many contrasts that are not necessarily independent are examined. The aim is to restrict the number of unwarranted conclusions of real differences that may be reached due to carrying out all possible tests. Such conclusions are often referred to as *false positives,* although that terminology is also used in a wider sense in diagnostic studies.

If 20 treatments are involved in an experiment, an overall parametric test such as the F-test in the analysis of variance, or a Kruskal–Wallis test, may indicate no acceptable evidence of treatment differences. In that situation, a comparison of the largest observed treatment mean with the smallest observed treatment mean, ignoring all others, might well produce a P-value sufficiently small to indicate strong evidence that these means differ, contrary to the conclusion in the overall test that treatments do not differ. Doing this runs the risk of misleading findings, and it is not advisable to carry out such a test purely because the data look as though that test might reveal something interesting.

Multiple comparison procedures should not be used blindly, especially when there is no obvious treatment structure such as that implied in factorial experiments or a structure where comparisons with a control are of prime interest.

A constructive approach is one where an experimenter, often in association with a statistician, elects (or nominates) comparisons that are of interest before the data are obtained and makes relevant assessments of evidence about these on the basis of what information is later provided by the data. The interpretation of P-values should be used as a tool for weighing evidence rather than as a decision tool in a way that is implicit in the assignment of pre-fixed formal significance levels.

One approach widely used for nominated pairwise comparisons is to compute what are known as *least significant differences*. Logically, these can only be justified if applied to comparisons chosen before the data are collected, or at latest before the data are examined. Even then, they are applied only if an overall Kruskal–Wallis, Friedman or other relevant test indicates strong evidence for differences. The critical P-value or significance level used in multiple comparison tests should be no less stringent than that in the overall test.

When using, for example, the Kruskal–Wallis test, the criteria for accepting a location difference between the ith and jth samples are that

1. The overall test indicates significance, and
2. If $r_{i.}$ and $r_{j.}$ are the mean ranks for these samples of size n_i, n_j, then

$$|r_{i.} - r_{j.}| / \sqrt{(S_r - C)(N-1-T)(n_i+n_j)/[n_i n_j (N-t)(N-1)]} > t_{N-t,\alpha}, \tag{8.9}$$

where T is given by (7.2) and $t_{N-t,\alpha}$ is the t-value required for significance at level α% in a t-test with $N-t$ degrees of freedom, other quantities being defined as in Section 7.2.2.

An analogous result holds for van der Waerden scores, with sample rank means replaced by sample score means and C now zero and T being the form of the statistic appropriate for these scores.

Nominated comparisons for the Friedman test may be based on either treatment rank means or totals since here there is equal replication. The algebraic expression in terms of rank totals is simpler. An analogue of (8.9) gives the requirement for a least significant difference between the treatment rank totals $R_{i.}, R_{j.}$:

$$|R_{i.} - R_{j.}| / \sqrt{2b(S_r - S_t) / [(b-1)(t-1)]} > t_{(b-1)(t-1),\alpha} \tag{8.10}$$

where $S_r = \sum_{i,j} r_{ij}^2$ and $S_t = \sum_i R_{i.}^2 / b$.

Many statisticians would agree with Pearce (1965, pp. 21–2) that it is reasonable to compare a subset of all samples (or treatments) if that subset is selected before the data are obtained (or at least before they are inspected) using formulae like (8.9) or (8.10) if one could sensibly anticipate that the specified pair are likely to show a treatment difference that may be of practical interest or importance.

Example 8.10

Uniform editions by each of three writers of detective fiction are selected. The numbers of sentences per page, on randomly selected pages in a work by each, are

C.E Vulliamy	13	27	26	22	26		
Ellery Queen	43	35	47	32	31	37	
Helen McCloy	33	37	33	26	44	33	54

For this example, assume that there is strong evidence for differences between the authors in terms of the numbers of sentences per page. The data suggest that Vulliamy may use longer sentences than Queen. It is reasonable to investigate this further, since the former was a British author with an academic background and limited fictional output, whereas Queen is the pseudonym for a pair of popular and prolific American authors. Use a least significant difference test to determine whether the difference is significant.

The data are ranked as shown below. The relevant quantities are calculated using the method explained in Example 7.2 and substituted into (8.9).

Vulliamy	Number	13	22	26	26	27		
	Rank	1	2	4	4	6		
Queen	Number	31	32	35	37	43	47	
	Rank	7	8	12	13.5	15	17	
McCloy	Number	26	33	33	33	37	44	54
	Rank	4	10	10	10	13.5	16	18

Relevant rank means are $r_1=17/5=3.4$ (Vulliamy), $r_2=72.5/6=12.1$ (Queen). Also, $S_r-C=$ 2,104.5 – 1,624.5=480.0, $T=9.146$, $N=18$, $n_1=5$, $n_2=6$, $t=3$. With the difference between rank means 12.1–3.4=8.7, the left-hand side of (8.9) is:

$$8.7 / \sqrt{(2,104.5 - 1,624.5)\,(18-1-9.164)(5+6)/[5 \times 6(18-3)(18-1)]} = 3.73.$$

Tables give the critical t-value at the 1% significance level with 15 degrees of freedom to be 2.95, so since 3.73>2.95 the difference is formally significant at the 1% level, i.e., there is strong evidence of a real difference. Note that the 1% level is used since the overall test indicated significance at this level.

Output from the ANSM5 package in R can be obtained using `kruskal.wallis.lsd(ch8$sentences, ch8$authors, c("Vulliamy", "Queen"))`.

Example 8.11

The description of the experiment and the data in the next example were kindly provided by Chris Theobald.

As part of a study of the feeding habits of the larvae of the Blue-tailed Damsel Fly, 6 of the larvae were collected along with six members of each of seven of the species of prey on which they usually fed. Each larva was placed on a cocktail stick in a glass beaker that contained water and one of each of the seven types of prey. Records were kept of the order in which each larva ate the prey. The results are shown in Table 8.4.

Some ties denoted by mid-ranks arose due to failure to observe the order when a larva ate two prey in quick succession. Is there evidence that the larvae prefer to eat some species of prey before others?

Evidence of preference requires a Friedman test analogous to that in Example 7.6. A multiple comparison test using (8.10) is one approach to examining the situation regarding *cyclops*.

There are $t=7$ types of prey (treatments) and $b=6$ larvae (blocks). Relevant ranks are given in Table 8.4, where the final column gives rank totals s_i for each prey. Squaring each rank and adding gives $S_r=1^2+2^2+1^2+\ldots+3^2+4^2+5^2=837$. The sum of the squared row totals is $9^2+41^2+28.5^2+23^2+25.5^2+15^2+26^2=4{,}654.5$. Hence $S_t=4{,}654.5/6=775.75$. Also, $C=6\times7\times(7+1)^2/4=672$.

Applying (7.6) gives $T=22.64$, and (7.7) gives $T_1=8.47$, indicating a P-value of less than 0.01, or probably less than 0.001. Indeed, the Monte Carlo estimate of the exact P-value from the `friedman` function of the ANSM5 package in R indicates almost certainly that $P<0.0005$. Setting $\alpha=0.01$, the appropriate t-value with $(b-1)(t-1)=30$ degrees of freedom is 2.75, so in (8.10) the least significant difference between treatment rank totals given in the last column of Table 8.4 that needs to be exceeded is $2.75\times\sqrt{2\times6(837-775.75)/30}=13.61$.

TABLE 8.4

Preferences of Blue-Tailed Damsel Fly Larvae for Various Prey

	Larva						
Prey	**1**	**2**	**3**	**4**	**5**	**6**	**Total**
Anopheles	1	2	1	1	3	1	9
Cyclops	7	7	7	6.5	6.5	7	41
Ostracod	5.5	5.5	5	5	5	2.5	28.5
Simocephalus (Wild)	3	4	4	6.5	1.5	4	23
Simocephalus (Domestic)	2	1	6	4	6.5	6	25.5
Daphnia magna	4	3	2	2	1.5	2.5	15
Daphnia longispina	5.5	5.5	3	3	4	5	26

The *cyclops* was usually the last to be eaten. Taking each rank total in turn, the differences relative to the rank total for *cyclops* exceeds 13.61 apart from the total for *ostracod*, so there is strong evidence that there is a low preference for *cyclops*.

Output from the ANSM5 package in R can be obtained using `friedman.lsd(ch8$prey.preference, ch8$prey, ch8$larva, c("Cyclops", "Anopheles"))`.

It is possible to devise a test comparing *cyclops* with the average rank response for all other prey, which would confirm the conclusion that *cyclops* was less favoured. A more sophisticated approach to multiple comparisons is given by Leach (1979, Section 6.2). A useful paper on this topic is that by Shirley (1987), and an important earlier paper is Rosenthal and Ferguson (1965).

A widely used way of allowing for the testing of multiple pairs is the *Bonferroni correction* (Bland and Altman, 1995). If an overall test involves k treatments, there are $k(k–1)/2$ paired comparisons between treatments. With this correction, pairwise comparisons are accepted as significant only if the associated

$$P \le \frac{\alpha}{k(k-1)/2}$$

Thus, if $k=6$ and $\alpha=0.05$, each pairwise test would require $P \le 0.05/15 \approx 0.0033$.

Depending on what procedure is relevant, each pairwise test may be carried out by an appropriate parametric procedure such as the *t*-test or a nonparametric one such as the WMW test with $P \le 0.0033$ as the criterion for significance. In the terminology of hypothesis testing, this method tends to be conservative. For $k=6$, the true overall significance level is $1-(1-0.05/15)^{15}$. $=1 - 0.9511=0.0489$ rather than 0.05. This difference is due to using $1-\alpha$ as an approximation for $(1-\alpha/n)^n$. Although possibly trivial here, this discrepancy can be important if k is large.

SUMMARY POINTS:

- The analysis of pairs of groups subsequent to the testing of all the groups together is described as post-hoc pairwise testing.
- Post-hoc testing should only be carried out if there is evidence against the original multi-sample null hypothesis.
- If several pairings are tested in turn, there is an increased probability of finding evidence against the null hypothesis even if the null hypothesis for all the groups together is true.
- Multiple treatment comparisons and specific treatment comparisons are often based on least significant differences.
- If the Bonferroni correction is applied, the *P*-value benchmark used for the pairwise testing is equal to the benchmark *P*-value used in the multi-sample test divided by the number of pairs tested.

8.7 Exercises

8.1 By modifying one of the variables in Table 8.1, show how this information can be presented as illustrating a negative interaction.

8.2 Researchers hypothesized that the time required for two players to complete a video game is longer for those with more experience, as their skills are more finely tuned. The data below show the time taken (minutes) to complete a video game by pairs of similarly skilled players. Experience of video game playing (factor A) – is rated as *high, intermediate, low* and each pair plays one of two different games (factor B) – I, II. Seven pairs compete in each category. Use suitable EDA techniques, such as boxplots, for each factor combination to study and highlight important features of the data.

High	I	25.1	20.2	13.5	11.3	10.9	8.7	7.8
High	II	48.6	40.9	31.5	24.9	20.8	18.5	14.8
Intermediate	I	18.0	12.9	11.4	10.7	9.1	8.8	5.6
Intermediate	II	27.1	20.9	20.1	16.3	15.3	13.8	10.4
Low	I	11.2	8.5	7.0	6.3	3.8	3.4	2.9
Low	II	12.6	8.3	7.7	7.0	5.4	4.7	3.8

8.3 For the data in Exercise 8.2, apply the Hettmansperger and Elmore analysis to investigate relationships between experience, the game used, and length of play. Is there evidence of an interaction between experience and the game played?

8.4 Show that if in (8.6) and (8.7) all d_{ij} are equal, then the expressions reduce to the degrees of freedom for interaction and for error in classic parametric ANOVA.

8.5 Show that if in (8.8) all d_{ij} are equal, then the expression reduces to the degrees of freedom for interaction in classic parametric ANOVA.

8.6 A crossover design is used to assess the writing skills of young children living in a seaside town. Fourteen children of the same age and taught in a similar manner each produce two pieces of creative writing. One is a factual description of a local tourist attraction (A), and the other is a short story set in their neighbourhood (B). Half the children write A in Period I and B in Period II (sequence AB), and the other half write B in Period I and A in Period II (sequence BA). For each child, the number of spelling mistakes is recorded as follows:

Child	**1**	**2**	**3**	**4**	**5**	**6**	**7**
Sequence	AB	AB	AB	AB	AB	AB	AB
Per I	7	4	11	9	0	8	2
Per II	5	3	8	7	2	6	3
Child	**8**	**9**	**10**	**11**	**12**	**13**	**14**
Sequence	BA	BA	BA	BA	BA	BA	BA
Per I	5	8	6	8	7	3	4
Per II	6	5	7	8	8	6	5

Do these data indicate differences in the number of spelling mistakes between A and B? Is there evidence that the number of mistakes depends upon whether the factual description or the story is written first?

8.7 A sportswear manufacturer produces a low-priced shoe (A) for athletes that it recommends for training runs and a higher-priced shoe (B) that it recommends for competitive events. The coach at an athletics club who takes a doubtful view of advertising suggests that performance may be no better with a change to the more expensive type for competitions, indeed that the reverse may be true. A crossover trial is conducted in which 16 members of the club of varying ability test them in 100 m sprints. Eight use shoe A in training runs, followed by shoe B in a competitive run (combination AB). The remaining 8 use shoe B in training and shoe A in the competitive run (combination BA). Do the experimental results below justify the coach's doubts? What recommendation should be made to members about which type of shoe to use?

Athlete	**1**	**2**	**3**	**4**	**5**	**6**	**7**	**8**
Sequence	AB	AB	AB	AB	AB	AB	AB	AB
Per I	10.7	11.4	13.2	12.2	12.9	11.1	15.1	13.7
Per II	11.1	11.3	14.9	12.1	14.2	16.1	14.9	14.3
Athlete	**9**	**10**	**11**	**12**	**13**	**14**	**15**	**16**
Sequence	BA	BA	BA	BA	BA	BA	BA	BA
Per I	11.3	13.2	13.5	14.4	15.8	11.3	14.1	15.1
Per II	11.1	13.3	12.7	13.9	10.3	11.0	13.1	14.9

8.8 The severity of seizures experienced by people with epilepsy can be assessed using the Chalfont Seizure Severity Scale (Duncan and Sander, 1991). The overall score can take a value between 1 (least severe) and 178 (most severe). Outpatients recruited from the neurology departments of three hospitals had the seizure severity scores:

Hospital A	4	11	35	41	49	62	80	104	114	128
Hospital B	2	10	27	36	38	79	103	119	130	
Hospital C	80	90	99	104	108	131	162	166		

At Hospital C, less severe patients are cared for in the community, whereas Hospital A is highly traditional, and outpatient care for all patients with epilepsy is regarded as important. Is there evidence for a difference between hospitals A and C?

8.9 For Exercise 8.8, assuming that post-hoc examination of pairs of hospitals is reasonable, apply multiple testing with the Bonferroni correction.

8.10 Mediaeval silver coins produced in Tunisia over a period of more than 800 years were analysed for percentage silver content, and for each coin, the dynasty during which it was produced was recorded (Ben Abdelouahed et al., 2010). This information is tabulated below. Given that the coins with the highest silver content were produced in the final dynasty, is there an indication that the percentage of silver content increased between dynasties? Is it reasonable to conduct a pairwise post-hoc test of the difference in average silver content between the first and final dynasties?

Dynasty	Period (AD)	Silver Content (%)					
Umayyad	702–748	98.7	96.1				
Abbasid	775–833	90.6	90.3				
Fatimid	909–975	75.8	74.3	69.3	68.8	77.2	64.2
Zirid	975–1159	57.6	70.6	63.0	54.9	59.8	73.0
Mowahid	1159–1230	91.0	99.8	99.7	99.8	97.4	96.3
Hafsid	1230–1574	90.8	99.8	99.8	99.9	99.9	99.4

9

Analysis of Survival Data

9.1 Special Features of Survival Data

9.1.1 Historical Context

The term survival in its literal sense concerns matters of life and death. Since the dawn of human civilization, society has been aware that length and quality of life can be increased through, for instance, appropriate selection and preparation of food, isolation of those with incurable infectious diseases from the remainder of the community, and the existence of peace rather than war. The Old Testament describes how the Hebrew people were expected to follow these principles.

It was not until the 1660s that John Graunt constructed a basic life table using the weekly Bills of Mortality (lists of deaths) recorded in London (Graunt, 1665). Starting from a specific number of births, Graunt estimated the number of individuals that would still be alive at specific ages (Table 9.1). These figures reflect the squalid conditions under which most Londoners lived at the time and, in particular, the impact of cholera epidemics (Thompson, 2011).

Developments in the construction of life tables led to interest in the possible factors that might influence human survival. For instance, Edward Wigglesworth, who was a professor at Harvard University during the late eighteenth century, noted that in America, the clergy and university professors experienced better survival than the general population. He applied his work on life expectancy to a proposal for the establishment of an annuity fund for the widows of the clergy in New England (Vinovskis, 1971).

TABLE 9.1

Estimated Survival Rates Based on Graunt's Data

Age (Years)	Percentage Alive
0	100
6	64
16	40
26	25
36	16
46	10
56	6
66	3
76	1
86	0

DOI: 10.1201/9780429326172-9

By linking the expectation of human life to carefully invested regular financial contributions from a large number of individuals, provision can be made for a particular individual's dependants in the event of his or her untimely death. For a highly detailed account of the origins of life assurance (now commonly referred to as life insurance), see Walford (1885), from which the social attitudes of the day speak for themselves. Actuarial science has led to the development and application of a wide range of quantitative techniques for modelling financial data (Promislow, 2015).

The modelling of survival data has evolved into a substantial field. This chapter is intended to serve as an introduction to the concepts in a nonparametric setting. The details can become technical, so the chapter should be read in conjunction with a paper describing a practical application of survival analysis in a relevant field, e.g., the article by Wolfe et al. (2005) on survival following a stroke. A more detailed text on survival analysis suitable for readers with some statistical background is Kleinbaum and Klein (2011).

9.1.2 Survival Data in Contemporary Research

The collection of survival data is no longer confined to studies of human longevity and may involve the comparison of two or more groups. A medical treatment could be compared with a placebo by randomly allocating consenting patients to one of the interventions. Measures of interest from the two groups might include times from diagnosis of the illness to discharge from hospital, duration of remission or the time from an operation to complete recovery. Here the word *treatment* is used in a very broad sense; for instance, it might be a drug, a surgical procedure, a lifestyle intervention such as cessation of smoking, or even different forms of a disease such as cancer.

Length of survival can also be compared by demographic variables such as gender or ethnic group. For instance, Wolfe et al. (2005) report a comparison of black and white ethnic groups on survival following a stroke.

Survival data are commonly displayed graphically as a survival curve. The horizontal axis represents time, and the vertical axis is the proportion of the sample for which the event of interest (e.g., death, hospital discharge, complete recovery) has not as yet occurred (Figure 9.1). In a medical context, the asymmetric nature of the distribution reflects a common tendency for many treatments to be largely ineffective, or only moderately effective, with some patients, while markedly prolonging the time to the event of interest for others.

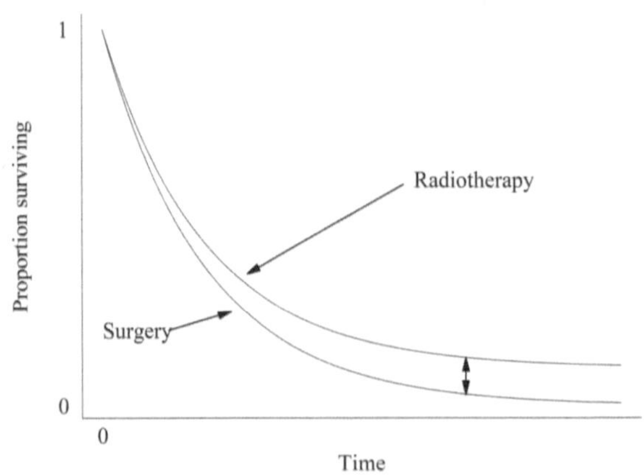

FIGURE 9.1
Survival curves for patients treated by either surgery or radiotherapy.

Two key features of survival data have stimulated the development of appropriate new tools. These features are that:

- Distributions of survival times of patients in clinical trials or of times to failure for components in an engineering or industrial context are typically long-tailed or skewed to the right.
- Some units are often lost to a study before the response of interest (e.g., death of a patient or complete recovery, or failure of a machine) occurs.

In medical studies, the progress of patients is often monitored for a limited time after treatment; this may be anything from a few weeks to 10 years or more. Consider a comparison of two drug treatments A and B, in which patients are compared on their survival in months. Suppose that the observations for Groups A and B are 1 3* 4* 5* 7 9 and 2 6* 8 12 15, respectively, where an asterisk * indicates that the patient left the study and their subsequent survival is unknown.

The precise survival time is therefore not known for three patients in Group A and one patient in Group B. The lengths of survival at exit from the study are recorded and used in analyses; the observations are described as *censored.*

Censored observations may be due to a variety of causes. The main ones are:

- The study is terminated before all subjects show the designated response.
- For one of the reasons given below, some subjects withdraw from the study before showing the response.
- Data limitations may preclude observation of the designated response.

How censored data are best analysed depends crucially on which of these causes come into play and how the censoring is related to a particular treatment when survival times for several treatments are compared.

The tests introduced in this chapter will only be valid if censoring is not caused by a factor that is treatment dependent. Valid analysis is possible when censoring is treatment dependent, but the methods are more advanced than those given here.

Survival analysis becomes more difficult to address when, given the choice of treatment, death and censorship are still correlated with each other. This issue is undecidable as there will always be a model that fits a set of data and assumes that times due to die and times due to be censored are statistically independent (Tsiatis, 1975).

One cannot tell if censoring is treatment dependent merely by looking at the data. At first sight, the data for Groups A and B appear to indicate treatment-dependent censoring since there is a greater proportion of censored observations in Group A (3 out of 6) than there is in Group B (1 out of 5). However, this difference might simply be due to a decision to terminate the study on a fixed date, particularly if most of the earlier patients were given treatment B.

On the other hand, when two or more treatments are used in a trial to compare the efficacy of two drugs to treat the same illness or disorder and one of these has serious side effects that cause, or even forces, many patients to withdraw before the condition of interest (recovery or whatever) is observed, censoring would be treatment dependent.

Causes of censoring, which often are not treatment dependent, include the subject leaving the district or dropping out of the study for other reasons (e.g., contracting another illness, death due to another illness or in an accident), failing to follow the treatment regime, forgetting to keep hospital appointments, etc. If it is reasonable to assume such factors are not treatment dependent, censoring might reasonably be regarded as random. As already indicated, the observations themselves cannot indicate whether dropout is random or treatment dependent. Those conducting the experiment are best placed to answer that question.

It is evident that censored items do not contain as much information as complete data. However, they contain some information. For example, an observation censored at 362 weeks clearly has a longer survival time than one that survives for only 251 weeks, or indeed shows the response of interest at any time less than 362 weeks.

Another area in which survival analysis is important is engineering. Interest is often in time to failure of some product or machine or machine component, or time to completion of some process, or time to attain some goal such as a specified speed or pressure. Davis and Lawrance (1989) give an example of censoring in the assessment of tyre failure, in which those that had not failed at the end of an experimental period were regarded as censored observations. In this context, many components may break down at or near the median time, while a few will have appreciably extended life spans.

The characteristic long right tail of survival time distributions reduces the role of the normal distribution. Parametric methods for survival analysis often assume that data are samples from exponential, Weibull or other long-tail distributions. Some nonparametric analogues will be explored, indicating how these take censorship into account.

Sometimes the exact time of occurrence of an event of interest is not known. For instance, in the United Kingdom, many people choose to undergo a routine dental examination every 6 months. If a patient has a filling placed on a particular tooth and subsequently loses the filling but does not report the loss until the next routine inspection, there is some uncertainty regarding the exact time that the filling became detached from the tooth.

Innes et al. (2006) report a dental crown survival study and discuss the estimation of the length of survival where the time of loss is not known exactly. In their study, values were censored at the date of the last inspection with the crown present, although dental researchers customarily estimate the date of the loss of a crown as the middle of the interval between the date of the last inspection with the crown in place and that of a subsequent visit when it is noted to be absent. In practice, the method of estimation may be a minor issue, as dental restorations typically survive well. For instance, Opdam et al. (2007) found that around 80% of amalgam fillings in adult teeth remain in place for at least 10 years.

There are situations in which left, rather than right, censorship can occur. For example, in a study that considers survival times from the first symptoms of a particular type of cancer, the symptoms may go unrecognized until a medical examination at a later date, and the time at which the symptoms first appeared will be unknown.

In the remainder of this chapter, only right censorship is considered.

SUMMARY POINTS:

- Survival analysis developed from an interest in human longevity (Section 9.1.1).
- Length of survival can be displayed as a survival curve (Section 9.1.2).
- Survival data are typically described by long-tail distributions (Section 9.1.2).
- If an individual or item is lost to a study before the event of interest has occurred, the time at leaving is recorded as a censored value (Section 9.1.2).

- Censored observations provide less information than complete observations, but various techniques enable us to use what information they contain (Section 9.1.2).
- Censored observations may or may not be treatment dependent (Section 9.1.2).

9.2 Estimating Survival Proportions with Censored Observations Present

For uncensored data, the proportion surviving at a particular time point is the number remaining at that time point divided by the sample size. Using this method where censored values are present treats the censored values as deaths and underestimates the true proportion remaining alive. Instead, *regard a censored value as a death once the next observed death has occurred.* One estimator of this corrected proportion is the *Kaplan–Meier estimator,* proposed by Kaplan and Meier (1958).

How the Kaplan–Meier method operates is shown in Table 9.2 for the observations in treatment Group B.

Group B starts the study with five patients (the *initial group*). As each patient leaves either by death or study exit, the number remaining in the study decreases by 1, and the proportion remaining in the study decreases by 1/5, as shown below.

At 2 months there is a single departure, a *death*. This reduces the proportion alive by 1/5 down to 4/5. At 6 months, a patient exits the study *alive*. Immediately following their study exit, this patient is still alive, so the proportion alive is assumed to continue at 4/5.

From the following day, survival proportions are based on the three remaining patients known to be alive at that point (call this the *core group*). At 8 months there is a single departure, a death. This reduces the proportion of the core group alive to 2/3. The death at 12 months reduces the proportion of the core group alive to 1/3. The overall proportion alive after 8 months, taking account of all five original patients, is estimated to be $4/5\times2/3=8/15$. At 12 months, the overall proportion alive is $4/5\times1/3=4/15$. The final death, at 15 months, reduces the survival proportion to zero.

This illustration is based on a tiny study group with just one patient who exits alive. For a study of a realistic size, calculations need to be performed using statistical software.

The Kaplan–Meier method is seen in a more general context in Section 9.4.

TABLE 9.2

Estimated Survival Proportions for Treatment B

Time Since Start of Study (Months)	Deaths	Exits	Remaining in Study	Proportion Remaining	Alive	Proportion Alive
Based on initial group (5 individuals)						
0	0	0	5	1	5	1
2	1	0	4	4/5	4	4/5
6*	0	1	3	3/5	4	4/5
Based on core group (3 individuals)						
8	1	0	2	2/5	2	4/5×2/3
12	1	0	1	1/5	1	4/5×1/3
15	1	0	0	0/5	0	4/5×0/3

SUMMARY POINTS:

- With censored data, the proportion surviving at each failure time can be estimated by using the Kaplan–Meier estimator for the expected survival time of censored observations.
- Censored values are not regarded as failures immediately but are assumed to be failures by the time the next genuine failure occurs.

9.3 Modified Wilcoxon Tests

9.3.1 The Gehan–Wilcoxon Test

Gehan (1965a, 1965b) proposed generalized Wilcoxon-type tests for both one- and two-sample problems with various types of censoring. The two-sample case is considered here. A censored observation clearly has a longer survival time than any unit that has failed at or before the time of censoring. However, for this test, any pair of censored observations are regarded as giving only partial information on survival times relative to each other. In addition, a censored observation is assumed to provide only partial information on whether that subject survives a later uncensored failure. These *partial information* cases are scored as ties. Example 9.1 illustrates the method.

Example 9.1

Dinse (1982) gives data for survival times in weeks for ten patients with symptomatic lymphocytic non-Hodgkin's lymphoma and 28 patients with asymptomatic lymphocytic non-Hodgkin's lymphoma. These data are given in Table 9.3.

Test the hypothesis H_0: *the survival times distributions are identical* against H_1: $G(x) \leq F(x)$ *with strict inequality for some x*, where $F(u)$, $G(v)$ are survival time cumulative distribution functions for the symptomatic and asymptomatic cases, respectively.

The alternative hypothesis is one-sided. A one-tailed test is reasonable since asymptomatic patients are generally at an earlier stage of the disease and hence have a better prognosis relative to those with symptoms. Censored observations are present, so Gehan–Wilcoxon scores are used.

The Gehan–Wilcoxon approach uses analogues of the statistics U_m or U_n in the Mann-Whitney formulation (Section 6.1.3). For convenience, the first (symptomatic) sample

TABLE 9.3

Survival Times in Weeks for 10 Symptomatic (S) and 28 Asymptomatic (A) Cases of Lymphocytic Non-Hodgkin's Lymphoma

S	49	58	75	110	112	132	151	276	281	362*
A	50	58	96	139	152	159	189	225	239	242
	257	262	292	294	300*	301	306*	329*	342*	346*
	349*	354*	359	360*	365*	378*	381*	388*		

An * represents a censored observation.

is referred to as Group S and the second (asymptomatic sample) as Group A. To obtain the modified form of U_n, a count is made of the number of uncensored Group A observations that exceed an uncensored Group S observation. To this is added the number of censored Group A observations greater than or equal to an uncensored Group S observation. In both of these cases, the Group A failure time is regarded as exceeding the Group S failure time.

Ties are scored as 1/2 as these give only partial information. Equal uncensored Group A and Group S values form a tie along with uncensored Group A observations that exceed a censored Group S observation. A censored Group A observation that occurs before an uncensored Group S observation is also scored as a tie. All censored Group A observations are scored as ties if the Group S observation is censored, as it is unclear as to which of the two events occurred first.

The statistic U_n is the sum of all such scores over all first sample observations.

For the first Group S observation (49), the score is 28 because all of the 28 Group A participants survived for more than 49 weeks. For the second Group S observation (58), there are 26 Group A values that are greater. In addition, the observation of 58 in Group A is a tie and given a score of 0.5. Hence the total score for the second Group S observation is 26.5. For the final entry in Group S (362*), the score is 6. No Group A participants definitely had a longer survival time than 362 weeks, but there are 12 censored observations in Group A. Although these participants were lost to the study prior to 362 weeks, they could have survived beyond 362 weeks. Each of these observations is regarded as a tie and scored as 1/2. The score for the largest Group S observation is therefore $12 \times 1/2 = 6$. Carrying out the process for all observations in Group S shows that

$$U_n = 28 + 26.5 + 26 + 25 + 25 + 25 + 24 + 16 + 16 + 6 = 217.5$$

An exact WMW test with sample sizes of 10 and 28, respectively, gives a one-tail $P=0.004$.

Using the ANSM5 package in R and the command `gehan.wilcoxon(ch9$symp.survtime, ch9$asymp.survtime, ch9$symp.censor, ch9$asymp.censor, alternative = "less")` gives a Monte Carlo $P=0.00372$ for a one-tailed test.

There is strong evidence against the null hypothesis indicating that on average survival is longer for the asymptomatic group. Examining the censored values, all of which are at least 300, highlights this difference. For the asymptomatic cases, survival times are sufficiently long for many patients to survive this time, whereas only one symptomatic case does.

Note that the Wilcoxon approach, in which the observations are combined and ranked, cannot be applied in this situation; it is not possible to give precise ranks to the censored observations.

The treatment of censored data in the Gehan–Wilcoxon test is conservative. Peto and Peto (1972) and others have suggested stronger, but realistic, assumptions about the life expectancy of units that provide only censored observations. These often lead to more powerful tests, especially if there is substantial censoring. One such test is described below.

9.3.2 The Peto–Wilcoxon Test

One advantage of the Gehan–Wilcoxon test is that it only makes mild assumptions about the life expectancy of censored units. Note, however, that when censoring is severe, there will be an appreciable loss of power relative to tests that give more weight to censored data. The effect of censoring will be to lower the U_n relative to that obtained had censoring not occurred, and a larger value of U_n corresponds to a smaller P-value. Ties are only scored as 1/2, and heavily censored data contribute many ties and hence half-scores to U_n.

To explore how the Gehan–Wilcoxon test might be modified to overcome this loss of power, the concept of a *survival function* is introduced here. If the distribution of survival times, T, is $F(u)$ this implies that for any individual, the probability of the response of interest (which for simplicity is called a *failure*) at or before time t is $\Pr(T \le t) = F(t)$. The opposite event is survival beyond t. The relevant probability is given by the survival function

$$H(t) = \Pr(T > t) = 1 - F(t).$$

The physical interpretation of $H(t)$ is that it is the proportion of the population that survives beyond time t. For survival time distributions, $F(t)=0$, if $t<0$. This is often indicated by the shorthand $F(0_-)=0$. Since $F(t)$ is a monotonic increasing function of t and $F(\infty)=1$, by implication $H(t)$ decreases monotonically from 1 to 0 as t increases from 0 to ∞.

In Section 4.2, the sample distribution for a sample of n distinct values was defined as

$$S_n(t) = \frac{\text{number of sample values } \le t}{n}.$$

The corresponding sample survival function is

$$R_n(t) = \frac{\text{number of sample values} > t}{n} = 1 - S_n(t).$$

Like $S_n(t)$, if there are no ties, $R_n(t)$ is a step function with steps of height $1/n$ at each sample value, but now at each step, the value decreases by $1/n$, remaining constant between each step. It represents the proportion of the sample surviving beyond time t.

It is sometimes a matter of practical convenience to modify the definition of $R(t)$ at the rth step t_r, $r=1, 2, \ldots, n$ by assigning to it the mean of the value immediately before and immediately after that step, i.e., setting

$$R(t_r) = \frac{1}{2}\left\{\left[1 - \frac{(r-1)}{n}\right] + \left[1 - \frac{r}{n}\right]\right\} = \frac{(n-r+1/2)}{n}. \tag{9.1}$$

It is reasonable to look at $R(t_r)$ defined this way as a score allocated to that observed t_r. This scoring system assigns to each observation the probability that it survives at least to time t_r. If this score is replaced by

$$W(r) = n + \frac{1}{2} - nR(t_r)$$

it follows from (9.1) that $W(r)=r$, the usual rank score used in computing rank-sum statistics.

Thus, for samples of size m, n with no ties and no censored observations, to test the hypothesis that the population survival distributions are the same against alternatives of a location shift or of dominance of survival times in one distribution, a Wilcoxon test is appropriate. The argument can be extended to cover tied survival times when the survival function has a step of height s/n if there are s tied values at a given t.

The above approach suggests a way to formulate a test when there are censored data. Steps in $R(t)$ are still confined to failure times, but if there is censoring between two *consecutive* failures, these steps are no longer equal. The modification to the WMW test proposed by Peto allocates scores to censored data in a different way than those allocated to known failure times. This is known as the *Peto–Wilcoxon test*.

The basic procedure is illustrated with simple numerical illustrations at key points. Denote by $R(t-)$, $R(t+)$ the values of $R(t)$ immediately before and immediately after time t, respectively. These are identical unless t corresponds to one or more failure times, t_r. If there are exactly s, $s \geq 1$ failures at time t_r, the Kaplan–Meier method (Section 9.2) for calculating $R(t_r+)$ gives

$$R(t_r+) = R(t_r-)\left[1-\frac{s}{r(t_r)}\right] \tag{9.2}$$

where $r(t_r)$ indicates the number of failures plus the number of observations censored at or after time t_r. It may be described as the *number of units at risk* immediately prior to any failures at time t_r.

Example 9.2

The study of Treatment B (Section 9.1.2) is repeated. The survival times obtained (months) are 2 5 9* 11 11 22* 29* 47 51*. Find the value of $R(t)$ before and after each failure. Use this scoring system to obtain appropriate scores for these data. An asterisk denotes a censored observation.

The Kaplan–Meier procedure is used to compute relevant $R(t_r-)$, $R(t_r+)$. The computation process is equivalent to that used for the observations with a single censored value in Section 9.2. Table 9.4 illustrates the stages of this procedure.

There are steps in the number of failures at 2, 5, 11, and 47. At 2, $R(2-)=1$. Thus by (9.2) $R(2+)=1(1-1/9)=8/9$ since $s=1$ and $r(t_r)=9$. This implies a single step of 1/9 at $r=2$. Noting that $R(2+)=R(5-)$ by similar reasoning, $R(5+)=(8/9)(1-1/8)=7/9$, again indicating a step of 1/9. Since steps only occur at observed failure times, $R(5+)=R(11-)$, whence $R(11+)=(7/9)(1-2/6)=14/27$ since there are 2 failures at $t=11$ and 6 at risk. Similarly at the next failure at $t=47$, so $R(47+)=(14/27)(1-1/2)=7/27$. All that can be said about $R(t)$ for values of $t>47$ is that at some unknown $t>51$ when the observation censored at 51 has failed, then $R(t)$ will fall to zero.

TABLE 9.4

Estimated Survival Proportions for Hospital 2, Treatment B. An * represents a censored observation.

Time Since Start of Study (Months)	Deaths	Exits	Remaining in Study	Prop Remaining	Alive	Prop Alive
Based on initial group (9 individuals)						
0	0	0	9	1	9	1
2	1	0	8	8/9	8	8/9
5	1	0	7	7/9	7	7/9
9*	0	1	6	6/9	7	7/9
Based on 6 assumed alive						
11	2	0	4	4/9	4	7/9×4/6
22*	0	1	3	3/9	4	7/9×4/6
29*	0	1	2	2/9	4	7/9×4/6
Based on 2 assumed alive						
47	1	0	1	1/9	1	7/9×4/6×1/2
51*	0	1	0	0/9	1	7/9×4/6×1/2

TABLE 9.5

Peto Scores for Hospital 2, Treatment B. An * represents a censored observation.

Time (t)	$R(t-)$	$R(t+)$	$R(t-)+R(t+)$	s_i
2	1	8/9	17/9	17/18
5	8/9	7/9	15/9	15/18
9*	7/9	0	7/9	7/18
11	7/9	14/27	35/27	35/54
11	7/9	14/27	35/27	35/54
22*	14/27	0	14/27	7/27
29*	14/27	0	14/27	7/27
47	14/27	7/27	21/27	7/18
51	7/27	0	7/27	7/54

In summary, the relevant values are

$$R(2-) = 1, R(2+) = R(5-) = 8/9, R(5+) = R(11-) = 7/9,$$

$$R(11+) = R(47-) = 7/9 \times 4/6 = 14/27, R(47+) = 7/9 \times 4/6 \times 1/2 = 7/27.$$

Remembering how scores are formed in the no ties, no-censoring situation, an obvious thing to do at the steps is again to take the mean of each relevant $R(t+)$ and $R(t-)$. All that is known about a censored observation is that it will fail between the time of censoring and infinity when $R(t)$ is zero. Peto and Peto suggested that a reasonable score to assign to a censored observation is the mean of $R(t)$ at censoring and the final value zero, i.e., half the value of $R(t)$ at censoring. For instance, with the observations censored at 22* and 29*, the value of $R(t)$ is (1/2)(14/27)=7/27.

Considering the Treatment B data, at each failure time t_r, the score s_i is the mean of the Kaplan–Meier scores $R(t_r-)$, $R(t_r+)$. For each censored observation, the score is one half of the then current $R(t)$. The details are shown in Table 9.5.

The transformation $W(r_i)=n+1/2-nR(t_r)=9+1/2-9s_i$ gives the scores: 1, 2, 6, 3.67, 3.67, 7.16, 7.16, 6, 8.33, respectively.

The $W(r_i)$ are appropriate scores for the given data if the data were relevant to a Peto–Wilcoxon test.

The sum of these scores is $W=1+2+6+3.67+3.67+7.16+7.16+6+8.33=45$, which is the same as the sum of the ranks in a combined sample of 9 in a WMW test had there been no censoring.

Example 9.3

For the data on patients with lymphocytic non-Hodgkin's lymphoma (Table 9.3), use the Peto–Wilcoxon method to test the hypothesis H_0: *the survival times distributions are identical* against H_1: $G(x) \le F(x)$ *with strict inequality for some x*, where $F(u)$, $G(v)$ are survival time cumulative distribution functions for the symptomatic and asymptomatic cases, respectively.

Using the ANSM5 package in R and the command `peto.wilcoxon(ch9$symp.survtime, ch9$asymp.survtime, ch9$symp.censor, ch9$asymp.censor,` `alternative = "less")`, the one-tailed test gives a Monte Carlo $P=0.0034$, showing strong evidence against the null hypothesis and greater survival in the asymptomatic group. This finding is in close agreement with that of the Gehan–Wilcoxon test.

An alternative formulation uses a scoring system proposed by Kalbfleisch and Prentice (2002, p. 147). The modification takes account of ties by averaging the scores that would be given if the ties had been arbitrarily split, a procedure often used in other contexts. For the data in Table 9.3, it gives similar results to the Peto–Wilcoxon test. In fact, the two tests usually give similar results unless censoring is severe.

In formulating the Gehan–Wilcoxon and Peto–Wilcoxon tests, there is an implicit assumption that under the null hypothesis, the survival time distribution for censored items is the same as that for uncensored ones. This may not be the case if some or all of the censored items have a characteristic unrelated to treatments that affect their life expectations. For example, if some have a history of drug addiction, their life expectancy may be shorter than that of those not so addicted. However, if the allocation of such patients to treatments is random, and under H_0 the failure time distributions for such patients are the same for each treatment but not necessarily the same as that for patients with no history of drug addiction, the analysis is still valid.

The above statement can appear to be contradicted if censored observations are considerably shorter than the corresponding true failure times.

Example 9.4

Consider the survival time data for two groups of patients receiving a medical treatment:

Sample I	9	25	35	38	41			
Sample II	10	12*	15*	22*	37	48	57	58

The one-tailed Peto–Wilcoxon test for dominance in the second population gives $P=0.0482$. The censored items suggest a lower life expectancy in Sample II. However, if it is later discovered that the actual failure times for these patients are, respectively, 26, 29 and 36 the WMW test gives a one-tail $P=0.218$. The finding based on the actual failure times provides no evidence for a survival difference between the two groups.

Output from the ANSM5 package in R can be obtained using `peto.wilcoxon (ch9$sampleI.survtime, ch9$sampleII.survtime, ch9$sampleI.censor, ch9$sampleII.censor, alternative = "less")` and `wilcoxon.mann.whitney (ch9$sampleI.survtime, ch9$sampleII.survtime.2, alternative = "less")`.

Even though all censored items appear in one sample, this need not imply censoring is treatment dependent, since with only three censored items, there is a probability of 0.25 that these would all be in the same sample if these were both of equal size and may even be greater for unequal sample sizes. In these circumstances, the Gehan–Wilcoxon or Peto–Wilcoxon tests in fact protect against making incorrect inferences when all censored observations having a different life expectancy fall in one sample.

If a measured variable has a potentially different distribution in some subgroups, the use of appropriate covariates may sometimes increase precision. If censored observations form a subgroup that might, with hindsight, have been measured by an appropriate covariate (e.g., duration of drug addiction), the censoring may act as a surrogate for such a covariate.

SUMMARY POINTS:

- If censoring is not treatment dependent, the Gehan–Wilcoxon test is a modification of the WMW test that makes mild assumptions about the information in censored observations (Section 9.3.1).
- The Peto–Wilcoxon test makes stronger assumptions about the information in censored observations (Section 9.3.2).

9.4 Savage Scores

The prevalence of long-tail distributions, often not unlike an exponential distribution, has given that distribution a role in survival and failure time studies almost as central as that of the normal distribution in many other types of problems. While methods of analysis based on the normal distribution are often moderately robust against many minor (and even some major) departures from assumptions, tests based on the exponential distribution are often sensitive to even small departures from that distribution.

This has led to the development of distribution-free methods to overcome such problems. Savage (1956) introduced a transformation – now often referred to in a slightly extended form as the *logrank* transformation – in which ranks are replaced by what are often known as *Savage* or *exponential* scores. The transformation is essentially one to expected values of the order statistics of an exponential distribution with parameter $\lambda=1$. This is sometimes referred to as the *unit* exponential distribution. If there are no ties and no censoring in two samples of m, n observations, then the Savage score corresponding to the observation in the combined sample having rank r, ($r=1, 2, \ldots, N$) is

$$s_r' = \frac{1}{N} + \frac{1}{N-1} + \ldots + \frac{1}{N-r+1}$$

where $N=m+n$.

It can be shown using routine algebra that the sum of all s_r' is N, and an equivalent score commonly used is

$$s_r = 1 - s_r' \qquad (9.3)$$

and these latter scores sum to zero. The sums of these scores for either sample are appropriate statistics in an exact permutation test, which may be carried out using either specific tests for these scores or they may be substituted for the original data in an arbitrary score permutation test program. Some computer programs can calculate scores automatically for a given data set, making appropriate modifications for ties or censored data. Several different approaches are given in the literature for calculating appropriate scores in these circumstances.

It can be shown that two commonly used approaches are equivalent. To do so, it is necessary to look more closely at the structure of Savage scores in the no-tie and no-censoring cases. The form (9.3) is used to look at the statistic S_m formed by the sum of these scores allocated to Sample I of size m. If Sample II has n observations, the combined sample has $N=m+n$ observations. From (9.3),

$$s_t = 1 - \sum_{j=1}^{t} \left[1/(m+n-j+1) \right]$$

so if S_m is the sum of all s_t in Sample I, it follows that S_m is given by

$$S_m = m - \sum_{j=1}^{m+n} \frac{m_j}{m+n-j+1}, \tag{9.4}$$

where m_j is the number of all Sample I failures having rank j or higher. This is the number at risk immediately before failure at time $t(j)$. In particular, $m_1 = m$; also $m_j = 0$ when j exceeds the highest ranked observation in Sample I.

The denominator in the summation terms in (9.4) is the number of survivors *in both* samples immediately before failure at time t_j. This means that if n_j is defined for Sample II in a way analogous to the definition of m_j for Sample I, it follows that the denominator term may be written $m_j + n_j$, whence (9.4) becomes

$$S_m = m - \sum_{j=1}^{m+n} \frac{m_j}{m_j + n_j}, \tag{9.5}$$

Careful examination of (9.5) shows that the terms in S_m may be arranged in two groups. The first consists of m terms, each corresponding to a failure in Sample I and each taking the form

$$s_{1j} = 1 - \frac{m_j}{m_j + n_j}. \tag{9.6}$$

The second consists of n terms, each corresponding to a failure in Sample II and each taking the form

$$s_{2k} = -\frac{m_k}{m_k + n_k}. \tag{9.7}$$

It can be shown that (9.5) is the sum of all s_{1j}, s_{2k}.

Example 9.5

Calculate S_m for the following data using (i) the Savage scores given by (9.3) and (ii) by using (9.4) directly.

Sample I	19	24	31		
Sample II	26	29	33	34	37

Tests are based on the ordering of observations, and it is convenient for (i) to work with overall rankings, though these are not needed for (ii).

Ranking all data in ascending order gives

Sample I	1	2	5		
Sample II	3	4	6	7	8

For Sample I, the Savage scores using (9.3) are $s_1=1-1/8$, $s_2=1-1/8-1/7$, $s_5=1-1/8-1/7-1/6-1/5-1/4$ whence

$$S_m = 3-(3/8-2/7-1/6-1/5-1/4) = 3-1{,}073/840 \approx 1.723.$$

Equation (9.4) can be used directly without the need for ranks if for each failure time the number of survivors in each sample immediately before that failure is known. For example, just before $t=29$, there is one survivor in Sample 1 and 4 in Sample II giving a contribution in (9.4) with $j=4$ and $m_j=1$, i.e., a contribution $1/(8-4+1)=1/5$ to the sum. The complete set of contributions to the sum starting with $j=1$ and concluding with $j=m+n$ are 3/8, 2/7, 1/6, 1/5, 1/4, 0/3, 0/2, 0/1, and it immediately follows that this leads to the same value of S_m as that using (9.3).

SUMMARY POINTS:

- Savage or logrank scores are particularly appropriate for survival data.
- They reflect the key role of the exponential distribution in simple survival models.

9.5 The Logrank Test

An assessment of whether or not two or more survival curves are the same can be performed using the logrank test. The null hypothesis is H_0: *the survival curves are the same* with the alternative H_1: *the survival curves differ.* This test is commonly performed using an asymptotic version based on the chi-squared distribution that is available in most statistical packages (Chapter 12). The section looks at the formulation of the logrank test in detail.

In the absence of ties or censoring, if the sample sizes are m, n, then under the null hypothesis for any time t the proportions surviving in the identical populations take the same value, so immediately before the jth failure time $t_{(j)}$, if the numbers at risk in each sample are, respectively, m_j, n_j, then the probability that the jth failure is in Sample I is $m_j/(m_j+n_j)$. A counter, d_j, is now introduced such that $d_j=1$ if the jth failure is in Sample 1 and $d_j=0$ if the jth failure is in Sample II. Then,

$$\Pr(d_j = 1 \mid m_j, n_j) = \frac{m_j}{m_j+n_j}$$

and

$$\Pr(d_j = 0 \mid m_j, n_j) = \frac{n_j}{m_j+n_j}$$

This in turn implies that

$$e_j = \mathrm{E}(d_j \mid m_j,\ n_j) = \frac{m_j}{m_j+n_j} = \Pr(d_j = 1 \mid m_j, n_j).$$

Here the d_j (either 0 or 1) are the observed number of failures in Sample I at each of the observed failure times for the combined samples, and e_j are the corresponding expected numbers. An intuitively reasonable test statistic is the sum of the differences between these observed and expected numbers, i.e.,

$$S_m = \sum_{j=1}^{m+n} d_j - \sum_{j=1}^{m+n} e_j = m - \sum_{j=1}^{m+n} \frac{m_j}{m_j + n_j}$$

This is the same as (9.5). The terms for which $d_j=1$ contribute components of the form (9.6) and those for which $d_j=0$ contribute components of the form (9.7).

Modification for censoring or ties is relatively straightforward. This requires the calculation of the conditional expectations relevant to each observed failure, replacing m_j, n_j by m'_j, n'_j, where these are now the numbers still certainly at risk immediately before the jth failure. That is, one excludes all earlier failures and any observations that have previously been censored as in the calculation of survival using the Kaplan–Meier method. If there are c_j tied failures at the time of the jth failure, the corresponding d_j is the number of those failures in Sample I, and the corresponding $e_j=c_jm'_j/(m'_j+n'_j)$. As the components of S_m are only computed at actual failure times, there are now fewer than $m+n$ components.

Example 9.6

Determine the logrank statistic S_m for testing H_0: *population survival distributions are the same* against the alternative H_1: *population survival distributions differ* given

Sample A	7	9	13*	16		
Sample B	8	10	14*	16	25*	27

All relevant m'_j, n'_j, c_j and e_j are computed in the way described above, and these values are used to compute the modified S_m. The expected values are calculated as $4/(4+6)$, $3/(3+6)$, $3/(3+5)$, $2/(2+5)$, $2\times1/(1+3)$, $0/(0+1)$.

It is convenient to exhibit the computations in tabular form, as in Table 9.6. The censored observations do not appear directly in the table. Their presence is reflected by changes in

TABLE 9.6

Computation of Logrank Statistic with Censoring and Ties

				Failures in Sample A	
Failure Time	Number in Sample A (m'_j)*	Number in Sample B (n'_j)*	Number of Failures (c_j)	Observed (d_j)	Expected (e_j)
7	4	6	1	1	0.4000
8	3	6	1	0	0.3333
9	3	5	1	1	0.3750
10	2	5	1	0	0.2857
16	1	3	2	1	0.5000
27	0	1	1	0	0.0000
Total				3	1.8940

* immediately prior to the failure.

the numbers remaining in the two samples. The two failures when $t=16$ only require one table entry because the number of failures recorded (2) takes account of the tie. Here

$$S_m = \sum d_i - \sum e_i = 3 - 1.8940 = 1.106.$$

Using the ANSM5 package in R and the command `logrank(ch9$samplesAB.survtime, ch9$samplesAB.censor, ch9$samplesAB, score.censored = FALSE)` gives an exact $P=0.3381$. There is no evidence against the hypothesis of identical survival distributions.

The method used above does not provide scores for censored observations. Peto and Peto derived logrank scores for individuals using an estimator of the survival function proposed by Altshuler (1970).

Details are omitted, but in the notation used here, their method leads to the following scores:

- If the jth ordered observation is a failure, or if there are c_j tied failures at that time, each is assigned the score

$$s_j = 1 - \sum_i \frac{c_i}{m_i + n_i}$$

  where the summation is over all failures at or before $t(j)$. When $c_i=1$, there is a single failure. Earlier censored items are not included in this count.
- For all censored items between a failure at time $t(j)$ and the next failure (if any), each is given the score

$$s_k = -\sum_i \frac{c_i}{m_i + n_i}$$

  where the summation is over all failure times up to and including that at time $t(j)$. Censored items prior to the first failure are recorded as zero.

Applying the process outlined above to the data in this example it can be shown that the scores are those given in Table 9.7.

TABLE 9.7

Logrank Scores with Censoring and Ties. An * represents a censored observation

Sample	Response	Score
A	7	1−1/10
B	8	1−1/10−1/9
A	9	1−1/10−1/9−1/8
B	10	1−1/10−1/9−1/8−1/7
A	13*	−1/10−1/9−1/8−1/7
B	14*	−1/10−1/9−1/8−1/7
A	16	1−1/10−1/9−1/8−1/7−1/2
B	16	1−1/10−1/9−1/8−1/7−1/2
B	25*	−1/10−1/9−1/8−1/7−1/2
B	27	1−1/10−1/9−1/8−1/7−1/2−1

From Table 9.7, the sum of the scores for the sample I units is given by:

$$S_m = 1 - 1/10 + 1 - 1/10 - 1/9 - 1/8 - 1/10 - 1/9 - 1/8 - 1/7$$
$$+ 1 - 1/10 - 1/9 - 1/8 - 1/7 - 1/2 = 3 - 4{,}773/2{,}520 = 1.106.$$

This is the same total as obtained using Table 9.6.

The scores in Table 9.7 may be used in an asymptotic test valid for large $m+n$, providing neither m nor n is particularly small. Some computer programs include such a test, but slightly different results may be obtained if the program deals differently with ties, assuming initially that ties are separated by a small amount and then allocating to each tied value the mean of these separate scores. This has a carryover effect to other scores, but the test results are generally very close to those obtained using the Altshuler scores, or using the differences between the sum of observed and expected scores at each failure time.

The basic scoring systems behind the Gehan–Wilcoxon test, the Peto–Wilcoxon test, and the logrank test all generalize to more than two samples. Desu and Raghavarao (2004, Section 4.7) give a concise formulation of these extensions. At their simplest, these are analogous to the Kruskal–Wallis test, with the relevant scoring system replacing ranks and proceeding along the general lines described in Section 7.2.4.

Modification to the scoring systems when there are more than two samples is usually relatively straightforward. For example, suppose that an analogous scoring system to that derived by Peto and Peto (Example 9.6) is used. Then in the formulae for the scores s_j, s_k given there, the only modification is that the denominator sum m_i+n_i is extended to cover the relevant observations in all k samples. This is illustrated for a simple example.

Example 9.7

Using the extension outlined above for the scoring system derived by Peto and Peto (Example 9.6), test whether there is any evidence that the following samples indicate that the corresponding population survival distributions differ.

Sample X	4	16		
Sample Y	9	19	23	51*
Sample Z	11	49*	63	

Scores for a logrank test are obtained in the manner indicated above, and an analogue of the Kruskal–Wallis test is applied using these scores.

The scores are given in Table 9.8. Using the ANSM5 package in R and the command `logrank(ch9$samplesXYZ.survtime, ch9$samplesXYZ.censor, ch9$samplesXYZ)` gives an exact $P=0.2976$. Hence, there is insufficient evidence to reject the hypothesis of identical populations. This is largely an indication that the samples are too small to detect any possible difference.

The use of alternative scoring systems with censored survival distributions often leads to somewhat different P-values. This is largely a consequence of differing assumptions made about the information given by a censored observation.

Several software packages provide programs using scoring systems suggested in this text — and also a range of alternative scoring systems not covered here. Many scoring systems used for procedures discussed here may not be identical to those in Table 9.8 but are usually equivalent. For example, the scores may have opposite signs to those given in Table 9.8, but they lead to identical conclusions. In the case of data ties, different ways of

TABLE 9.8

Logrank Scores with Censoring and Ties for Three Samples. An * represents a censored observation.

Sample	Response	Score
X	4	1−1/9=0.889
Y	9	1−1/9−1/8=0.764
Z	11	1−1/9−1/8−1/7=0.621
X	16	1−1/9−1/8−1/7−1/6=0.454
Y	19	1−1/9−1/8−1/7−1/6 − 1/5=0.254
Y	23	1−1/9−1/8−1/7−1/6 − 1/5 − 1/4=0.004
Z	49*	−1/9−1/8−1/7−1/6−1/5 −1/4=−0.996
Y	51*	−1/9−1/8−1/7−1/6−1/5 −1/4=−0.996
Z	63	1−1/9−1/8−1/7−1/6 − 1/5 −1/4 − 1=−0.996

handling these may result in slight differences in allocated scores, but these usually have only a small effect on computed *P*-values.

Tarone and Ware (1977) generalized many of the procedures covered in this and the previous section. A full account of their test, highlighting its relationship to those described here, is given by Desu and Raghavarao (2004, Section 2.6.3).

Akritas and Brunner (1997) give an alternative nonparametric approach to the analysis of factorial designs with censored data. A wide-ranging review of developments in nonparametric analysis of survival data is given by Akritas (2004).

SUMMARY POINTS:

- The logrank test is used with two or more groups to examine whether population survival distributions differ.
- For two samples, it is based on the observed and expected numbers of failures for the first sample at each of the failure times.
- The test can be generalized to three or more samples using an analogue of the Kruskal–Wallis test.

9.6 The Proportional Hazards Model

In survival analysis, an important concept is that of *hazard*, which in the context of survival is concerned with the chance of the event of interest occurring in the short term. For a population without an effective health care system, such as in the seventeenth-century London experienced by John Graunt (see Thompson, 2011), beyond early childhood hazard is almost unrelated to age and the survival distribution is broadly exponential. For healthy individuals in contemporary developed societies, the risk of death in the short term is extremely small, so, as discussed below, the concept of hazard is more useful in *comparing* survival between two or more groups.

The concept of hazard is defined more formally as follows. Given a continuous distribution of survival times $F(u)$, the *hazard rate* $h(t)$ at time t is the limiting probability of failure in the small interval $(t, t+\delta t)$ given survival to time t. If $f(u)$ is the probability density function corresponding to $F(u)$, this conditional probability is

$$h(t) = \frac{f(t)}{1-F(t)}$$

The denominator is the survival function $H(t)$.

The idea of the hazard rate can be considered informally in the context of comparing the probability that one individual will die within the next specified short period of time relative to that of a second individual of a similar age. The ratio of these probabilities reflects the chance of death in the short term for one individual relative to that of the other.

In the context of statistical methods, the hazard rate is an important concept in the comparison of survival for two groups. Survival for one group relative to that in the other can be assessed by the ratio of their respective hazard rates, which is generally referred to as the *hazard ratio (HR)*.

The HR is key in the comparison of survival time cumulative distribution functions that are *Lehmann alternatives*, i.e., if the cumulative distribution functions $F(u)$, $G(v)$ have survival functions that satisfy the relationship

$$1 - F(t) = [1 - G(t)]^{\alpha} \tag{9.8}$$

for some $\alpha > 0$. Differentiating (9.8) with respect to t gives

$$f(t) = \alpha[1 - G(t)]^{\alpha-1} g(t)$$

and writing $j(t)$ for the hazard function corresponding to $G(t)$, this implies

$$h(t) = \alpha j(t) \tag{9.9}$$

It follows immediately that the ratio of the hazard rates is a constant α; hence the name *proportional hazards*.

One-parameter exponential survival time distributions satisfy (9.8) and (9.9). This follows since, if the distributions have means λ_1, λ_2, then

$$F(t) = 1 - \exp(-t / \lambda_1) \text{ and } G(t) = 1 - \exp(-t / \lambda_2)$$

Whence

$$1 - F(t) = \exp(-t / \lambda_1) \text{ and } 1 - G(t) = \exp(-t / \lambda_2).$$

It follows that (9.8) holds with $\alpha = \lambda_2/\lambda_1$.

If the use of the logrank test shows that survival time distributions differ, the proportional hazards model can then be used to examine differences in terms of hazard ratios. For a proportional hazards model, it is reasonable to test H_0: $\alpha = 1$, which implies identical survival time distributions or equal hazard rates in the population under consideration, against an alternative indicating a different value of α. This approach may be extended to deal with ties and censoring. Peto and Peto (1972) showed this to be a locally most powerful test for proportional hazard alternatives.

If the proportional hazards assumption is valid in a comparison of survival for two or more groups, then the survival curves will not cross and will touch only at time 0 [e.g., Wolfe et al. (2005)]. However, the absence of survival curve crossing is *not* sufficient by itself to infer a proportional hazard relationship between the curves.

The assumption of proportional hazards is often appropriate in the analysis of survival data. For instance, in the study of stroke survival by Wolfe et al. (2005), the HR was approximately 0.7 across a period of 9 years, in favour of individuals of black ethnicity (see Example 9.8). However, this can be an oversimplification. Examples of *non-proportional hazards* are well documented. Ogle (1887) reported that younger clergymen experienced better survival relative to men in the "higher professions" but from around the age of 60, this advantage was reversed.

Survival analysis using the proportional hazards assumption can be extended to take covariates, such as gender and smoking status, into account (Cox, 1972). This multivariable technique is known as *proportional hazards regression* or *Cox regression*. Estimates of the HR based on survival time only are known as *unadjusted* estimates; those from models that include covariates are called *adjusted* estimates. As with the reporting of most statistical findings, it is good practice to give HR estimates with 95% confidence intervals.

Example 9.8

An investigation was conducted in south London into the possible effect of ethnicity on the length of survival following a stroke (Wolfe et al., 2005). There were 1,721 individuals of white ethnicity and 414 individuals of black ethnicity, followed up for a maximum of 9 years. Survival in the two groups was compared using proportional hazards modelling.

State an appropriate null hypothesis H_0 regarding survival in the two ethnic groups and suggest an alternative H_1. The observed HR was 0.72 in favour of those of black ethnicity, with a 95% confidence interval adjusted for age, socio-economic status, and stroke subtype from 0.61 to 0.79, and $P < 0.001$. Interpret this information.

A suitable null hypothesis might be H_0: *For the population of London residents, the HR α for those of black ethnicity relative to those of white ethnicity is equal to 1.* A one-tailed alternative should be avoided, as comparisons between groups do not always yield findings in the direction that the researchers envisage. A two-tailed H_1: $\alpha \neq 1$ is more realistic.

The sample HR indicates that at given time points, the hazard function for those of black ethnicity takes a lower value than the hazard function for those of white ethnicity. This indicates that on average those of black ethnicity survive longer. With 95% confidence, the HR in the population is between 0.61 and 0.87. This interval does not contain the null hypothesis value of 1, showing strong evidence against H_0. The P-value indicates that the probability of observing a HR of less than 0.72 or greater than 1/0.72, assuming H_0 is true, is less than 0.001. This confirms the strong evidence against H_0.

SUMMARY POINTS:

- The logrank test for any survival difference is usually performed prior to using the proportional hazards model.
- The null hypothesis for the proportional hazards model is that for the population, the HR is equal to 1.

- If the proportional hazards assumption is appropriate, the survival curves for the groups do not cross.
- Data demonstrating non-proportional hazards are commonplace.
- Use of the proportional hazards model with covariates is known as proportion hazards or Cox regression.

9.7 Median Tests for Sequential Data

In Section 6.2, caution was advised regarding the use of the median test for small samples, even in situations where its Pitman efficiency was relatively high. However, this test may result in an appreciable saving of resources in survival studies when observations are made sequentially over time.

In many survival experiments, observations are collected sequentially and are automatically ordered as times to failure not only within each of the samples of size m, n but also in the combined samples. If it is sufficient to know whether the median times to failure appear to differ for the populations providing the samples, Gastwirth (1968) showed that it may be possible to terminate the experiment before all $m+n$ failure times have occurred and still get the same P-value as one would get when the experiment was completed where either the median or the control median test is used. This may lead to an appreciable saving in experimental resources, especially if a control median test is used.

For the median test, it follows from the arguments used in Section 6.2.2 that the critical value of the test statistic is the number in one of the samples either above or below the combined sample median, M, under H_0. Without loss of generality, one might make this the number below M in Sample I. Further, for either the median or the control median test, when m, n are fixed, one may work out prior to the start of the experiment the critical value for the number of Sample I values that must be below M for any assigned significance level P. This means that if the median test is used, the experiment may be terminated when this number is exceeded in Sample I, because in that case H_0 is then accepted. If, on the other hand, the critical number has *not* been exceeded by the time $(m+n)/2+1$ observations have been obtained, it is then possible to terminate the experiment with rejection of H_0. This follows because the observations are ordered, and therefore any further observations in Sample I will be greater than M.

A similar argument holds for the control median. Taking Sample I as the control, the stopping rule is now that the experiment is stopped if the critical value in Sample II is exceeded, at which point H_0 is accepted. Otherwise, the experiment is stopped when $m/2+1$ values have been obtained; for then, because the observations are ordered, no more Sample II values will be below the control median.

It is clear that the decision would not, in these circumstances, be changed if the experiment were continued until all items had failed since the test only requires numbers below the combined median M or the control median M_1 to be known.

Gastwirth showed that asymptotically, the average number of failures that had to be observed for any given m, n before a decision is made is less for the control median test than it is for the median test. He also proposed a slight modification to the median test that reduces even further the expected number of observations needed before a decision can be made.

SUMMARY POINTS:

- The median and the control median test allow early termination of an experiment where results are obtained sequentially in time.
- For these tests, the only interest is in possible median differences between the two populations.

9.8 Choice of Tests

Sometimes, especially with small data sets, it is difficult to decide which of a number of possible test procedures is optimal. In the case of two independent samples, there might be a choice between a WMW test or one based on, e.g., van der Waerden scores or logranks. Using one test when another is appropriate often results in loss of Pitman efficiency and of power. Gastwirth (1985) proposed tests applicable in a wide range of circumstances where it may be reasonable to assume that one of a finite set of models is appropriate without clear evidence of which is optimal. The tests he proposed are robust over the family of possible models in the sense that no other test has asymptotically a higher minimum efficiency relative to the optimum test for each model in the family. Gastwirth called such tests *maximin efficiency robust tests* – now often abbreviated to MERT.

The theory involves some advanced ideas, and the reader should consult Gastwirth's paper for a comprehensive theoretical discussion. In essence, the tests are based on linear functions of test statistics used in the competing optimal tests for the different models. The appropriate linear function takes into account the correlations between the different statistics or the scores used in them. The method provides protection against a poor choice of test and is especially useful with censored data. MERT procedures generally have a relatively high efficiency when compared with the optimal test that would be used if the true model were known. Computer packages may provide facilities for the calculation and use of MERT scores. A related robust procedure, based on the maximum of the standardized optimum statistics, referred to as MX, is described by Freidlin, Podgor and Gastwirth (1999).

Software may allow tests to be carried out quickly with a wide range of scores, so the safety net feature of MERT is perhaps less useful than it would be without these facilities. In trying many tests, one must bear in mind the caution given in Section 1.7 about unethical statistical behaviour. MERT and MX procedures are, however, more than just a safety net, and their robustness can prove invaluable when there is appreciable uncertainty about an appropriate model. Andersen et al. (1982) also discuss tests applicable to alternative model situations.

SUMMARY POINTS:

- Using one test when another is appropriate often results in loss of Pitman efficiency and of power.
- MERT tests provide a useful safety net when it is not clear which of several tests is most appropriate.

9.9 Exercises

For these questions, an * indicates a censored value.

9.1 The hospital study in Section 9.1.2 gives survival times for treatment A as

1 3* 4* 5* 7 9

Construct a table showing the estimated survival probabilities for this treatment.

9.2 The observations below indicate the ages at first tooth filling for a group of young people.

Boys	3	5	6	8	11	12	15	16
Girls	4	7	9	10	14	17		

What is meant by the term "failure" here? Is there a difference in "survival" times for this event between boys and girls?

9.3 In Example 9.1, is there a case for using a two-sided alternative hypothesis? Explain your answer.

9.4 Use the data for treatment A in Exercise 9.1 to obtain Peto scores.

9.5 Peto et al. (1977) give the following data for survival times in days of two groups of cancer patients under two different treatment regimes, A and B. Investigate the evidence for a difference in survival times using the Gehan–Wilcoxon test.

A	8	852*	52	229	63	8	1,976*	1,296*
	1,460*	63	1,328*	365*				
B	180	632	2,240*	195	76	70	13	1,990*
	18	700	210	1,296	23			

9.6 Analyse the data in Exercise 9.5 using the logrank test. Do the findings differ from those for the Gehan–Wilcoxon test?

9.7 An investigation was conducted in Dallas County, Texas, into the possible effect of hyponatraemia (low sodium level in the blood) on the length of survival (Gankam-Kengue et al., 2013). There were 3,266 individuals with a serum sodium level of at least 135 mmol/L and 285 individuals with a serum sodium level of less than 135 mmol/L, followed up for a median of 8.4 years.

State an appropriate null hypothesis for a comparison of survival in the two groups and suggest an alternative hypothesis. The paper states: "The unadjusted hazard ratio (HR) for deaths in individuals with hyponatraemia compared with those without hyponatraemia was 1.94 (95% CI: 1.31–2.87, $P = 0.001$)". Interpret this information.

9.8 A manufacturer of light bulbs claims that their standard (type A) bulbs have an average life of 3,500 hours. A long-life bulb (type B) is developed, which they claim is considerably superior. A researcher takes a sample of type A bulbs, switches them on at the same time, and records the running period until failure in units of 100 hours. The process is similarly applied to a sample of type B bulbs. Use a control median test approach to determine the minimum number of observations needed before a decision can be reached to either accept or reject the hypothesis of identical medians in a one-tailed test where the alternative hypothesis is that bulbs of type B are from a population with a median greater than the control, type A, and the test is performed using a nominal significance level $\alpha = 0.05$. What decision is then reached?

A 2 5 9 14 21 27 33 37 42 49 53 59

B 22 29 47 56 73 82 87 94 95 95 98 102 105 109
112 115 116 117

10

Correlation and Concordance

10.1 Correlation in Bivariate Data

So far the main focus has been on outcomes that can be represented as a single distribution, such as height, weight, and seizure severity score. However, an outcome may consist of measurements from two random variables such that data from the two variables are paired. For example, pairs of observations are recorded in this way if both height and weight are measured for each individual in a sample of children. The two variables can then be represented by what is known as a bivariate distribution. Exploring the strength of relationships between pairs of variables is facilitated by the use of correlation coefficients; these measure the strength of association or dependence. Note that the algebra involved can be more complex than that encountered up to this point.

10.1.1 The Pearson Product-Moment Correlation Coefficient

The most widely used correlation measure, the Pearson product-moment correlation coefficient, was developed from a parametric context and estimates the degree of linear association between two variables (Pearson, 1895). The coefficient takes values between –1 and +1. The extremes are attained only when points lie exactly on a straight line. The extreme –1 implies a negative slope, and +1 a positive slope.

If one of the variables tends to be large when the other is large and small when the other is small, the correlation is positive. If large values of one variable are associated with small values of the other, the correlation is negative. If the two variables are independent of each other, the value of the correlation is zero. A nonlinear relationship can also produce a correlation value close to zero. A scatter diagram illustrating a positive relationship between two variables is shown in Figure 10.1.

It is prudent to plot a scatter diagram before calculating any correlation coefficient.

If measurements have been taken on two variables (such as height and weight) for a sample of n individuals, these paired observations can be written $(x_1, y_1), (x_2, y_2), \ldots, (x_n, y_n)$ where the population distributions are X and Y. The sample *Pearson product-moment correlation coefficient*, r, is defined as

$$r = \frac{\sum_{i=1}^{n}(x_i - \bar{x})(y_i - \bar{y})}{\sqrt{\sum_{i=1}^{n}(x_i - \bar{x})^2 \sum_{i=1}^{n}(y_i - \bar{y})^2}} \tag{10.1}$$

DOI: 10.1201/9780429326172-10

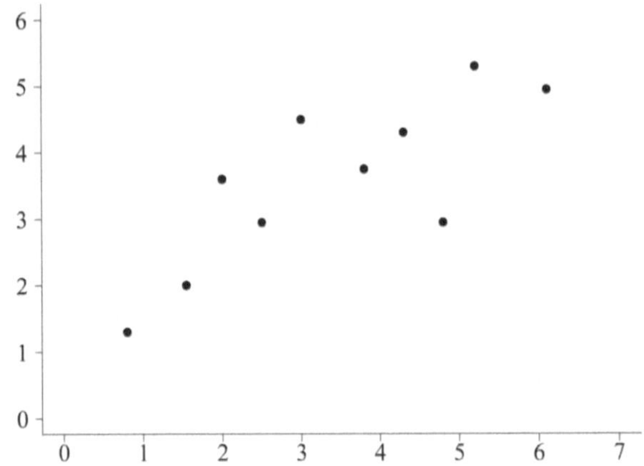

FIGURE 10.1
Scatter diagram for a positive correlation between two variables.

which, for simplicity and computational convenience, is usually rearranged and written

$$r = \frac{c_{xy}}{\sqrt{c_{xx}c_{yy}}}$$

where

$$c_{xy} = \sum_i (x_i y_i) - \left(\sum_i x_i\right)\left(\sum_i y_i\right) / n,$$

and

$$c_{xx} = \sum_i x_i^2 - \left(\sum_i x_i\right)^2 / n, \ c_{yy} = \sum_i y_i^2 - \left(\sum_i y_i\right)^2 / n.$$

Example 10.1

In a study of social conditions, five families are asked to state the number of vehicles owned (variable X) and the number of bedrooms in their home (variable Y). If the responses are (0, 2) (1, 1) (2, 3) (2, 4) and (3, 5), calculate the sample Pearson correlation coefficient r.

The quantities required are $\sum_i x_1 = 0+1+2+2+3 = 8$, $\sum_i y_1 = 2+1+3+4+5 = 15$, $\sum_i x_1 y_1 = 0+1+6+8+15 = 30$, $\sum_i x_i^2 = 0+1+4+4+9 = 18$, and $\sum_i y_i^2 = 4+1+9+16+25 = 55$.

Hence, $c_{xy} = 30-(8\times 15/5) = 6$, $c_{xx} = 18-\left(8^2/5\right) = 5.2$, $c_{yy} = 55-\left(15^2/5\right) = 10$.

This gives $r = \frac{6}{\sqrt{5.2\times 10}} = 0.8321$.

This demonstrates that there is a positive association between the number of bedrooms in the home and the number of vehicles owned, which is unsurprising.

It can be verified from (10.1) that r is unaltered by what are called location and scale changes (see also location and scale parameters, Section 6.4). That is, r is unaltered if all x_i are replaced by $(x_i-k)/s$ and all y_i are replaced by $(y_i-m)/t$ where k, m are any constants that represent location changes and s and t are any positive constants that represent scale changes. Another way of putting this is to say that r is invariant under linear transformations of the x and y. This property extends to the alternative correlation coefficients described in Sections 10.1.3–10.1.5. It also has important practical implications in some regression methods considered in Chapter 11.

In parametric inference, the Pearson coefficient is particularly relevant to a bivariate normal distribution where the sample coefficient r is an appropriate estimate of the population correlation coefficient ρ. If $\rho=0$ for a bivariate normal distribution, this implies X and Y are independent and values of r close to zero support that hypothesis. Even when the sample $(x_1, y_1), (x_2, y_2), \ldots, (x_n, y_n)$ is assumed to come from a bivariate normal distribution, making inference about an unknown population correlation coefficient ρ based on the sample coefficient r given by (10.1) is more difficult than that for inferences about the means or variances of the distributions of X or Y. If the joint distribution of X and Y is not bivariate normal, parametric inference about ρ is even harder.

Basic nonparametric correlation inference does not require the assumption of bivariate normality and can be applied to both paired observations of continuous data and data consisting of ranks. These ranks may be the original data or they may be derived from continuous measurements.

Pitman (1937b) introduced a permutation test based on the Pearson coefficient for continuous data that may be used for inference. Like other permutation tests with continuous data such as those for the one- and two-sample problems, the test is a conditional test and fresh computation of the permutation distribution is required for each data set. Modern computer software makes the computation feasible for small samples. For larger samples, Monte Carlo or asymptotic approximations are available in, for example, the ANSM5 package in R. The appropriate test is based on the fact that if the order of the x values is fixed, then all $n!$ permutations of the y values are equally likely under a hypothesis of independence or lack of association between X and Y. Permutations giving values of r near to –1 or +1 indicate strong evidence against the hypothesis H_0: $\rho=0$. Lack of robustness means the test, although of historical interest, is little used in practice. When normality assumptions hold, the Pitman efficiency of the test is 1. A simple numerical example is given in Sprent (1998, Section 9.1).

10.1.2 Other Measures of Bivariate Association

Desirable properties of any correlation coefficient are that like the Pearson coefficient its values should be confined to the interval [–1, 1] and that lack of association implies a value near zero. Values near +1 should imply a strong positive association, and values near –1 a strong negative association. For the Pearson coefficient, $r=\pm1$ implies linearity, but for the rank coefficients introduced below values of ±1 need not, and usually do not, imply linearity in continuous data from which the ranks may have been derived. Rather, interest is in what, for continuous data, is known as *monotonicity*. If x and y increase together, this is a monotonic increasing relationship. If y decreases as x increases the relationship is monotonic decreasing. For rank correlation, the value +1 implies strictly increasing monotonicity, and the value –1 strictly decreasing monotonicity.

Rank correlation coefficients are also relevant to, and indeed were originally developed for use in, situations where there is no underlying continuous measurement scale, but where the ranks simply indicated the order of preference expressed by two assessors for a group of objects, e.g., contestants in a diving contest, or different brands of tomato soup in a tasting trial.

Some basic concepts in nonparametric correlation are illustrated using the artificial, but realistic, data in Table 10.1. This gives, for 12 candidates, the scores achieved in an examination paper consisting of a short question marked out of 20 and a long question marked out of 60. To a certain extent, the two marks are positively associated. However, some students concentrated their revision on the material for Question 1 (Q1), which they answered well, and performed less well on Question 2 (Q2). One student only revised the material for Q1 and did not even attempt Q2. Figure 10.2 is a scatter diagram for these data. It shows the tendency for the Q2 marks to increase with the Q1 marks up to a certain point, beyond which the Q2 marks tend to decrease.

Other examples of this type of relationship can be found in a biological context. A measured response often increases up to a maximum as the level of a stimulus increases. After reaching this maximum, the response may level off or decrease. This might be due to a toxic effect associated with an overdose of a potentially beneficial treatment. For the data in Table 10.1, the Pearson coefficient is $r=0.373$. This is not significantly different from zero at the 5% level even in a one-tailed test (Monte Carlo one-tail $P\approx 0.117$) if the sample is assumed to be from a bivariate normal distribution. That assumption is not realistic for these data. This lack of significance of the Pearson coefficient is counter to an intuitive feeling that there is evidence of a reasonably high degree of association in the data. Not unexpectedly, because it lacks robustness, the Pitman permutation test for these data gives an exact one-tail $P=0.112$, close to the value assuming normality.

Output from the ANSM5 package in R can be obtained using `pearson(ch10$q1, ch10$q2, alternative = "greater")`.

TABLE 10.1

A Data Set for Scores x, y for Two Questions Q1 and Q2 on a Paper

Q1	x	1	3	4	5	6	8	10	11	13	14	16	17
Q2	y	13	15	18	16	23	31	39	56	45	43	37	0

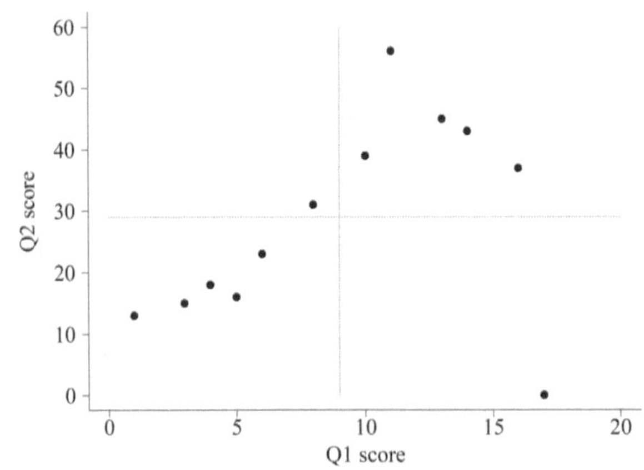

FIGURE 10.2
Scatter diagram for data in Table 10.1. The broken lines at right angles represent axes when the origin is transferred to the x, y sample medians.

In Sections 10.1.3–10.1.5, three nonparametric measures of correlation are considered and applied to the data in Table 10.1. The coefficients are presented in a way that shows relationships between these measures of correlation and certain concepts introduced in earlier chapters.

Exact tests are based on appropriate permutations of data, ranks or counts. For tests based on ranks, the simplest permutation procedure is to fix the rank order associated with one variable, conventionally x, and to calculate the value of the chosen coefficient under all equally likely permutations of ranks of the other variable, y. The probabilities of the observed or more extreme values of the chosen coefficient, or some linear function of it, determine the size of the relevant critical region for testing for zero correlation.

10.1.3 The Spearman Rank Correlation Coefficient

Spearman (1904) proposed a correlation coefficient that is calculated from the ranks of the data. It is computationally equivalent to the Pearson coefficient. Ranks replace the original continuous data if these are given. The coefficient is often denoted by the Greek letter ρ (rho) and referred to as *Spearman's rho.* Because ρ is also used for the correlation coefficient in bivariate normal distributions (estimated by the Pearson coefficient r), to avoid confusion the Spearman coefficient is denoted by ρ_s and the corresponding sample value by r_s. Formula (10.1) may be used to calculate r_s if the pairs of observations (x_i, y_i) are replaced by their ranks (r_i, s_i). When there are no ties, a simpler formula can be obtained by straightforward algebraic manipulation using properties of sums of ranks and sums of squares of ranks. Each within-pair difference is found and squared. These squared values are then added together to give $T = \sum_{i=1}^{n}(r_i - s_i)^2$. The coefficient is obtained as

$$r_s = 1 - \frac{6T}{n(n^2 - 1)} \tag{10.2}$$

If the x, y ranks are equal for each pair, i.e. if $r_i = s_i$ for all i, then the differences are zero so $T=0$ and $r_s=1$. For a complete reversal of ranks, elementary algebra establishes that $r_s=-1$. If there is no correlation between ranks, it can be shown that $\mathrm{E}(T)=n(n^2-1)/6$, so that r_s has expected value of zero. If the observations are a random sample from a bivariate distribution with X, Y independent near-zero values for r_s are expected.

Tables of critical values at nominal 5% and 1% significance levels for testing H_0: $\rho_s=0$ against one- and two-sided alternatives are widely available (see, e.g., Gibbons and Chakraborti, (2020, Table M) or Neave (1981, p. 40)). For small samples, the ANSM5 package in R provides exact P-values for this test with the choice of an asymptotic or a Monte Carlo estimate of P for larger samples. Since r_s is a monotonic function of T when n is fixed, T itself may be used as a test statistic and critical values of T have been tabulated for small n. It is preferable to use r_s because it gives a more easily appreciated indication of the level of association. It complies with the convention that possible values of a correlation coefficient should be in the interval [–1, 1].

Both this coefficient and one developed in Section 10.1.4 are often used to test whether there is broad equivalence between ranks, or orderings, assigned by different assessors. Do two examiners concur in their ranking of candidates? Do two judges agree on the places awarded to each of the n competitors in a diving contest? Do job applicants' ranks for reasoning skills based on a psychological test show a relationship to their rankings in a further test for mathematical skills?

These coefficients are also appropriate for tests of trend. Does Y increase (or decrease) as X increases? Such tests are often more powerful than the Cox–Stuart test described in Section 4.5.3.

Example 10.2

Compute the Spearman coefficient for the data in Table 10.1. Use the result to test H_0: $\rho_s=0$ against the one-sided alternative H_1: $\rho_s>0$.

The coefficient is computed using (10.2), and tables or appropriate software are used to assess the evidence.

To compute T in (10.2), the data in Table 10.1 are replaced by the ranks and the rank differences d are computed as given in Table 10.2.

T is calculated by squaring the difference, d, and adding the squares. Thus

$$T=(-1)^2+(-1)^2+(-2)^2+\ldots+(11)^2=1+1+4+\ldots+121=162,$$

whence $r_s=1-(6\times 162)/(12\times 143)=0.4336$. One-tail critical values for significance at nominal 5% and 1% levels given by tables are 0.503 and 0.678. For this example using the ANSM5 package in R and the command `spearman(ch10$Q1, ch10$Q2, alternative = "greater")` gives a Monte Carlo one-sided $P=0.082$. Using an asymptotic approximation (see Section 10.1.6), $P=0.080$. Since $r_s=0.4336$ is less than the value 0.503 required for significance at a nominal 5% level and corresponds to a $P\approx 0.08$, there is not strong evidence against a hypothesis of lack of monotonic association.

Section 10.1.2 indicates that the test based on the Pearson coefficient gave an associated $P\approx 0.117$, slightly greater than that obtained here. This again reflects the influence of the curved nature of the relationship between x and y for high Q1 marks. The conceptual relationship between the Spearman and the Pearson coefficients often results in both coefficients lacking robustness when a few observations are markedly out of line with the general pattern.

Many general statistical packages have programs that compute r_s or T or an equivalent statistic, but often either leave one to look up tables or else give only an asymptotic result to establish whether there is strong evidence against H_0. If no specific program is included for the Spearman coefficient, one might use the paired ranks as data for a Pearson test. This will give the correct numerical value of the coefficient, but if such a program gives only an asymptotic P-value, this will usually be calculated on the assumption that the data are a sample from a bivariate normal distribution and so it is not relevant.

The exact permutation distribution of r_s for small samples is not hard to compute. R will do this. The basic idea is similar to that outlined in Section 10.1.2 for the permutation distribution of the Pearson coefficient. In practice, one would be likely to use a facility like that in R to compute the distribution, but if it were needed, the procedure is illustrated for four paired ranks in Example 10.3.

TABLE 10.2

Paired Ranks and Their Differences d for the Data in Table 10.1

x-Rank	1	2	3	4	5	6	7	8	9	10	11	12
y-Rank	2	3	5	4	6	7	9	12	11	10	8	1
Difference	−1	−1	−2	0	−1	−1	−2	−4	−2	0	3	11

Example 10.3

Given four paired observations calculate the exact permutation distribution of the Spearman correlation coefficient r_s. Assume there are no ties.

Write the ranks for the first variable as 1 2 3 4, and calculate the number of ways in which the ranks for the second variable can be matched to these four observations. There are 4!=24 possible data pairings. For each such pairing, the value of r_s is computed. Since all pairings are equally likely under a hypothesis of no association, each has an associated probability of 1/24. Table 10.3 shows the orderings with its associated Spearman correlation coefficient.

The possible values of r_s and the number of times each occurs, n_s, are those given in Table 10.4. Since each of the 24 cases is equally likely to occur division of these numbers by 24 gives P_r, the probability of occurrence of each possible value of r_s under the null hypothesis H_0: $\rho_s=0$.

In summary, the distribution of r_s under the null hypothesis is discrete and takes only 11 possible values in the interval [–1, 1] with the given probabilities.

The distribution is symmetric about zero, and this property holds whatever the size of the sample.

TABLE 10.3

Spearman Rank Correlation Coefficients for the Possible Arrangements of the y-Ranks Corresponding to x-Ranks in the Order 1, 2, 3, 4

Order of y-Ranks				Spearman Rank Correlation Coefficient
1	2	3	4	1.0
1	2	4	3	0.8
1	3	2	4	0.8
1	3	4	2	0.4
1	4	2	3	0.4
1	4	3	2	0.2
2	1	3	4	0.8
2	1	4	3	0.6
2	3	1	4	0.4
2	3	4	1	–0.2
2	4	1	3	0
2	4	3	1	–0.4
3	1	2	4	0.4
3	1	4	2	0
3	2	1	4	0.2
3	2	4	1	–0.4
3	4	1	2	–0.6
3	4	2	1	–0.8
4	1	2	3	–0.2
4	1	3	2	–0.4
4	2	1	3	–0.4
4	2	3	1	–0.8
4	3	1	2	–0.8
4	3	2	1	–1.0

TABLE 10.4

Frequencies for the Spearman Rank Correlation Values

r_s	−1.0	−0.8	−0.6	−0.4	−0.2	0	0.2	0.4	0.6	0.8	1.0
n_s	1	3	1	4	2	2	2	4	1	3	1
P_r	1/24	1/8	1/24	1/6	1/12	1/12	1/12	1/6	1/24	1/8	1/24

The program for the Spearman coefficient in R will generate the complete distribution for any specified sample that is not too large. For 10 observations, the discontinuities are appreciably smaller than those in this example. The statistic takes 166 distinct values in the interval [−1, 1].

Ties in data are replaced by mid-ranks. To obtain exact *P*-values with ties requires a program like that in R, although the value of r_s may be computed by applying the Pearson coefficient formula (10.1) to these mid-ranks. Thomas (1989) showed that if (10.2) is used to compute r_s when there are ties represented by mid-ranks, it gives a value that is always greater than or equal to the correct value given by the Pearson formula. An asymptotic approximation is considered in Section 10.1.6.

10.1.4 The Kendall Correlation Coefficient

Methods reminiscent of the Mann–Whitney scoring procedure for the Wilcoxon–Mann–Whitney (WMW) test are now used to obtain another measure for monotonic association. Inferences involve a coefficient known as the *Kendall correlation coefficient*, or *Kendall's tau*, which is closely related to the Jonckheere–Terpstra test statistic.

In practice, as in the Jonckheere–Terpstra test, continuous data do not need to be replaced by ranks. Computation is simpler if it is assumed that the x data are arranged in ascending order. Data can be analysed either in a form like that in Table 10.1 where the x data are already in ascending order, or by using the corresponding ranks. It is convenient to use the labels "x-ranks" and "y-ranks" to refer, respectively, to the ranks associated with the first and second members of a bivariate pair (x, y), even when the basic data are themselves ranks only, and there is no specific underlying continuous variable. The latter is the case for taste preferences or for two judges ranking a number of competitors performing a skill such as gymnastics.

Denoting the ranks of x_i, y_i by r_i, s_i, respectively, the Kendall coefficient is based on the principle that if there is an association between the ranks of x and y, then, if the x-ranks are arranged in ascending order (i.e., so that $r_i = i$), the s_i should show an increasing trend if there is positive association and a decreasing trend if there is negative association. Kendall (1938) therefore proposed that after arranging observations in increasing order of x-ranks, each paired difference $s_j - s_i$ is calculated for $i = 1, 2, \dots, n-1$ and $j > i$. A difference is then scored as +1 if $s_j - s_i$ is positive, and as −1 if negative. Kendall called positive and negative differences *concordances* and *discordances* respectively. Denoting the numbers of concordances and discordances by n_c, n_d, respectively, Kendall's tau, which is denoted here by t_k, although historically it has usually been denoted by the Greek letter τ, is

$$t_k = \frac{n_c - n_d}{n(n-1)/2} \tag{10.3}$$

The corresponding population value is denoted by τ_k.

In a comparison of ranks assigned by two judges to a finite set of objects such as ten different brands of tomato soup, this is essentially the entire population of interest, so the calculated coefficient is a measure of agreement between the judges in relation to that population. On the other hand, for a sample of n bivariate observations from some continuous distribution (of measurements, say) if t_k is calculated using (10.3), this is an estimate of some underlying τ_k that is a measure of the degree, if any, of monotonic association, or dependence between the variables X, Y in the population. Since there are $n(n-1)/2$ pairs s_j-s_i, if all are concordances $n_c=n(n-1)/2$ and $n_d=0$, whence $t_k=1$. Similarly, if all are discordances, then $t_k=-1$. If the rankings of x and y are independent, there should be a fair mix of concordances and discordances and t_k should be close to zero.

When there are no ties in the rankings, $n_c+n_d=n(n-1)/2$. Although the coefficient is often referred to as a rank correlation coefficient, in contrast to the Spearman coefficient, its computation requires only the order of the relevant x and y and not the actual ranks.

Example 10.4

Compute Kendall's tau for the data in Table 10.1.

As the values of x are ranked in ascending order, n_c may be computed from the y-ranks in Table 10.2. The value of n_d may be deduced from $n_d=n(n-1)/2-n_c$.

To count the number of concordances, the y-ranks in Table 10.2 are taken in turn from left to right. For each, the number of succeeding ranks that are greater is noted. The total number of concordances is equal to the sum of these counts. In this example, the first y-rank is 2 and this is succeeded by 10 greater ranks, namely, 3, 5, 4, 6, 7, 9, 12, 11, 10, 8. Similarly, the next y-rank, 3, is succeeded by the nine greater ranks 5, 4, 6, 7, 9, 12, 11, 10, 8. Proceeding in this way, the total number of concordances is $n_c=10+9+7+7+6+5+3+0+0+0+0=47$. Note that from the eighth y-rank onwards, all ranks to the right are smaller so the final four terms contribute nothing to the total.

There are $n=12$ observations, so $n(n-1)/2=66$. It follows that $n_d=(12\times 11)/2-47=19$. From (10.3), $t_k=(47-19)/66=0.4242$.

Since the x are ordered in Table 10.1, the order of the corresponding y values is a one-to-one ordered transformation of the ranks in Table 10.2. Thus, one could equally well compute the number of concordances and discordances by noting the signs of all differences y_j-y_i in Table 10.1 as these signs correspond to those of the s_j-s_i above.

The counting procedure for concordances is reminiscent of the Mann–Whitney counts for the WMW test, and equivalence to the Jonckheere–Terpstra test is shown below.

The values of x_i are in ascending order in Table 10.1. This simplifies computation. More generally, the number of concordances is the number of positive c_{ij} among all $n(n-1)/2$ of the $c_{ij}=(y_j-y_i)/(x_j-x_i)$, $i=1, 2, \ldots, n-1$ and $j>i$. The number of negative c_{ij} is the number of discordances.

Equivalence to a Jonckheere–Terpstra test is evident if the values of x_i are arranged in ascending order, as in Table 10.1, and the x_i (or the associated ranks) are regarded as indexing n ordered samples, each of 1 observation, where that observation is the corresponding y value. The number of concordances is the sum of the Mann–Whitney sample pairwise statistics used in the Jonckheere–Terpstra test in Section 7.2.4. Thus, in theory, any program for the exact Jonckheere–Terpstra test may be used for Kendall's coefficient to test H_0: *no association between x and y ranks* against either a one- or two-sided alternative. In practice, however, most programs for the Jonckheere–Terpstra test will, in this situation, only give exact tail probabilities for very small n. Asymptotic results, or Monte Carlo estimates, of the exact tail probability must be used with this approach if the samples are moderate to large. More specific asymptotic results are considered in Section 10.1.6.

Programs in general statistical packages often calculate t_k directly, some indicating significance levels (usually asymptotic), others leaving the user to determine these at a nominal level from tables. For example, tables can be found in Neave (1981, p. 40) and Gibbons and Chakraborti (2020, Table L). Some tables give equivalent critical values of $n_c - n_d$ for various n. However, calculating t_k requires little extra effort and gives a better feel for the degree of association implied by any correlation coefficient. For the no-tie situation tests may be based on n_c only, for it is clear from (10.3) that if n is fixed, t_k is a linear function of n_c. The ANSM5 package in R includes a program that computes exact P-values for the test of no association based on this coefficient for small sample sizes.

In passing, it is worth noting one difference between the usual scoring systems for the Kendall coefficient and that used for Jonckheere–Terpstra. Whereas the former uses 1, 0, –1 for concordances, ties and discordances, the latter uses 1, 1/2, 0. This is not a fundamental difference, because providing certain obvious adjustments are made to relevant formulae, the scoring systems may be interchanged without altering conclusions.

Example 10.5

For the data given in Table 10.1 use Kendall's tau to test H_0: *no association* against H_1: *positive association between x and y*.

The appropriate test is a one-tailed test of H_0: $\tau_k = 0$ against H_1: $\tau_k > 0$.

Example 10.4 showed that $t_k = 0.4242$, and $n_c - n_d = 47 - 19 = 28$. Tables give nominal one-tail 5% and 1% critical values for significance when $n = 12$ as 0.3929 and 0.5455. Corresponding values for $n_c - n_d$ are 26 and 36. More precisely, using the ANSM5 package in R and the command `kendall.tau(ch10$Q1, ch10$Q2, alternative = "greater")` gives an approximate Monte Carlo estimate of the exact one-tail $P = 0.032$. Hence, there is moderately strong evidence against the hypothesis of no association. In a formal significance test, one would reject H_0 at a nominal 5% significance level and conclude a positive association between the ranks of x and y.

Had the alternative hypothesis been H_1: $\tau_k \neq 0$, requiring a two-tailed test, H_0 would not have been rejected at the 5% significance level.

The evidence of association provided by Kendall's tau is stronger than that provided by Spearman's rho. A likely explanation is given in Section 10.1.7.

As indicated above, most general statistical programs do not give exact P-values or significance levels for relevant hypothesis tests. That in the ANSM5 package in R gives exact results for small samples, but a Monte Carlo approximation is needed for moderate to large samples.

For small samples, the exact distribution of t_k can be calculated in a similar way to that used for the Spearman coefficient in Example 10.3.

Example 10.6

Calculate the exact permutation distribution of Kendall's t_k when $n = 4$, under the null hypothesis of no association.

If the x-ranks are in ascending order, then for each of the $4! = 24$ equally likely permutations of the y-ranks the number of concordances, n_c, is counted. For $n = 4$, $n(n-1)/2 = 6$. It follows that $n_c + n_d = 6$ whence n_d can be obtained for any given n_c. The value of t_k is computed using (10.3), and the number of times each value of t_k occurs is recorded. Dividing each number of occurrences by 24 gives the probability of observing each value (see Table 10.5).

From Table 10.5, the probabilities associated with each of the seven possible values of t_k are

t_k	1	0.667	0.333	0	−0.333	−0.667	−1
Probability	1/24	1/8	5/24	1/4	5/24	1/8	1/24

When $n=4$, a P-value less than 0.05 can only be obtained in a one-tailed test corresponding to $t_k=1$ for which $P=1/24 \approx 0.042$. Samples of this size are of little use in practice for inference. Note that two-tail P-values can be obtained by doubling one-tail probabilities since the distribution of t_k is symmetric.

The ANSM5 package in R will calculate complete distributions for small samples. The Monte Carlo option may be used to estimate P-values rapidly and with good accuracy.

Modifications are needed to calculate and use t_k when there are rank ties. With ties (10.3) no longer takes values +1 or −1 even when all mid-rank pairs lie on a straight line. Ties among the values of x or y are both scored zero in Kendall's coefficient because they are neither concordances nor discordances. Now (10.3) can no longer take the value ±1 because

TABLE 10.5

Kendall Rank Correlation Coefficients for the Possible Arrangements of the y-Ranks Corresponding to x-Ranks in the Order 1, 2, 3, 4

Order of y-Ranks				Kendall Rank Correlation Coefficient
1	2	3	4	1.000
1	2	4	3	0.667
1	3	2	4	0.667
1	3	4	2	0.333
1	4	2	3	0.333
1	4	3	2	0.000
2	1	3	4	0.667
2	1	4	3	0.333
2	3	1	4	0.333
2	3	4	1	0.000
2	4	1	3	0.000
2	4	3	1	−0.333
3	1	2	4	0.333
3	1	4	2	0.000
3	2	1	4	0.000
3	2	4	1	−0.333
3	4	1	2	−0.333
3	4	2	1	−0.667
4	1	2	3	0.000
4	1	3	2	−0.333
4	2	1	3	−0.333
4	2	3	1	−0.667
4	3	1	2	−0.667
4	3	2	1	−1.000

$|n_c+n_d|$ is necessarily less than $n(n-1)/2$. It can be shown (see e.g., Kendall and Gibbons, 1990, Chapter 3) that a coefficient that takes values in the interval [–1, 1] can be obtained by replacing the denominator in (10.3) by $[(D-U)(D-V)]$ where

$$D = n(n-1)/2,\ U = \sum u(u-1)/2,\ V = v(v-1)/2$$

and u, v are the number of consecutive ranks in a tie within the x- and the y-ranks, respectively, and the summations in U, V are over all sets of tied ranks. If there are no tied ranks, $U=V=0$ and the modified denominator reduces to that in (10.3). This modified statistic is denoted by t_b. It is often called *Kendall's tau-b.* If there is an exact linear relationship between mid-ranks t_b takes either the value +1 or –1. A simple example with heavy tying illustrates some of the above points.

Example 10.7

Two motoring experts are each asked to assess 12 sport-utility vehicles (SUVs) in order of preference. Both agree on which vehicle they rank 1 (or best). There are four vehicles that they both agree are almost as good, but neither can separate within that group, so each gives these a tied rank 3.5, while there are a further five vehicles that they both classify as unremarkable but neither expert can separate the vehicles within that group. Each of these vehicles is thus given a mid-rank 8. Both also agree the remaining two vehicles are equally disappointing and thus are ranked 11.5. Thus the mid-ranks allocated are

Expert A	1	3.5	3.5	3.5	3.5	8	8	8	8	8	11.5	11.5
Expert B	1	3.5	3.5	3.5	3.5	8	8	8	8	8	11.5	11.5

There is a linear relationship between these mid-ranks pairs since they all lie on a straight line through the origin with a unit slope. The ranks awarded by Expert A (i.e., x-ranks) are ordered, so n_c, the total number of concordances, is obtained by counting the number of positive differences s_j-s_i, $j>i$. This is found by a direct, but careful, counting process to be 49, whence (10.3) gives $t_k=0.7424$, since there are no discordances. Also $U=V=(4\times3+5\times4+2\times1)/2=17$, whence

$$t_b = 49/\sqrt{(66-17)^2} = 1$$

Here t_b exhibits the desirable property of a correlation coefficient, i.e., $t_b=1$, when there is complete ranking agreement, whereas t_k does not.

The next example involves ties but to a less extreme extent.

Example 10.8

Life expectancy showed a general tendency to increase during the nineteenth and twentieth centuries as standards of health care and hygiene improved. The extra life expectancy varies between countries, communities and even families. Table 10.6 gives the year of death and age at death for 13 males in clan McDelta buried in the Badenscallie burial ground in Wester Ross, Scotland (see Appendix 1). Is there an indication that life expectancy is increasing for this clan in more recent years?

TABLE 10.6

Year of Death and Ages of 13 McDeltas

Year	1827	1884	1895	1908	1914	1918	1924
Age	13	83	34	1	11	16	68
Year	1928	1936	1941	1964	1965	1977	
Age	13	77	74	87	65	83	

If there is an increase in life expectancy those dying later will tend to be older than those dying earlier. Years of death, x, are already arranged in ascending order. The number of concordances and discordances may be counted by examining all pairs of ages at death, y, and scoring each as appropriate (1 for concordance, –1 for discordance, 0 for a tie).

There are no ties in the years of death. The first age at death is 13. In the following entries, this is exceeded in nine cases (concordances); there are two discordances and one tie. Similarly, for the next age at death (83), there is one concordance, nine discordances and one tie. Proceeding this way one gives

$$n_c = 9 + 1 + 6 + 9 + 8 + 6 + 4 + 5 + 2 + 2 + 0 + 1 = 53$$

and

$$n_d = 2 + 9 + 4 + 0 + 0 + 1 + 2 + 0 + 2 + 1 + 2 + 0 = 23.$$

Since $n=13$, (10.3) gives

$$t_k = \frac{53 - 23}{(13 \times 12)/2} = 0.3846.$$

There are two tied values at each of $y=13$ and at $y=83$. Using the formula for the adjustment of the denominator gives $D=78$, $U=0$ and $V=(2\times 1)/2+(2\times 1)/2=2$.

Hence

$$t_b = \frac{53 - 23}{\sqrt{78 \times (78 - 2)}} = 0.3896.$$

This small difference between t_k and t_b does not affect the conclusion. Using the ANSM5 package in R and the command `kendall.tau(ch10$death.year, appl$McDelta, alternative = "greater")` gives a Monte Carlo estimate of $P=0.038$. Hence, there is reasonable evidence against the hypothesis H_0: *no increasing trend of life expectancy* and favouring the one-tail alternative that expectancy has increased during the period covered.

If there are only a few ties and these are ignored, the test that makes use of (10.3) is conservative, for if ties are allowed for by using t_b the denominator is less than that in (10.3).

A one-tailed test is justified because an increase in life expectancy has become an established trend in most developed countries.

Care is needed in counting concordances if there are ties in both x and y. For example, for the data set:

x	2	5	5	7	9
y	5	1	3	8	1

the first pair (2, 5) gives one concordance (7, 8) and three discordances; the second pair (5, 1) gives one concordance (7, 8). There are no discordances due to the tied values in x at (5, 3) and in y at (9, 1). The following pair (5, 3) gives one concordance (7, 8) and one discordance (9, 1).

With more than a few ties, a program that gives at least a Monte Carlo estimate of the relevant tail probabilities is useful.

For moderately large samples with only a few ties, use of (10.3) and tables of critical values for the no-tie situation, or the use of asymptotic results given in Section 10.1.6 should not be seriously misleading.

So far it has been established only that it is intuitively reasonable to expect Kendall's t_k to take values near zero when the (x_i, y_i) are sample values from a bivariate population where the random variables X and Y associated with that population are independent, implying a population $\tau_k=0$ (though the converse may not be true). This leaves open the question of what precisely the sample Kendall coefficient is estimating when $\tau_k \neq 0$.

The following gives a probabilistic interpretation of τ_k that leads naturally to t_k as a reasonable sample estimator of τ_k irrespective of whether X and Y are independent and establishes the parameter τ_k as a reasonable measure of a certain type of dependence between X and Y. Further, the measure is distribution-free as it makes only the assumption that sampling is from some unspecified bivariate distribution.

The argument is based on the fact that if (X_s, Y_s) and (X_t, Y_t), are independent samples from a population with some joint cumulative distribution function $F_{XY}(x, y)$, then each has that same distribution.

Consider a variable $B_{ts}=(Y_s-Y_t)/(X_s-X_t)$. If $B_{ts}>0$, this implies that either the joint event ($Y_s>Y_t$ and $X_s>X_t$) or the joint event ($Y_s<Y_t$ and $X_s<X_t$) has occurred. These are mutually exclusive, so this in turn implies that

$$\Pr(B_{ts} > 0) = \Pr[(Y_s - Y_t)/(X_s - X_t) > 0] = \Pr(Y_s > Y_t \text{ and } X_s > X_t) + \Pr(Y_s < Y_t \text{ and } X_s < X_t).$$

If also X and Y are independent, then

$$\Pr(Y_s > Y_t \text{ and } X_s > X_t) = \Pr(Y_s > Y_t) \times \Pr(X_s > X_t). \tag{10.4}$$

Because Y_s and Y_t both have the marginal distribution of Y, it follows that $\Pr(Y_s>Y_t)=0.5$. Similarly $\Pr(X_s>X_t)=0.5$. Thus from (10.4), independence implies $\Pr(Y_s>Y_t$ and $X_s>X_t)=0.5\times0.5=0.25$. Similarly, under independence $\Pr(Y_s<Y_t$ and $X_s<X_t)=0.25$, which in turn implies that under independence

$$\Pr\left(B_{ts} > 0\right) = \Pr\left[\frac{Y_s - Y_t}{X_s - X_t} > 0\right] = \frac{1}{4} + \frac{1}{4} = \frac{1}{2}.$$

Defining the population Kendall τ_k to be

$$\tau_k = 2\Pr(B_{ts} > 0) - 1 \tag{10.5}$$

it follows that under independence $\tau_k=2\times0.5-1=0$. This indicates that if the departures from independence are such that high values of Y tend to be associated with high values of X and low values of Y with low values of X, then

$$\Pr\,(B_{ts} > 0) = \Pr\left[\frac{Y_s - Y_t}{X_s - X_t} > 0\right]$$

will be greater than 0.5, and will equal 1 if $Y_s > Y_t$ always implies $X_s > X_t$ in which case, from (10.5), $\tau_k = +1$. A similar argument shows that if high values of Y tend to be associated with low values of X and low values of Y with high values of X, then τ_k will be negative. For a complete reversal of rank orders, $\tau_k = -1$.

The discussion following Example 10.4 showed that if for all pairs of n sample values (x_i, y_i), (x_j, y_j) one formed quotients $b_{ij} = (y_j - y_i)/(x_j - x_i)$, $i = 1, 2, \ldots, n-1$, for all $j > i$, then a positive value of this quotient corresponds to a concordance and a negative value to a discordance. In the notation used in (10.3) $n_c/(n_c + n_d)$ provides a sensible estimate of

$$\Pr[(Y_s - Y_t)/(X_s - X_t) > 0]$$

in (10.5). Further, in the no-tie case since $n_c + n_d = n(n-1)/2$, it can be verified that the substitution of $n_c/(n_c + n_d)$ in (10.5) leads to (10.3). Kerridge (1975) gives an interesting example of an estimation problem with this probabilistic interpretation.

It should be realized that the type of dependence described above is not the only possible kind. For example, it is not uncommon for high values of Y to go with both high values of X and low values of X while intermediate values of Y go with intermediate values of X. At its simplest, this might represent a near quadratic relationship between X and Y. The correlation measures described in this chapter are generally unsuitable for detecting this, or many other, possible kinds of nonlinear dependence.

10.1.5 A Median Test for Correlation

Blomqvist (1950, 1951) proposed a test that embodies similar concepts to the median test considered in Section 6.2. Suppose the medians of the marginal distribution of X, Y are, respectively, θ_x, θ_y. In a sample about half the observed x should be below θ_x and about half the y below θ_y. If X and Y are independent a good mix of high, medium and low values of Y should be associated with the various values of X. If θ_x and θ_y were known, the origin in a scatter diagram of the observed (x_i, y_i) could be shifted to (θ_x, θ_y). Then, under the null hypothesis, H_0: X, Y *are independent* one would expect about one-quarter of all points to be in each of the four quadrants determined by the new axes.

Usually θ_x and θ_y are unknown. Blomqvist proposed replacing them with the sample medians M_x, M_y as reasonable estimates. In Figure 10.2, axes with a new origin at (M_x, M_y) are shown using broken lines. Working anticlockwise from the first (or top right) quadrant the numbers of points in the respective quadrants determined by the new axes are 5, 1, 5, 1. The concentration of points in the first and third quadrants suggests dependence between X and Y. One may formalize a test of H_0 against a one-sided alternative of either positive or negative association, or a two-sided alternative of some association, following closely the procedure in Section 6.2.2. In general, if the sample size is even, denoting the sample medians for X, Y by M_x, M_y and assuming no values coincide with these medians, a count is made of the numbers of points a, b, c, d in each of the four quadrants determined after this shift of the scatter plot origin to (M_x, M_y).

From sample median properties, a 2×2 contingency table of the form shown in Table 10.7 can be constructed.

The marginal totals follow from the properties of the sample median. The table assumes an even sample size and no observations at either M_x or M_y. If the sample size is odd, at least one sample value for x and one for y will coincide with the medians, resulting in points at or on axes through the sample medians. Such points are omitted from the count. This necessitates the modifications given at the end of this section.

TABLE 10.7

Contingency Table for the Median Test for Correlation

	Above M_x	**Below M_x**	**Row Total**
Above M_y	a	b	$n/2$
Below M_y	c	d	$n/2$
Column Total	$n/2$	$n/2$	n

For the case shown in Table 10.7, if X and Y are independent a, b, c and d will each have an expected value of $n/4$. Marked departures from this value indicate association between X, Y.

The appropriate procedure is then to use the Fisher exact test. This is illustrated in Example 10.9.

First consider what is an appropriate correlation measure having the desirable properties that it takes values in the interval [–1, 1] with values near ±1 indicating near-monotonic dependence. Independence results in values near zero (although the converse may not be true).

Two possible coefficients are

$$r_d = (a + d - b - c)/n \tag{10.6}$$

and

$$r_m = 4(ad - bc)/n^2 \tag{10.7}$$

It can be verified that in the special circumstances of Table 10.7, where the marginal totals imply that $a=d$, $b=c$ and $a+b=n/2$, both (10.6) and (10.7) have the desirable properties for a correlation coefficient and are equivalent, each taking the same set of $n/2+1$ possible values with the same associated probabilities under the null hypothesis of independence. While (10.6) has arithmetic simplicity, (10.7) generalizes more readily to other contexts, including the case n odd, so it is usually preferred.

Example 10.9

Calculate r_m (or r_d) for the data in Table 10.1 and test the hypothesis of independence against that of positive association between X and Y.

First find the sample medians M_x, M_y and deduce the entries in Table 10.7. The relevant coefficient values are computed using (10.6) or (10.7) and the test of significance is based on the Fisher exact test as in Example 6.7. A one-tailed test is appropriate since the alternative is in a specified (positive) direction.

For the data in Table 10.1, the sample medians are $M_x=9$, $M_y=27$, from which the values of a, b, c, d in Table 10.7 can be deduced. Proceeding clockwise around the quadrants in Figure 10.2 and counting the numbers of observations gives $a=5$, $b=c=1$ and $d=5$. The sample size $n=12$ so $r_m=4(ad-bc)/n^2=4\ ((5\times5)\ -\ (1\times1))/12^2=2/3$. Alternatively, $r_d=2/3$. The Fisher exact test is then used, the relevant tail probabilities being associated with the observed value $a=5$ and the more extreme $a=6$. The relevant probability is $P=0.0400$. There is moderately strong evidence against the hypothesis of independence and favouring that of positive association. The numerical value of the appropriate median correlation coefficient is $r_m = 0.667$.

Output from the ANSM5 package in R can be obtained using `blomqvist(ch10$Q1, ch10$Q2, alternative = "greater")`.

The Pearson correlation coefficient was not significant for these data. It is not robust against departures from normality of the type implicit in the observation (17, 0). However, r_m is robust as it depends only on the number of observations in each quadrant. In the example, the same conclusion would have been reached had the point (17, 0) been replaced by any point in the fourth quadrant (positive x, negative y) keeping the other observations the same. In general, the median correlation test is robust against a few major departures from a monotonic or near-monotonic trend. Unfortunately, it has low Pitman efficiency compared to the Pearson coefficient test when the relevant normality assumption holds.

The Blomqvist test is applicable no matter what the marginal distributions of X, Y may be.

For all but very small samples, a computer program such as that in R or many other packages is desirable for the Fisher exact test. The asymptotic result based on the chi-squared test is generally reliable for large samples.

If the sample size is odd, or generally if there are values equal to one or both sample medians, the marginal totals in a table like Table 10.7 will no longer be $n/2$. The Fisher exact test is then used with the marginal totals observed. If a correlation coefficient is required, it is best to use a modification of r_m, namely,

$$r_m = \frac{ad - bc}{\sqrt{(a+b)(c+d)(a+c)(b+d)}}$$

where a, b, c, d are the observed cell values. This reduces to (10.7) when all marginal totals are $n/2$.

The coefficient r_m is easy to calculate, but in practice, it is used less often than the Spearman or Kendall correlation coefficients.

The concept of partitioning bivariate data by the median that is inherent in Blomqvist's approach has been extended to partitioning based on empirical or sample quantiles by Borkowf et al. (1997).

Other nonparametric tests for correlation have been suggested. Gideon and Hollister (1987) proposed a measure that is also relatively easy to calculate, although this advantage is partly negated for small samples by the fact that the coefficient takes only relatively few possible values and the permutation distribution under the null hypothesis is less easily established than that for the Kendall or Spearman coefficients. The authors give fairly extensive tables of nominal critical values.

Dietz and Killeen (1981) extended the concept of Kendall's tau to a multivariate test for trend applicable to a pharmaceutical problem.

10.1.6 Asymptotic Results

For large samples, tests using the Spearman and Kendall coefficients may be based on the distribution of functions of r_s and t_k that have an asymptotic standard normal distribution.

Detailed discussions of the distribution of r_s under H_0: $\rho_s = 0$ are given by Gibbons and Chakraborti (2020, Section 11.3) and by Kendall and Gibbons (1990, Section 4.14). Convergence to normality tends to be slow, so the asymptotic result should not be used for values of samples of less than 30, and some writers suggest an even higher value. For large n, it is usually assumed that $z = r_s\sqrt{n-1}$ has approximately a standard normal distribution under H_0. This approximation is identical in form to one often used for the Pearson coefficient. A better approximation that is often reasonable for values of n as low as 10 for both the Pearson and Spearman coefficients suggested in Kendall and Gibbons and elsewhere is

$$t_{n-2} = \frac{r\sqrt{n-2}}{\sqrt{1-r^2}}.$$

This has approximately a t-distribution with $n-2$ degrees of freedom under H_0. This approximation is used in R for asymptotic estimation of a P-value for both the Pearson and Spearman coefficients.

For Kendall's tau, the exact distribution of t_k under H_0: $\tau_k=0$ is difficult to obtain. A full discussion is given in Gibbons and Chakraborti (2020, Section 11.2) and by Kendall and Gibbons (1990, Section 4.8) who show under H_0 that $E(t_k)=0$ and $Var(t_k)=2(2n+5)/[9n(n-1)]$ and that for reasonably large n

$$z = \frac{3t_k\sqrt{n(n-1)}}{\sqrt{2(2n+5)}} \qquad (10.8)$$

has approximately a standard normal distribution. This is the basis of the asymptotic P-value determination in R and in many general statistical programs. Modifications are needed for ties and Gibbons and Chaktraborti discuss these, although a few ties may make little difference. The ANSM5 package in R uses Kendall's tau-b. Some writers suggest that the asymptotic result for the Kendall coefficient should not be used if $n<15$ so it would not be recommended for the data in Example 10.8. However, in that example where $t_b=0.3896$, it is found that $z=1.84$ implying that for a one-tailed test $P=0.0331$, broadly in line with the Monte Carlo approximations to the exact P-value in Example 10.8 despite the small sample size.

Output from the ANSM5 package in R can be obtained using `kendall.tau(ch10$death.year, app1$McDelta, alternative = "greater", do.asymp = TRUE, do.exact = FALSE)`.

One difficulty in obtaining confidence intervals for the population coefficients, ρ_s and τ_k using asymptotic results is that the limits may lie outside the closed interval [–1, 1]. Another is that the distributions of the sample statistics are not simple for non-zero ρ_s or τ_k. In particular, the variances are not the same as those under a zero-value hypothesis. Kendall and Gibbons (1990, Chapters 4 and 5) discusses the problem further and gives useful approximations.

10.1.7 A Comparison of the Spearman and Kendall Coefficients

The foregoing discussion raises the question of whether the Spearman coefficient is preferable to the Kendall coefficient or vice versa. The two coefficients seldom lead to markedly different conclusions, though Examples 10.2 and 10.5 show this is possible when one ranked pair is markedly out of line with the general trend. For instance, in Example 10.2, one pair of ranks, i.e. (12, 1), contributes $(12-1)^2=121$ to a total $T=162$. This suggests r_s may be sensitive to such an outlier.

An alternative way of calculating the coefficients clarifies the relationship between them. It is assumed there are no ties and that the pairs of ranks are arranged in ascending order of x-ranks as in Tables 10.1 and 10.2. Writing s_k for the rank of y_k define $s_{ij}=s_j-s_i$, $i=1, 2, 3, \ldots, n-1$ for all $j>i$. It can be seen that the number of concordances in Kendall's t_k equals the number of positive s_{ij}, and the number of discordances equals the number of negative s_{ij}.

In addition, denoting by n_{cs} the sum of the values of the positive s_{ij} (the values of differences that are concordant) and by n_{ds} the sum of the values of the negative s_{ij} (the values of differences that are discordant), it can be shown that in the no-tie case $n_{cs}+n_{ds}=n(n^2-1)/6$ and r_s is given by

$$r_s = \frac{n_{cs} - n_{ds}}{n\left(n^2-1\right)/6} \tag{10.9}$$

This has a formal similarity to (10.3). The s_{ij} may be written as an upper triangular matrix

$$\begin{array}{cccccc}
s_2 - s_1 & s_3 - s_1 & s_4 - s_1 & \cdot\ \cdot & \cdot & s_n - s_1 \\
 & s_3 - s_2 & s_4 - s_2 & \cdot\ \cdot & \cdot & s_n - s_2 \\
 & & \cdot & \cdot\ \cdot & \cdot & \cdot \\
 & & & \cdot\ \cdot & \cdot & \cdot \\
 & & & & s_{n-1} - s_{n-2} & s_n - s_{n-2} \\
 & & & & & s_n - s_{n-1}
\end{array}$$

For the data in Table 10.2, the matrix is

$$\begin{array}{rrrrrrrrrrr}
1 & 3 & 2 & 4 & 5 & 7 & 10 & 9 & 8 & 6 & -1 \\
 & 2 & 1 & 3 & 4 & 6 & 9 & 8 & 7 & 5 & -2 \\
 & & -1 & 1 & 2 & 4 & 7 & 6 & 5 & 3 & -4 \\
 & & & 2 & 3 & 5 & 8 & 7 & 6 & 4 & -3 \\
 & & & & 1 & 3 & 6 & 5 & 4 & 2 & -5 \\
 & & & & & 2 & 5 & 4 & 3 & 1 & -6 \\
 & & & & & & 3 & 2 & 1 & -1 & -8 \\
 & & & & & & & -1 & -2 & -4 & -11 \\
 & & & & & & & & -1 & -3 & -10 \\
 & & & & & & & & & -2 & -9 \\
 & & & & & & & & & & -7
\end{array}$$

To compute t_k using this array, the number of positive entries, $n_c=47$, and the number of negative entries, $n_d=19$, are counted. These were the values obtained in Example 10.4. To obtain n_{cs}, all positive entries are added together. Starting with the first row and working along columns from left to right gives $n_{cs}=1+3+2+4+\ldots+3+2+1=205$. Similarly, adding all negative entries ignoring signs gives $n_{ds}=1+2+1+4+\ldots+10+2+9+7=81$. Substitution in (10.9) gives

$$r_s = 6 \times \frac{205-81}{12 \times 143} = 0.4336,$$

agreeing with the value in Example 10.2.

Examining the triangular array helps explain why, in this example, r_s plays down any correlation relative to that indicated by t_k. A few negative s_{ij} in the last column make relatively large contributions to the sum n_{ds}. This is attributable to the fact that the low-rank value of the 12th and last observation, $s_{12}=1$, is out of line with the high ranks assigned to its near neighbours.

As illustrated above, Kendall's tau is more easily interpretable, as a difference between probabilities of concordance and discordance, than the Spearman correlation coefficient, which is usually explained only as a measure of correlation on the scale from –1 to 1. On the other hand, the Spearman coefficient is easier to calculate without a computer. This was a crucial advantage in 1904, when Spearman introduced the coefficient, and was still important as late as the 1940s; Daniels and Kendall (1947) published a confidence interval formula for Kendall's tau and made a mistake in the worked example with a sample size of 30.

Many other aspects of parametric correlation have nonparametric equivalents, including that of partial correlation. A brief introduction to that topic is given in Gibbons and Chakraborti (2020, Section 12.6).

10.1.8 Efficiency, Power and Sample Size

Various results concerning efficiency, power and sample sizes needed to ensure a required power when specific alternative hypotheses hold are scattered throughout the literature for both the Kendall and Spearman coefficients. One important result is that when sampling from the bivariate normal distribution both the Kendall and Spearman coefficients have the same Pitman efficiency of 0.912 relative to the Pearson coefficient. The Kendall coefficient tends to do rather better than the Pearson coefficient when sampling from long-tailed symmetric distributions and has Pitman efficiency of 1.266 relative to the Pearson coefficient for a bivariate double exponential distribution. Noether (1987) gives the following asymptotic formula for the sample size needed to ensure power $1-\beta$ in a one-tailed test of H_0: $\tau_k=0$ against H_1: $\tau_k=\tau_1$ where τ_1 has some fixed nonzero value

$$n \approx \frac{(z_\alpha + z_\beta)^2}{9\tau^2}.$$

Here z_α, z_β have the standard normal tail probability meanings assigned to them in (5.3). Some further results on power for the Kendall coefficient are given by Hollander, Wolfe and Chicken (2014).

SUMMARY POINTS:

- The Pearson product-moment correlation estimates the degree of linear association between two variables (Section 10.1.1).
- If points lie on a straight line, the coefficient values are 1 and –1 for positive and negative association, respectively (Section 10.1.1).
- A value close to zero indicates either independence between the variables or a non-linear relationship (Section 10.1.1).
- Rank coefficient values depend on monotonicity. Values of 1 and –1 indicate strictly increasing and decreasing associations, respectively (Section 10.1.2).

- The Spearman rank correlation coefficient may be calculated by ranking the data and using the formula for the Pearson coefficient (Section 10.1.3).
- The test for the Kendall coefficient is equivalent to one for a Jonckheere–Terpstra statistic (Section 10.1.4).
- The median test for correlation is based on the number of observations located in the four quadrants formed from horizontal and vertical lines passing through the point representing the sample medians (Section 10.1.5).
- Asymptotic tests should be avoided with samples of less than 30 (Section 10.1.6).
- The Spearman and Kendall coefficients usually give the same conclusions (Section 10.1.7).
- Kendall's tau is easier to interpret than the Spearman coefficient (Section 10.1.7).
- The Spearman coefficient is easier to calculate by hand (Section 10.1.7).
- When sampling from the bivariate normal distribution, the Kendall and Spearman coefficients have the same Pitman efficiency of around 0.91 (Section 10.1.8).

10.2 Ranked Data for Several Variables

If more than two observations are ranked for each of a set of n experimental units, it may be of interest to test for evidence of concordance between rankings of the units. As indicated in Section 7.3.1, the Friedman test may be applied to rankings of objects (e.g., preferences for different varieties of raspberry, placings in a gymnastics contest by different judges, ranking of candidates by different examiners) to test whether there is evidence of consistency between those making the rankings.

Kendall, independently of Friedman, proposed the use of a function of the Friedman statistic, which is often referred to as the *Kendall coefficient of concordance* and tabulated some small-sample critical values relevant to testing the hypothesis that the rankings were essentially random against the alternative of evidence of consistency. Kendall regarded his coefficient of concordance as an extension of the concept of correlation to more than two sets of rankings. In the bivariate case, the Kendall correlation coefficient may be used to measure both agreement (positive association) and disagreement (negative association). Concordance, whether measured by Kendall's original statistic, or the Friedman modification, is one-sided in the sense that rejection of the null hypothesis indicates positive association. For example, if four judges A, B, C, D rank five objects in the order given in Table 10.8, the hypothesis of no association would be retained because the Friedman statistic (7.5) would take the value zero.

There is here complete agreement between judges A and C and between judges B and D. However, judges B and D disagree completely with judges A and C. The Kendall and the Friedman statistics do not detect such patterns.

An example indicates Kendall's approach in developing his concordance test.

TABLE 10.8

Ranking of Five Objects by Four Judges

	Judge			
Object	A	B	C	D
I	1	5	1	5
II	2	4	2	4
III	3	3	3	3
IV	4	2	4	2
V	5	1	5	1

Example 10.10

There are six contestants in a diving competition. Three judges each independently rank their performance in order of merit (1 for best, 6 for worst). The rankings allocated by each judge are given in Table 10.9. Is there evidence of consistency between judges?

Clearly the judges are not in complete agreement, but there is a reasonable consensus that competitor III is a good performer. There is fairly strong support for competitor I, whereas competitors IV and VI are thought to perform poorly. A low-rank total (last column) indicates a performance considered by the judges overall to be better than that of a competitor who attains a high-rank total. Had the judges shown only chance agreement, effectively equivalent to allocating ranks at random, one would expect all rank sums to be nearer to the average value of $63/6=10.5$. Had there been complete agreement between judges it can be verified that the rank sums for the six competitors would be some permutation of 3, 6, 9, 12, 15, 18.

Kendall's coefficient of concordance is based on the sum of squares of deviations of the competitors' rank sums from the mean or expectation under random allocation.

If one denotes this sum of squares by S, then it is intuitively reasonable, and can indeed be shown, that S has a maximum when the judges are in complete agreement, and a minimum of zero when all rank sums equal the mean of 10.5 (which would imply some tied ranks in the situation considered in Table 10.9). Arguing along these lines, Kendall proposed the statistic

$$W = \frac{S}{\text{Max}(S)}.$$

TABLE 10.9

Ranks Awarded to Six Competitors by Three Judges in a Diving Contest

	Judge			
Competitor	A	B	C	**Rank Total**
I	2	2	4	8
II	4	3	3	10
III	1	1	2	4
IV	6	5	5	16
V	3	6	1	10
VI	5	4	6	15

He called W a coefficient of concordance. This takes the maximum value 1 when there is complete agreement, and the minimum value zero when there is no agreement. It will also take the value zero in a situation like that in Table 10.8 where there are contrary opinions held by pairs of judges. Tables of critical values are available, but any program that gives exact P-values for the Friedman test may be used also to obtain P for the Kendall statistic.

From Table 10.9 it can be seen that

$$S=(8-10.5)^2+(10-10.5)^2+\ldots+(15-10.5)^2=99.5.$$

If there are no ties, it can be shown that $\text{Max}(S)=m^2n(n^2-1)/12$ if there are m judges and n competitors, which in this case gives $\text{Max}(S)=157.5$, whence $W=99.5/157.5=0.632$. Using the ANSM5 package in R and the command `kendall.concordance(ch10$diving.rank, ch10$competitors, ch10$judges)` gives a Monte Carlo $P=0.0615$. There is no strong evidence of concordance between the judges.

The weakness of the evidence indicates the fairly low power of the test against alternatives other than a very high level of agreement. The judges are particularly erratic in their assessment of competitor V.

It can be shown that the statistic T for the Friedman test given by (7.5) and W satisfy the relationship

$$W=\frac{T}{m(n-1)}. \tag{10.10}$$

Some modifications to the procedure in the above example are needed for ties. These are similar to those for ties in the Friedman test. If T is computed for the tied Friedman case, as described in Section 7.3.2, the correct value for W may be derived using (10.10).

An intuitively reasonable measure of concordance between rankings of competitors by several judges is one based on the mean of pairwise rank correlation coefficients. It can be shown that for m judges, the mean, R_s, of all $m(m+1)/2$ such Spearman coefficients is such that $R_s=(mW-1)/(m-1)$. It follows that when $R_s=1$ (i.e., when all pairwise rankings are in complete agreement) then $W=1$. When $W=0$, then $R_s=-1/(m-1)$ which can be shown to be the least possible value for R_s. Examination of the overall pattern of pairwise rank correlation coefficients is useful for detecting patterns of the type illustrated for an extreme case in Table 10.8 where the pairwise Spearman coefficients can be shown to take the value +1 twice and the value −1 four times, thus exhibiting a definite pattern where $W=0$. This indicates that, like rank correlation in the bivariate situation, $W=0$ when there is no association pattern, but the converse may not hold.

SUMMARY POINTS:

- The Kendall coefficient of concordance can be applied to observations ranked by three or more judges.
- For multivariate ranked data, Kendall's coefficient of concordance is equivalent to the Friedman test statistic for ranked data in randomized blocks given in Section 7.3.1.

10.3 Agreement

10.3.1 Assessing Observer Variation

All measures so far described in this chapter relate to the concept of association between variables. With positive association between two variables, if one variable takes a large value then so will the other. This does not necessarily mean, however, that the scores on some scale for both variables are identical. If, in addition to positive association, scores on both variables tend to be the same, this is described as a high level of agreement. Agreement is not the same as association. It is possible for a correlation coefficient to be very high while at the same time the agreement is very low. For instance, scores for one variable may be consistently higher than the score for the other, with no zero differences. Agreement in this sense can be assessed for categorical variables.

This concept of agreement is particularly important in assessing observer variation. Suppose that two dentists examine the X-ray of a particular tooth for the presence or absence of dental caries on two separate occasions. Although the information available from the X-ray remains the same, there are two possible types of variation with such examinations. A dentist re-examining the X-ray of a tooth may make a different judgement (e.g., caries absent rather than caries present). This is known as *intra-observer* variation. Also, a colleague may make a different decision about the same tooth, leading to *inter-observer* variation. For both types of variation, the same question can be asked, i.e., how well do the two sets of data agree with each other?

If there are relatively few possible categories (ordered or nominal), the most obvious measure of agreement between two sets of data is the proportion of cases in which agreement occurs. This has the clear drawback that a substantial amount of agreement can occur by chance alone, so the level of agreement should be *chance-corrected*. Myrick Doolittle developed a chance-corrected measure of association in the context of tornado forecasting (Doolittle, 1888). It was rediscovered as the *kappa statistic* by the psychologist Jacob Cohen, and published in a highly cited paper (Cohen, 1960). For convenience, it is referred to as Cohen's kappa statistic in this chapter.

10.3.2 The Kappa Statistic for Agreement between Two Assessors

Two observers allocate each of n patients to one of k possible categories. Denote the number of patients allocated to category i by Observer A and to category j by Observer B by n_{ij}, the total number of patients allocated to category i by Observer A by n_{i+}, and the total number of patients allocated to category j by Observer B by n_{+j}. Table 10.10 shows the data in a $k \times k$ contingency table.

Dividing the number of observations, n_{ij}, for the cell in the ith row and jth column by the total number of patients, n, gives the proportion p_{ij} of patients for that cell. The observed proportion of agreement is the proportion of patients for which i and j are equal (i.e., on the diagonal of the table) which is given by

$$p_o = (n_{11} + n_{22} + \ldots + n_{kk}) / n = \sum_{i=1}^{k} p_{ii}$$

The expected proportion of agreement p_e is determined using the appropriate row and column totals in a similar manner to that used for expected numbers in what is known as the chi-squared test for contingency tables (Sections 12.1.1 and 12.2.2) so that

TABLE 10.10

A General Contingency Table for Assessing Agreement between Two Observers

Obs B		1	2	.	.	.	k	Total
	1	n_{11}	n_{12}	.	.	.	n_{1k}	n_{1+}
	2	n_{21}	n_{22}	.	.	.	n_{2k}	n_{2+}
	.	.	.	.	.	.	.	.
Obs A	.	.	.	.	.	.	.	.
	.	.	.	.	.	.	.	.
	k	n_{k1}	n_{k2}	.	.	.	n_{kk}	n_{k+}
		n_{+1}	n_{+2}	.	.	.	n_{+k}	n

$$p_e = \sum_{i=1}^{k} p_{i+} + p_{+i}$$

where $p_{i+}=n_{i+}/n$ and $p_{+i}=n_{+i}/n$.

The kappa statistic is defined as:

$$\frac{\text{Observed proportion} - \text{Expected proportion}}{1 - \text{Expected proportion}},$$

or symbolically as

$$\kappa = \frac{p_o - p_e}{1 - p_e}$$

The kappa coefficient takes values between –1 and 1, with the value 1 for perfect agreement, zero for the level of agreement that would be expected by chance, and negative values for less than chance agreement, i.e. apparent disagreement. Landis and Koch (1977) suggested benchmarks for kappa, e.g. a score over 0.8 indicates good agreement, 0.6–0.8 indicates substantial agreement and 0.4–0.6 moderate agreement. Assuming that the patients are assessed independently of each other and that the assessors operate independently, one can test the null hypothesis that for the population $\kappa=0$. An exact test for small samples is available in the ANSM5 package in R together with a Monte Carlo approximation and an asymptotic result for larger samples. Fleiss, Lee and Landis (1979) give an asymptotic formula for the standard error of kappa assuming H_0 holds, namely

$$\text{se}(\kappa) = \frac{\sqrt{p_e + p_e^2 - \sum_{i=1}^{k} p_{i+}p_{+i}(p_{i+} + p_{+i})}}{(1-p_e)\sqrt{n}}$$

The null hypothesis is tested by assuming that if it holds, then $Z=\kappa/\text{se}(\kappa)$ has a standard normal distribution.

Fleiss, Levin and Paik (2003) give an asymptotic formula for a confidence interval for κ based on the estimated standard error of its maximum-likelihood estimate. The ANSM5 package in R gives a Monte Carlo confidence interval.

Example 10.11

Two dentists inspected 100 patients and classified them as either requiring treatment or not requiring treatment. Table 10.11 shows the decisions by the two dentists as to whether they thought treatment was required. Use the kappa statistic to calculate the chance-corrected agreement for this sample and test the null hypothesis that $\kappa=0$.

It is assumed that patients are classified independently of each other and that the dentists do not confer. Cohen's kappa is calculated using the formulae for observed and expected proportions of agreement given above. The null hypothesis may be tested using either an exact test if relevant software is available or the asymptotic formula for the standard error.

First display the data in terms of observed proportions:

	Dentist B		
Dentist A	**Treatment Needed**	**Treatment not Needed**	**Total**
Treatment Needed	0.40	0.05	0.45
Treatment not Needed	0.25	0.30	0.55
Total	0.65	0.35	1.00

With only two assessors and two categories, the formulae for the observed and expected proportions of agreement are straightforward. The observed proportion of agreement is 0.4+0.3=0.7. However, some of this agreement could have been expected by chance, and this is calculated from the row and column totals. Under chance agreement, the expected proportion of cases in the *yes/yes* cell of the table is given by 0.65×0.45=0.2925. The expected proportion of cases in the *no/no* cell is given by 0.35×0.55=0.1925. The total proportion of *expected* agreeing cases is therefore 0.2925+0.1925=0.485. Hence

$$\kappa = (0.7 - 0.485)/(1 - 0.485) = 0.417.$$

The asymptotic formula for the standard error of kappa under the null hypothesis gives:

$$\text{se}(\kappa) = \frac{\sqrt{0.485 + 0.485^2 - 0.45 \times 0.65(0.45 \times 0.65) - 0.55 \times 0.35(0.55 \times 0.35)}}{(1 - 0.485)\sqrt{100}} = 0.09215,$$

whence

$$Z = \kappa / \text{se}(\kappa) = 0.417 / 0.09215 \approx 4.53.$$

Reference to a standard normal distribution indicates $P<0.0001$. Hence, applying the benchmarks of Landis and Koch, there is moderate chance-corrected agreement ($\kappa=0.417$) between the two dentists. The very small P-value shows overwhelming evidence against the null hypothesis.

TABLE 10.11

Need of Treatment for a Series of Patients as Assessed by Two Dentists

	Dentist B		
Dentist A	**Treatment Needed**	**Treatment not Needed**	**Total**
Treatment Needed	40	5	45
Treatment not Needed	25	30	55
Total	65	35	100

Output from the ANSM5 package in R can be obtained using `cohen.kappa(ch10$dentistA, ch10$dentistB)`.

The strong evidence against the null hypothesis should come as no surprise. One would expect dentists, who receive several years of training, to have considerable agreement with each other. However, the fact that kappa is only 0.417 should be a cause for concern. The observed agreement, which appears satisfactory, can be misleading. Once chance agreement is taken into account, the level of agreement is less impressive.

A Monte Carlo 95% confidence interval for kappa is (0.254, 0.576). This is rather wide, indicating that even a sample of 100 patients does not provide a great deal of information about the population value for kappa. Studies of observer agreement frequently involve complex and time-consuming methods for assessing patients or specimens. As a consequence, samples are generally small (typically no more than 50) and the benchmarks given by Landis and Koch (1977) are in practice therefore not always helpful.

Output from the ANSM5 package in R can be obtained using `cohen.kappa(ch10$dentistA, ch10$dentistB, do.exact = FALSE, do.CI = TRUE)`.

10.3.3 Extensions of Cohen's Kappa Statistic

If there are more than two categories, some types of disagreement between the two assessors might be particularly serious. For instance, in assessing a patient's state of health, disagreement between the responses of "well" and "poor" is more serious than disagreement between "well" and "moderately well". In the calculation of weighted kappa (Cohen, 1968), different types of disagreement are given different weights.

The conditional form of the kappa statistic is generally accredited to Light (1971). However, this measure is equivalent to the centesimal scale proposed by Francis Galton for measuring the degree of matching for fingerprint pattern types on the left and right hand (Galton, 1892).

The kappa statistic has been extended to allow for more than two assessors, see Conger (1980) and Posner et al. (1990). A straightforward method for calculating a multiple kappa statistic is to take the arithmetic mean of the kappa values obtained by taking pairs of observers in turn. However, the formulae for hypothesis testing and the calculation of confidence intervals are complicated (Fleiss, Levin and Paik, 2003).

Kappa has also been developed for ordinal data by Fleiss (1978) and continuous data by Rae (1988). In certain situations, kappa is equivalent to the intraclass correlation coefficient (Fleiss and Cohen, 1973; Rae, 1988). Sample size calculation for the case of two assessors has been investigated by Cantor (1996).

10.3.4 Testing the Equality of Cohen's Kappa Statistic in Several Populations

A multicentre study of health care usually involves the selection and assessment of a group of patients at several different locations. The measurement of inter-observer agreement may indicate considerable variation between centres (Crane et al., 2003). If the level of agreement is summarised by the kappa statistic, it is appropriate to test the null hypothesis that the population kappa values are equal.

Donner, Eliasziw and Klar (1996) describe methods of dealing with this problem for the case of two categories. These include the large sample approach sketched out here, which is based on the chi-squared distribution.

Suppose that there are k centres and at each a pair of doctors assess the need for a particular treatment for a series of patients. The contingency table for the hth centre ($h=1, 2, \dots, k$) will take the form shown in Table 10.12.

TABLE 10.12

Need of Treatment for a Series of Patients as Assessed by Two Doctors at Centre h

	Doctor B		
Doctor A	**Treatment Needed (1)**	**Treatment not Needed (0)**	**Total**
Treatment Needed (1)	a_{11h}	a_{12h}	$a_{1.h}$
Treatment not Needed (0)	a_{21h}	a_{22h}	$a_{2.h}$
Total	$a_{.1h}$	$a_{.2h}$	$a_{..h}$

TABLE 10.13

Frequency of Rating Combinations in the k Centres

Category	Ratings	Centre 1	Centre 2	…	Centre k	Total
1	(1, 1)	n_{11}	n_{12}	…	n_{1k}	$n_{1.}$
2	(1, 0) or (0, 1)	n_{21}	n_{22}	…	n_{2k}	$n_{2.}$
3	(0, 0)	n_{31}	n_{32}	…	n_{3k}	$n_{3.}$
	Total	N_1	N_2	…	N_k	N

Writing $n_{1h}=a_{11h}$, $n_{2h}=a_{12h}+a_{21h}$, $n_{3h}=a_{22h}$ and $a_{..h}=N_h$ the information from the k centres can be displayed as in Table 10.13.

Now define for the hth centre

$$\hat{\kappa}=1-\frac{n_{2h}}{2N_h\hat{\pi}_h\left(1-\hat{\pi}_h\right)}$$

where

$$\hat{\pi}=\frac{2n_{1h}+n_{2h}}{2N_h}$$

For the test of the null hypothesis H_0: $\kappa_1=\kappa_2=\ldots=\kappa_k$ one calculates for each centre ($h=1, 2, \ldots, k$)

$$\hat{V}_h=\frac{1-\hat{\kappa}_h}{N_h}\left[\left(1-\hat{\kappa}_h\right)\left(1-2\hat{\kappa}_h\right)+\frac{\hat{\kappa}_h\left(2-\hat{\kappa}_h\right)}{2\hat{\pi}_h\left(1-\hat{\pi}_h\right)}\right]$$

to obtain

$$\hat{W}_h=\frac{1}{\hat{V}_h}$$

The test statistic required is

$$\chi_V^2=\sum_{i=1}^{k}\hat{W}_h\left(\hat{\kappa}_h-\tilde{\kappa}\right)^2$$

where

$$\tilde{\kappa}=\frac{\sum_{h=1}^{k}\hat{W}_h\hat{\kappa}_h}{\sum_{h=1}^{k}\hat{W}_h}$$

This is referred to a chi-squared distribution with $k-1$ degrees of freedom.

Example 10.12

A group of 601 young adults living in Limache, central Chile, were investigated regarding their experience of symptoms of asthma (Smeeton et al., 2006). Initially the participants completed an asthma questionnaire that included an item about whether they had experienced symptoms of wheeze in the previous 12 months (*yes* or *no*). Around 18 months later, the participants answered the same series of questions, but on this occasion, a trained interviewer gave a physical demonstration of how symptoms of wheeze (and other characteristics of asthma) would manifest themselves. For each individual, the response regarding wheeze recorded from the initial questionnaire was compared with that given following the demonstration. The level of agreement between the two sets of responses was assessed using the kappa statistic.

One issue of interest was whether the level of agreement was influenced by the individual's standard of living. Each participant indicated whether they owned the following household items: a gas fuelled water heating device, personal computer, refrigerator, washing machine, microwave oven. Individuals were categorized by how many of these items they possessed into *low* (0–1), *intermediate* (2–3) or *high* (4–5). The contingency tables for the question regarding wheeze experienced in the previous 12 months stratified by the number of items owned are shown in Table 10.14.

It is assumed that individuals did not confer with each other and with an 18-month period between the two assessments any memory effect will be minimal. There are $k=3$ levels for the standard of living and Cohen's kappa is calculated by the method shown in Example 10.10. The asymptotic chi-squared test of H_0: $\kappa_1=\kappa_2=\kappa_3$ is then performed.

TABLE 10.14

Questionnaire and Demonstration Responses Regarding Personal Symptoms of Wheeze in the Last 12 months, Stratified by Number of Specified Items Owned*

		0–1 Items	
Demonstration	**Symptoms (1)**	**No Symptoms (0)**	**Total**
Questionnaire			
Symptoms (1)	14	23	37
No Symptoms (0)	5	54	59
Total	19	77	96
		2–3 Items	
Demonstration	**Symptoms (1)**	**No Symptoms (0)**	**Total**
Questionnaire			
Symptoms (1)	45	104	149
No Symptoms (0)	23	213	236
Total	68	317	385
		4–5 Items	
Demonstration	**Symptoms (1)**	**No Symptoms (0)**	**Total**
Questionnaire			
Symptoms (1)	5	34	39
No Symptoms (0)	9	71	80
Total	14	105	119

*Household item information was incomplete for one adult, resulting in a sample size of 600.

From the information in Table 10.14, κ_1=0.323, κ_2=0.227, κ_3=0.019, which indicates that there might be lower agreement for the group owning the most items. Performing the test of H_0: $\kappa_1=\kappa_2=\kappa_3$ using the information in Table 10.14 gives:
$n_{11} = 14, n_{21} = 28, n_{31} = 54, N_1 = 96; n_{12} = 45, n_{22} = 127, n_{32} = 213, N_2 = 385; n_{13} = 5, n_{23} = 43,$ $n_{33} = 71, N_3 = 119.$

Hence, $\hat{\pi}_1 = 0.292$, $\hat{\pi}_2 = 0.282$, $\hat{\pi}_3 = 0.223$ from which, after lengthy calculation, it follows that $\hat{W}_1 = 90.370$, $\hat{W}_2 = 351.755$, $\hat{W}_3 = 130.032$, finally giving χ_V^2=7.191. This value is referred to a chi-squared distribution with two degrees of freedom, which indicates a *P*-value of 0.027.

Output from the ANSM5 package in R can be obtained using `cohen.kappa (ch10$questionnaire, ch10$demonstration, ch10$items)`.

There is moderate evidence against the null hypothesis. For the less prosperous groups, some agreement may exist between responses from the two formats, but for the group having a high standard of living, there is no agreement whatsoever.

Donner, Eliasziw and Klar (1996) showed using Monte Carlo simulation that the large sample test applied here is appropriate if the size of each group exceeds 100. The smallest group in this example is very close to this threshold, so the approximate test should give reasonable results.

There is some evidence of an association between the prevalence of asthma and deprivation in the medical literature. It could be reasoned that individuals from a less prosperous background might know more people who suffer from asthma and so have a clearer understanding of the symptoms involved. It would be difficult to generalize these findings to more economically developed countries, including contemporary Chile, where most of the population would have access to all these items.

It might be argued that the level of agreement between the two formats was influenced by treatment received during the intervening 18 months. However, at the time of the study appropriate treatment for asthma was virtually unavailable in this area of Chile so once asthma was diagnosed (usually during childhood) symptoms would generally have continued.

Most statistical packages do not contain procedures for assessing the equality of population kappa values. However, the *parmhet* package in Stata can perform this type of test with any asymptotically normal statistics, including kappa, if the estimates and standard errors are given for each population.

SUMMARY POINTS:

- Agreement is a more stringent concept than association (Section 10.3.1).
- Measures of agreement can be used to assess intra-observer agreement (responses given on two occasions by the same observer) and inter-observer agreement (responses given by two observers) (Section 10.3.1).
- Cohen's kappa statistic is a chance-corrected measure of agreement between two raters that can be used for categorical data (Section 10.3.1).
- Kappa takes a value between 1 (perfect agreement) and –1 (perfect disagreement). A value of 0 indicates a level of agreement that would occur by chance alone (Section 10.3.2).
- Strong evidence against the null hypothesis of chance agreement only does not necessarily imply that the strength of agreement between two raters is satisfactory (Section 10.3.2).

- For more than two categories, the severity of disagreement can be taken into account with the use of weighting (Section 10.3.3).
- For large samples, the equality of Cohen's kappa in several independent populations can be tested using an asymptotic method based on a chi-squared distribution (Section 10.3.4).

10.4 Exercises

10.1 The Epilepsy Risk Awareness (ERA) scale and the Epilepsy Self-Management Scale (ESMS) are two methods for scoring how an individual with epilepsy manages the risks involved. High scores indicate good awareness/ management. For a sample of 10 participants in the study reported in Braun et al. (2017), the scores were

ERA	67	80	85	94	68	71	111	105	107	108
ESMS	132	146	147	156	119	142	162	164	175	165

Find the value of Spearman's correlation between the two scales.

10.2 Calculate the exact permutation distribution of the Spearman correlation coefficient r_s for (i) three and (ii) five paired observations. Assume there are no ties. Compare your findings with those from Example 10.3.

10.3 For the data in Exercise 10.1, calculate Kendall's tau and test the null hypothesis H_0: *no association* against H_1: *positive association between ERA and ESMS scores.*

10.4 The table for the McDelta clan deaths is shown below:

Year	1827	1884	1895	1908	1914	1918	1924
Age	13	83	34	1	11	16	68
Year	1928	1936	1941	1964	1965	1977	
Age	13	77	74	87	65	83	

Reanalyse these data using Spearman's rho. How do your conclusions compare with those based on Kendall's tau?

10.5 Use the Cox–Stuart trend test (Section 4.5.3) to investigate a possible time trend in the life spans of the McDelta clan (Exercise 10.4). Do the results agree with those based on the Kendall and on the Spearman coefficients?

10.6 Perform the median test for correlation on the epilepsy risk awareness and management data given in Exercise 10.1.

10.7 In the study reported in Braun et al. (2017), participants completed the Chalfont Seizure Severity Scale (for details see Example 3.1). Higher scores indicate more severe seizures. For a sample of 10 participants in Exercise 10.1, the ERA and SSS scores were

ERA	67	80	85	94	68	71	111	105	107	108
SSS	104	103	60	38	2	27	50	35	108	35

Perform the median test for correlation and comment on your findings.

10.8 A china manufacturer is investigating market response to seven designs of dinner set. The main markets are the British and American. To get some idea of preferences in the two markets, a survey of 100 British and 100 American individuals is carried out and each participant is asked to rank the designs in order of preference from 1 for favourite to 7 for least acceptable. For each country, the 100 rank scores for each design are totalled. The design with the lowest total is assigned rank 1 that with the next lowest total rank 2, and so on. Overall rankings for each country are

Design	A	B	C	D	E	F	G
British Rank	1	2	3	4	5	6	7
American Rank	3	1	2	5	4	7	6

Calculate the Spearman and Kendall correlation coefficients. Is there evidence of a positive association between orders of preference?

10.9 The manufacturer in Exercise 10.8 later decides to assess preferences in the Canadian and Australian markets by a similar method and the rankings obtained are

Design	A	B	C	D	E	F	G
Canadian Rank	5	3	2	4	1	6	7
Australian Rank	3	1	4	2	7	6	5

Calculate the Spearman and Kendall correlation coefficients. Is there evidence of a positive association between orders of preference?

10.10 Perform an analysis of the ranked data for all four countries in Exercises 10.8 and 10.9 to assess the evidence for any overall concordance. Comment on the practical implications of your result.

10.11 Paul (1979) discusses marks awarded by 85 different examiners to each of 10 scripts. The marks awarded by six of these examiners were

	Script									
Examiner	**1**	**2**	**3**	**4**	**5**	**6**	**7**	**8**	**9**	**10**
1	22	30	27	30	28	28	28	28	36	29
2	20	28	25	29	28	25	29	34	40	30
3	22	28	29	28	25	29	33	29	33	27
4	24	29	30	28	29	27	30	30	34	30
5	30	41	37	41	34	32	35	29	42	34
6	27	27	32	33	33	23	36	22	42	29

Use rank tests to determine (i) whether the examiners show reasonable agreement on ordering the scripts by merit and (ii) whether some examiners tend to give consistently higher or lower marks than others.

10.12 Timber veneer panels are classified by experts as either deluxe (DL), standard (S) or reject (R) depending upon the number and type of faults detected. To check for consistency of classification between two observers, each is asked to allocate the same set of 150 panels to the categories they consider relevant. Use Cohen's kappa to assess whether the following findings for a pair of observers indicate reasonable agreement between them. Comment critically on the interpretation of your results.

		Observer A			
	Grade	**DL**	**S**	**R**	**Total**
Observer B	**DL**	7	2	1	10
	S	4	96	18	118
	R	1	10	11	22
	Total	12	108	30	150

10.13 For the data in Example 10.12 the information for agreement between the two methods of questioning with regard to wheeze in the last 12 months was tabulated by gender and is shown below.

Questionnaire and demonstration responses regarding personal symptoms of wheeze in the last 12 months, stratified by gender.

	Males		
Demonstration	***Symptoms* (1)**	***No Symptoms* (0)**	***Total***
Questionnaire			
Symptoms (1)	27	58	85
No Symptoms (0)	15	136	151
Total	42	194	236
	Females		
Demonstration	**Symptoms (1)**	**No Symptoms (0)**	**Total**
Questionnaire			
Symptoms (1)	37	104	141
No Symptoms (0)	22	202	224
Total	59	306	365

Calculate the two kappa values. Is there evidence that the population kappa values differ between males and females?

11

Bivariate Linear Regression

11.1 Regression and Correlation

Correlation is concerned with the direction and strength of possible associations between two variables. As discussed in Chapter 10, associations may be positive or negative, strong or weak, linear or non-linear, or there may be no association whatsoever. By contrast, regression is concerned with the quantitative aspects of relationships, such as determining the slope and intercept of a straight line that, in some sense, provides the best fit to given data.

Correlation and regression are closely related. For example, in straight-line regression, a gradient of zero (i.e. horizontal regression line) corresponds to a correlation of zero. Values of +1 or –1 for the Pearson product moment correlation coefficient indicate that all the observed points lie on a straight line.

For a pair of variables, it is important to identify which of the two could be described as an *outcome* variable, the other measure being the *explanatory* variable. For example, with the height and weight of boys, individuals are regarded as increasing in weight as they grow taller, so height is the explanatory variable and weight is the outcome variable. In presenting the observations on a scatter diagram, the horizontal (x) axis is used for the explanatory variable and the vertical (y) axis for the outcome variable. The terms *dependent* and *independent* are sometimes used instead of outcome and explanatory. The fitted relationship can then be used to *predict* the weight of a boy given that his height is known.

Example 11.1

An investigation is conducted into the amount of time spent by adolescents playing video games on weekends (Saturday/Sunday) and parental attitudes towards video gaming. The young people participating keep a diary of their weekend video game playing, and their parents state the maximum length of weekend playing time that they would regard as reasonable. Observations for the seven participating families, in hours, are shown below.

Upper Limit Stated by Parents (x):	0	1	2	3	4	5	6
Reported Gaming Time (y):	2.5	3.1	3.4	4.0	4.6	5.1	11.1

Illustrate this information using a scatter diagram.

DOI: 10.1201/9780429326172-11

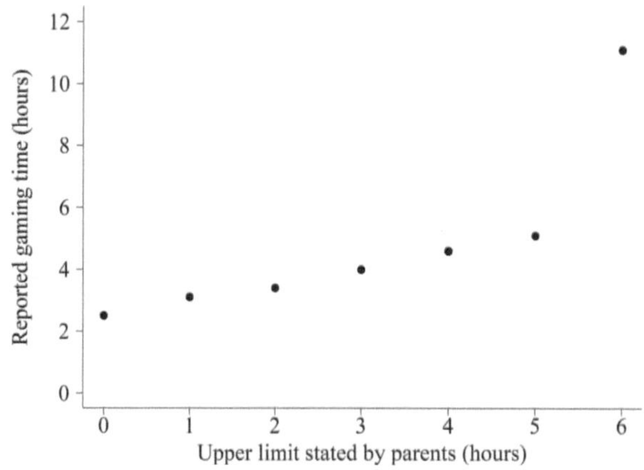

FIGURE 11.1
Scatter diagram for the data in Example 11.1.

Here, the outcome variable is the length of gaming time reported by the young people and the explanatory variable is the time that his or her parents would regard as a reasonable upper limit. The scatter diagram is shown in Figure 11.1. There is a positive relationship, with a higher parental limit being associated with longer playing time. The largest reported value (11.1 hours) is either an outlier or an indication that the relationship between reported gaming time and parental limit does not follow a straight line.

SUMMARY POINTS:

- In a straight-line regression, a gradient of zero corresponds to a correlation of zero.
- If all points lie on a straight line, the Pearson correlation is either equal to 1 (positive slope) or –1 (negative slope).
- Bivariate relationships are between an outcome (y-axis) variable and an explanatory (x-axis) variable.

11.2 Regression Associated with Pearson's Correlation

11.2.1 The Least Squares Method

If the scatter diagram suggests that there might be a linear relationship between two variables, it is useful to fit a straight line to the data for predictive purposes. An appealing initial approach might be to draw a line through the points using a ruler and pencil. Unfortunately, this method is highly subjective and does not lead to a single replicable solution. A conceptually straightforward objective approach is to fit a line using the method of *least squares*.

Unless the association is perfect, any fitted straight line cannot pass through all of the points, and there will be differences between the observed values for the outcome variables and those expected according to the fitted line. These differences are known as *errors*

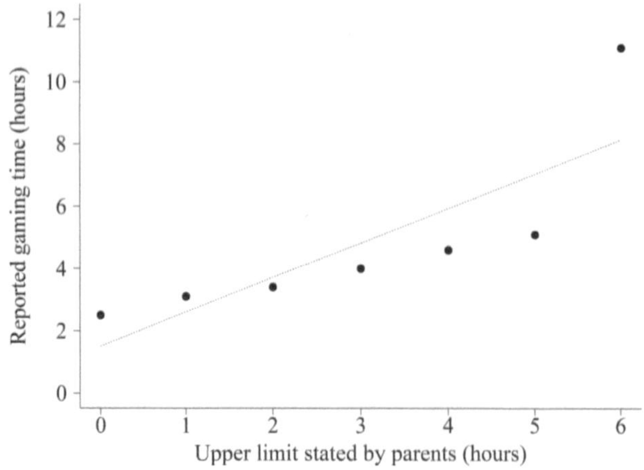

FIGURE 11.2
Scatter diagram and classic least squares line of best fit to data in Example 11.1.

or *residuals*. The method of least squares finds the straight line that minimizes the sum of the squares of these errors. Thus, relatively large errors are penalized heavily. Figure 11.2 shows the video gaming scatter diagram with the least squares line fitted.

The least squares method is the classic method of fitting a straight line to bivariate data. The method has optimal properties subject to well known independence and homogeneity assumptions. Under certain normality assumptions, there are standard procedures for hypothesis testing and estimation. Regression provides useful tools for forecasting and prediction. The description given here closely follows the approach by Maritz (1995, Chapter 5), but much of the theory given there is omitted.

It is important to note that there are many cases where the assumptions needed to validate least squares procedures for bivariate data do not hold. This may result in misleading or invalid inferences. The least squares method is relatively insensitive to some types of departures from basic assumptions, but it is strongly influenced by others. This has resulted in the development of a group of techniques known as regression diagnostics. Although this approach is not pursued here, Atkinson (1985) and Cook and Weisberg (1982) provide comprehensive treatments. McKean, Sheather and Hettmansperger (1990) describe regression diagnostics for rank-based methods. An alternative to classic least squares with regression diagnostics is to use distribution-free methods.

In the classic parametric approach, consider n values denoted by x_i, $i=1, 2, \dots, n$. These may be observed from one or more random X variables or consist of a set of fixed values that may or may not be chosen in advance. It is assumed that each value x_i is associated with a corresponding value y_i of a random variable, Y_i, which has the property that its mean depends upon (i.e., is conditional upon) the value of x_i in such a way that

$$E(Y_i \mid x_i) = \alpha + \beta x_i \tag{11.1}$$

while the variance of Y_i is independent of x and, for all x_i, has the value

$$\text{Var}(Y_i) = \sigma^2,$$

where α, β and σ^2 are unknown. The straight-line relationship between $E(Y|x)$ and x of the form $E(Y|x)=\alpha+\beta x$ between the conditional mean of Y and any x defines the *regression* of Y on x. The line has slope β and intercept α on the y-axis. The notation $Y_i|x_i$ is the conventional notation for an event or variable Y_i having some property conditional upon a specified x_i.

For many inference purposes, it is also assumed that the conditional distribution of $Y_i|x_i$ takes a normal distribution with mean $\alpha+\beta x_i$ and variance σ^2. A classic regression problem is to estimate α, β and sometimes also σ^2, given a set of n paired observations (x_1, y_1), (x_2, y_2), … , (x_n, y_n) where, for a given x_i, the y_i is an observed value of the random variable Y_i featured in (11.1). These conditions hold if each (x_i, y_i) satisfies a relationship

$$y_i = \alpha + \beta x_i + \varepsilon_i,$$

where each ε_i is an unobserved value of a $N(0, \sigma^2)$ random variable and the ε_i are independent of each other and also of x_i. The portion $\alpha+\beta x_i$ is the systematic, or *deterministic*, part of the model for y_i, and the ε_i is the *random* element.

The least squares method obtains as estimates of α, β, the values a, b that minimize

$$S = \sum_{i=1}^{n} (y_i - a - bx_i)^2 \tag{11.2}$$

Denoting these estimates by $\hat{\alpha}$, $\hat{\beta}$, the straight line $y = \hat{\alpha} + \hat{\beta}x$ is called the *least squares regression* of y on x.

If, for any line $y=a+bx$, the y-coordinate corresponding to $x=x_i$ is denoted by $\hat{y}_i$ (i.e., $\hat{y}_i = a + bx_i$), then the differences between the observed and predicted values, i.e., $e_i = y_i - \hat{y}$, $i=1, 2, \ldots, n$, are called the *residuals* with respect to that line. Equation (11.2) may be written $S = \sum_{i=1}^{n} e_i^2$. The least squares estimators of α, β are so called because they minimize the sum of squares of residuals. To obtain $\hat{\alpha}$, $\hat{\beta}$, the expression S is differentiated separately with respect to a and b. These derivatives are set equal to zero to obtain the so-called *normal equations* that are solved for a, b.

The solutions are $\hat{\alpha} = \bar{y} - \hat{\beta}\bar{x}$ where $\bar{y} = \left(\sum_i y_i / n\right)$, $\bar{x} = \left(\sum_i x_i / n\right)$ and

$$\hat{\beta} = \frac{\sum_i (x_i - \bar{x})(y_i - \bar{y})}{\sum_i (x_i - \bar{x})^2} \tag{11.3}$$

With the assumptions above, classic least squares regression is a generalization of classic tests for comparison of treatment means. In particular, a test of H_0: $\beta=0$ is equivalent to a test of equality of a set of sample means. This follows from (11.1) if the values x_i are regarded as indicators or labels attached to samples. For n paired observations (x_i, y_i), each y_i corresponding to a particular x_i may be looked upon as an observed value from the sample labelled by that x_i. The total number of samples, m, may be any number between 2 and n, and the numbers of observations, n_j, in sample j, $j=1, 2, \ldots, m$ are subject to the constraint $n_1+n_2+\ldots+n_m=n$. In particular, if no two x_i are equal, there are n samples, each with one observation, and at the other extreme, if there are only two distinct x_i, there are two samples with n_1 and $n_2=n-n_1$ observations, respectively. In this last case, the first sample may be labelled by $x=0$ and the second sample by $x=1$, and the first sample values are

$$y_{11}, y_{12}, \ldots y_{1n_1}$$

with mean m_0, and the second sample values are

$$y_{21}, y_{22}, \ldots y_{2n_2}$$

with mean m_1. It can then be shown using (11.3) that, in this case $\hat{\beta} = m_1 - m_0$, the sample mean difference used in the familiar *t*-test for equality of two treatment means (Exercise 11.1). Thus, the test of equality of means is in this case identical to testing H_0: $\beta=0$.

In the case of $m>2$ samples, the appropriate regression model is a multivariate regression model with a single intercept equal to the mean for the baseline sample and multiple slopes equal to the regression coefficients with respect to indicators for the remaining samples. The appropriate test would be an *F*-test for the joint null hypothesis asserting that *all* the slopes are zero. In practice, confidence intervals for the multiple slopes, also known as differences between the non-baseline and baseline means, would be of more interest. Either way, there will be *multiple slopes*, not a single slope.

It is well known that for the classic least squares model, the estimator of β does not depend upon α, but that of α depends upon β through $\hat{\beta}$ since $\hat{\alpha} = \bar{y} - \hat{\beta}\bar{x}$. From this expression for $\hat{\alpha}$, it can be seen that the fitted equation may be written, without reference to the intercept, as

$$y = \bar{y} + \hat{\beta}(x - \bar{x}) \tag{11.4}$$

implying that the line passes through the point $(\bar{x}, \bar{y})$.

A graphical interpretation is that the slope of the fitted line is unaltered by a change of origin. In particular, if the origin is shifted to the bivariate mean $(\bar{x}, \bar{y})$ and writing $x' = x - \bar{x}$, $y' = y - \bar{y}$, equation (11.4) becomes

$$y' = \hat{\beta}x'$$

and (11.3) reduces to

$$\hat{\beta} = \frac{\sum_i x_i' y_i'}{\sum_i x_i'^2} \tag{11.5}$$

The form (11.5) still holds if the origin is only moved to the mean of the x_i and the original y_i are used.

From Section 10.1, the Pearson correlation coefficient is unchanged by linear transformations of (x, y) of the form $x'=(x-k)/s$ and $y'=(y-m)/t$, where s and t are both positive. By contrast, in regression, although the estimate of α is affected by a change in origin, the estimate of β is not, i.e., transformations of the form $x'=(x-k)$, $y'=(y-m)$ do not affect the estimate of β. However, it can be established using (11.3) that the transformation $x'=(x-k)/s$ and $y'=(y-m)/t$ alter both the true value of β and its least squares estimate by the same scale factor s/t, changing β to $\beta'=s\beta/t$. However, scale changes are not made in this chapter.

11.2.2 Permutation Tests

This section makes use of the fact that since all other quantities in (10.1) remain constant under permutation, permutation tests may be based on the statistic $T = \sum_i x_i y_i$ in the case of the Pearson coefficient as an alternative to using the sample correlation coefficient r (Section 10.1.1). A corresponding property carries over to the Spearman coefficient, with ranks replacing the x_i, y_i.

Dropping the normality assumption for the Y_i makes joint inferences about α and β difficult, so initially, the estimation of β only is considered. The problem then reduces to making inferences about differences in medians or means of samples labelled by the different x_i. For any independent Y_i, Y_j associated with distinct x_i, x_j, the normality assumptions on the conditional distributions of $Y|x$ are dropped, and it is assumed now only that for any x_i, x_j, they have distributions $F_i(Y_i|x_i)$, $F_j(Y_j\ |x_j)$ that differ only in their location measure. Here, this will be taken to be the *median*, which will coincide with the mean for symmetric conditional distributions for which the mean exists.

For the straight line regression model, the median of $F_i(Y_i|x_i)$ is $\text{Med}(Y_i|x_i)=\alpha+\beta x_i$ and that of $F_j(Y_j\ |x_j)$ is $\text{Med}(Y_j\ |x_j)=\alpha+\beta x_j$. These are the analogues of (11.1), and the difference between the medians is

$$\text{Med}[(Y_j - Y_i)\,|\,x_j, x_i] = (\alpha + \beta x_j) - (\alpha + \beta x_i) = \beta(x_j - x_i).$$

Assuming that differences between distributions are confined to a median difference implies that for all i

$$D_i(\beta) = Y_i - \beta x_i$$

are identically and independently distributed with median α. Since the D_i are therefore independent of the x_i, it follows that they are uncorrelated with the x_i. As pointed out already, the x_i can be altered by the addition or subtraction of a constant without affecting either the slope or its estimate. In particular, it will often simplify the algebra if the x_i are adjusted by adding an appropriate constant to make $\sum_i x_i = 0$. In graphical terms, this implies shifting the origin to a point on the x-axis corresponding to the mean $\bar{x}$.

Writing $d_i=y_i-\beta_0 x_i$, an intuitively reasonable test of the hypothesis H_0: $\beta=\beta_0$ against a one- or two-sided alternative is a test for zero correlation between the x_i and the d_i since it is known that if H_0 holds then $D_i(\beta_0)$ is uncorrelated with the x_i and the d_i are then observed values of the variable $D_i(\beta_0)$. Whether the most appropriate test is based on the Pearson, Spearman, Kendall or some other coefficient will depend upon what assumptions are made about characteristics of the distribution $F(Y|x)$, (e.g., whether long-tailed, symmetric or asymmetric). Here, the d_i are referred to as *residuals* for convenience, but this is an unorthodox use of the term, which more conventionally refers to the $e_i=y_i-\alpha_0-\beta_0 x_i$ where α_0 is some hypothesized value of α. An appropriate statistic based on the Pearson coefficient for the test of zero correlation is

$$T(\beta_0) = \sum_i x_i d_i = \sum_i x_i (y_i - \beta x_i) \tag{11.6}$$

To estimate β, an intuitively reasonable estimation procedure takes the form of solving for b the equation

$$T(b) - \sum_i x_i \sum_i d_i = 0 \tag{11.7}$$

This follows from the form of c_{xy} given below (10.1), since equating c_{xy} to zero ensures a zero value for the sample correlation coefficient because it makes the numerator in (10.1) zero.

Recalling that the x_i may be adjusted to have mean zero, i.e., so that $\sum_i x_i = 0$ without affecting the estimate of β, assume here that this has been done so that (11.7) simplifies to

$$T(b) = \sum_i x_i(y_i - bx_i) \tag{11.8}$$

with solution $\hat{\beta} = \left(\sum_i x_i y_i\right) / \left(\sum_i x_i^2\right)$. This is identical to the solution given by (11.3) with the constraint $\bar{x} = 0$.

No assumptions have been made about the distribution of the D_i except that they are identical for all i, so that for hypothesis testing a permutation test for a zero Pearson correlation coefficient based on the sample values (x_i, d_i) is appropriate. As noted in other contexts, a difficulty with inferences based on raw data using a Pitman permutation test is that the results tend to be similar to those based on equivalent theory assuming normality, even when the normality assumption is clearly violated. In other words, the method lacks robustness.

This is illustrated for a small data set in Example 11.2. There are many situations where data like these arise, although usually in larger data sets where the complexity of computation may mask a clear understanding of what is happening.

Example 11.2

In Example 11.1, the lengths of time spent by adolescents playing video games on weekends and the preferred upper limits stated by their parents are as follows:

Upper Limit Stated by Parents (x):	0	1	2	3	4	5	6
Reported Gaming Time (y):	2.5	3.1	3.4	4.0	4.6	5.1	11.1

Use the method of least squares to fit a straight line to the data and a permutation test based on the Pearson correlation coefficient to test the hypothesis that reported gaming time parallels the parental upper limit.

The estimates of β for classic least squares and permutation-based least squares are identical and may be obtained using (11.3), or from (11.5) after replacing x by $x'=x-3$, or from any least squares regression program. If reported playing time differs from the parental upper limit by a constant amount, then an increase of 1 hour in the time limit stated by parents will be associated with an increase of 1 hour in reported playing time. Hence, the appropriate null hypothesis is that the gradient of the true regression line is 1, i.e. H_0: $\beta=1$. A two-tailed test is appropriate with the alternative H_1: $\beta\neq1$. This hypothesis test is performed by computing the P-value associated with the Pearson coefficient between the x_i and $d_i=y_i-\beta x_i$ after setting $\beta=1$.

For these observations, $n=7$, $\sum_i x_i = 0+1+2+3+4+5+6=21$, $\sum_i y_i = 2.5+3.1+3.4+4.0+4.6+5.1+11.1=33.8$.

Also,

$$\sum_i x_i y_i = (0\times 2.5)+(1\times 3.1)+(2\times 3.4)+(3\times 4.0)+(4\times 4.6)+(5\times 5.1)+(6\times 11.1)$$

$$= 3.1+6.8+12.0+18.4+25.5+66.6=132.4$$

$$\sum_i x_i^2 = 0+1+4+9+16+25+36=91$$

$$\left(\sum_i x_i\right)^2 / n = 21^2/7 = 63.$$

Using (11.3) gives

$$\hat{\beta} = \frac{132.4-(21\times 33.8/7)}{91-63} = \frac{31}{28} = 1.107$$

and

$$\hat{\alpha} = \bar{y} - \hat{\beta} = (33.8/7) - (1.1071\times 21/7) = 1.507.$$

To test the hypothesis H_0: $\beta=1$ against a two-tailed alternative, it is necessary to calculate the differences $d_i=y_i-x_i$ for each data pair and compute the correlation coefficient between the x and d values. Ideally, for such a small sample, an exact permutation test program should be used to compute the P-value appropriate to the hypothesis H_0: $\rho=0$, implying that the x_i and the d_i are uncorrelated. Relevant values of the x_i and d_i are as follows:

x_i	0	1	2	3	4	5	6
$d_i=y_i-x_i$	2.5	2.1	1.4	1.0	0.6	0.1	5.1

Using the ANSM5 package in R and the command `pearson.beta(ch11$reportedtime, ch11$parentlimit, H0 = 1)` gives $r=0.1392$ for the sample Pearson coefficient and the exact two-tail $P=0.7925$. Hence, the least squares line of best fit is $y=1.507+1.107x$. Since H_0: $\beta=1$ has an associated permutation test $P=0.7925$ the data certainly do not provide evidence against H_0. Based on an uncritical analysis of these data, playing time is approximately the length of the parental upper limit plus 1.5 hours showing that time spent in video gaming over a weekend broadly parallels the parental upper limit.

Figure 11.2 (Section 11.2.1) shows the data points and fitted line. The fit is unsatisfactory, and the conclusion above is questionable because the point (6, 11.1) seems to be an outlier relative to the other points. If that point is omitted, the remaining points are well fitted by the least squares regression line $y=2.490+0.517x$. If this is the case, one can conclude that the maximum length of play stated by parents has limited influence on the reported length of playing time.

With the point (6, 11.1) omitted the P-value associated with the test of H_0: $\beta=1$ against H_1: $\beta\neq1$ reduces to $P=0.0028$, providing strong evidence (as one would expect from an inspection of Figure 11.2) against H_0. Given the impact of a single outlier in this example, a much larger study would be required for a serious investigation.

Unfortunately, both the theory and the practice of obtaining exact permutation-based confidence intervals for β present difficulties. Possible procedures are discussed by Maritz (1995, Section 5.2) and Sprent (1998, Section 10.2). Using a Monte Carlo approach with the ANSM5 package in R and the command `pearson.beta(ch11$reportedtime, ch11$parentlimit, H0 = 1, do.CI = TRUE)` gives a 95% confidence interval for β of (0.497, 1.948), which is not very different from the interval (0.231, 1.984) given by classic parametric least squares. Both clearly demonstrate the tendency for the upper limit (as is the point estimate of β itself) to be pulled upward by the final data point.

The extreme playing time associated with a high parental limit of 6 hours requires careful examination. There could have been a genuine recording error on the part of the young person. Alternative possible explanations include exaggeration on the part of the young person and incorrect recording of the parental response. The latter is less likely as 6 hours play would be considered to be more than sufficient by most parents. In-depth interviews might help to clarify these issues.

This type of data can arise from time series measurements, especially following incidents, natural or accidental, that have an environmental impact over an extended period. For example, each y reading could be parts per million of a toxic gas in the atmosphere x hours after an explosion at a chemical plant, or the y could be some measure of radioactivity recorded at daily (or weekly) intervals after an accident at a nuclear power station.

Even simple tests based on the Pearson coefficient require suitable computing facilities. Asymptotic tests may be unreliable if normality assumptions break down seriously. Bootstrapping methods (a topic discussed in Section 14.3) are often used in this context.

Better procedures for dealing with data like that in Example 11.1 are often based on ranks.

SUMMARY POINTS:

- The method of least squares minimizes the sum of the squares of the differences between the observed and expected values (residuals) (Section 11.2.1).
- In linear regression, a change in origin affects the estimate of α but not that of β (Section 11.2.1).
- Although inferences may be based on the relevant permutation distribution, there is a lack of robustness against certain departures from homogeneity of error distributions (Section 11.2.2).

11.3 Regression Associated with Spearman's Rho

In practice, the procedures described in Section 11.4 are easier to apply and more widely used than those in this section. However, the method introduced here illustrates some key features of distribution-free regression and helps to clarify the relationship between procedures.

The obvious analogue to the Wilcoxon rank sum test procedure is to replace the d_i in (11.6) by rank(d_i). The exact test and estimation procedures when the x_i are equally spaced are simplified, for inferences may then be based on the distribution of Spearman's rho, since, if the x_i are equally spaced, they are linear functions of the ranks and hence the statistic

$$T_s(\beta_0) = \sum_i x_i \text{rank}(d_i) = \sum_i x_i \text{rank}(y_i - \beta_0 x_i) \tag{11.9}$$

is appropriate for a test of zero Spearman correlation. It is convenient for estimation problems, though not essential, to assume also that any origin change has been made to make the mean of the x_i zero. The test is not conditional on the observed x or y because for any hypothesized β_0 and fixed n, only the paired set of ranks (or linear transformations of these ranks) from 1 to n is needed for the permutation reference set.

Example 11.3 shows that a program providing an exact test for Spearman's $\rho_s=0$ may be used to test H_0: $\beta=\beta_0$ against a one- or two-sided alternative.

Point estimation of β is now less straightforward than it is for least squares; however, it is still possible, and confidence intervals based on the permutation test are relatively easy to obtain if appropriate software is available.

Estimation is based on an analogue of (11.8) derived from (11.9), and the solution b is sought to the equation

$$T_s(b) = \sum_i x_i \text{ rank}(y_i - bx_i) = 0 \tag{11.10}$$

A difficulty is that there is, in general, no value of b for which (11.10) holds exactly because $T_s(b)$ is a step function in b. This is because the ranking of the d_i only changes as b passes through a value where two or more d_i, d_j are equal, i.e., for b such that for some i, j, it is the case that $y_i-bx_i=y_j-bx_j$. Denoting this value of b by b_{ij}, it follows that $b_{ij}=(y_j-y_i)/(x_j-x_i)$. Thus, $T_s(b)$ does not change in value for any b lying between successive values of b_{ij} for $i=1, 2, \ldots, n-1, j>i$. It can be shown [see, e.g., Sprent (1998, Section 10.3)] that $T_s(b)$ is a nonincreasing step function in b and that if for different (i, j) the corresponding b_{ij} are equal, this implies either three or more observations are collinear, or the lines joining the relevant pairs of points have identical slopes. The steps will then not all be of equal height.

Some implications of the properties of $T_s(b)$ for hypothesis testing and estimation, including confidence intervals, are best demonstrated by an example.

Example 11.3

The data in Example 11.1 with the x_i adjusted to have mean zero become

x	−3	−2	−1	0	1	2	3
y	2.5	3.1	3.4	4.0	4.6	5.1	11.1

It is convenient to record the pairwise slopes b_{ij} as a triangular array or upper triangular matrix similar to that used for the s_{ij} in Section 10.1.7. A typical entry is

$$b_{26} = \frac{5.1-3.1}{2-(-2)} = 0.500$$

shown in bold in the array below which are pairwise slopes in the order of the columns in an upper triangular matrix.

0.600	0.450	0.500	0.525	0.520	1.433
	0.300	0.450	0.500	**0.500**	1.600
		0.600	0.600	0.567	1.925
			0.600	0.550	2.367
				0.500	3.250
					6.000

Repeated values among the b_{ij} imply collinearity or parallelism of some pairwise joins. Table 11.1 is easily formed using (11.9), although the complete table is not needed for most inferences. Since $T_s(b)$ remains constant between successive values of b_{ij}, it is only necessary to evaluate (11.9) for one value β_0 of b in each interval between successive values. When b coincides with some b_{ij}, there are ties in the ranks of the d_i, and using mid-ranks for these gives a value of the corresponding $T_s(b)$ that is the mean of its values for b immediately

TABLE 11.1

Intervals for b-Values Corresponding to Each Possible $T_s(b)$ Arising from the Data in Example 11.3

b-Interval	Mid-b Value	$T_s(b)$	r_s
−∞ to 0.300		28	1.000
0.300+ to 0.450−	0.375	27	0.964
0.450+ to 0.500−	0.475	23	0.821
0.500+ to 0.520−	0.510	12	0.429
0.520+ to 0.525−	0.5225	7	0.250
0.525+ to 0.550−	0.5375	3	0.107
0.550+ to 0.567−	0.5583	1	0.036
0.567+ to 0.600−	0.5835	−2	−0.071
0.600+ to 1.433−	1.0165	−7	−0.250
1.433+ to 1.600−	1.5165	−13	−0.464
1.600+ to 1.925−	1.7625	−18	−0.643
1.925+ to 2.367−	2.146	−22	−0.786
2.367+ to 3.250−	2.8085	−25	−0.893
3.250+ to 6.000−	4.625	−27	−0.964
6.000+ to ∞		−28	−1.000

The values of $T_s(b)$ when $b=b_{ij}$ are the means of the values in adjacent b intervals. Values of r_s corresponding to each $T_s(b)$ are also given.

greater than and immediately less than the b_{ij} concerned. As an illustration of the computation of $T_s(b)$ for a given b consider that for the interval between $b_{16}=0.520$ and $b_{15}=0.525$. It suffices to set $b=0.523$, say, and to evaluate

$$T_s(0.523) = \sum_i x_i \text{ rank}(y - 0.523x_i)$$

The relevant values from the given data are

x_i	−3	−2	−1	0	1	2	3
y_i	2.5	3.1	3.4	4.0	4.6	5.1	11.1
$d_i=y_i-0.523x_i$	4.07	4.15	3.92	4.0	4.08	4.05	9.53
rank(d_i)	4	6	1	2	5	3	7

whence

$$T_s(0.523) = (-3)\times 4 + (-2)\times 6 + (-1)\times 1 + 0\times 2 + 1\times 5 + 2\times 3 + 3\times 7 = 7.$$

Now consider the slightly higher b value of 0.533. This is contained in the interval from $b_{15}=0.525$ to $b_{46}=0.550$.

The relevant values needed for calculating T_s are now

x_i	−3	−2	−1	0	1	2	3
y_i	2.5	3.1	3.4	4.0	4.6	5.1	11.1
$d_i=y_i-0.533x_i$	4.10	4.17	3.93	4.0	4.07	4.03	9.50
rank(d_i)	5	6	1	2	4	3	7

Note that the ranks associated with $x_i=-3$ and $x_i=1$ have been interchanged giving

$$T_s(0.533) = (-3)\times 5 + (-2)\times 6 + (-1)\times 1 + 0\times 2 + 1\times 5 + 2\times 3 + 3\times 7 = 3.$$

The remaining values of $T_s(b)$ in column 3 of Table 11.1 are obtained in a similar manner. Since there is no unique b corresponding to each possible value of $T_s(b)$, it is sometimes convenient to associate with each such value, the mid-b value. This is the midpoint of the interval of b-values associated with that $T_s(b)$. For example, as shown above, $T_s(b)=7$ is associated with b in the interval (0.520, 0.525), so the associated mid-b value is 0.5225. These mid-b-values are given in column 2 of Table 11.1.

The maximum value, 28, of $T_s(b)$ corresponds to a Spearman sample coefficient $r_s=1$. Since T_s is a linear function of r_s dividing any $T_s(b)$ by 28 gives a corresponding value of the equivalent r_s. The relevant values are shown in column 4 of Table 11.1. This is useful for determining confidence intervals for β, but first consider the point estimation of β. Maritz (1995) proposed an estimate obtained by treating the step function $T_s(b)$ as an approximation to a continuous function. He then used linear interpolation between the mid-b-values for the closest $T_s(b)$ on either side of zero. In this example, interpolation is performed between $T_s=1$, $b=0.5583$ and $T_s=-2$, $b=0.5835$ to obtain a point estimate corresponding to $T_s=0$. This is

$$b = 0.5583 + (0.5835 - 0.5583)/3 = 0.567.$$

It is fortuitous that this is identical with the median of the b_{ij}, although an estimate obtained in this way will usually be close to that median. In practice, if only a point estimate of β is required, one need calculate only a few entries in Table 11.1 for values of b in intervals close to med(b_{ij}), then use linear interpolation.

If an exact test program for the Spearman coefficient is available, hypothesis tests about β are straightforward. For example, to test H_0: $\beta=1$ against H_1: $\beta\neq1$, the values of x_i and d_i computed in Example 11.2 may be inserted in a Spearman test program. Not surprisingly (as can be verified from Table 11.1), this gives $r_s=-0.250$ and the corresponding two-tail $P=0.5948$, so there is no evidence against H_0. If no program is available for an exact test, statistical tables indicate that only values of r_s greater than about 0.7 in magnitude provide any acceptable evidence against H_0.

Output from the ANSM5 package in R can be obtained using `spearman.beta (ch11$reportedtime, ch11$parentlimit, H0 = 1)`.

Given a facility for generating the complete distribution of r_s under the null hypothesis H_0: $\rho_s=0$, it is relatively easy to obtain confidence intervals for β. Using this distribution, one finds that for $n=7$, the one-tail P associated with $r_s\geq0.786$ is $P=0.024$.

From Table 11.1, note that because tied ranks occur for certain values of b, there is no upper tail b exactly corresponding to $r_s=0.786$, but in the lower tail when $r_s=-0.786$, the corresponding mid-b value is 2.146. The smallest value of b that will produce an $r_s=-0.786$ is a value slightly above 1.925, say 1.926. In the upper tail of r_s values, the largest value of b that will produce a value of r_s of at least 0.786 is b slightly less than 0.500, say 0.499. Thus, the b-interval (0.500, 1.925) is an appropriate $(1-2\times0.024)100=95.2\%$ confidence interval.

Output from the ANSM5 package in R can be obtained using `spearman.beta (ch11$reportedtime, ch11$parentlimit, H0 = 1, do.CI = TRUE)` which, in order to calculate the confidence interval, uses the exact distribution for small sample sizes, such as in this example, and an asymptotic distribution for larger samples.

The interval obtained here is similar to those given in Example 11.2. More interestingly, the point estimate $\hat{\beta}=0.567$ is much smaller than the full-data least squares estimate and reasonably close to the least squares estimate 0.517 obtained if the rogue point (6, 11.1) is omitted. This indicates that the suspect point is less influential on the point estimate of β, although it still exerts upward pressure on the confidence limit. Here, the confidence limits, unlike those in classic parametric least squares estimation, are no longer symmetric about the point estimator, but they reflect more realistically the upward pressure of the point (6, 11.1) on estimates of β.

If confidence intervals are required at or near conventional levels, only a few values of $T_s(b)$ corresponding to large values of $|T_s(b)|$ need be calculated.

An asymptotic formula leading to approximate confidence intervals is discussed briefly in Sprent (1998, Section 10.3). This usually works well for larger samples. Since inference procedures discussed in this section are generalizations of the Wilcoxon–Mann–Whitney test, it is not surprising that they have the same Pitman efficiencies.

If the x_i are not equally spaced, $T_s(b)$ given by (11.9) is no longer a linear function of Spearman's rho. Then, the method given in the next section is generally preferable. Some writers have suggested that inferences about β might be based on a statistic that replaces x_i by rank(x_i) in (11.9), i.e.,

$$S(b)=\sum_i \text{rank}(x_i)\,\text{rank}(y_i-bx_i). \tag{11.11}$$

This is equivalent to using $T_s(b)$ for equally spaced x for hypothesis testing, but care is needed for estimation since the mean of the ranks of the x_i is no longer zero, but adjustments can be made for this. However, for unequally spaced points, transformation to ranks is not a linear transformation of the x_i. This precludes the use of $S(b)$ for making reliable inferences about the population value β for the original data.

SUMMARY POINTS:

- Regression based on the Spearman coefficient is generally performed if the points along the explanatory variable x-axis are equally spaced.
- This type of regression uses the slope values for all possible combinations of pairs of points.
- If the x-values are equally spaced, this method may be more robust to outliers than classic least squares.

11.4 Regression Associated with Kendall's Tau

11.4.1 Linkage of Points and Regression

Least squares regression focuses on the differences between the observed y-values and the y-values that would be expected according to the fitted regression line. Although it is good practice to plot the calculated regression line on a scatter diagram, connecting individual points by straight lines is discouraged.

An alternative approach to the least squares method is based on connecting pairs of points on the scatter diagram by straight lines and examining their gradients (see Example 11.3). If most of the gradients are positive and few are negative, this suggests a positive association between the two variables. A negative association is indicated if most of the gradients of the lines connecting points are negative. In estimating a regression line using this approach, the median of the slope values for the individual lines is an intuitively reasonable estimate for the gradient of the regression line.

11.4.2 The Theil–Kendall Method

Section 11.3 demostrated that regression inference procedures based on Spearman's rho are complicated if the x_i are not equally spaced. Equal spacing is not needed for inferences based on Kendall's tau, since τ_k depends only upon the order of the observations, not on magnitudes of differences between data values. To simplify presentation when the x_i are all different, it is assumed that $x_1 < x_2 < \ldots < x_n$. The case of some x_i being equal is considered briefly in Section 11.5.

The statistic used for inference about β is

$$T_t(b) = \sum_i \text{sgn}[d_{ij}(b)] \qquad (11.12)$$

where

$$d_{ij}(b) = (y_j - a - bx_j) - (y_i - a - bx_i) = y_j - y_i - b(x_j - x_i),$$

and summation of the signs of the d_{ij} is over all $i=1, 2, \ldots, n-1$ and $j>i$. Since the x_i are all different and in ascending order, it follows that $T_t(b)$ is the numerator in the expression (10.3) for Kendall's tau composed of the numbers of concordances minus the number of discordances in the data pairs (x_i, d_i) where $d_i=y_i-bx_i$.

The use of a statistic equivalent to $T_t(b)$ was first proposed by Theil (1950). The procedure is widely known as Theil's method. Sen (1968) highlighted the relationship to Kendall's tau, so it is referred to here as the *Theil–Kendall method.*

$T_t(b)$ is a linear function of the sample estimator of Kendall's tau, and using arguments similar to those in Sections 11.2 and 11.3, an appropriate point estimator of β is obtained by setting $T_t(b)=0$. Since the x_i are in ascending order, it is apparent that $T_t(b)$ is unchanged if $d_{ij}(b)$ is replaced by $b_{ij}-b$ where, as in Section 11.3, $b_{ij}=(y_j-y_i)/(x_j-x_i)$. Then, $T_t(b)=0$ with $b=\text{med}(b_{ij})$, for in that case the number of positive and the number of negative d_{ij} will be equal. As was the case for $T_s(b)$, it can be seen that $T_t(b)$ only changes in value when b passes through a value of b_{ij}, and since the statistic is the numerator term in Kendall's tau, it follows that if all b_{ij} are distinct as b increases from $-\infty$ to ∞, $T_t(b)$ is a step function decreasing by steps of 2 from $n(n-1)/2$ to $-n(n-1)/2$. Division of $T_t(b)$ by $n(n-1)/2$ yields Kendall's t_k which may be used as an equivalent statistic. If the b_{ij} are not all distinct, some of the steps will be multiples of 2. In particular, if a b_{ij} occurs r times, it induces a step of height $2r$. The value of $T_t(b)$ at any b_{ij} may be regarded as the mean of the values immediately above and below that b_{ij}.

Tests of hypotheses about β are straightforward, using either Kendall's t_k or the equivalent $T_t(b)$ as the test statistic. If the exact distribution of the sample Kendall coefficient when $\tau_k=0$ is known, a confidence interval for β may be obtained. The procedures are illustrated in Example 11.4.

Example 11.4

Given the data in Example 11.1, i.e.,

x	0	1	2	3	4	5	6
y	2.5	3.1	3.4	4.0	4.6	5.1	11.1

use the Theil–Kendall method (i) to test the null hypothesis H_0: $\beta=1$ against the alternative H_1: $\beta\neq 1$ and (ii) to obtain a point estimate of β and a confidence interval giving at least 95% coverage.

Testing the hypothesis in (i) requires the computation of $T_t(1)$ and the use of an exact test program to assess evidence for or against H_0. To obtain a point estimate of β, the median of the b_{ij} (pairwise slopes) defined above is determined. For a confidence interval, one then calculates $T_t(b)$ and the corresponding t_k for all b. Using the exact distribution of t_k when $\tau_k=0$, or appropriate tables if the exact distribution when $n=7$ is not available, the relevant confidence interval for β is computed.

The b_{ij} may conveniently be arranged in a triangular matrix with the first row elements $b_{12}\, b_{13} \ldots b_{1n}$, second row elements $b_{22}\, b_{23} \ldots b_{2n}$ and so on, to give

0.600	0.450	0.500	0.525	0.520	1.433
	0.300	0.450	0.500	0.500	1.600
		0.600	0.600	0.567	1.925
			0.600	0.550	2.367
				0.500	3.250
					6.000

which is identical to the array introduced in Example 11.3. Inspection shows that the point estimate of β, i.e., med(b_{ij}), is 0.567 because there are 10 greater and 10 lesser b_{ij}. To test the hypothesis $\beta=1$, subtract $b=1$ from each b_{ij}. This implies 15 discordances (negative values) and 6 concordances when $T_t(1)=6-15=-9$, and when $n=7$, this implies $t_k=-9/21=-0.4286$. Corresponding to this value of t_k, using the ANSM5 package in R and the command `theil.kendall(ch11$reportedtime, ch11$parentlimit, H0 = 1)` gives an exact two-sided $P=0.0817$, so there is no convincing evidence against H_0.

The values of T_t can be established for all values of b using the method described above, noting that there are, respectively, 2, 4 and 4 tied observations of b_{ij} at 0.450, 0.500 and 0.600. Table 11.2 shows the T_t values together with corresponding values of t_k obtained by dividing each $T_t(b)$ by 21.

To obtain an approximate 95% confidence interval for β, it is necessary first to determine a value of $|T_t|$ with a one-tail P-value as close as possible to, but not exceeding, $P=0.025$. The exact distribution when $n=7$ indicates that $P=0.015$ when $T_t=15$, while $P=0.035$ when $T_t=13$. From Table 11.2, it can be deduced that the shortest $(1-2\times0.015)100=97\%$ interval

TABLE 11.2

Intervals for b-Values Corresponding to Each Possible $T_t(b)$ or the Corresponding t_k for the Data in Example 11.4

b-Interval	$T_s(b)$	t_k
$-\infty$ to 0.300	21	1.000
0.300+ to 0.450–	19	0.905
0.450+ to 0.500–	15	0.714
0.500+ to 0.520–	7	0.333
0.520+ to 0.525–	5	0.238
0.525+ to 0.550–	3	0.143
0.550+ to 0.567–	1	0.048
0.567+ to 0.600–	–1	–0.048
0.600+ to 1.433–	–9	–0.428
1.433+ to 1.600–	–11	–0.524
1.600+ to 1.925–	–13	–0.619
1.925+ to 2.367–	–15	–0.714
2.367+ to 3.250–	–17	–0.810
3.250+ to 6.000–	–19	–0.905
6.000+ to ∞	–21	–1.000

The values of $T_t(b)$ when $b=b_{ij}$ are the means of the values in adjacent b intervals.

based on $|T_t| = 15$ will be (0.500–, 1.925+) or the open interval (0.500, 1.925). Hence, the point estimate of β is 0.567, and a nominal 95 (actual 97)% confidence interval for β is (0.500, 1.925).

Output from the ANSM5 package in R can be obtained using `theil.kendall (ch11$reportedtime, ch11$parentlimit, do.CI = TRUE)`, which gives a Monte Carlo 95% confidence interval of (0.500, 2.367). Appropriately, this is a slightly wider interval than the sample-based 97% confidence interval, above.

Note that finding a confidence interval does not require the complete Table 11.2, since the observation $T_t=15$ implies $n_c-n_d=15$, and since $n_c+n_d=21$, this implies $n_c=18$ and $n_d=3$. Thus, one should reject values of β greater than or equal to the three largest b_{ij} or less than or equal to the three smallest. Inspection of the matrix of the b_{ij} values shows that the third largest and third smallest b_{ij} are, respectively, 2.367 and 0.450. Rejecting these values along with those more extreme implies the limits 0.500 and 1.925 given above.

The point estimate of β is identical to that obtained in Example 11.3 using Spearman's r_s, and the nominal 95% confidence interval is virtually the same despite minor differences in the exact coverage.

The Theil–Kendall procedure is easier to compute than that based on Spearman's correlation. More importantly, in practice (though not relevant to this specific example), the restriction to equally spaced x_i is no longer needed.

The close relationship between Kendall's t_k and T_t makes it possible to carry over asymptotic approximations between them. Using results for τ_k, it can be shown (see, e.g., Maritz, 1995, Section 5.2.5) that $E(T_t)=0$ and that $\text{Var}(T_t)=n(n-1)(2n+5)/18$. For large n, the distribution of $Z = T_t / \sqrt{\text{Var}(T_t)}$ is approximately standard normal. Asymptotic inferences can then be made in the usual way. Even for $n=7$, the approximation is sometimes not seriously misleading (Exercise 11.4), but caution is advisable with so small a sample.

Section 10.1.4 showed that the Kendall procedure is a special case of the Jonckheere–Terpstra method. This, in turn, is a generalization of the Mann–Whitney formulation of the WMW procedure, whereas the method using Spearman's rho is a generalization of the Wilcoxon formulation of the WMW procedure. Although the generalizations are not exactly equivalent, it is not unreasonable when applying them to regression problems to expect them to lead to broadly similar conclusions, as seen in Examples 11.3 and 11.4.

11.4.3 The Abbreviated Theil Procedure

Having noted that the Theil–Kendall procedure is a generalization of the WMW method, it is interesting to explore the possibility of procedures that generalize those of a sign test. Theil proposed one that is usually called the *abbreviated Theil method.* It uses a small independent subset of the b_{ij} and is not recommended for sample sizes smaller than about 14. Even for larger sample sizes, the full Theil–Kendall method is preferable if adequate computing facilities or tables are available.

For an even sample size, the only b_{ij} used are those of the form $b_{i,i+n/2}$, where $i=1, 2, \ldots, n/2$. If n is odd, only the b_{ij} of the form $b_{i,i+(n+1)/2}$ are needed, where $i=1, 2, \ldots, (n-1)/2$. These estimators all involve different data pairs and hence are independent, so test and estimation procedures based on the sign test may be used. The point estimator of β is the median of the reduced set of b_{ij} indicated above. Tests and confidence intervals are easily derived from the corresponding sign-test procedures.

Example 11.5

Given below is the modal length (y cm) of samples of Greenland turbot of various ages (x years) based on data obtained from Kimura and Chikuni (1987). Fit a straight line using the abbreviated Theil method and obtain a 95% confidence interval for β.

Age (x)	4	5	6	7	8	9	10	11	12
Length (y)	40	45	51	55	60	67	68	65	71
Age (x)	13	14	15	16	17	18	19	20	
Length (y)	74	76	76	78	83	82	85	89	

For the 17 points, the subset of pairwise slopes $b_{10,1}, b_{11,2}, \ldots, b_{17,8}$ is calculated and their median obtained. The 10th observation is matched with the 1st, the 11th with the 2nd and so on. The differences in modal length are as follows: $y_{10}-y_1=74-40=34$, $y_{11}-y_2=76-45=31$, … , $y_{17}-y_{10}=89-65=24$. The required b_{ij} are obtained by dividing each y_j-y_i by its $x_j - x_i$: 34/9, 31/9, … , 24/9 respectively. The median b_{ij} gives an estimate for β of $\hat{\beta} = 23.5/9 = 2.61$.

Note that for the abbreviated Theil method, the x_j-x_i do not need to be equal (see Exercise 11.6), but in this example, all differences take the same value (9).

A confidence interval for β is obtained in a manner analogous to that for a median based on the sign test. Following the argument in Section 3.2.2 for a nominal 95% confidence interval for the above set of 8 differences between the y-values, appropriate B(8, 0.5) distribution probabilities can be used to find that an exact 93% confidence interval for the median of these differences is the interval (17, 31). Dividing the limits by 9 gives the corresponding interval for β, i.e., (1.89, 3.44). Similarly, the interval (1.67, 3.78) is a 99.2% interval. With so small a sample, the discontinuities in coverage of possible intervals are quite marked. In summary, an appropriate estimate for β is $\hat{\beta} = 2.61$ and a 93% confidence interval for β is (1.89, 3.44).

Output from the ANSM5 package in R, including a 95% Monte Carlo confidence interval of (1.88, 3.77), can be obtained using `theil.kendall(ch11$length, ch11$age, do.abbreviated = TRUE, do.CI = TRUE)`.

Generating all pairwise slopes for these data (using the default `do.abbreviated = FALSE` in the above command), one can show that the full Theil–Kendall procedure gives an estimator $\hat{\beta}=\text{med}(b_{ij})=2.65$ and a Monte Carlo 95% confidence interval of (2.00, 3.27). It may also be established that a 95.8% confidence interval is (2.2, 3.1) following the procedure outlined in Example 11.4. This follows because the one-tail $P=0.021$ when $n=17$ corresponds to $T_t=50$, implying $n_c-n_d=50$. Hence, to establish the relevant confidence interval, the 43 smallest and 43 largest b_{ij} are rejected. Inspection of the paired values establishes the above interval. If one has a facility to generate all b_{ij}, the full rather than the abbreviated Theil–Kendall method is recommended for samples of this size. A comparison of the lengths of the confidence intervals given by the two methods indicates an appreciable loss of efficiency when using the abbreviated method. This is not surprising because the abbreviated method only uses 8 of the 136 items of information (the b_{ij}) involved in the full method. It does, however, use an independent set, whereas there are correlations in the complete set.

A problem arises with both the Theil–Kendall and abbreviated Theil methods when the values of x_i and x_j within a pair are tied. The slope value corresponding to such a pair becomes infinite. For only a small proportion of ties, there will be little loss of efficiency if such points are replaced by one data entry with the y-value set equal to the mean of the y-values for all the points with that tied x-value. This is usually preferable to artificial tie-breaking devices such as splitting the ties by making arbitrary small changes to tied x-values to separate them. That process may lead to large or even bizarre b_{ij} associated with such splits because of the small denominators. An alternative with greater appeal is to exclude all comparisons between points with a common x_i value but to consider all joins of each of these points to points with other values x_j, i.e., where $x_j \neq x_i$. An extreme example is when there are only two distinct x-values, x_1 and x_2, and repeated y at each. The problem is then equivalent to the two-sample problem for means or medians of the y. By considering the slopes of all pairwise joins between y-values associated with the two x, it becomes apparent that these are similar to the differences computed for the Hodges–Lehmann estimator in Section 6.1.4. Indeed, they are simply these differences divided by the constant x_2-x_1, so the procedure is exactly equivalent to the Wilcoxon–Mann–Whitney test.

SUMMARY POINTS:

- For methods based on Kendall's tau, the gradient estimate for the regression line is the median of the pairwise slope values (Section 11.4.1).
- The Theil–Kendall method is straightforward to apply even if the x-values are not equally spaced (Section 11.4.2).
- Confidence intervals based on the Theil–Kendall method depend on the distribution of Kendall's tau (Section 11.4.2).
- An abbreviated Theil procedure uses only a limited subset of slopes that join independent pairs of points and is not recommended for small samples (Section 11.4.3).
- Confidence intervals for the abbreviated procedure depend upon the sign test (Section 11.4.3).
- Ties in x-values within pairs are problematic as the corresponding pairwise slope values will be infinite (Section 11.4.3).

11.5 Some Alternative Approaches

Many variants of procedures considered in this chapter for slope estimation appear in the literature. Another sign-test analogue related to the Spearman coefficient rather than the Kendall coefficient is briefly discussed by Sprent (1998, Section 10.5). As alternatives to procedures involving correlation between x_i and rank(d_i) such as that discussed in Section 11.3, Adichie (1967) considered possible transformations of the rank(d_i) such as that to van der Waerden or to normal scores. These have not been widely used in practice, and one might expect their performance to show little improvement on the direct use of ranks.

The examples so far show that rank-based techniques depend heavily on the b_{ij}. This is particularly true for those such as the Theil–Kendall methods. The use of median estimators based on these pairwise slope estimators introduces robustness to point estimators in the presence of rogue observations, or observations that might indicate inadequacy of the model. In Example 11.3, for instance, the observation (6, 11.1), which is so obviously out of line with the other observations, influences only the b_{ij} that involve that point, i.e., those in the last column of the triangular matrix of b_{ij}, and these have little influence on med(b_{ij}) as a point estimator of β. They do, however, exert an influence on the confidence limits and raise the upper limit above what it would be with no such out-of-line observation.

If there are no suspect observations, one might feel that more weight should be given to those b_{ij} associated with larger values of $x_j - x_i$. This is what the classic least squares method does, for it can be shown that that estimator is a weighted mean of the b_{ij} with weights proportional to $(x_j - x_i)^2$. Jaeckel (1972) recommended taking as a point estimator of β the median of the weighted b_{ij} with weights $w_{ij} = (x_j - x_i)/\left[\sum_{i<j}(x_j - x_i)\right]$. The procedure has some optimum properties when there are no outliers, but these and other weighting schemes that have been proposed may be less satisfactory than, for instance, the Theil–Kendall method if there are rogue observations. In simulation studies, Hussain and Sprent (1983) found that the Theil–Kendall method performed almost as well as least squares when relevant assumptions held for the latter and that it showed a marked improvement in efficiency for long-tail error distributions, whereas in the latter situation weighted medians performed no better than, or sometimes less well than, Theil–Kendall.

SUMMARY POINTS:

- Outliers have little impact on the median estimators used in the Theil–Kendall and abbreviated Theil methods.
- Outliers have a greater influence on the confidence limits, e.g., extremely large y-values raise the upper limits.
- The classic least squares estimator is a weighted mean of the pairwise slope values.

11.6 Joint Estimation of Slope and Intercept

Estimating slope is often the main aim of a regression analysis, but if the fitted line is to be used for forecasting or prediction, the intercept α must also be estimated. It is then possible to predict using the estimate of $E(Y|x)=\alpha+\beta x$ for some specified x. The intercept corresponds to the choice $x=0$. Whereas the estimation of β is unchanged if x is replaced by $x'=x-h$, where h is any constant, this is not the case for α.

Even in classic parametric least squares, estimates of α and β are in general correlated. Only in the special case when the mean of the x_i is zero and the estimator of α reduces to $\hat{\alpha} = \bar{y}$ is this estimate uncorrelated with that for β and then the customary normality assumption implies independence, so in that case inferences can be made about α without having to worry about what is now a *nuisance* parameter β. Unfortunately, when moving to distribution-free methods, this simplification does not hold because even when the mean of the x_i is zero, the statistics used to estimate α and β are correlated.

Many of the implications are discussed in detail by Maritz (1995, Section 5.3) and more informally by Sprent (1998, Section 10.6). A broad class of statistics relevant to the joint estimation of α and β are discussed by Adichie (1967). These lead to inferences that in practice call for iterative methods of estimation starting with an estimate of β that is used to obtain an estimate of α. This estimate of α is used to get a revised estimate of β and so on. The process usually converges after a few cycles. Some of the statistics used have links with, and similarities to, those used for the estimation of slopes in Sections 11.3 and 11.4.

Methods used in practice often give satisfactory estimates of α, providing a reasonably good estimate of β is available. For example, for the data in Example 11.1, the estimate $\beta = 0.567$ obtained by the methods used in Sections 11.3 or 11.4 is a reasonable slope estimate for a line through all points other than (6, 11.1) if that is what is of interest.

The assumptions in Section 11.3 imply that all $D_i = y_i - \beta x_i$ (as defined in Section 11.2) have identical distributions with median α. The value of β is unknown, but if $\hat{\beta}$ can be accepted as a good estimate of β, it is not unreasonable to assume that $d_i = y_i - \hat{\beta} x_i$ will have a distribution with a median close to α. An obvious procedure is to compute all n values of d_i and to take the median of these d_i as an estimate of α. This is equivalent to the optimum estimator based on the sign test. If one makes the further assumption that the D_i are symmetrically distributed about α, the Hodges–Lehmann estimator based on the Walsh averages of the d_i is arguably more appropriate. It is important to note, however, that these results are only approximate and will obviously be influenced by the choice of $\hat{\beta}$.

Example 11.6

Given the data in Example 11.1, i.e.,

Upper Limit Stated by Parents (x):	0	1	2	3	4	5	6
Reported Gaming Time (y):	2.5	3.1	3.4	4.0	4.6	5.1	11.1

and assuming a reasonable estimate of β is $\hat{\beta} = 0.567$, obtain an estimate of the intercept α.

All the values of $d_i = y_i - 0.567x_i$ are computed. The estimate is then either (i) the median of the d_i or (ii) the Hodges–Lehmann estimator, which is the median of the Walsh averages.

The d_i are 2.500, 2.533, 2.266, 2.299, 2.332, 2.265 and 7.698. The median of these seven values is 2.332, there being three smaller and three larger values. The Hodges–Lehmann estimator can be obtained from any program that computes the Walsh averages, including the ANSM5 package in R. For these data, the estimate (the median of the Walsh averages) is 2.399. Hence, two possible estimates of α are 2.332 or 2.399.

Output from the ANSM5 package in R can be obtained using `theil.kendall (ch11$reportedtime, ch11$parentlimit, do.alpha = TRUE)`.

One may argue that the median of the d_i should be the preferred estimator because the extreme value $d_7 = 7.698$ suggests that the symmetry assumption needed to justify Hodges–Lehmann estimation may not hold. The line of best fit using the Theil–Kendall estimator of slope and the median of the d_i as intercept estimator is shown in Figure 11.3.

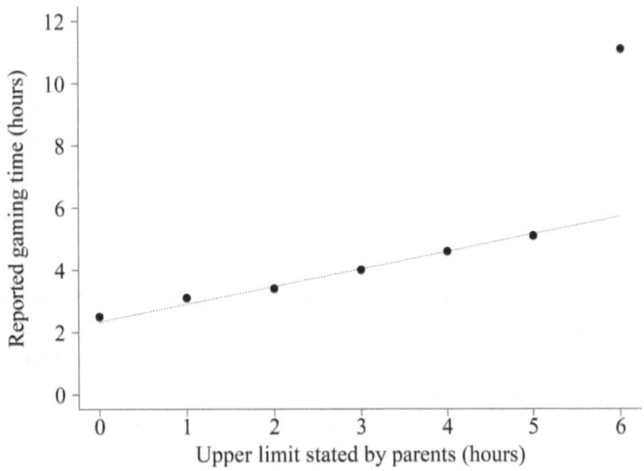

FIGURE 11.3
The Theil–Kendall regression line fitted to the data in Example 11.6.

It is tempting to apply either the sign-test method or the Hodges–Lehmann procedure to obtain a confidence interval for α. However, this may not be appropriate because only an estimate of β is used, as its true value is not known. An alternative choice of the estimate of β may profoundly influence both the pattern of the d_i and the estimate based upon them. For instance, in Example 11.2, if the last point is ignored, the least squares estimate of β becomes $\hat{\beta} = 0.517$. For this, the values of the d_i are 2.500, 2.583, 2.366, 2.449, 2.532, 2.515 and 7.998 with a median of 2.515 reflecting a general upward shift in the d_i.

The pattern changes even more markedly if taking $\hat{\beta} = 0.5$ when the d_i, apart from the last, are even more concentrated, being 2.5, 2.6, 2.4, 2.5, 2.6, 2.6 and 8.1. Indeed, to obtain reasonable approximate confidence intervals for α, a more sophisticated approach must be used. As indicated above, some that are available require an iterative approach for the joint estimation of α and β. One possible approach is given by Sprent (1998, Examples 10.7 and 10.8) where it is shown that in this example an approximate interval based on the estimate $\beta = 0.567$ may not be unreasonable, as iteration does not improve the joint estimates of slope and intercept.

Some alternative proposals for the estimation of α and the more general problem of estimating med($Y|x$) are discussed by Maritz (1995, Section 5.3) and by Sprent (1998, Section 10.6).

SUMMARY POINTS:

- If the fitted regression line is to be used for forecasting or prediction, the intercept must be estimated.
- Methods for estimating the intercept generally involve iteration.
- Obtaining confidence intervals for the estimate of the intercept requires sophisticated methods.

11.7 Comparison of Slopes for Several Lines

Suppose there are k mutually independent sets of paired observations of variables (x, y) where the ith set contains n_i observations $(i=1, 2, \ldots, k)$ and $k \geq 2$. Also assume that the jth observation $(j=1, 2, \ldots, n_i)$ in the ith set [i.e., (x_{ij}, y_{ij})] satisfies the relationship

$$Y_{ij} = \alpha_i + \beta_i x_{ij} + \varepsilon_{ij}$$

where the ε_{ij} are realized values of a N(0, σ^2) random variable, and the α_i, β_i and σ^2 are all unknown. This model specifies k regression lines all with the same error distribution. The model has been widely studied in parametric analysis, particularly in the context of testing the hypothesis that all the β_i are equal against an alternative of some inequality. Equality of the β_i implies all k lines are parallel. An extended hypothesis of equality of all α_i in addition to equality of all β_i implies the k regression lines are identical.

If the normality assumptions on the distribution of the ε_{ij} are dropped, assuming only that they all have median zero and are otherwise identically distributed appropriate methods of estimation of β within each of the k data sets include those given in Sections 11.3–11.5. The importance of the pairwise slopes denoted in the methods in those sections by b_{ij} (not to be confused with the i, j subscripts as in x_{ij} in this section) here suggests that a comparison of these for the different data sets might produce a useful base for tests for parallelism.

Unfortunately, this intuitively reasonable approach is generally complicated by the lack of independence between pairwise slopes within any one set. This complication does not arise, however, in the case of the abbreviated Theil method. If this is applied to each data set, the pairwise slopes involved are independent within that set, and the independence of observations in different sets implies they are independent of the pairwise slopes obtained by the same method in the other sets. If, in addition, the x are equally spaced, all pairwise slopes have the same distribution within a particular set. If the slopes of all sets are the same, they will have identical distributions in all sets. If the β_i are not all equal (i.e., if the lines are not all parallel), the distributions of the pairwise slopes for each set will differ only in their medians. Thus, in this very restricted case, an appropriate test for parallelism against the alternative of at least one different slope is the Kruskal–Wallis test (or the WMW test when $k=2$).

The Pitman efficiency is not known for the above procedure, but it may not be high because the abbreviated Theil procedure does not make full use of all relevant information in the data.

More sophisticated tests for parallelism are discussed by Sen (1969). A very general test that is asymptotically distribution-free under broad assumptions developing some of Sen's ideas in combination with proposals due to Adichie (1984) is described with an example by Hollander, Wolfe and Chicken (2014, Section 9.5).

SUMMARY POINTS:

- In a comparison of the regression lines for several groups, equality of the population slope values implies that all lines are parallel.
- For the full Theil–Kendall method, the lack of independence between pairwise slopes within the groups makes comparison of groups difficult.
- Straightforward comparison is possible for the abbreviated Theil method, as pairwise slopes are independent.

11.8 Efficiency, Power and Sample Size

It is well known in classic least squares straight-line regression that the standard error of the estimator $\hat{\beta}$ of β is highly dependent upon the positioning of the x_i. If one has a choice of the x_i before data are collected, one aim should be to select these to make this standard error as small as possible. If this is done, the confidence interval for β at any chosen level will be the shortest available. It is also well known that for least squares, the shortest confidence interval is obtained in the case when the sample size is even by setting half the x_i at the least possible value and the other half at the greatest possible value.

In practice, this ideal design for an experiment to estimate β may be impossible because of experimental constraints. Also, other factors may have to be taken into consideration. One of these could be the need to test whether a straight line is adequate to describe the systematic component of the model. To test this, one or more x-values intermediate between the two extreme values must be included.

In many practical situations, there are major restrictions on the available x-values. For example, if the study is designed to determine whether a straight line provides a reasonable relationship between weight and age of chickens from shortly after hatching to maturity, one may have to select a sample from available chickens. There may be doubt about whether a straight-line relationship is adequate, so one would be unwise to choose only day-old chicks and fully mature birds. However, it would be helpful if one could reduce the standard error of the estimate of β by choosing, if possible, relatively large numbers of young, and of mature, birds and just a few intermediate in size. There may well be practical limitations on the extent to which this can be done. For example, there may be only a few mature birds available. For valid inferences if more than one bird is available at a given x and only some of these are included, those included should be chosen at random.

The often quite profound influence of design (in the sense of choosing x) on the precision of estimates for β carries over to nonparametric situations.

Example 11.7

Suppose that the data in Example 11.1 had been generated by adding errors to some Y generated by the systematic relationship $Y=x+2$.

Upper Limit Stated by Parents (x):	0	1	2	3	4	5	6
Reported Gaming Time (y):	2.5	3.1	3.4	4.0	4.6	5.1	11.1
Systematic Relationship ($Y=x+2$):	2.0	3.0	4.0	5.0	6.0	7.0	8.0
Error ($y-Y$):	0.5	0.1	−0.6	−1.0	−1.4	−1.9	3.1

Thus, for instance, for the first point (0, 2.5), an error of 0.5 has been added to the systematic $Y=0+2=2$, the other errors being as shown above. Now consider the situation in which the systematic relationship $Y=x+2$ is retained as part of the model, but the values $x=4$, 5, 6 in Example 11.1 are replaced by the unlikely but not impossible values of $x=24$, 25, 26. If the errors from the systematic relationship remain as those shown above, this leads to the table

Upper Limit Stated by Parents (x):	0	1	2	3	24	25	26
Reported Gaming Time (y):	2.5	3.1	3.4	4.0	24.6	25.1	31.1
Systematic Relationship ($Y=x+2$):	2.0	3.0	4.0	5.0	26.0	27.0	28.0
Error ($y-Y$):	0.5	0.1	−0.6	−1.0	−1.4	−1.9	3.1

For example, the point (5, 5.1) is replaced by the point (25, 25.1) since, when $x=25$, $Y=25+2=27$ and if the associated error is −1.9, this leads to $y=27-1.9=25.1$.

Use the Theil–Kendall method to obtain a point estimate of β and a confidence interval giving at least 95% coverage for the gaming times with the values $x=4, 5, 6$ replaced by x=24, 25, 26.

The b_{ij} arranged in a triangular matrix with the first row elements $b_{12}\ b_{13} \ldots b_{1n}$, second row elements $b_{22}\ b_{23} \ldots b_{2n}$ etc. are

0.600	0.450	0.500	0.921	0.904	1.100
	0.300	0.450	0.935	0.917	1.120
		0.600	0.964	0.943	1.154
			0.981	0.959	1.178
				0.500	3.250
					6.000

From Table 11.3, the point estimate of β given by the Theil–Kendall method is 0.935 and a Monte Carlo 95% confidence interval for β is (0.50, 1.178). The corresponding results in Example 11.3 were 0.567 and (0.50, 1.925). Although the lower limit is the same as that obtained in the earlier example, the upper limit is appreciably lower, and the point estimator is now near the top end of the interval. The reduction in width of the interval implies an increase in power attributable to the design change of altering dramatically three of the x-values.

Output from the ANSM5 package in R can be obtained using `theil.kendall (ch11$reportedtime.2, ch11$parentlimit.2, do.CI = TRUE)`.

It may be beneficial to think carefully about the implications of these findings. In particular, one should consider whether the model with systematic part $Y=x+2$ and the given errors seems reasonable in relation to the amended data values, and why the point estimate of β changes so dramatically between that in Example 11.3 and that obtained here.

In earlier sections, attention was drawn to a relationship between some linear regression models and the procedures used to analyse them and methods such as WMW and the sign test. Where relevant, results for Pitman efficiency under various error structures carry over to regression. However, power studies become very complicated not only because, as in the simpler situations, they are often highly dependent upon what distributional assumptions are made, but as just indicated, the power is strongly influenced by design factors in choosing x, as well as by the size of the sample. Further, changes in sample size may well introduce new x-values that may or may not in themselves enhance the power. In view of the complexity, this matter is not pursued further here. Hollander, Wolfe and Chicken (2014, Section 9.8) give other useful comments on efficiency.

TABLE 11.3

Intervals for *b*-Values Corresponding to Each Possible $T_t(b)$ or the Corresponding t_k for the Data in Example 11.7

b-Interval	$T_s(b)$	t_k
−∞ to 0.300	21	1.000
0.300+ to 0.450–	19	0.905
0.450+ to 0.500–	15	0.714
0.500+ to 0.600–	11	0.524
0.600+ to 0.904–	7	0.333
0.904+ to 0.917–	5	0.238
0.917+ to 0.921–	3	0.143
0.921+ to 0.935–	1	0.048
0.935+ to 0.943–	−1	−0.048
0.943+ to 0.959–	−3	−0.143
0.959+ to 0.964–	−5	−0.238
0.964+ to 0.981–	−7	−0.333
0.981+ to 1.100–	−9	−0.429
1.100+ to 1.120–	−11	−0.524
1.120+ to 1.154–	−13	−0.619
1.154+ to 1.178–	−15	−0.714
1.178+ to 3.250–	−17	−0.810
3.250+ to 6.000–	−19	−0.905
6.000+ to ∞	−21	−1.000

SUMMARY POINTS:

- Concentrating on low and high *x*-values reduces the standard error of the slope estimate.
- As a result, confidence intervals for the slope estimate are shorter with a consequential increase in the power of the design.

11.9 Exercises

11.1 Show that if a sample of n_1 observations has mean m_1 and another sample of $n_2=n-n_1$ has mean m_2 and the first sample values are indexed by $x=0$, and the second sample values by $x=1$, then $\hat{\beta}$ given by (11.3) equals m_2-m_1.

11.2 For the data in Example 11.1, verify that the sum of residuals e_i and the sum of the products of residuals with the corresponding x_i is zero. Explain why this is an inevitable consequence of the least squares estimation procedure.

11.3 The numbers of rotten oranges (y) in 10 randomly selected boxes from a large consignment are counted after they have been kept in storage for a stated number of days (x). Use the Theil–Kendall method to compute the slope of a straight line fitted to these data and obtain an appropriate estimate of the intercept.

x	3	5	8	11	15	18	20	25	27	30
y	2	4	7	10	17	23	29	45	59	73

Plot these data and the fitted line. Does the fit seem reasonable?

11.4 Use the asymptotic result relevant to the Theil–Kendall method given in Section 11.4.2 to obtain an approximate 95% confidence interval for the data used in Example 11.4. Compare this with the exact permutation-based interval obtained in that example.

11.5 Verify the confidence limits quoted in Example 11.5 using the abbreviated Theil method.

11.6 The Epilepsy Risk Awareness (ERA) scale and the Epilepsy Self-Management Scale (ESMS) are similar scales developed in the United Kingdom and United States, respectively. For a sample of 10 participants in the study reported in Braun et al. (2017), the scores were as follows:

ERA	67	68	71	80	85	94	105	107	108	111
ESMS	132	119	142	146	147	156	164	175	165	162

Use the abbreviated Theil method to estimate the slope between the ERA (x-variable) and ESMS (y-variable) scores and find a 95% confidence interval. Comment.

11.7 Compare the abbreviated Theil estimator and the Cox–Stuart test for trend (Section 4.5.3) in terms of similarities and differences.

11.8 Gat and Nissenbaum (1976) provide ammonia concentration (y mg/L) at various depths (x m) in the Dead Sea. Fit a linear regression for concentration on depth using the Theil–Kendall method and obtain an approximate 95% confidence interval for β.

x	25	50	100	150	155	187	200	237	287	290	300
y	6.13	5.51	6.18	6.70	7.22	7.28	7.22	7.48	7.38	7.38	7.64

Do you consider a straight-line fit satisfactory for these data?

11.9 Katti (1965) provides data for weight of food eaten (x) and weight gain (y) for 10 pigs fed on one type of food (A) and for 10 fed on a second type (B). Use the abbreviated Theil method to estimate the slope of each regression line and test whether the hypothesis that the true slopes β_1, β_2 are equal is supported.

A	x	575	585	628	632	637	638	661	674	694	713
	y	130	146	156	164	158	151	159	165	167	170
B	x	625	646	651	678	710	722	728	754	763	831
	y	147	164	149	160	184	173	193	189	200	201

11.10 The Phenological Reports (1891–1948) of the Royal Meteorological Society give the average observation date (denoted by January 1st=1, January 2nd=2, etc.) for the first flowering of various plants in several locations in the United Kingdom. Given below are the records for 10 plants in Southwest England (x) and in Northern

Scotland (y) [Jeffree (1960), Table 4]. Draw a scatter diagram for these data and in the light of any evidence as to a possible relationship and the presence of any outliers fit what you consider an appropriate linear regression and comment on the adequacy of fit or otherwise. If a further species has an average flowering date of 122 in Southwest England, what estimate would you give for the average flowering date in Northern Scotland?

	SW England	N Scotland
Hazel	38	56
Coltsfoot	57	76
Wood Anemone	85	103
Hedge Garlic	108	131
Hawthorn	129	150
White Ox-Eye	145	160
Dog Rose	156	168
Greater Bindweed	185	201
Harebell	193	195
Ivy	265	292

12

Categorical Data

12.1 Categories and Counts

12.1.1 Some Basic Concepts

Data may consist of counts of the number of units (people, institutions, towns, countries, items, etc.) with given attributes. It is often convenient to present these in one-, two-, three- or higher-dimensional tables usually referred to as one-way, two-way, three-way, etc., *contingency tables.* Each dimension, or way, corresponds to a classification into categories representing attributes.

Attributes may be explanatory (e.g., dose levels of a drug, the names of several different drugs, gender, psychiatric diagnoses, ethnic groups, income levels). Alternatively, they may be responses (e.g., side-effects of drugs classified as none, slight, moderate, severe; blood pressure levels after administration of a drug; examination grades). Attributes are often qualitative. Recalling Section 1.3, if there is no natural ordering, they are described as *nominal* (e.g., psychiatric diagnoses, ethnic groups). Attributes that may be arranged in a natural order are described as *ordered* or *ordinal* (e.g., reactions to a drug classified as slight, moderate, severe; grouping by age under 50 and age 50 and over).

In Table 12.1, the row categories are nominal and explanatory. Had each row referred to different dose levels of the same drug, they would have been ordered in explanatory categories. The column categories are ordered responses that reflect increasing seriousness.

Situations have already been met in which contingency tables have been formed from measurement data by counting the numbers of measurements falling into defined categories, e.g., with the median test (Section 6.2.2, Section 7.2.5). For instance, Table 7.5 involves counts from six samples of numbers in each of two categories (above and below an overall sample median M). In the column categories, there were ordered responses ($<M$ or $>M$). The row categories (samples) were explanatory and nominal.

TABLE 12.1

Numbers of Patients Showing Side-Effects for Two Drugs

	Side-Effect Level					
	None	**Slight**	**Moderate**	**Severe**	**Fatal**	**Total**
Drug A	23	8	9	3	2	45
Drug B	42	8	4	0	0	54
Total	65	16	13	3	2	99

DOI: 10.1201/9780429326172-12

TABLE 12.2

The Data in Example 6.1 Presented in a Contingency Table

Time	17	18	19	21	22	23	24	25	26	27	28
Group A	1	1	1	0	1	1	0	1	1	0	0
Group B	0	0	0	1	0	0	1	0	0	1	1
Time	**29**	**30**	**31**	**32**	**33**	**34**	**35**	**36**	**39**	**41**	
Group A	1	0	1	0	1	0	0	0	0	0	
Group B	0	1	0	1	0	1	1	1	1	1	

For display purposes, the table is divided, but it should be read as a 2×21 table.

TABLE 12.3

Outcome for Patients Who Receive a Prescription for a Particular Antibiotic

	Otitis Media			
Bronchitis	**Worse**	**The Same**	**Better**	**Total**
Worse	13	5	6	24
The Same	1	19	4	24
Better	4	0	8	12
Total	18	24	18	60

The data in Example 6.1 could be expressed as a 2×21 contingency table. The rows correspond to Group A and Group B. Each column corresponds to one of the times taken to carry out the calculations, times being arranged in ascending order. The numbers in the cells of this table are either 0 or 1. An entry 1 is placed in the cell in the first row of a time column if someone from Group A took that time to complete the calculations, and similarly, a 1 is entered in the appropriate cell in the second row for those in Group B. All other cell entries are zero. This leads to Table 12.2. Why this is useful is explained in Sections 12.3.1 and 13.2.6.

In Example 6.5, a contingency table format was used for what were for practical purposes tied data. A contingency table was a natural way to present the data.

The correspondence between tables of measurements and counts of units or participants exhibiting each measurement is a useful tool for showing relationships between methods.

Many tests for lack of association applicable to nominal categories in both rows and columns may still be applied when either or both row and column categories are ordered. Generally, however, more powerful procedures exist to take account of ordering.

Sometimes both rows and columns represent response categories. For instance, a particular antibiotic can be used to treat both bronchitis and otitis media (infection of the middle ear) in the same patient (Hornbrook, Hurtado and Johnson, 1985). Table 12.3 shows a data set where the antibiotic is prescribed to 60 patients who have both health problems. Its effect is noted in changes (worse, the same, better) in *bronchitis* (rows) and *otitis media* (columns). Such a variation in outcome might occur in practice, as a substantial minority of the patients may not take the prescribed antibiotic, leading to a worsening in their condition. Since, for each illness, the three outcomes are mutually exclusive, all subjects can be classified into one and only one combination of row and column categories.

This may not be the case in a different study; in asthma, the outcomes of interest might be no symptoms, cough and wheeze. It is possible for individuals to have both cough and wheeze so a fourth category of "both" would be needed to take this into account.

In Table 12.3 both row and column categories are ordered; in the study of cough and wheeze in asthma outlined above there is no natural ordering and these categories are nominal.

This chapter is mainly about two-way contingency tables consisting of two or more rows and columns. The *cells* of the table represent the different pairings of the categories associated with the two variables, and the *cell entries* indicate the frequency of each combination of categories within the sample. A complete contingency table includes the *marginal totals*. The *row totals* form a margin to the right of the cell entry columns, and the *column totals* form a margin beneath the cell entry rows. The *grand total* is shown in the bottom right corner of the table. For example, Table 12.1 has two rows, five columns and hence ten cells. The two-row totals, five-column totals and the grand total are also shown.

It may be of interest to investigate whether, in comparing the rows of the table, the proportions of observations in each column category take the same profile. The null hypothesis of independence is often expressed as one of *no association between row and column categories*. For Table 12.1, the null hypothesis is that the corresponding population proportions for the various outcomes (none, slight, moderate, severe, fatal) are equal. If categories are merged so that the comparison becomes "side-effects experienced" vs. "no side-effects", the null hypothesis is that in the population the proportion of individuals experiencing side-effects is equal for drugs A and B.

Hypothesis testing in the context of contingency tables typically involves a comparison of the *observed* frequencies in the cells of the table and the values *expected* for those frequencies were the null hypothesis to be true. The chi-squared test for contingency tables uses this approach.

12.1.2 An Overview of the Chi-Squared Test for Contingency Tables

This nonparametric test, formulated by Pearson (1904), is well known and widely used. Expected frequencies are calculated for each cell by multiplying together the totals for the intersecting row and column, and dividing by the grand total. For each cell, the squared difference between the observed and expected frequencies is divided by the expected frequency. These terms are then added together to give the *chi-squared statistic*. Under the null hypothesis, the observed frequencies are equal to those expected on the grounds of chance alone, and this statistic takes the value zero. Evidence against the null hypothesis is found if there are large differences between observed and expected values that result in a high value for the chi-squared statistic.

If the table has r rows and c columns, it is referred to as an $r \times c$ table. The number of degrees of freedom for such a table is given by $(r-1)(c-1)$. This can be thought of as the minimum number of cell entries required in order to deduce the complete contingency table if the row and column totals are known (Fisher, 1934). For example, with a 2×2 table, only one cell entry is required, so the table has one degree of freedom. Under the null hypothesis, the chi-squared statistic asymptotically takes a chi-squared distribution with $(r-1)(c-1)$ degrees of freedom.

This approximation should not be relied upon with small samples; a longstanding recommendation is that for 2×2 tables, all of the expected values should be at least 5 (e.g., Fisher, 1934). For larger tables, at least 80% of the entries should have an expected value of 5 or greater, and no expected value should be less than 1 (Armitage, Berry and Matthews, 2001).

TABLE 12.4

Reported Frequency of Receiving/Giving Feedback on PPI Comments

	Representatives	Researchers	Total
Frequency			
Never	12	4	16
Sometimes	40	17	57
Always	10	17	27
Total	62	38	100

TABLE 12.5

Expected Frequencies for Receiving/Giving Feedback on PPI Comments

	Representatives	Researchers	Total
Frequency			
Never	9.92	6.08	16
Sometimes	35.34	21.66	57
Always	16.74	10.26	27
Total	62	38	100

Example 12.1

It is considered good practice for researchers engaged in projects with patient and public involvement (PPI) to provide feedback on the advice that they receive. Mathie et al. (2018) asked PPI representatives and researchers to indicate how often they received/gave feedback, using the categories never, sometimes and always (Table 12.4). Is there any evidence that perceived feedback frequency differs between representatives and researchers?

The null hypothesis is that the true profile of frequencies is the same for representatives and researchers. Given a table with 3 rows and 2 columns, the chi-squared test for contingency tables might be considered. In order to calculate the chi-squared statistic, it is first necessary to find the expected frequencies for this table from the row and column totals (Table 12.5). For a 3×2 table, only 2 cell entries are required. For representatives "never" receiving feedback, the expected frequency is $\frac{62 \times 16}{100} = 9.92$. For the representatives who report receiving feedback "sometimes", the expected frequency is $\frac{62 \times 57}{100} = 35.34$. The remaining expected frequencies can be calculated by subtraction from the row and column totals, whence the chi-squared statistic is given by

$$\frac{(12-9.92)^2}{9.92}+\frac{(4-6.08)^2}{6.08}+\frac{(40-35.34)^2}{35.34}+\frac{(17-21.66)^2}{21.66}+\frac{(10-16.74)^2}{16.74}+\frac{(17-10.26)^2}{10.26}=9.906.$$

Referring the chi-squared statistic to a chi-squared distribution with (3 − 1)(2 − 1) or 2 degrees of freedom gives $P=0.007$. There is strong evidence against the null hypothesis.

Output from the ANSM5 package in R can be obtained using `chisqtest.ANSM(ch12$feedback.freq, ch12$PPI.person, do.asymp = TRUE)`.

Comparing the observed cell entries with those expected, the proportion of representatives who report "always" receiving feedback is noticeably less than the proportion of researchers who stated that they "always" give feedback. Given that some of the representatives and researchers were attached to the same projects, it appears that the representative perception of feedback frequency is less positive than that of the researchers.

12.1.3 Some Models for Counts in $r \times c$ Tables

Books on categorical data analysis in both parametric and nonparametric contexts include Bishop, Fienberg and Holland (1975), Breslow and Day (1980), Plackett (1981), Agresti (2010, 2013, 2019), Christensen (1997), Everitt (1992), Lloyd (1999) and Simonoff (2003).

It is convenient to represent a general $r \times c$ table in the form of Table 12.6. The entry n_{ij} at the intersection of the ith row and jth column, referred to as cell (i, j), is the number of items having that combination of row and column attributes. The total count for row i is denoted by n_{i+}, i.e., $n_{i+} = \sum_j n_{ij}$. Similarly, n_{+j} is the total count for column j, and

$$N = \sum_i n_{i+} = \sum_j n_{+j} = \sum_{i,j} n_{ij} = n_{++}$$

is the total count over all cells. Some writers use $n_{i.}$ and $n_{..}$ instead of n_{i+} and n_{++}.

The methods developed here are nonparametric in the sense that, although several parametric models may give rise to an observed distribution of counts in a contingency table, the inference procedures often do not depend upon which model is relevant or upon the precise values of certain parameters since these are not specified in the hypotheses. The use of *conditional inference* avoids the problem of specifying parameter values. This draws on properties of sufficient and ancillary statistics, a topic that is beyond the scope of this text. Readers familiar with the method of maximum likelihood will find an excellent account of inference theory for categorical data in Agresti (2013, Chapter 3). Agresti (2019) deals at a less technical level with both classic and modern methods of categorical data analysis.

When making inferences, row and column totals are nearly always regarded as fixed. This is the conditional feature implied by conditional inference. It is not, in general, a consequence of the experimental setup. Sometimes, by the nature of the experiment, neither row nor column totals are fixed *a priori*. In other cases, one may be fixed, the other not. Sometimes both are fixed.

TABLE 12.6

A General $r \times c$ Contingency Table

		Column Categories						
		1	**2**	**3**	**.**	**.**	***c***	**Total**
	1	n_{11}	n_{12}	n_{13}	.	.	n_{1c}	n_{1+}
	2	n_{21}	n_{22}	n_{23}	.	.	n_{2c}	n_{2+}
Row Categories	.	.	.	.	.	.	.	.
	.	.	.	.	.	.	.	.
	r	n_{r1}	n_{r2}	n_{r3}	.	.	n_{rc}	n_{r+}
	Total	n_{+1}	n_{+2}	n_{+3}	.	.	n_{+c}	N

For example, in a comparison of two drugs, I and II, that will lead to either an improvement (a success, denoted by *S*) or no improvement (a failure, denoted by *F*), there may be 19 patients, with 9 of these allocated at random to receive drug I. The remaining 10 receive drug II. Suppose the results are:

	S	*F*	Total
Drug I	7	2	9
Drug II	5	5	10
Total	12	7	19

(12.1)

Row totals are fixed by the numbers allocated to each drug. If the 19 patients are the entire population of interest, then under the null hypothesis H_0 that the drugs are equally effective, one is justified in regarding the total number of successes, 12, as fixed irrespective of how the patients to the groups of 9 and 10 corresponding to the two drugs are allocated. Just as in Example 2.3, it is extreme outcomes under randomization like none (or 9) of the 12 successes being among the 9 patients allocated to drug I that cast doubt on the hypothesis of equal effectiveness. The probability of each possible outcome under H_0 when the marginal totals are fixed can be computed.

The situation is different if the patients are sampled as two groups of 9 and 10 from a larger population. Even under H_0, taking another pair of samples of 9 and 10 (the fixed row totals representing the designated sample sizes) is unlikely to result in the combined number of successes for these different patients again being 12. The outcome might be as in (12.2).

	S	*F*	Total
Drug I	6	3	9
Drug II	3	7	10
Total	9	10	19

(12.2)

Note that the hypothesis of independence does not involve the actual probability of success with each drug but stipulates only that the probability of success, π, say, is the same for each drug. In practice, the value of π is generally not known.

However, if it is assumed that there is such a common value for both drugs, an estimate can be obtained based on the total number of successes, 12 from 19, for the sample of 19 in the observed situation, i.e., that in (12.1). This information can be used to compute the expected number of successes in sub-samples of 9 and 10 from the total of 19 under the assumption that H_0 is true. Again, big discrepancies between observed and expected numbers of successes suggest an association, i.e., that the probability of success for drug I is not the same as that for drug II. In effect, a permutation distribution is once more under examination, but only in possible samples that have fixed column totals (in this case 12, 7, respectively) as well as fixed row totals. Such conditional inference is perfectly valid. It is the entries in the body of any such table with fixed marginal totals that provide the evidence for or against independence. As indicated above, the column marginal totals in this example provide information about the likely value of π, the probability of success if H_0 is true. They contain little information as to whether there is really a common probability, π, of success for each drug. The individual cell values contain virtually all the information on this latter point. Similar arguments apply regarding all marginal totals in an $r \times c$ table as fixed for nearly all the models considered in this and the next chapter.

Methods for inference that do not require all marginal totals to be fixed have been proposed and are used. However, there are many statisticians who feel strongly that conditional inference with fixed marginal totals is the most appropriate way to test for independence (although independence is not the only thing that might be tested in contingency tables).

Conditional inference is convenient, valid and overcomes technical problems with hypothesis tests in this context. For enlightened discussions about this approach and those that do not assume marginal totals are fixed, see Yates (1984), Upton (1992), a wide-ranging paper by Agresti (1992), and Freidlin and Gastwirth (1999).

Experimentally, the outcome in (12.1) might be reached without pre-fixing even the row totals. For example, the data might have come from what is called a retrospective study, by simply going through hospital records of all patients treated with the two drugs during the last 12 months and noting the responses. In this context, the method of testing for no association using conditional inference would be the same, being conditional upon the observed row and column totals.

For elucidation, several models used to describe counts are described briefly, indicating where each might be appropriate and showing that each leads to the same conditional inference procedures. This introductory discussion is largely confined to 2×2 tables. Readers who are not familiar with the multinomial and Poisson distributions and the concept of conditional and marginal distributions may wish to omit the remainder of this section. Applications are described in Section 12.2.

A typical model for the situation exemplified by (12.1), where the row classifications are explanatory and the column classifications are responses, is to assume that each row has associated with it independent binomial distributions and that the cell entries are the numbers of occurrences of a response I and a response II for a $B(n_{1+}, p_1)$ distribution in the first row and for a $B(n_{2+}, p_2)$ distribution in the second row. In the case of (12.1), these numbers are given by $n_{1+}=9$ and $n_{2+}=10$. Under this model, for each cell, probabilities p_{ij} are obtained for the observed numbers n_{ij} as follows:

$$p_{11} = \Pr(n_{11} = 7) = \binom{9}{7} p_1^7 (1-p_1)^2, \; p_{12} = 1 - p_{11},$$

$$p_{21} = \Pr(n_{21} = 5) = \binom{10}{5} p_2^5 (1-p_2)^5, \; p_{22} = 1 - p_{21}.$$

The hypothesis of no association implies H_0: $p_1=p_2$ but the associated common value π, say, is not specified. Independence of response from patient to patient and the additive property of binomial variables imply that under H_0 the probability of observing a total of 12 favourable outcomes in 19 observations (the column total line) is

$$p_{+1} = \binom{19}{12} \pi^{12} (1-\pi)^7.$$

Since the classifications of individuals are independent, it follows from standard definitions of conditional and independent probabilities that the probability of observing 7 successes ($= n_{11}$) in 9 ($= n_{1+}$) trials, conditional upon having observed a total of 12 successes in 19 trials, is

$$P^* = p_{11}p_{21}/p_{+1}.$$

Thus, when $\pi = p_1 = p_2$

$$P^* = \left[\binom{9}{7}\pi^7(1-\pi)^2 \times \binom{10}{5}\pi^5(1-\pi)^5\right] \Big/ \left[\binom{19}{12}\pi^{12}(1-\pi)^7\right]$$

$$= \left[\binom{9}{7} \times \binom{10}{5}\right] \Big/ \left[\begin{matrix}19\\12\end{matrix}\right]$$

$$= [(9!)\times(10!)\times(12!)\times(7!)]/[(19!)\times(7!)\times(2!)\times(5!)\times(5!)] = 0.180.$$

As indicated in Section 6.2.2, for a general 2×2 table with cell entries

a	b
c	d

this generalizes to

$$P^* = \frac{(a+b)!(c+d)!(a+c)!(b+d)!}{(a+b+c+d)!a!b!c!d!} \tag{12.3}$$

and this does not involve π. Equation (12.3) represents the frequency, or probability mass, function for the *hypergeometric distribution*. In Section 6.2.2, it was seen that when the marginal totals are fixed, P^* depends essentially only on a, since, given a, then b, c and d are automatically fixed to give the correct marginal totals. The fact that fixing a determines all cell entries is summarized by saying that a 2×2 contingency table has *one degree of freedom*. The value of P^* given by (12.3) is relevant to testing for independence (no association). As explained in Section 6.2.2, the test based on the hypergeometric distribution is the *Fisher exact test*. Its application to a 2×2 table was described in Example 6.7. By calculating P^* for all possible values of a, one may set up critical regions in the tail of the distribution relevant to assessing the strength of evidence against H_0 in either one- or two-tailed tests, as in Example 6.7.

Sometimes estimates of the expected values of each n_{ij} under H_0 are needed. If π is known, exact expected values can be obtained, but when π is unknown, the estimates are based on the marginal totals, using these to estimate π. Since n_{+1} has an expected value $N\pi$, one can estimate π by $\hat{\pi} = n_{+1}/N$, and the expected value of n_{11} is $m_{11} = n_{1+}\hat{\pi} = (n_{1+}n_{+1})/N$. In the same way, the general result $m_{ij} = (n_{i+}n_{+j})/N$ can be established. These expected values sum to the marginal totals.

More generally, if there is an association in a 2×2 table, then in (12.1), for example, if the binomial probabilities were known to be p_1, p_2, the expected values in that case would be

$$\begin{array}{ll} 9p_1 & 9(1-p_1) \\ 10p_2 & 10(1-p_2) \end{array} \tag{12.4}$$

The quotients, $p_1/(1-p_1)$ and $p_2/(1-p_2)$, represent the odds of the outcome *success* for each row. A quantity often used as an indicator of association is the *odds ratio*, defined as

$$\theta = \frac{p_1/(1-p_1)}{p_2/(1-p_2)} = \frac{p_1(1-p_2)}{p_2(1-p_1)}$$

From (12.4) the odds ratio is also the ratio of cross products of the cell expected frequencies m_{ij}, i.e., $\theta=(m_{11}m_{22})/(m_{12}m_{21})$. If $p_1=p_2$, then $\theta=1$. This is also true for the estimated expected values calculated under H_0, i.e., $m_{ij}=(n_{i+}n_{+j})/N$. If, as usual, p_1, p_2 are not known, θ may be estimated by the sample odds ratio, which is defined as $\theta^*=(n_{11}n_{22})/(n_{12}n_{21})$. Section 13.2 gives tests for association, including a test for whether θ^* is consistent with H_0: $\theta=1$.

If $p_1=3/4$ and $p_2=1/2$, it is intuitively obvious that more successes will be expected with drug I than with drug II, and from (12.4) it is found that $m_{11}=6.75$ and $m_{21}=5$. These expected values no longer add to the column total, as do the expected values under H_0. This should cause no surprise, as the probabilities of success are known and different for each drug. For these p_1, p_2 values $\theta=[(3/4)\times(1/2)]/[(1/2)\times(1/4)]=3$. If the probabilities are interchanged, so that $p_1=1/2$ and $p_2=3/4$, it is found that $\theta=1/3$. Both these cases reflect the same degree of association, but in opposite directions. This is demonstrated more clearly by using $\phi=\log\theta$ as a measure of association, for now

$$\log 3 = -\log(1/3)$$

and generally

$$\log\theta = -\log(1/\theta).$$

Possible values of ϕ range from $-\infty$ to ∞. When there is no association $\phi=0$. The above relationships are true for logarithms to any base, but most formulae for testing and estimation use natural logarithms with the exponential base *e*. These are sometimes denoted by *ln* rather than *log*, but the latter is used here.

The binomial model given above is not appropriate if data are for two response classifications, like those in the table (12.5), which gives numbers of patients responding in specified ways to a drug.

	Blood Cholesterol Level		
Blood Pressure	**Increased**	**Unchanged**	**Total**
Higher	14	7	21
Lower	11	12	23
Total	25	19	44

(12.5)

Here all four cells are on a par in the sense that all one can say is that, with an unknown probability p_{ij}, any of the $N=44$ subjects may be allocated to cell (i, j). The multinomial distribution is an extension of the binomial distribution with independent allocation of N items to $k>2$ categories, with specified probabilities for each category. Here there are four categories, one corresponding to each cell. The probabilities are subject to the constraint $p_{11}+p_{12}+p_{21}+p_{22}=1$, since each person is allocated to one and only one cell. The notation p_{1+},

p_{2+} is used for the marginal probabilities that items will fall in row 1 or row 2. Likewise, marginal probabilities p_{+1}, p_{+2} are specified for the columns. For independence between rows and columns, standard theory gives the relationship $p_{ij}=p_{i+}p_{+j}$ for $i, j=1, 2$.

Even if there is independence, the values of the p_{ij} are still unknown, but this presents no problem if conditional inference is used. Assuming no association (i.e., independence), estimates are based on marginal totals, estimating p_{i+} by n_{i+}/N and p_{+j} by n_{+j}/N. Since $p_{ij}=p_{i+}p_{+j}$ if there is independence, p_{ij} is estimated by $\hat{p}_{ij} = n_{i+}n_{+j}/N$. It follows from the properties of the multinomial distribution that expected estimated counts can be obtained as $m_{ij} = N\hat{p}_{ij} = n_{i+}n_{+j}/N$, exactly as in the two-sample binomial model. Tests are performed as in that model. Under independence, the odds ratio, here defined as $\theta=(p_{11}p_{22})/(p_{12}p_{21})$, is unity (hint: put $p_{11}=p_{1+}p_{+1}$, etc.). If the p_{ij} are estimated under H_0 in the way indicated above, using these estimates, or the estimated expected frequencies m_{ij}, gives $\theta=1$. The sample odds ratio is $\theta^*=(n_{11}n_{22})/(n_{12}n_{21})$.

Another model often relevant is one where counts in each cell have independent Poisson distributions with a mean (Poisson parameter) m_{ij} for cell (i, j). If $\theta=(m_{11}m_{22})/(m_{12}m_{21})=1$, there is no association between rows and columns. In general, the m_{ij} are not known, but it can be shown that, again under H_0, these may be estimated from marginal totals in the way used for the row-wise binomial or the overall multinomial models.

Extension of the three models above to $r \times c$ tables is reasonably straightforward. For $c>2$, binomial sampling within rows generalizes to multinomial sampling within rows. For this, or the overall multinomial sample, or the Poisson count situation, under the null hypothesis of no association, the expected cell frequencies are estimated in the notation of Table 12.5 by $m_{ij}=n_{i+}n_{+j}/N$, $i=1, 2, \ldots, r$ and $j=1, 2, \ldots, c$. An odds ratio may be defined for any two rows i, k and any two columns j, l as

$$\theta_{ijkl} = (m_{ij}m_{kl})/(m_{il}m_{kj}),$$

and $\theta_{ijkl}=1$ for all i, j, k, l if there is no association.

The sample odds ratios are $\theta^*_{ijkl} = (n_{ij}n_{kl})/(n_{il}n_{kj})$.

The Fisher exact test extends to $r \times c$ tables in a way described in Section 12.2.1.

SUMMARY POINTS:

- Two-way contingency tables consist of two or more rows and two or more columns (Section 12.1.1).
- Hypothesis testing usually involves a comparison of the observed frequencies in the cells of the table and the values expected under the null hypothesis (Section 12.1.1).
- The number of degrees of freedom for an $r \times c$ table is given by $(r - 1)(c - 1)$ (Section 12.1.2).
- The asymptotic chi-squared test for contingency tables should not be used if expected values in cells are small (Section 12.1.2).
- Testing for independence between row and column classifications in contingency tables is usually based on conditional inference, regarding the marginal totals as fixed (Section 12.1.3).
- For 2×2 tables, an indicator of association is the odds ratio (Section 12.1.3).

12.2 Nominal Attribute Categories

Tests developed in this section are appropriate when row and column categories are both nominal. They are often used for tests of no association when attributes are ordered, but tests described in Section 12.3, or some in Chapter 13, are then often more appropriate.

Three commonly used procedures are described here. Confusion exists in the literature because each often (but not invariably) leads to essentially equivalent permutation distributions for 2×2 tables but to different asymptotic test statistics even in this case. For general $r \times c$ tables, both the exact permutation distribution of the test statistics and the asymptotic test statistics differ. One should not exaggerate the importance of these differences; unless the choice of a precise significance level is critical, it is seldom that the three possible exact tests lead to markedly different conclusions. However, in some circumstances, there may be considerable differences between exact and asymptotic tests. Numerical values of the three test statistics often differ widely, even for moderately large samples, but each statistic has asymptotically a chi-squared distribution with $(r-1)(c-1)$ degrees of freedom under the hypothesis of independence (no association) between row and column classifications.

The three tests are

- The *Fisher exact test* (also known as the Fisher–Irwin or as the Fisher–Freeman–Halton test).
- The *Pearson chi-squared test* (introduced in Section 12.1.2).
- The *likelihood ratio test.*

Practice has often been to use the Fisher exact test when asymptotic theory is clearly inappropriate and either of the others when asymptotic results are thought to be acceptable. Despite differences in the numerical value of the relevant test statistics and the fact that exact permutation distribution theory differs in each case, there seems no compelling reason to depart from this practice.

Tests based on exact permutation theory, except in the case of 2×2 tables (where the Fisher exact test is commonly used), are not practical without relevant software, such as that available in the ANSM5 package in R. This means there is no longer the need to rely on asymptotic theory when it is of doubtful validity. This applies particularly to sparse contingency tables in which there are many low cell counts, even though N may be large.

This last situation is common in pioneering medical research where responses may be recorded on a limited number of diseased patients because resources are in short supply or responses in certain categories are few in number.

12.2.1 The Fisher Exact Test

The Fisher exact test for 2×2 tables based on the hypergeometric distribution specified in (12.3) was first described by Fisher in the 5th edition of his landmark book *Statistical Methods for Research Workers* (1934). Freeman and Halton (1951) extended the test to $r \times c$ tables. In the notation of Table 12.5, the generalized hypergeometric distribution relevant to an $r \times c$ table gives the probability of an observed configuration under the null hypothesis H_0, conditional upon fixed marginal totals as

$$P^{*} = \frac{\prod_{i}(n_{i+}!)\prod_{j}(n_{+j}!)}{n!\prod_{i,j} n_{ij}!} \tag{12.6}$$

The P-value for testing no association is the sum of all probabilities for configurations with the given marginal totals having the same, or lower, probability.

Example 12.2

Scarlet fever is a childhood infection that, among other symptoms, gives rise to severe irritation of the nose, throat and ears. In a developing country where infection is a problem, six districts A–F were chosen. In each district, patients were located, and parents were asked to state the site at which they thought their child's irritation was worst. Numbers in each response category were:

	District						
	A	B	C	D	E	F	**Total**
Nose	1	1	0	1	8	0	11
Throat	0	1	1	1	0	1	4
Ears	1	0	0	0	7	1	9
Total	2	2	1	2	15	2	24

Is there evidence of an association between districts and sites of greatest perceived irritation?

The Fisher exact test for independence is used and the command `fishertest.ANSM(ch12$infection.site, ch12$district)` from the ANSM5 package in R.

The probability of observing this configuration under H_0: *no association* is 0.0001, and the probability of this or a less probable configuration is $P=0.0261$, hence there is fairly strong evidence against H_0.

It is evident that the main cause of association is the relatively high number of responses of greatest irritation at the nose or ears (but not throat) in district E. This district has responses out of line with the general trend, whereas the other districts produce reasonably consistent results for all three sites of irritation. It may be that in district E more importance is placed on visual appearance compared with the other five districts.

The finding of dependence is not surprising, so it is relevant to note that the asymptotic theory here does not indicate significance at a conventional 5% level. The more commonly used asymptotic statistics are discussed in Sections 12.2.2 and 12.2.3.

This example does not consider whether incidence levels differ between districts, but whether the proportions at irritation sites differ between districts (or equivalently, whether the proportions for districts differ between irritation sites). Had there, in addition, been eight cases of greatest irritation in the throat in district E, little evidence would have been found against the hypothesis of no association, although clearly the numbers of cases at all three irritation sites would be much higher for district E.

The test of association is essentially two-tailed, the situation being broadly analogous in this aspect to that in the Kruskal–Wallis test, where testing is for differences in location regardless of direction, or in different directions in populations corresponding to different samples.

If the table has many cells and the sample size is large, the computation of the P-value may take several minutes. Hence, many general packages include the Fisher exact test for 2×2 tables only.

12.2.2 The Pearson Chi-Squared Test

An alternative statistic for testing the independence of row and column categories is the Pearson chi-squared statistic introduced in Section 12.1.2. It has been widely used by researchers, partly because of the ease of its computation. It also has fairly rapid convergence to the chi-squared distribution as the sample size increases.

An exact test is based on the permutation distribution of this statistic over all sample configurations having fixed marginal totals. Computation of P-values is no easier than it is for the Fisher exact test, and only for 2×2 tables are the exact tests often equivalent.

Denoting, as before, the expected number in the cell (i, j) by $m_{ij}=n_{i+}n_{+j}/N$, then $(m_{ij}-n_{ij})^2$ is a measure of departure from expectation under the null hypothesis. It is reasonable to weight this measure inversely by the expected count. This is because a departure of 5 from an expected count of 100 is less surprising than a departure of 5 from an expected count of 6 for any of the models given in Section 12.1.3. Pearson (1904) proposed the statistic

$$X^2 = \sum_{i,j}\left[\frac{(m_{ij}-n_{ij})^2}{m_{ij}}\right] \tag{12.7}$$

as a test for evidence of association.

For computational purposes, it is easier to use the equivalent form

$$X^2 = \sum_{i,j}\left[\frac{(n_{ij})^2}{m_{ij}}\right] - N \tag{12.8}$$

Example 12.3

For the data in Example 12.2 calculate X^2. Test for significance using the asymptotic result that X^2 has a chi-squared distribution.

Each $m_{ij}=n_{i+}n_{+j}/N$ is calculated, following which the formula (12.8) is used for X^2.

Using the row and column totals for the data in Example 12.2, the following table of expected frequencies is obtained. An explanation of how to get these is given in Example 12.1.

	District						
	A	B	C	D	E	F	Total
Nose	11/12	11/12	11/24	11/12	55/8	11/12	11
Throat	1/3	1/3	1/6	1/3	5/2	1/3	4
Ears	3/4	3/4	3/8	3/4	45/8	3/4	9
Total	2	2	1	2	15	2	24

Is there evidence of association between the location of the rash and the district?
Use of (12.8) gives

$$X^2 = [1^2/(11/12) + 1^2/(11/12) + \cdots + 7^2/(45/8) + 1^2/(3/4)] - 24 = 14.96.$$

With 10 degrees of freedom, chi-squared tables show that significance at the 5% level requires $X^2 \geq 18.31$. In fact, $\Pr(\chi^2 \geq 14.96) = 0.1335$. As a result, there is little evidence against the null hypothesis that there is no association between the location of rash and district. Standard statistical packages include a program for tests based on X^2 using the asymptotic result. Using the ANSM5 package in R and the command `chisqtest.ANSM(ch12$infection.site, ch12$district)` gives a Monte Carlo $P=0.1094$. This not only shows that the Fisher exact test and the Pearson chi-squared tests are not equivalent but suggests that the tests may differ in power.

For 2×2 tables, the Fisher exact test and the exact tests based on X^2 are often equivalent, though the asymptotic test statistics generally differ. It is customary to use the latter in asymptotic tests, and when doing so, a better approximation is obtained by using a continuity correction commonly known in this context as Yates' correction (Yates, 1934). With this correction, the computation is

$$X^2 = \sum_{i,j} \left[\frac{(|n_{ij} - m_{ij}| - 1/2)^2}{m_{ij}} \right],$$

i.e., the quantity 1/2 is subtracted from the magnitude of each difference in the numerator before squaring.

The correction is used only for 2×2 tables, with an alternative form useful for computation being

$$X^2 = \frac{N[(|n_{11}n_{22} - n_{12}n_{21}|) - N/2]^2}{n_{1+}n_{+1}n_{2+}n_{+2}}. \tag{12.9}$$

Omitting the term $N/2$ gives the X^2 value without continuity correction.

Example 12.4

A new drug (Y) is only available in a pilot experiment for 6 patients who are chosen at random from a total of 50 patients; the remaining 44 receive the standard drug (Z). Of those receiving drug Y, only one shows a side-effect, while 38 of those receiving drug Z do. Test the hypothesis of no association between side-effects and the drug administered.

Using the information above, the following 2×2 table is formed:

	Side-Effect	None	Total
Drug Y	1	5	6
Drug Z	38	6	44
Total	39	11	50

Equation (12.9) is used to compute X^2. The chi-squared test with 1 degree of freedom is then applied.

Substituting in (12.9) gives

$$X^2 = \frac{50\left[\left|(1\times 6 - 5\times 38)\right| - 25\right]^2}{6\times 39\times 44\times 11} = 11.16.$$

Tables, or software for the chi-squared distribution, indicate $P = \Pr(\chi^2 \geq 11.16) = 0.0008$. Hence, there is strong evidence against H_0: *no association*.

The permutation X^2 test indicates $P = 0.0012$. Without Yates' correction $X^2 = 14.95$, which overestimates the evidence against H_0.

Most general statistical packages include a program for the Pearson chi-squared test for $r\times c$ tables, often using Yates' correction or providing it as an option for 2×2 tables. Output from the ANSM5 package in R can be obtained using `chisqtest.ANSM(ch12$drugYZ, ch12$side.effect)` with the additional argument `cont.corr = FALSE` being used if Yates' correction is not wanted.

12.2.3 The Likelihood Ratio Test

An alternative statistic for a test of association is the likelihood ratio

$$G^2 = 2\sum_{i,j} n_{ij} \log(n_{ij}/m_{ij}) \tag{12.10}$$

where log is the natural logarithm, i.e., to the base e. Asymptotically, G^2 has a chi-squared distribution with $(r-1)(c-1)$ degrees of freedom if there is no association. For the data in Example 12.2, $G^2 = 17.33$. Although this does not indicate significance at a conventional 5% level, it is a better approximation to the Fisher exact test than that given by X^2. The test has some optimal properties for the overall multinomial sampling model, and convergence to the chi-squared distribution under H_0 is often faster than that of X^2.

An exact permutation theory test may be based on G^2 and, except for 2×2 tables, the results will generally differ from those for exact tests based on F^2 or X^2. For the data in Example 12.2, the test based on G^2 gives a Monte Carlo $P = 0.048$.

Output from the ANSM5 package in R can be obtained using `lik.ratio(ch12$infection.site, ch12$district)`.

In each of the three tests for the data in Example 12.2, the asymptotic test is an unsatisfactory approximation. Although the three exact permutation distribution tests are not equivalent, two give strong indications of evidence that there is an association, and all give lower P-values than those for asymptotic theory. The sample has a preponderance of sparse cell entries. Although such situations do arise in practice, in many contingency tables very few cells have low expected values under H_0.

There is a variety of conflicting advice in the literature about when it is safe to use asymptotic results with G^2 or X^2. Generally speaking, larger values of r and c reduce the need to worry about a small proportion of sparse cell counts. Programs such as the ANSM5 package in R that provide exact Monte Carlo and asymptotic P-values so that these may be compared are invaluable for those who frequently meet contingency tables with sparse data.

Since the tests in Section 12.2 apply to nominal categories, reordering rows or columns does not affect the value of the test statistic. As already indicated, for ordered categories there are more appropriate tests that take account of any natural ordering in rows and columns.

SUMMARY POINTS:

- For $r \times c$ tables with nominal categories, exact tests for association are commonly based on the Fisher exact test (Section 12.2.1).
- Asymptotic tests are usually based on the Pearson chi-squared statistic, X^2 (Section 12.2.2).
- A continuity correction devised by Yates can be used for small samples (Section 12.2.2).
- The likelihood ratio statistic, G^2, is also used for asymptotic tests (Section 12.2.3).
- Exact permutation tests may also be based on X^2 and G^2. While the permutation distributions differ from that for the Fisher exact test (except in 2×2 tables), conclusions for all three approaches are often broadly in line (Section 12.2.3).

12.3 Ordered Categorical Data

The case of nominal row explanatory categories and ordered column response categories is considered first; then ordered row explanatory and ordered column response classifications, or both ordered row and ordered column response categories.

12.3.1 Nominal Row and Ordered Column Categories

In Table 12.1, the explanatory row categories, type of drug, are nominal but column responses are ordinal. Here the Wilcoxon–Mann–Whitney (WMW) test is appropriate. However, the data are heavily tied, so, e.g., the 65 responses tied as *none* are allocated the mid-rank 33, the 16 responses tied as *slight* are given the mid-rank of ranks 66 to 81, i.e., 73.5, and so on. Using the Mann–Whitney formulation obviates the need to specify mid-ranks. The procedure becomes a count of the number of recipients of drug B showing the same or more severe side-effects than that for each recipient of drug A, scoring an equality as 0.5 and an excess as 1. Thus, the 23 responses *none* for drug A each give rise to 42 ties, scoring 21 and 8+4 higher ranks, contributing a total score of $23(0.5 \times 42+8+4)=759$ to the Mann–Whitney U_m.

Example 12.5

For the data in Table 12.1, compute the Mann–Whitney statistic and test for evidence of a difference in levels of side-effects.

Either U_m or U_n are computed in the way described above, which was also illustrated in Example 6.5. Since the sample is large, an asymptotic result might be used, but if a suitable computer program is available with so many ties, an exact test is preferable.

Choosing U_m the calculation is

$$U_m = 23(0.5 \times 42 + 8 + 4) + 8(0.5 \times 8 + 4) + 9(0.5 \times 4) = 841.$$

Using (6.2) (i.e., ignoring ties) leads to $Z=2.63$, whereas the appropriate modification of (6.6) gives $Z=3.118$. Using the ANSM5 package in R, the test gives a two-tail Monte Carlo $P=0.0019$, equal to that using $Z=3.118$ at this number of decimal places. This shows that with sample sizes 45 and 54, having allowed for ties, an asymptotic result should be reasonable, and this indicates strong evidence against H_0 since $P<0.01$. This is confirmed by the Monte Carlo $P=0.0019$.

Output from the ANSM5 package in R can be obtained using `wilcoxon.mann.whitney(ch12$side.effect.level[ch12$drugAB == "Drug A"], ch12$side.effect.level[ch12$drugAB == "Drug B"], do.asymp = TRUE)`.

If ordering is ignored and the Fisher exact test is used, this gives an exact $P=0.0127$, while for the asymptotic Pearson chi-squared test $P=0.0193$. These results are in line with the general rule of thumb that power is gained by taking account of ordering. Nevertheless, for some configurations, marginally smaller P-values may be observed when ordering is ignored.

Output from the ANSM5 package in R can be obtained using `fishertest.ANSM(ch12$drugAB, ch12$side.effect.level)`.

Since asymptotic results appear reasonable here, any statistical package for the WMW test that uses mid-ranks and gives the significance level based on the asymptotic theory that takes account of ties should be satisfactory.

A program such as the one in the ANSM5 package in R does, however, give additional confidence in that it compares not only exact and asymptotic results for the WMW test but also allows the Fisher exact test or a Pearson chi-squared test (exact or asymptotic) to be performed and results compared.

In Section 6.2.2, the Fisher exact test was recommended when using a median test to compare treatments. Since the categories were ordered, this raises the question as to why the WMW test was not recommended given a program for exact permutation distribution tests. It turns out that it does not matter which is used for 2×2 tables, for the exact tests are then equivalent, although the asymptotic forms of the two are not quite the same.

The position is more complicated for the median test in Section 7.2.5, where there are more than two samples. This is discussed in Section 12.3.2.

Nearly all tests for location in earlier chapters can be formulated as tests involving contingency tables with fixed marginal totals. This is a key to the efficient computation of exact permutation distributions and is how many computer programs handle data for these tests. An example is given in Table 12.2 for a data set encountered earlier. In Example 2.3, a permutation test was developed for what is a specific example of the WMW test. In that example, there were sets of 2 and 10 sailors, receiving citrus fruits and not receiving citrus fruits, respectively, ranked 1 to 12. Those who received citrus fruits were ranked 1 and 2. In the contingency table format, the citrus fruits and no citrus fruits categories are regarded as the row classifications and the ordered ranks 1 to 12 as column (response) classifications. The number of items allocated to each row and column classification is counted; e.g., this count will be 0 in row 1, column 4 because in Example 2.3, no sailor given citrus fruits was ranked 4. It would be 1 in row 2, column

5 had a sailor given citrus fruits been ranked 5. There were no ties in Example 2.3, so had the ranks 1, 5 been observed for the two sailors given citrus fruits (implying those receiving no citrus fruits were ranked 2, 3, 4, 6, 7, 8, 9, 10, 11, 12), a contingency table may be constructed as

	Rank											
	1	**2**	**3**	**4**	**5**	**6**	**7**	**8**	**9**	**10**	**11**	**12**
Citrus Fruits	1	0	0	0	1	0	0	0	0	0	0	0
No Citrus Fruits	0	1	1	1	0	1	1	1	1	1	1	1

In this table, the row totals are 2, 10 (the sample sizes). In this no-tie situation, each column total is 1. In the Wilcoxon formulation of the WMW test, the statistic is $S_i = \sum_j jn_{ij}$, where either $i=1$ or $i=2$. In this context, j is a *column score* that here is the rank, and the n_{ij} take the value 0 or 1.

The permutation distribution in Example 2.3 is obtained by permuting all possible cell entries in the above table consistent with the fixed row and column totals and calculating S_i for each permutation. Rank is only one possible column score. The score j in column j may be replaced by other appropriate ordered scores s_j. One can then obtain in the form $S_i = \sum_j s_j n_{ij}$, where $i=1$ or $i=2$, various other statistics discussed in Chapter 6 for testing location. For example, the s_j may be the van der Waerden scores or the ordered raw data, the latter giving rise to the raw scores Pitman permutation test. Graubard and Korn (1987) discuss choices of column scores in $2 \times c$ tables for many well-known tests.

In the case of ties in the WMW test, cell entries correspond to the number of ties in the relevant sample (row) at any given mid-rank, and the column scores are the appropriate mid-rank values. Each column total is the total number of ties assigned to that mid-rank score. Permutation is over all configurations, giving the fixed row and column totals.

In a no-tie situation, under the null hypothesis, a good mix of ones and zeros is expected over the cells of the $2 \times c$ table. A concentration of ones at opposite corners of the table indicates association and possible rejection of H_0. For example, a configuration with the marginal totals above that indicates rejection is that in Example 2.3

	Rank											
	1	**2**	**3**	**4**	**5**	**6**	**7**	**8**	**9**	**10**	**11**	**12**
Citrus Fruits	1	1	0	0	0	0	0	0	0	0	0	0
No Citrus Fruits	0	0	1	1	1	1	1	1	1	1	1	1

The approach used here is examined in more detail in Section 13.2.6.

When there are more than two nominal explanatory categories, each corresponding to a row and the column responses are ordinal, the obvious extension from the WMW test is to the Kruskal–Wallis test (with ties if appropriate). Unless some samples are very small or there is a concentration of many entries in one or two columns (heavy tying), asymptotic results are usually satisfactory. Except for fairly small samples, many computer programs cannot cope with the exact permutation distribution for the Kruskal–Wallis test, but the Monte Carlo approximation available in the ANSM5 package in R works well in practice.

Example 12.6

Data from a study of the association between the cause of car tyre failure and recorded failure time are given below, courtesy of Timothy P. Davis. The failure times are grouped by cause:

- *A* – Open joint on the inner lining
- *B* – Rubber chunking on the shoulder
- *C* – Loose casing low on the sidewall
- *D* – Cracking on the sidewall
- *E* – Other causes

Depending on one's viewpoint, the row categories *A, B, C, D, E* might be regarded as either nominal explanatory or as nominal response categories. In the sense that they are different types of failure, they represent different responses. On the other hand, each tyre failing for a particular reason might be regarded as one of a sample of tyres doomed to fail in that way. Then, if the focus is on times of failure at which different faults occur, the various faults might be regarded as explanatory variables.

For illustrative purposes in this example, the latter stance is taken, although the analysis would still be appropriate if the faults themselves were regarded as distinct response categories.

The column classification is time to failure recorded to the nearest hour, grouped into intervals <100, 100–199, 200–299, 300+ hours.

	Time to Failure (hours)			
Cause	**<100**	**100–199**	**200–299**	**300+**
A	2	6	9	0
B	0	1	7	0
C	4	10	6	0
D	1	6	5	0
E	0	8	10	2

A Kruskal–Wallis test with five samples is performed. Mid-ranks are calculated regarding all units in the same column as ties. Thus, the seven ties in the column labelled <100 are all given the mid-rank 4, and so on. The Kruskal–Wallis statistic is calculated as in Section 7.2.1, and the statistic has asymptotically a chi-squared distribution with $r-1=4$ degrees of freedom. Computation is laborious without a computer program.

The computation outlined above gives the value 10.44 for the Kruskal–Wallis statistic, and tables or software indicate that for 4 degrees of freedom $P=\Pr(\chi^2 \geq 10.44)=0.0336$. Hence there is reasonably strong evidence of association between rows and columns, implying that different types of fault do not all have the same median times of occurrence.

Despite heavy tying, in the light of findings in earlier chapters, there is some justification for assuming that the asymptotic result is reasonable here. This was confirmed by the ANSM5 package in R's Monte Carlo procedure for estimating the exact *P*-value. This gave an estimated $P=0.0286$. This suggests that the asymptotic result is conservative, falling just above the upper confidence limit for the permutation distribution result.

Output from the ANSM5 package in R can be obtained using `kruskal.wallis(ch12$time.to.failure, ch12$cause)`.

Many packages compute the Kruskal–Wallis statistic, so the asymptotic test is readily available. The ANSM5 package will compute exact *P*-values with small numbers of rows and columns but may prove very time consuming for examples with a moderate number of ties. The Monte Carlo method is relatively fast. The estimate above $P=0.0286$ was based on 100,000 simulations.

In a no-tie situation with r samples and a total of N observations, the Kruskal–Wallis test may be formulated in contingency table terms as an $r \times N$ table with rows corresponding to samples and all cell entries 1 or 0. The columns are ordered by ranks, and the observation of rank j is recorded as 1 in the cell in the jth column in the row corresponding to the sample in which it occurs. All other entries in that column are zero in the no-tie case. The row totals are the sample sizes n_i, and each column total is 1. Ranks are the scores used to generate the exact permutation distribution. The statistic is a function of the rank sums for each row. Permutation is over all cell entries in the $r \times N$ table, consistent with row and column totals.

12.3.2 Ordered Row and Column Categories

The case where row categories are ordered explanatory categories and the columns are responses is considered first. One appropriate test of no association is then the Jonckheere–Terpstra test, with each row corresponding to one of the ordered samples. Asymptotic results may be reasonable if allowance is made for ties. The need to allow for ties is illustrated here in a context not uncommon in drug testing, where side-effects may not occur often but may be serious if they do.

Example 12.7

Side-effects (if any) experienced by patients at increasing dose levels of a drug are classified as none, mild, moderate or severe. Do the following data indicate that the incidence or severity of side-effects increases with dose level?

	Side-Effects			
Dose (mg)	**None**	**Slight**	**Moderate**	**Severe**
100	50	0	1	0
200	60	1	0	0
300	40	1	1	0
400	30	1	1	2

The Jonckheere–Terpstra statistic is relatively easy to compute as an extension of the procedure used for the Mann–Whitney formulation of the WMW test in Example 12.5, scoring the relevant ties in any column as 1/2. Those in the first column make a substantial contribution to the total.

Without a computer program, it is laborious but not difficult to compute the relevant statistic, denoted by U in Example 7.3. It is

$$U = 30\left[(40+60+50)/2+4\right]+40\left[(60+50)/2+2\right]+60(50/2+1)+3+1.5+1+1=6,217.$$

The asymptotic test in Section 7.2.4, with no correction for ties, gives $\Pr(U<6{,}217)=0.2288$. The two-tail probability is $P=0.4576$, so there is no evidence of association.

The hypothesis H_0: *side-effect levels are independent of dose* is acceptable.

Output from the ANSM5 package in R can be obtained using `jonckheere.terpstra(ch12$dose.side.effect, ch12$dose, alternative = "less", do.asymp = TRUE, do.exact = FALSE, do.asymp.ties.adjust = FALSE)`.

The null hypothesis would not be rejected here, even in a one-tailed test, which might be justified on the general grounds that increasing dose levels of a drug are more likely to produce side-effects if indeed there are any.

A clinician is unlikely to be satisfied with this finding. Side-effects are not common, but there is to a clinician (and even to the layperson) an indication that these are more likely and more severe at the highest dose level. Either the evidence is too slender to reach a firm conclusion or the chosen test is inappropriate. In particular, is it reasonable to justify an asymptotic result that does not allow for ties when there is so high an incidence of these?

The maximum safe dose of a drug is often determined by giving the drug to independent groups of healthy volunteers in increasing doses, with the first group receiving the smallest dose. Once a serious side-effect has been observed, the study is stopped; to continue the investigation with healthy volunteers would be unethical. Higher doses of the drug may have to be used with genuine patients in order to achieve an improvement. The potential gains would then have to be balanced against the likely side-effects.

Some general statistical packages include the asymptotic Jonckheere–Terpstra test without allowance for ties. Correcting for ties may have a dramatic effect as shown below. The ANSM5 package in R includes an asymptotic test that allows for ties.

Example 12.8

Use the exact Jonckheere–Terpstra test on the data in Example 12.7 to determine if there is acceptable evidence that side-effects become more common as the dose increases.

The permutation test requires computation of probabilities for the Jonckheere–Terpstra statistic for all permutations of the contingency table in Example 12.7 subject to the same marginal totals and with the same or a lesser probability than that observed.

The ANSM5 package in R provides a program for these computations and also, with a data set of this size, for Monte Carlo calculations which indicate $\Pr(U<6{,}217)$ or equivalently $\Pr(U>6{,}834)=0.0176$, providing clear evidence against the null hypothesis.

Output from the ANSM5 package in R can be obtained using `jonckheere.terpstra(ch12$dose.side.effect, ch12$dose, alternative = "less")`.

This is an illustration of a situation where an exact permutation test leads to a markedly different conclusion to a commonly used asymptotic test despite a large sample size.

In Section 7.2.4, an amended formula for $\text{Var}(U)$ which allows for ties was mentioned. The formula is adapted from one given in Kendall and Gibbons (1990, p. 66) and also in Lehmann (1975, 2006). The notation here matches that in Table 12.5.

The calculation required is

$$\text{Var}(U) = U_1/d_1 + U_2/d_2 + U_3/d_3 \tag{12.11}$$

where

$$U_1 = N(N-1)(2N+5) - \sum_i n_{i+}(n_{i+}-1)(2n_{i+}+5) - \sum_j n_{+j}(n_{+j}-1)(2n_{+j}+5)$$

$$U_2 = \left[\sum_i n_{i+}(n_{i+}-1)(n_{i+}-2)\right]\left[\sum_j n_{+j}(n_{+j}-1)(n_{+j}-2)\right]$$

$$U_3 = \left[\sum_i n_{i+}(n_{i+}-1)\right]\left[\sum_j n_{+j}(n_{+j}-1)\right]$$

$$d_1 = 72,\ d_2 = 36N(N-1)(N-2),\ d_3 = 8N(N-1).$$

When there are no ties, all $n_{+j}=1$ and $U_2=U_3=0$, and Var(U) reduces, after algebraic manipulation, to the form given in Section 7.2.4. In Example 12.8, the relevant values are $N=188$, $n_{1+}=51$, $n_{2+}=61$, $n_{3+}=42$, $n_{4+}=34$, $n_{+1}=180$, $n_{+2}=n_{+3}=3$, $n_{+4}=2$. Substituting in (12.11) reduces Var(U) from 172,522.25 (ignoring ties) to 21,078.33. The corresponding $Z=-2.125$, and standard normal distribution tables give $\Pr(Z<-2.125)=0.0168$, in close agreement with the Monte Carlo result.

Output from the ANSM5 package in R can be obtained using `jonckheere.terpstra(ch12$dose.side.effect, ch12$dose, alternative = "less", do.asymp = TRUE)`.

In clinical trials, situations like that in Examples 12.7 and 12.8 are common. They arise not only for relatively rare side-effects of drugs but also in other situations where some responses are rare, e.g., the incidence of lung cancer in populations exposed to adverse environmental factors. Several thousand people may be observed in each explanatory category, yet only a few develop lung cancer.

In Section 7.2.5, the Fisher exact test was recommended for the median test with $r>2$ samples. This ignores the fact that the responses $<M$ and $\geq M$ are ordered. If the sample classifications are nominal, an alternative is the Kruskal–Wallis test. If the samples are ordered so that the natural alternative hypothesis is some systematic increase or decrease in the median, one may do better using the Jonckheere–Terpstra procedure. With only two columns (ordered response categories), there will often be little difference between results using the Fisher test or Kruskal–Wallis or (if appropriate) the Jonckheere–Terpstra procedure. If only asymptotic results are used in this last test, a correction for ties is important. In the special case of only two ordered response categories, it is shown below that the Jonckheere–Terpstra test is equivalent to the WMW test.

Example 12.9

Exercise 7.6 provided data from Chris Theobald for blood platelet counts and spleen sizes in 40 patients. The investigators were interested in whether the proportion of patients with an abnormally low platelet count increased with increasing spleen size. If a count below 120 is abnormal, the data in Exercise 7.6 give the contingency table below. Does the Jonckheere–Terpstra test indicate an association between spleen size and platelet count?

	Blood Platelet Count	
Spleen Size	Abnormal	Normal
0	0	13
1	2	10
2	5	6
3	3	1

Exercise 7.6 indicated that platelet counts tend to decrease as spleen size increases. Ties were minimal as the actual counts were given. Here ties must be taken into account in an asymptotic test.

Proceeding as in Examples 12.7 and 12.8 gives $U=183$ and $E(U)=287.5$. Using (12.11), the standard deviation of U is $\sqrt{Var(U)} = 30.67$, whence $Z=-3.407$. Tables of the standard normal distribution give

$$\Pr(Z \le -3.407) = \Pr(Z \ge 3.407) = 0.0003.$$

There is strong evidence that the proportion of individuals with abnormal counts increases as spleen size increases.

Computing Z by the formula in Section 7.2.4 ignoring ties gives $Z=-2.551$, indicating a one-tail $P=0.0054$ (compared with $P=0.0003$ obtained above). A Monte Carlo test using the ANSM5 package in R with these data gave $P=0.0002$ for a one-tailed test, in close agreement with the asymptotic result taking ties into account.

Output from the ANSM5 package in R can be obtained using `jonckheere.terpstra(ch12$platelet.count, ch12$spleen.size, alternative = "greater", do.asymp = TRUE)`.

In general, ties have a greater effect on Var(U) when they occur *between* samples (in columns) than if they occur largely *within* one sample (in a row).

Regarding counts of less than 120 as abnormal is arbitrary. Using an extended data set, Bassendine et al. (1985) considered counts below 100 as abnormal. With such arbitrary decisions about categories, one choice may lead to a result significant at a formal level such as $\alpha=0.05$ or $\alpha=0.01$ while another does not. In Section 12.4.4, a similar problem arising with grouped continuous data is discussed.

Since there are only two columns, they remain ordered if they are interchanged.

The *abnormal* and *normal* cases could be viewed as samples from two populations, with the spleen sizes as rankings for those populations. The WMW test could then be applied as in Example 12.5. Doing so gives exactly the same results as with the Jonckheere–Terpstra procedure.

The expression for Var(U) in (12.11) is unaltered if rows and columns are interchanged in an $r \times c$ table. However, the statistic U and the mean E(U) are affected by this change. By writing U_1 for the Jonckheere–Terpstra statistic for the original $r \times c$ table and U_2 for the corresponding statistic when rows and columns are interchanged, it can be shown that

$$U_2 = U_1 + \left(\sum_i n_{i+} - \sum_j n_{+j}\right)/4 \text{ and } E(U_2) = E(U_1) + \left(\sum_i n_{i+} - \sum_j n_{+j}\right)/4.$$

The asymptotic tests using U_1 and U_2 are identical. Although U_1 and U_2 differ by the constant value given above, their permutation distributions are otherwise identical, i.e., for all k

$$\Pr(U_1 = k) = \Pr\left[U_2 = k + \left(\sum_i n_{i+} - \sum_j n_{+j}\right)/4\right]$$

This equivalence may seem surprising, as the Jonckheere–Terpstra test was introduced as a test for a monotone trend between samples. However, if there are no ties in p samples with $n_1, n_2, \ldots, n_p$ observations, respectively, the problem may be stated in a contingency table format with p rows and $N = \sum_i n_i$ columns. In each column, all cells except one have zero entries. The remaining cell has an entry unity. The entries in row i are constrained so that the sum of the cell entries that are unity (all others being zero) is n_i. If the rows and columns are interchanged, the equivalent test problem is one in which there are N samples each consisting of one observation, and these take only p different values of which there are n_1 tied at the lowest value, n_2 tied at the next lowest, and so on.

An extreme case is that in which there are N samples, each of one observation and none of these is tied. This situation may be represented by an $N \times N$ table in which all column totals and all row totals are equal to 1. This implies that in any row or column, there is one cell value of unity and all others are zero. In this case, $U_2 = U_1$. As pointed out in Section 10.1.4, the Jonckheere–Terpstra test procedure is here equivalent to that for Kendall's tau; U_1 or U_2 are the numbers of concordances in this no-tie situation, and this is sufficient to determine Kendall's t_k.

Any situation with ties, and where the concept of linear rank association implicit in Kendall's tau is relevant, may be looked upon as generating an $r \times c$ matrix which shrinks an $N \times N$ matrix of zeros and ones by super-imposing rows with identical x values and super-imposing columns with identical y. In particular, if the ith sample has n_i observations, one shrinks n_i adjacent rows to a single row by adding all column entries in those n_i adjacent rows. The x variable values in this situation are effectively labels that distinguish the (ordered) samples. It has been pointed out that in the no-tie case, the Jonckheere–Terpstra statistic is equivalent to the number of concordances in Kendall's t_k. With ties, the two statistics use slightly different counting systems. The Kendall statistic counts concordances as unity, ties as zero and discordances as –1. For the Jonckheere–Terpstra statistic, concordances count as 1, ties as 1/2 and discordances as zero. With appropriate adjustments to certain formulae, the two counting systems are interchangeable and, used correctly, lead to the same conclusions.

If rows and columns are both response categories, there may be interest in whether there is a patterned association (akin to a correlation) between row and column classifications in that high responses in one classification tend to be associated with high responses in the other, or high responses in one are associated with low responses in the other. Again the Jonckheere–Terpstra test is relevant.

Several statistics are used as measures of association, but none is completely satisfactory in all circumstances. For some of those suggested, an exact Jonckheere–Terpstra test is appropriate to test for significant association in the form of a monotonic trend. However, for nonmonotonic associations, such tests may fail to detect patterns of association. For example, in an ordered categories contingency table with cell entries

$$\begin{matrix} 0 & 17 & 0 \\ 5 & 0 & 5 \\ 3 & 0 & 3 \end{matrix} \qquad (12.12)$$

there is a clear – but obviously not monotonic – association. It can be verified that the asymptotic Jonckheere–Terpstra statistic Z takes the value 0, giving $P=0.50$ for a one-tailed test. Apart from a small discontinuity effect, the exact test confirms this probability.

A desirable property of a measure of monotonic association, akin to that of a correlation coefficient, is that it takes the value 1 when there is complete positive association, the value zero if there is no rank association, and the value –1 if there is inverse association, e.g., high category row classifications are associated with low category column classifications and vice versa. An appropriate test for linear rank association using such statistics is often the Jonckheere–Terpstra test, but to measure the degree of association, there is a case for having a statistic that behaves like a correlation coefficient. One such coefficient is discussed here. Kendall and Gibbons (1990, Chapter 3) and Siegel and Castellan (1988, Chapter 9) describe this and other measures of association in considerable detail. These and other authors give guidance on relevant asymptotic tests that allow for the tied nature of the data.

The *Goodman–Kruskal gamma statistic* is usually denoted by G. The number of concordances and discordances between row and column classifications is counted as for Kendall's t_k, but no allowance is made for ties. When both row and column classifications are ordered, for each count in cell (i, j), there is concordance between that count and the count in any cell below and to the right. Thus, if N_{ij}^{+} denotes the sum of all counts below and to the right of cell (i, j), which itself has count n_{ij}, the total number of concordances is

$$C=\sum_{i,j} n_{ij}N_{ij}^{+},\; 1\le i\le r-1,\; 1\le j\le c-1.$$

This resembles the count for the Jonckheere–Terpstra test, except that for counts in cell (i, j) ties in column j are ignored. Each of these is counted as 1/2 in Jonckheere–Terpstra. Similarly, denoting the sum of all elements to the left of and below the cell (i, j) by N_{ij}^{-}, the total number of discordances is

$$D=\sum_{i,j} n_{ij}N_{ij}^{-},\; 1\le i\le r-1,\; 2\le j\le c.$$

The Goodman–Kruskal statistic is $G=(C-D)/(C+D)$. Hence, if $D=0$, then $G=1$. If $r\ge c$, then $G=1$ if all entries in any column except the last are zero, unless $i=j$; in the last column they must be zero if $i<c$. If $c\ge r$, $G=1$ if all entries in any row except the last are zero, unless $i=j$. In the last row, they must be zero if $j<r$. If $r=c$, $G=1$ if all entries for which $i\ne j$ are zero.

One can obtain analogous conditions for $G=-1$. When there is a good scatter of counts over all cells, C and D will not be very different and G will be close to zero. Near-zero values of G are also possible if the association is not monotonic. It can be verified that $C=D$, hence $G=0$, in the highly patterned contingency table in (12.12).

Like Kendall's tau (as distinct from Kendall's tau-b), G takes no account of ties in both x and y, yet unlike Kendall's tau, it may equal 1 even when there is heavy tying because the denominator is $C+D$. This falls well short of the denominator in Kendall's tau for the no-tie case. In Section 10.1.4 Kendall's t_b is used in preference to t_k when there are ties. This is a compromise between ignoring them (as in t_k) and deleting their effect if in the same row or column as is done in G. The calculation of t_b for contingency tables is straightforward and follows the method outlined in Section 10.1.4.

Other measures of association include Somers' D, proposed by Somers (1962). It is closely related to the Goodman–Kruskal statistic and to Kendall's tau. This and some other appropriate procedures for measuring association are described by Siegel and Castellan (1988, Chapter 9). Other approaches to exploring association in contingency tables are described in Chapter 13.

SUMMARY POINTS:

- For tables with the explanatory variable represented as two nominal rows and the outcome variable as ordered columns, the Mann–Whitney formulation of the Wilcoxon–Mann–Whitney test is an appropriate test of association (Section 12.3.1).
- If the explanatory variable has multiple nominal rows and the outcome variable has ordered columns, the Kruskal–Wallis test adjusted for ties can be used (Section 12.3.1).
- If the rows represent ordered explanatory categories and the columns are ordered, the Jonckheere–Terpstra test is appropriate (Section 12.3.2).
- The Goodman–Kruskal gamma statistic, G, is closely related to Kendall's tau as a measure of association (Section 12.3.2).

12.4 Goodness-of-Fit Tests for Discrete Data

12.4.1 Introduction

The Pearson chi-squared test and the likelihood ratio test (Sections 12.2.2 and 12.2.3) may be looked upon as tests of goodness-of-fit of data in a contingency table to a model of independence (i.e., no association) between row and column classifications. The asymptotic chi-squared test is also widely used as a goodness-of-fit test of data to any discrete distribution. This procedure, first published by Pearson (1900), is covered in most standard courses in statistics. Only a brief resume is given here, drawing attention to a few points that are sometimes not covered specifically in introductory treatments.

Tests of goodness-of-fit to any discrete distribution (often appropriate for counts) are usually based on the Pearson chi-squared statistic, X^2, in (12.7).

Expected values are calculated for a hypothesized distribution that may be binomial, negative binomial, Poisson, uniform or some other discrete distribution. Sometimes one or more parameters must be estimated from the data.

If there are r counts or cells, and no parameters to be estimated, the test will have $r-1$ degrees of freedom. If a parameter (e.g., p for the binomial or λ for the Poisson distribution) is estimated from the data, one further degree of freedom is lost for each parameter estimated. While tests of the Kolmogorov–Smirnov type are sometimes also applied for discrete distributions, generally the test criteria are no longer exact and the tests are often inefficient.

12.4.2 Goodness-of-Fit with No Parameter to Be Estimated

The most straightforward goodness-of-fit situation, where no parameter is involved, is illustrated by the example below.

Example 12.10

It has been suggested that recorded ages at death can be influenced by misstatement of age, since elderly people may be imprecise about giving their age, tending to round it; e.g., if they are in the mid-seventies, they tend to say they are 75 if anywhere between about 73 and 77. Similarly, a stated age of 80 may correspond to a true age a year or two above or below. If this is the case, the final digits in recorded ages at death would not be uniformly distributed. Table 12.7 gives the final digits at age of death for the 117 males recorded at the Badenscallie burial ground (see Appendix 1). Is the null hypothesis H_0: *any digit is equally likely* acceptable?

For 117 deaths, the expected number of occurrences of each of the 10 digits, if all are equally likely, is 11.7. Since the denominator in X^2 is always the expected number, $m=11.7$, one may calculate X^2 by taking the differences between the observed numbers n_i and 11.7, squaring these, adding and finally dividing the total by 11.7.

The differences $n_i - m$ are, respectively,

$$-4.7,\ -0.7,\ 5.3,\ 7.3,\ -2.7,\ 1.3,\ -2.7,\ -0.7,\ 1.3,\ -3.7.$$

These differences sum to zero – a well-known property of deviations from the mean. The sum of their squares is 136.1. Division by the expected number $m=11.7$ gives $X^2=136.1/11.7=11.63$. This is well below the value of 16.92 for χ^2 that corresponds to $P=0.05$ when there are $10-1=9$ degrees of freedom. Indeed, it corresponds to $P=0.235$. The ANSM5 package in R gives a Monte Carlo $P=0.239$. Hence there is no real evidence against H_0.

Output from the ANSM5 package in R can be obtained using `chisqtest.ANSM(table(ch12$last.digits), do.asymp = TRUE)`.

In posing the problem, a possible build-up of digits 0, 1 and perhaps also 5 was suggested. The observed variation in digit frequency is not in line with such a pattern. Any build-up is around 2 and 3. As the overall result is not significant, it is important not to read too much into this.

In this data set, ages at death ranged from less than 1 to 96 years. In particular, there is some evidence of an increased risk of death for very young children. In all, 41 of the 117 recorded ages at death were less than 60 years. The influence of elderly people is therefore diluted. It might also be reasoned that a formal record of age at death, as in these data, is unlikely to be incorrect.

TABLE 12.7

Recorded Last Digit of Age at Death, Badenscallie Males

Digit	0	1	2	3	4	5	6	7	8	9
Freq	7	11	17	19	9	13	9	11	13	8

Misstatement of age occurs across a wide range of ages for a variety of reasons, as discussed in detail by Cox (1978).

12.4.3 Goodness-of-Fit with an Estimated Parameter

Data are often encountered that might belong to a particular distribution, such as the Poisson distribution with an unknown mean. The statistic X^2 may be used to test this, although, for the Poisson and some other distributions, alternative parametric tests may be more powerful.

For the asymptotic test using X^2, a difficulty arises if some expected numbers are small. Traditional advice is to group such cells to give a group expected number close to 5 (but this is only a guide). There is a corresponding reduction in degrees of freedom.

Example 12.11

A factory employs 220 people. The numbers experiencing 0, 1, 2, 3, … accidents in a given year are recorded

Number of Accidents	0	1	2	3	4	5	≥6
Number of People	181	9	4	10	7	4	5

Are these data consistent with a Poisson distribution?

The maximum likelihood estimate, $\hat{\lambda}$, of the Poisson parameter λ, is the mean number of accidents per person. Expected numbers having r accidents are $E(X = r) = 220\hat{\lambda}^r \exp\left(-\hat{\lambda}\right)/r!$. This follows because, for a Poisson distribution with parameter λ, $\Pr(X{=}r){=}\lambda^r \exp(-\lambda)/r!$.

To obtain the mean number of accidents per person, each number of accidents is multiplied by the number of people having that number of accidents. These products are added together, and the sum is divided by 220. This gives

$$\lambda = [(0 \times 181) + (1 \times 9) + (2 \times 4) + (3 \times 10) + (4 \times 7) + (5 \times 4) + (6 \times 5)]/220 \approx 0.568.$$

This is approximate because "six or more" is treated as exactly six. In practice, this has little influence on the result, but it is a limitation that should be recognized. To obtain expected numbers, an iterative algorithm may be used, noting that

$$\mathrm{E}\left(X = r + 1\right) = \mathrm{E}\left(X = r\right) \times \hat{\lambda}/(r + 1).$$

The expected numbers are

Number of Accidents	0	1	2	3	4	5	6 or more
Number of People	124.74	70.82	20.12	3.81	0.54	0.06	0

Grouping results for 3 or more accidents gives an associated total expected number of 4.4. The value of X^2 is calculated using (12.8) as

$$X^2 = \left(181^2/124.74\right) + \left(9^2/70.82\right) + \left(4^2/20.12\right) + \left(26^2/4.42\right) - 220 = 197.7.$$

There are 2 degrees of freedom, since there are four cells in the final test and one parameter, λ, is estimated.

It is clear, even if using only conventional chi-squared tables, that there is extremely strong evidence that the data are not consistent with a Poisson distribution since, with 2 degrees of freedom, $\Pr(\chi^2 \geq 13.81) = 0.001$.

Output from the ANSM5 package in R can be obtained using the following commands:

```
lambda <- sum(0:6 * table(ch12$accidents)) / length(ch12$accidents)
poisson.probs <- c(dpois(0:2, lambda), 1 - sum(dpois(0:2, lambda)))
chisqtest.ANSM(table(ch12$accidents.reduced), p = poisson.probs)
```

Accident data seldom follow a Poisson distribution. The number of accidents experience by an individual would follow this distribution if accidents occurred entirely at random, each person had the same probability of experiencing an accident, and having one accident did not alter the probability of that person having another. In practice, a better model is one that allows people a differing degree of accident proneness. The negative binomial distribution assumes that individual proneness has a gamma distribution within the population. This is very flexible and has been shown to give a good fit to the number of episodes of mental illness experienced by adults in a fixed period of several years (Smeeton, 1986). If an individual experiences an accident, the probability of a further accident may be reduced if he or she becomes more careful or increased if concentration or confidence is lowered, making that person more accident-prone. These factors may act differently from person to person or for individuals exposed to different risks. Multiple accidents, which involve more than one individual in a single incident, also affect the distribution of the number of accidents per person in some accident data.

Standard statistical packages include a program for chi-squared goodness-of-fit tests.

12.4.4 Goodness-of-Fit for Grouped Data

The chi-squared test is sometimes applied to test the goodness-of-fit of grouped data from a specified continuous distribution, commonly a normal distribution with specified mean and variance, or with these estimated from the data. This is not recommended unless the grouped data are all that are available and grouping is on some natural basis. If the grouping is arbitrary, one arbitrary grouping may result in the rejection of a hypothesis while some other grouping may not.

A situation where a test for normality based on Pearson's X^2 might be used is that for sales of clothing of various sizes. A large retailer might note the number of sales of ready-made trousers with nominal leg lengths (cm) 76, 78, 80, etc.

The implication is that customers requiring some ideal leg length between 75 and 77 cm will purchase trousers with leg length 76, those requiring leg lengths between 77 and 79 will purchase trousers of length 78, and so on. The sizes are the centre value for each group. To test whether sales at nominal lengths are consistent with a normal distribution with an unspecified mean and variance, these parameters must be estimated from the grouped data and 2 degrees of freedom deducted to allow for this.

Given complete sample data from a continuous distribution, it is better to use goodness-of-fit tests of the Kolmogorov/Lilliefors type as appropriate, or other tests relevant to particular continuous distributions. If grouping is used, and tests are to be based on X^2, Kimber (1987) points out that when both grouped and ungrouped data are available, anomalies may arise unless parameter estimates are based on the grouped data.

SUMMARY POINTS:

- The chi-squared goodness-of-fit test, which uses the Pearson X^2 statistic, is appropriate for examining discrete distributions (Section 12.4.1).
- If no parameter needs to be estimated, the number of degrees of freedom in the test is one less than the number of counts (Section 12.4.1).
- If one or more parameters need to be estimated, for the degrees of freedom, the number of parameters involved is additionally subtracted (Section 12.4.1).
- The uniformity of final digits in a series of numbers can be tested using this method as applied to a uniform distribution (Section 12.4.2).
- For counts within a group of individuals, a Poisson distribution assumes that the event of interest occurs at random, each individual has the same probability of experiencing the event, and the occurrence of the event does not change the probability that the individual will have another (Section 12.4.3).
- The chi-squared goodness-of-fit test can be applied to grouped continuous data (Section 12.4.4).

12.5 Extension of McNemar's Test

There is an asymptotic chi-squared approximation equivalent to the normal approximation to the binomial for McNemar's test (Section 5.2). For the data in Table 5.3, for example, the test was, in effect, whether 9 successes and 14 failures (or vice versa) are consistent with a binomial distribution with $p=0.5$ and $n=9+14=23$. A goodness-of-fit test may be performed using X^2. The expected number of successes or failures is 11.5, and the test statistic is $X^2=(2.5)^2/11.5+(2.5)^2/11.5=1.09$. This is compared with the relevant critical value for chi-squared with 1 degree of freedom. Recalling that the relevant cells in a 2×2 table for the McNemar test are the off-diagonal cells n_{12}, n_{21}, it follows that with these numbers of successes or failures ($n_{12}=9$ and $n_{21}=14$ in Table 5.3), then

$$X^2 = (n_{12} - n_{21})^2 / (n_{12} + n_{21}).$$

In Section 7.4, Cochran's test was introduced for comparing several treatments with binary responses scored as 0 or 1. When $t=2$, it can be shown that Cochran's Q is equivalent to the form given above for X^2. In this sense, the Cochran test is a generalization of McNemar's test.

Another generalization of McNemar's test is from a 2×2 contingency table to an $n \times n$ table in which off-diagonal symmetry is tested. Bowker (1948) proposed a relevant test. The test might be used if there is a new formulation of a drug that it is hoped will reduce side-effects. If a number of patients have been treated with the old formulation and records are available of any side-effects one might, if there are not ethical reasons to preclude this course, now treat each of these patients with the new formulation and note the incidence of side-effects. Table 12.8 shows a possible outcome of such an experiment.

This table shows that each off-diagonal count below the diagonal exceeds that in the symmetrically placed cell above the diagonal, e.g., $n_{31}=4$ is greater than $n_{13}=3$, etc. This gives

TABLE 12.8

Side-Effects with Old and New Formulations of a Drug

Side-Effect Levels – Old Formulation	Side-Effect Levels – New Formulation			
	None	Slight	Severe	Total
None	83	4	3	90
Slight	17	22	5	44
Severe	4	9	11	24
Total	104	35	19	158

the impression that the trend is towards less severe side-effects with the new formulation, although a few who suffered no side-effects with the old formulation do show some with the new formulation.

Bowker proposed a test statistic to determine whether at least one pair of probabilities associated with the symmetrically placed off-diagonal cells differs in an $n \times n$ table. It is a generalization of the X^2 statistic used in McNemar's test and takes the form

$$X^2 = \sum \frac{\left(n_{ij} - n_{ji}\right)^2}{n_{ij} + n_{ji}} \tag{12.13}$$

where the summation is over all i from 1 to $n-1$ and $j>i$. Under the null hypothesis of symmetry, the asymptotic distribution of X^2 in (12.13) is chi-squared with $n(n-1)/2$ degrees of freedom.

Example 12.12

Do the data in Table 12.8 provide evidence that side-effects are less severe with the new formulation of the drug?

The Bowker statistic given by (12.13) is appropriate. Here $n_{12}=4$, $n_{13}=3$, $n_{21}=17$, etc. The null hypothesis is that for all off-diagonal counts in the table, the associated probabilities are such that all $p_{ij}=p_{ji}$. The alternative is that for at least one such pair $p_{ij} \neq p_{ji}$.

Substitution of the relevant values in (12.13) gives

$$X^2 = \left(4-17\right)^2/21 + \left(3-4\right)^2/7 + \left(5-9\right)^2/14 = 9.33.$$

Under the null hypothesis, X^2 has a chi-squared distribution with 3 degrees of freedom, and appropriate tables or software establish that with 3 degrees of freedom $P=\Pr(\chi^2 \geq 9.33)=0.026$. There is evidence of a differing incidence rate for side-effects under the two formulations. Table 12.8 shows that this difference is towards less severe side-effects under the new formulation.

Output from the ANSM5 package in R can be obtained using `bowker(ch12$side.effect.new, ch12$side.effect.old)`.

Marascuilo and McSweeney (1977, Section 7.6) discuss an alternative test based on marginal totals proposed by Stuart (1955, 1957). Agresti (2013, Chapter 10) discusses other generalizations of McNemar's test.

SUMMARY POINTS:

- McNemar's test for off-diagonal symmetry in 2×2 tables may be carried out using a statistic which has an asymptotic chi-squared distribution.
- This may be extended to Bowker's test for off-diagonal symmetry.

12.6 Exercises

12.1 For the data presented in Table 12.3, use the chi-squared test for contingency tables to investigate whether there is an association between the outcomes for bronchitis and otitis media following a prescription of the antibiotic under consideration.

12.2 In an investigation of the factors associated with perseverance in adults learning the Welsh language (Newcombe and Newcombe, 2001), students who had completed a particular course in the Welsh language were asked regarding their use of Welsh and whether they had regular opportunities for practice. Of the 25 who reported having regular opportunities, 22 were using Welsh compared to 13 of the 22 who had less frequent opportunities. Is there an association between the use of the Welsh language and the frequency of opportunity to practice?

12.3 Using the data from a study of retired footballers and dementia reported by Ling et al. (2017), participants have been grouped according to position played (centre forward (CF), centre half (CH), other) and final clinical diagnosis (Alzheimer's disease (AD), Alzheimer's disease and Parkinson's disease (AD/PD), other). Is there evidence for an association between position played and final clinical diagnosis? Is the size of the sample an issue here?

	Position Played			
	CF	CH	Other	Total
AD	0	4	2	6
AD/PD	1	0	2	3
Other	2	0	3	5
Total	3	4	7	14

12.4 Mathie et al. (2018) asked PPI representatives and researchers to indicate their level of satisfaction with the PPI feedback received/given, using the categories very satisfied, fairly satisfied, neither, fairly unsatisfied, very unsatisfied. Is there any evidence that satisfaction with feedback differs between representatives and researchers?

	Satisfaction with PPI Feedback					
	Very Satisfied	Fairly Satisfied	Neither	Fairly Unsatisfied	Very Unsatisfied	Total
Representatives	12	20	7	7	2	48
Researchers	5	19	8	2	0	34
Total	17	39	15	9	2	82

12.5 Prior to an England vs. Scotland football match 80 English, 75 Scottish and 45 Welsh supporters are asked who they think will win. Do the numbers responding each way indicate that the proportions expecting each side to win are influenced by nationality?

	English	Scottish	Welsh
English Win	55	38	26
Scottish Win	25	37	19

12.6 The medical records of a group of patients who had experienced a haemorrhagic stroke (brain haemorrhage) were searched for evidence of diabetes prior to their stroke. Tabulated by ethnic group (white or Black Caribbean) and by diabetic status (no record of diabetes, treated for diabetes by insulin or tablet, untreated diabetes), the numbers were

	White	Black Caribbean
No Diabetes	279	52
Treated Diabetes	6	4
Untreated Diabetes	15	5

Investigate whether there is a relationship between ethnicity and history of diabetes (if any) for these patients treating the categories as (i) unordered and (ii) ordered, with untreated diabetes being of most concern. Discuss whether the assumption of ordering is appropriate.

12.7 It has been claimed that the probability of a horse winning a race on a circular or elliptic track is influenced by its starting position. Do the following data for the starting positions of the winners of 140 eight-horse races support this claim? Position number 1 is on the inside of the track, and with number 8 on the outside.

Position	1	2	3	4	5	6	7	8
Wins	24	27	21	16	18	12	8	14

12.8 The starting positions of the winning drivers in 70 Formula One races are given by Muehlbauer (2010). Assess and comment on any association between starting position and winning.

Position	1	2–4	5–8	9–20
Wins	38	26	4	2

12.9 In a study of weekly variation in stroke occurrence over an 11-year period in Kyoto, Japan, Shigematsu, Watanabe and Nakano (2015) found the numbers of strokes by day of the week to be:

Day of Week	Sun	Mon	Tues	Wed	Thu	Fri	Sat
Number	1,863	2,151	1,986	1,943	1,962	1,963	1,920

Test whether the strokes appear to be uniformly spread over days of the week.

12.10 Kim et al. (2018) investigated recurrent accident and emergency visits made by patients aged less than 16 years in South Korea. For three emergency departments, the distribution of the number of recurrent visits over 1 year is shown below. In addition, the mean number of recurrent visits for the 33,765 patients concerned was 0.61. Are these data consistent with a Poisson distribution?

No. of Recurrent Visits	0	1	2	3	4 or more
No. of Patients	23,384	5,610	2,353	1,172	1,246

The mean number of recurrent visits showed a decline with age from 0.78 for infants to 0.25 for those aged 15 years. Interpret your findings in the light of this trend.

12.11 To measure the abrasive resistance of cloth, 100 samples of fabric are each subjected to a 10-minute test under a series of 5 scourers, each of which may or may not produce a hole. The number of holes (0–5) is recorded for each sample. (i) What proportion of the tests produce a hole? (ii) Are the data consistent with a binomial distribution with $n=5$ and p estimated from the data?

No. of Holes	0	1	2	3	4	5
No. of Samples	42	36	14	3	4	1

12.12 Marascuilo and Serlin (1979) report a survey in which individuals were asked to state as true or false whether they thought that the most important qualities of a partner are determination and ambition. The respondents answered the same question at a later date. Numbers making the possible responses were as follows:

First Response	Second Response	Numbers
True	True	523
True	False	345
False	True	230
False	False	554

Is there evidence that experience significantly alters the attitudes of individuals towards the truth of the statement?

13

Association in Categorical Data

13.1 The Analysis of Association

In analysing a contingency table, if the hypothesis of independence is rejected, it may be useful to investigate the nature of the association. In Section 12.3.2, one aspect of association in tables with ordered row and column categories was considered, recommending the Jonckheere–Terpstra test for monotonic association. Further aspects of association are addressed in this chapter. The treatment is indicative rather than comprehensive. The approaches used to assess association range from parametric modelling to distribution-free methods. Parametric or semiparametric approaches are usually formulated to reflect knowledge about the mechanism generating the counts. One such method is the logistic regression model for binary responses such as success and failure, where the proportions responding depend upon levels of one or more explanatory variables. Agresti (2013, Chapter 5 and 2019, Chapter 5) gives general descriptions of this method, and a detailed treatment covering many applications is presented in Cox and Snell (1989).

In Chapter 12, two-way $r \times c$ tables were considered. Three or more way tables arise in practice. These may require detailed and sometimes subtle analysis to elucidate patterns of association. The techniques used range from extensions of some of the nonparametric models outlined for a few special cases in this chapter to analyses based on generalized linear models. A detailed account of the latter is given in McCullagh and Nelder (1989); Dobson and Barnett (2018) give a more elementary treatment illustrated by many practical applications.

A common way of presenting data from a three-way classification is by means of two-way cross-sectional tables. Table 13.1 is an illustration. The counts cover 197 patients in a study of coronary heart disease (CHD). The first categorization is into the presence or absence of CHD. The second classification is at three blood cholesterol levels (A, B, C), and the third is into five systolic blood pressure (SBP) groups (I–V).

Table 13.1 shows two cross sections of the second and third classifications determined or *stratified* by each of the two levels of the first classification (CHD present or absent). Marginal totals relevant to each cross section and a set of totals for each blood pressure group across the two strata are also given. The data may be presented in other ways (Exercise 13.1).

Tests for complete independence extend from two- to multi-way tables and may also be applied to two-way cross-sectional tables, but these analyses are usually insufficient (and often inefficient) because the nature of the associations is of greater interest. In particular, it is often productive to look upon such tables as representing k strata, each stratum consisting of a 2×2 table, as discussed in Sections 13.2.2–13.2.5.

DOI: 10.1201/9780429326172-13

TABLE 13.1

Categorization of 197 Patients in a Study of Coronary Heart Disease

		Systolic Blood Pressure (SBP) Group					
		I	II	III	IV	V	Total
Cholesterol							
CHD	A	2	3	1	0	4	10
	B	2	1	5	3	0	11
	C	4	7	8	6	2	27
	Total (1)	8	11	14	9	6	48
No CHD	A	16	14	11	8	6	55
	B	22	18	5	3	2	50
	C	15	13	10	5	1	44
	Total (2)	53	45	26	16	9	149
	Total (1+2)	61	56	40	25	15	197

A particular type of three-way classification that has received considerable attention is the $k\times2\times2$ table. The analysis method depends on the questions of interest and the nature of the categories in each classification.

For 2×2 tables, an appropriate measure of association is the *odds ratio* or, often better, the logarithm of the odds ratio, which was introduced in Section 12.1.2.

SUMMARY POINTS:

- Data from three-way tables can be presented using two-way cross-sectional tables.
- In the analysis of multi-way tables, they can be regarded as representing a number of strata, each stratum consisting of a two-way table.

13.2 Some Models for Contingency Tables

13.2.1 The Loglinear Model

In any $r\times c$ table if there is independence, the expected frequency in cell (i, j) is $m_{ij}=n_{i+}n_{+j}/N$, where n_{i+}, n_{+j} are the ith row and jth column totals and N is the total of all counts in the table (Section 12.1.1). Taking logarithms

$$\log m_{ij} = \log n_{i+} + \log n_{+j} - \log N \tag{13.1}$$

This means that under independence, for each cell, the logarithm of the expected number is a linear function of the logarithms of the row, column and grand totals.

This is reminiscent of the additive linear model (i.e., one without interactions) described in Section 8.3, with the parameters replaced by logarithms of certain contingency table marginal totals. With the model in Section 8.3, the observed yield for any unit differed from the expected yield by an added amount, usually regarded as a random error or departure from expectation.

By analogy, even when there is no association (i.e., independence) between row and column classifications in an $r \times c$ contingency table in general, the expected count m_{ij} will not equal the observed count n_{ij} for cell (i, j). For the model based on (13.1), the difference $\log n_{ij} - \log m_{ij} = \log(n_{ij}/m_{ij})$ may be regarded as an additive error or departure term when the model for no association is accepted as adequate. This quantity appears in the likelihood ratio statistic G^2 given in (12.10).

Association in contingency tables has analogies with interaction in factorial treatment structures in the analysis of variance, and these analogies are exhibited in extensions of (13.1) that allow for association. A detailed treatment is beyond the scope of this book, but the basic ideas for 2×2 and 2×2×2 contingency tables are illustrated, along with a brief indication of some extensions to simple cases for $r \times c$ tables.

In Section 12.1.2, it was established that for a 2×2 contingency table, the condition for no association (independence) between row and column categories was that the odds ratio $\theta = (m_{11}m_{22})/(m_{12}m_{21}) = 1$, or equivalently, that $m_{11}m_{22} = m_{12}m_{21}$. Taking logarithms, it can be seen that for the loglinear model

$$\log m_{11} + \log m_{22} - \log m_{12} - \log m_{21} = 0 \tag{13.2}$$

is the criterion for no association.

For observed counts n_{ij}, the empirical odds ratio $(n_{11}n_{22})/(n_{12}n_{21})$ does not usually take the value 1. A test of association (or interaction) in a contingency table amounts to determining whether this ratio differs sufficiently from 1 to provide strong evidence against the null hypothesis H_0: $\theta = 1$, or equivalently, whether $\log n_{11} + \log n_{22} - \log n_{12} - \log n_{21}$ differs significantly from zero, i.e., whether the alternative hypothesis

$$H_1 : \log m_{11} + \log m_{22} - \log m_{12} - \log m_{21} = I, I \neq 0 \tag{13.3}$$

is preferable to H_0: $I = 0$. Here (13.3) is the analogue of (8.2).

This is called a first-order, or two-factor, interaction model and represents the only possible kind of interaction with a 2×2 factorial treatment structure. The models in Section 8.1 and that in (13.3) extend respectively to 2×2×2 factorial experiments and to 2×2×2 contingency tables.

In the context of a loglinear model, the m_{ijk} are the expected counts based on marginal totals in a way described below in Example 13.1. For 2×2×2 tables, the no association (no-interaction) model corresponds to that of independence between classifications, and in terms of odds ratios, it may be written

$$\frac{m_{111}m_{221}}{m_{121}m_{211}} = \frac{m_{112}m_{222}}{m_{122}m_{212}} = 1$$

Dependence, or association, may be first or second order (first- or second-order interaction in the loglinear sense). For a first-order interaction

$$\frac{m_{111}m_{221}}{m_{121}m_{211}} = \frac{m_{112}m_{222}}{m_{122}m_{212}} = k, \quad k \neq 1 \tag{13.4}$$

and for a second-order interaction

$$\frac{m_{111}m_{221}}{m_{121}m_{211}} \neq \frac{m_{112}m_{222}}{m_{122}m_{212}}$$

By taking logarithms, analogies with the factorial treatment structure discussed in Section 8.1 can be seen. Specialist texts such as Bishop, Fienberg and Holland (1975), Fienberg (1980), Plackett (1981), Agresti (2010, 2013, 2019) or Everitt (1992) give a detailed discussion of loglinear models and their use in measuring association.

The assessment of association is now considered for a set of k cross-sectional tables each $2{\times}2$ with $k{\ge}2$, which may be formed from or combined as a $k{\times}2{\times}2$ table. How to answer some typical questions that may be asked will be illustrated using two specific data sets, the first given in Tables 13.2 and 13.3 and the second in Table 13.4. The data in Tables 13.2 and 13.3 are from the Health Behaviour in School-aged Children England 2010/2011 study of adolescents, and they give the self-reported time spent watching television during a weekday (2 hours or less, more than 2 hours) by gender and number of days per week in which physical activity (PA) is undertaken (Brooks et al., 2014). In Table 13.2, the results are stratified by *gender*, since the component $2{\times}2$ tables (left and right) are for boys and girls, respectively. In Table 13.3, stratification is by *engagement in PA*.

In Table 13.4, a $4{\times}2{\times}2$ table is in effect broken down into four $2{\times}2$ tables, where the $k{=}4$ strata correspond to different age groups. In the $2{\times}2$ tables within each age group, the

TABLE 13.2

Television Viewing (TV) (2 hours or Less, More than 2 hours Daily) by Gender and Engagement in Physical Activity (PA)

	Boy		Girl	
	TV≤2 hours	TV>2 hours	TV≤2 hours	TV>2 hours
Daily PA	167	320	123	157
Less than Daily PA	388	937	604	1,114

TABLE 13.3

TV (2 hours or Less, More Than 2 hours Daily) by Engagement in PA and Gender

	Daily PA		Less than Daily PA	
	TV≤2 hours	TV>2 hours	TV≤2 hours	TV>2 hours
Boy	167	320	388	937
Girl	123	157	604	1,114

TABLE 13.4

Side-Effects of Two Drugs Grouped by Ages

		Side-Effect Status	
Age Group (years)	Drug	None	Some
20–29	A	8	1
	B	11	1
30–39	A	14	2
	B	18	0
40–49	A	25	3
	B	42	2
50+	A	39	3
	B	22	6

explanatory variables are two drugs designated as A and B. The responses to each are the column categories *side-effect* or *no side-effect*. The k age groups forming the strata are also explanatory variables, often referred to as *covariates*.

For the data in Tables 13.2 and 13.3, the Fisher exact test, the Pearson chi-squared test or the likelihood ratio test all indicate association in each cross-sectional table.

A question of interest about association in the basic $2 \times 2 \times 2$ table is whether there is evidence of only first-order association or whether there might be second-order association. Remember that for the 2×2 cross-sectional tables, independence implies that the expected odds ratios must each take the value 1, while first-order association implies they should be equal, but the common value is not 1, and second-order association indicates the expected odds ratios should not be equal.

For the data in Table 13.4, the possibilities are more varied. However, the ones likely to be of most interest are whether there is evidence of association, and if there is any association, whether it appears to be similar in nature, i.e., implies a common odds ratio, over all strata.

Further questions of interest arise if the odds ratio appears to change between strata. For instance, in Table 13.4, do the odds ratios of the cross-sectional 2×2 tables relating drugs and side-effects tend to increase with the stratification covariate *age*? Sometimes, with small numbers in individual stratum, a test for independence applied separately in each stratum may give no evidence to indicate association. However, it is then wise to consider whether, collectively, there may be some evidence of association if appropriate tests are used. Section 13.2.2 considers the case of only two strata; the more general case of $k \geq 2$ strata is covered in Sections 13.2.3–13.2.5.

13.2.2 Analysis of Two Cross-Sectional 2 × 2 Tables

Although some more general methods given in the next section apply to the case $k=2$, once it has been established that there is evidence of association, the main point of interest is often whether a first-order model adequately describes the association. When $k=2$, this implies that (13.4) holds. The values of the expectations m_{ijk} for (13.4) are not known, but estimates $\hat{m}_{ijk}$ of the expected counts may be obtained from the data, where these are chosen to satisfy the condition

$$\frac{\hat{m}_{111}\hat{m}_{221}}{\hat{m}_{121}\hat{m}_{211}} = \frac{\hat{m}_{112}\hat{m}_{222}}{\hat{m}_{122}\hat{m}_{212}} = 1 \tag{13.5}$$

The $\hat{m}_{ijk}$ no longer have the simple values that they have for the separate cross-sectional tables under the hypothesis of independence. However, if all marginal totals in the complete $2 \times 2 \times 2$ table are fixed, there is only one degree of freedom, and once one of the $\hat{m}_{ijk}$ in (13.5) is known, all the others can be deduced.

After these expected values are determined, an associated P-value may be found using either a likelihood ratio test or a Pearson chi-squared test. Exact permutation tests are possible using these statistics, but the asymptotic chi-squared distribution results may be used for large samples. In practice, the likelihood ratio statistic G^2 has some advantages when the case $k=2$ is generalized to $k>2$. Then there is more than one degree of freedom. It is shown in Section 13.3.2 that this latter statistic may be partitioned into additive single degree of freedom components in a sense that is not in general true for the Pearson X^2 statistic.

Example 13.1

Determine whether the data in Table 13.2 are consistent with a model that specifies only first-order association.

A first-order association model (13.4) is postulated after first establishing that a model of no association or independence is not appropriate. If (13.5) holds the condition that the marginal totals are fixed means that the relationship can be used to calculate a value x for $\hat{m}_{111}$ under H_0: *association is first-order only*. The G^2 statistic may then be calculated using (12.10) and summing over all 8 cells of the two 2×2 subtables in Table 13.2.

For convenience, the data in Table 13.2 are reproduced below. The Fisher exact test, a Pearson chi-squared test or a likelihood ratio test applied to the two 2×2 constituent tables shows clear evidence of association. For the boys, all $P<0.045$, and for the girls, all $P<0.007$.

Output from the ANSM5 package in R can be obtained using. `fishertest.ANSM(ch13$physical.activity[ch13$gender == "Boy"], ch13$tv.viewing[ch13$gender == "Boy"])`, `chisqtest.ANSM(ch13$physical.activity[ch13$gender == "Boy"], ch13$tv.viewing[ch13$gender == "Boy"])` and `lik.ratio(ch13$physical.activity[ch13$gender == "Boy"], ch13$tv.viewing[ch13$gender == "Boy"])` for the boys, with `ch13$gender == "Boy"` being replaced by `ch13$gender == "Girl"` for the girls.

	Boy		Girl	
	TV≤2hours	**TV>2hours**	**TV≤2hours**	**TV>2hours**
Daily PA	167	320	123	157
Less than Daily PA	388	937	604	1,114

If the estimate $\hat{m}_{111}$ is denoted by x it can be shown using the fixed marginal totals that $\hat{m}_{121}=487-x$, $\hat{m}_{211}=555-x$, $\hat{m}_{221}=x+770$, $\hat{m}_{112}=290-x$, $\hat{m}_{122}=x-10$, $\hat{m}_{212}=x+437$, $\hat{m}_{222}=1,281-x$ and thus (13.5) is satisfied if x is chosen so that

$$\frac{x(x+770)}{(487-x)(555-x)}=\frac{(290-x)(1,281-x)}{(x-10)(x+437)}$$

This cubic equation in x may be solved numerically with an appropriate computer algorithm. The relevant solution is one that gives expected numbers that are all positive and turns out to be $x=\hat{m}_{111}=171.6$. The remaining expectations may be calculated using the relationships given above. As there is only 1 degree of freedom, the differences between observed and expected values are either 4.6 or –4.6. The expected values are summarized as

	Boy		Girl	
	TV≤2hours	**TV>2hours**	**TV≤2hours**	**TV>2hours**
Daily PA	171.6	315.4	118.4	161.6
Less than Daily PA	383.4	941.6	608.6	1,109.4

The likelihood ratio statistic obtained from (12.10), where the summation is over all 8 cells, is given by

$$\begin{aligned}G^2 = {} & 2\times[167\times\log(167/171.6)+320\times\log(320/315.4)+388\times\log(388/383.4)\\&+937\times\log(937/941.6)+123\times\log(123/118.4)+157\times\log(157/161.6)\\&+604\times\log(604/608.6)+1,114\times\log(1,114/1,109.4)]=0.6269.\end{aligned}$$

With 1 degree of freedom, this implies $P=0.4285$. The Pearson chi-squared statistic is $X^2=0.62733$ giving $P=0.4284$.

Output from the stats package in R can be obtained using the following commands:

```
fitted.model <- loglin(table(ch13$physical.activity,
  ch13$tv.viewing, ch13$gender), list(c(1, 2), c(1, 3),
  c(2, 3)))
pchisq(fitted.model$lrt, fitted.model$df, lower.tail =
  FALSE)
pchisq(fitted.model$pearson, fitted.model$df, lower.tail =
  FALSE).
```

There is insufficient evidence to reject the hypothesis that there is only a first-order interaction.

Accepting that there is only a first-order interaction is equivalent to saying that the odds ratios for the cross-sectional tables do not differ significantly. This might seem surprising as the association between television viewing and participation in PA is stronger for girls than boys. The sample odds ratios for boys and girls are 1.260 and 1.445, respectively. Despite the initial impression of a possible difference between genders, this finding could easily have arisen by chance.

The odds ratio, or its logarithm, contains all the information on association in a 2×2 table. Example 13.1 showed one way of using that information. As pointed out in Section 12.2.3 for asymptotic tests, normality is approached more rapidly if use is made of $\phi=\log\theta$ rather than θ itself.

Making use of natural logarithms and denoting the empirical log odds ratio by ϕ^*, i.e., $\phi^*=\log[(n_{11}n_{22})/\ (n_{12}n_{21})]$ then (see, e.g., Agresti, 2013, Chapter 3) ϕ^* is asymptotically normally distributed under the hypothesis of independence with mean zero and

$$\mathrm{Var}(\phi^*) = 1/n_{11} + 1/n_{12} + 1/n_{21} + 1/n_{22}. \tag{13.6}$$

If any $n_{ij}=0$, both ϕ^* and $\mathrm{Var}(\phi^*)$ become infinite. This difficulty is avoided if all n_{ij} $(i, j=1, 2)$ are replaced by $n_{ij}+0.5$ in (13.6). Differences in inference based on this modification are small when most n_{ij} are moderate or large. Association implies $\phi=\phi_0\neq0$ and that ϕ^* is asymptotically normally distributed with mean ϕ_0 and variance given by (13.6).

With two 2×2 tables having expected odds ratios θ_1 and θ_2 stating the null hypothesis H_0: $\theta_1=\theta_2$ implies $\theta_2/\theta_1=1$ or $\phi_2-\phi_1=0$. Thus, under H_0 it follows that asymptotically

$$Z = \frac{\phi_2^* - \phi_1^*}{\sqrt{\mathrm{Var}(\phi_1^*) + \mathrm{Var}(\phi_2^*)}} \tag{13.7}$$

has a standard normal distribution.

Example 13.2

For the data in Table 13.2 show that the hypothesis H_0: $\theta_1=\theta_2$ implying equal association (or independence if both equal 1) is plausible.

It is necessary to evaluate $\phi_k^* = \log\theta_k^*$ for $k=1, 2$ and base the test on (13.6) and (13.7)

From Table 13.2

$$\phi_1^* = \log[167 \times 937/(388 \times 320)] = 0.2314,$$

$$\phi_2^* = \log[123 \times 1{,}114/(604 \times 157)] = 0.3681$$

$$\text{Var}(\phi_1^*) + \text{Var}(\phi_2^*) = \left(1/167 + 1/320 + 1/388 + 1/937 + 1/123 + 1/157 + 1/604 + 1/1,114\right)$$

$$= 0.0298$$

whence

$$Z = \frac{0.3681 - 0.2314}{\sqrt{0.0298}} = 0.7917.$$

Since $Z=0.7917$ this corresponds to a two-tail $P=0.4285$, and there is no strong evidence against H_0.

Under H_0 the distribution of Z^2 is chi-squared with 1 degree of freedom. The value $Z^2=0.7917^2=0.6268$, which is in line with the result in Example 13.1 where it was found that $G^2=0.6269$.

In view of the large counts in each cell, having accepted the hypothesis that the odds ratios do not differ, it is reasonable to use the normal approximation to test for each table the hypothesis $\phi_i=0$, $i=1, 2$ implying independence. The standard test statistic $Z = \phi_i^* / \sqrt{\text{Var}(\phi_i^*)}$ shows evidence for an association with girls but not boys (Exercise 13.2).

The method used here makes it unnecessary to estimate the expected values $\hat{m}_{ijk}$ needed in Example 13.1.

An asymptotic confidence interval for the true difference $\phi_2-\phi_1$ may be obtained in the usual way. For example, the approximate 95% interval is of the form

$$\phi_2^* - \phi_1^* \pm 1.96\sqrt{\text{Var}\left(\phi_1^*\right) + \text{Var}\left(\phi_2^*\right)}$$

Since $\text{Var}(\phi_1^*)=(1/167+1/320+1/388+1/937)=0.0128$ and $\text{Var}(\phi_2^*)=(1/123+1/157+1/604+1/1{,}114)=0.0171$, for the above data the 95% interval is (−0.2017, 0.4751). The value of zero stated in H_0 is in the interval so a hypothesis of equal odds ratios is, as found above, acceptable.

Output from the ANSM5 package in R can be obtained using `oddsratio.2x2diff(ch13$physical.activity, ch13$tv.viewing, ch13$gender, do.asymp = TRUE)` which also gives an exact 95% confidence interval of (−0.3482, 0.3459).

13.2.3 Analysis of 2 × 2 Tables for Several Strata

From Section 12.1.2, binomial sampling with two levels of an explanatory variable is a model often associated with a 2×2 table. A typical situation is one where the presence or absence of side-effects (responses) is noted for each of two medicines (explanatory categories). Another example arises in a production process where two types of machines (explanatory categories) are being investigated in manufacturing runs, from each of which the numbers of satisfactory and flawed items (responses) are noted. Further explanatory variables, or covariates, may be introduced, leading to situations like that exemplified in Table 13.4, where each of the $k=4$ covariates is an age group. These groupings are usually referred to as strata. In that table, each stratum consists of a 2×2 table of counts classified by drugs and side-effects. An informal inspection of the table suggests that side-effect incidence tends to increase slightly with age for drug B, but there is no such tendency for drug A. Apart from this, there looks to be little difference in side-effect incidence between the drugs.

Two obvious questions that may be of interest are:

- Is there an association between drugs and side-effects?
- If there is an association does it change with age, i.e., does the odds ratio change between strata?

A simplistic approach is to test for evidence of association within each stratum. If there is no evidence, the hypothesis of independence is accepted. A disadvantage is that if the counts in some cells are small, the tests may have low power against alternatives that specify a slight association. Intuitively, one feels that if the information could be combined in some way for all strata, the larger sample size may detect such associations if these are similar for all strata. Care is needed in such analyses, for they may give misleading results, especially if the degree of association, or type of association, differs between strata, a matter explored further in Section 13.3.1.

Examples 13.1 and 13.2 considered the special case of only two strata (i.e., $k=2$). The large numbers in each cell enabled the detection of evidence of association within each stratum.

The situation is different for the data in Table 13.4, where exact tests applied to each of the four strata specified by age show no evidence of association. However, it will be seen in Example 13.4 that the suspicion aroused by a visual inspection of the data that there are more side-effects for drug B among older patients is not without justification, although the evidence for this is not strong.

One should begin by assessing whether there is evidence that the odds ratios are equal for all strata. If there is strong evidence that these ratios are not all equal, a test for independence across all strata (i.e., that the within-strata odds ratios $\theta_1, \theta_2, \dots, \theta_k$ have a common value of unity) is then not appropriate, since a decision they are probably not all equal has already been made.

Thus, a sensible procedure is to test first whether the hypothesis of equality of odds ratios holds. Only if there is little evidence against this should the possibility of a common value of unity be tested.

The methods used in Examples 13.1 and 13.2 for the case $k=2$ do not generalize easily to situations in which $k>2$. Further, only details of asymptotic methods were given, and these are not appropriate if there are many cells with small counts. These are unlikely to be satisfactory for the data in Table 13.4, for example.

The approaches using asymptotic and exact results are somewhat different, so they will be considered separately.

13.2.4 Asymptotic Tests About Odds Ratios in k 2×2 Tables

For any $k\geq 2$, three common questions are

- Is there evidence of association (i.e., $\theta_s \neq 1$, $s=1, 2, \dots, k$)?
- If there is association are all θ_s equal?
- If all θ_s are equal, how is the common value estimated?

Several asymptotic tests help answer these questions. When these are not adequate, exact or Monte Carlo permutation tests are possible. The basic ideas behind these procedures are outlined here, quoting relevant formulae without derivation. Some asymptotic tests are provided in general statistical packages and may even be conducted manually in simple cases. Many inference procedures in this, and the next, section rely heavily on the fact that once the observed or expected count for cell (1, 1) in any 2×2 table is known, this fixes the observed, or expected, odds ratio if marginal totals are fixed.

As already pointed out, it is desirable to test whether a hypothesis of equality of odds ratios is tenable before testing whether that common ratio may be unity. If the first possibility is rejected, the second is automatically rejected.

Data often suggest that if there is any association, then it may be similar within all strata; e.g., all the empirical odds ratios $\theta_s=(n_{11s}n_{22s})/(n_{12s}n_{21s})$ might take values slightly greater than 1.

Care is needed in choosing an estimate of a common odds ratio appropriate for testing the hypothesis that there is indeed such a common ratio. With different stratum sizes as in Table 13.4, and perhaps different allocations of numbers to rows or columns within strata, a simple average of all stratum odds ratios is a naïve approach.

Mantel and Haenszel (1959) proposed the estimator

$$\theta^* = \frac{\sum_s (n_{11s}n_{22s}/N_s)}{\sum_s (n_{12s}n_{21s}/N_s)} \tag{13.8}$$

This is a weighted mean of the strata odds ratios θ_s with weights $n_{12s}n_{21s}/N_s$ where N_s is the total count for the cells in stratum s. The motivation for choosing these weights is that when all θ are close to unity, they are nearly the reciprocals of the variances of the θ_s.

Breslow and Day (1980) proposed a test for whether data were consistent with an estimated common odds ratio θ^* given by (13.8). It is first necessary to compute the expected count m_{11s} in cell (1, 1, s) for each of the k strata conditional upon the common odds ratio being given by (13.8). This amounts to choosing for each stratum s an $x=m_{11s}$ that is the positive root not exceeding n_{1+s} or n_{+1s} of the equation

$$\frac{x(N_s - n_{1+s} - n_{+1s} + x)}{(n_{1+s} - x)(n_{+1s} - x)} = \theta^* \tag{13.9}$$

If odds ratios are equal across all strata, the differences $|m_{11s}-n_{11s}|$, $s=1, 2, \ldots, k$ should all be small and Breslow and Day (1980, Chapters 4 and 5) discuss appropriate tests for homogeneity of odds ratios in detail. They proposed a statistic

$$M_{BD}^2 = \sum_{s=1}^{k} \frac{(m_{11s} - n_{11s})^2}{\mathrm{Var}(n_{11s})} \tag{13.10}$$

where

$$\mathrm{Var}(n_{11s}) = \left[\frac{1}{m_{11s}} + \frac{1}{m_{12s}} + \frac{1}{m_{21s}} + \frac{1}{m_{22s}}\right]^{-1} \tag{13.11}$$

Asymptotically, M_{BD}^2 has a chi-squared distribution with $k-1$ degrees of freedom under H_0: *all θs are equal*. The expected frequencies in the expression for $\mathrm{Var}(n_{11s})$ are subject to the fixed marginal totals for stratum s given m_{11s}.

If the null hypothesis H_0 is accepted, a confidence interval for the common value θ, which was estimated by θ^*, may be obtained. Asymptotically, the distribution of $\phi^* = \log\theta^*$ approaches normality more rapidly than that of θ^*. Also, $E(\phi^*) = \log\theta$. Several estimates have been proposed for $\text{Var}(\phi^*)$. Robins, Breslow and Greenland (1986) proposed the estimate below. It has been shown to work well for small k and large counts or for large k even if some of the 2×2 tables are fairly sparse. The expression for this estimate is complicated, but it is given here for completeness. For each stratum s define

$$a_s = (n_{11s} + n_{22s}) / N_s,\ b_s = (n_{12s} + n_{21s}) / N_s,\ c_s = n_{11s}n_{22s} / N_s,\ d_s = n_{12s}n_{21s} / N_s,$$

then write

$$c_+ = \sum_s c_s,\ d_+ = \sum_s d_s$$

and

$$P = \left[\sum_s (a_s c_s)\right] \Big/ \left(2c_+^2\right),\ Q = \left[\sum_s (a_s d_s + b_s c_s)\right] \Big/ \left(2c_+ d_+\right),\ R = \left[\sum_s (b_s d_s)\right] \Big/ \left(2d_+^2\right)$$

then

$$\text{Var}(\phi^*) = P + Q + R.$$

Approximate 95% asymptotic confidence limits for ϕ are

$$\phi^* \pm 1.96\sqrt{\text{Var}(\phi^*)}. \tag{13.12}$$

For all but small data sets (where asymptotic procedures are not appropriate), a computer program is highly desirable for testing and estimation, but the salient features of the computation are illustrated in Example 13.3. Approximate confidence limits for θ can be obtained by back-transformation. Everitt (1992, Section 2.8.1) develops an alternative formula, basing the interval on θ^* directly.

Example 13.3

For the data in Table 13.5 explore whether the hypothesis of equality of odds ratios is acceptable. If it is, obtain an approximate 95% confidence interval for the common odds ratio θ.

TABLE 13.5

Production Count for Two Machines with Three Raw Material Sources

		Output Status	
Material Source	**Machine**	**Satisfactory**	**Faulty**
A	Type I	42	2
	Type II	33	9
B	Type I	23	2
	Type II	18	7
C	Type I	41	4
	Type II	29	12

The material sources form three strata (k=3), and the two machine types are explanatory categories. The output status categories represent two responses. The solution requires the following steps: (i) estimation of a common odds ratio using (13.8); (ii) calculation of M^2_{BD} from (13.10) for use in an asymptotic test of H_0: *all* θ_i *are equal* against H_1: $\theta_i \neq \theta_j$ for at least one $i \neq j$; (iii) if H_0 is accepted, a confidence interval for the common value θ is obtained by back-transformation from the interval for ϕ given by (13.12).

The widely available software is recommended for this test, but the main steps are outlined in the computation. Using (13.8) gives

$$\theta^* = \frac{(42\times 9)/86+(23\times 7)/50+(41\times 12)/86}{(33\times 2)/86+(18\times 2)/50+(29\times 4)/86} = 4.702.$$

The expected value m_{111} is the relevant root of (13.9), which here is

$$\frac{x(86-75-44+x)}{(75-x)(44-x)} = 4.702.$$

This gives $x = m_{111} = 41.68$. The other root $x = 100.54$ exceeds both row and column marginal totals. Similar calculations indicate $m_{112} = 23.07$ and $m_{113} = 41.25$. Then (13.11) gives

$$\text{Var}(n_{111}) = \left[\frac{1}{41.68}+\frac{1}{2.32}+\frac{1}{33.32}+\frac{1}{8.68}\right]^{-1} = 1.6660,$$

and in like manner $\text{Var}(n_{112}) = 1.3181$, $\text{Var}(n_{113}) = 2.4550$, whence (13.10) gives

$$M^2_{\text{BD}} = 0.32^2/1.6660 + 0.07^2/1.3181 + 0.25^2/2.455 = 0.09.$$

Assuming an asymptotic chi-squared distribution with $k-1=2$ degrees of freedom, there is clearly no evidence against H_0. Indeed, the fit is remarkably good (P=0.96).

To obtain the relevant confidence interval, note that $\phi^* = \log\theta^* = 1.5480$. Straightforward computation gives $\text{Var}(\phi^*) = 0.184557$ whence (13.12) gives the 95% confidence limits for ϕ as

$$1.5480 \pm 1.96\sqrt{0.184557}$$

leading to an interval (0.70598, 2.39002). Taking natural antilogarithms gives an interval (2.03, 10.92) for θ. Hence, the hypothesis of a common odds ratio is accepted, and it is estimated to be θ=4.70 with a 95% confidence interval (2.03, 10.92).

Output from the ANSM5 package in R can be obtained using `breslow.day(ch13$machine, ch13$output.status, ch13$material.source)`.

The confidence interval is wide and is not symmetric about the point estimate. This reflects the markedly skew distribution of θ referred to in Section 12.1.2. Note also that since independence, or lack of association, implies θ=1, there is strong evidence of association since θ=1 is well outside the above confidence interval.

Other software may give a different confidence interval to that obtained here, because, as indicated above, there are alternative estimates of variance to that given by the Robins, Breslow and Greenland formula.

The method in Example 13.3 tests whether a common odds ratio across all strata is an acceptable hypothesis. If this is the case, it is reasonable to assess, with the aid of a confidence interval, whether that common value may be θ=1, implying independence.

Another widely used test for independence is the Mantel–Haenszel test proposed by Mantel and Haenszel (1959), although Cochran (1954) proposed an effectively identical test. The latter is described by Everitt (1992, Section 2.7.2). The test is presented here in the Mantel–Haenszel formulation due to its similarities with the Breslow–Day test given above.

The test requires the calculation in each stratum of the cell (1, 1) expectations under the hypothesis of independence in that stratum, i.e., $m_{11s}=n_{1+s}n_{+1s}/N_s$. The statistic is

$$M^2 = \frac{(m_{11+} - n_{11+})^2}{\sum_s \mathrm{Var}(n_{11s})} \qquad (13.13)$$

where

$$m_{11+} = \sum_s m_{11s}, \quad n_{11+} = \sum_s n_{11s}$$

and

$$\mathrm{Var}(n_{11s}) = \frac{n_{1+s}n_{+1s}n_{2+s}n_{+2s}}{N_s^2(N_s - 1)}$$

Asymptotically (13.13) has a chi-squared distribution with 1 degree of freedom under H_0: *all* $\theta_s=1$.

If the expected and observed values in the cell of each table differ appreciably, and these differences are all in the same direction, this suggests odds ratios are consistently above, or consistently below, 1 and large values of M^2 are likely. Circumstances may arise where differences between expected and observed values in cell (1, 1) for each table are in an opposite direction in some tables to that in others. This may lead to small values of M^2 even when there is clear evidence of association – albeit in opposite directions – within some strata. For this reason, the Mantel–Haenszel test is not recommended without first using a test such as the Breslow–Day test for equality of odds ratios. However, if the asymptotic theory is accepted for the Breslow–Day test, one may, as shown above, compute approximate confidence intervals for a common odds ratio and accept or reject a hypothesis of independence on the basis of these, making the Mantel–Haenszel test somewhat superfluous. How it works is illustrated below.

Example 13.4

For the data in Table 13.4, compute M^2, and assuming all $k=4$ strata have the same expected odds ratio θ is it reasonable to accept the hypothesis H_0: $\theta=1$?

Assuming all θ_s are equal, one can compute M^2 using (13.13). A *P*-value can be obtained assuming a chi-squared distribution with 1 degree of freedom.

Calculation gives $m_{111}=9\times19/21=8.14$, $m_{112}=15.06$, $m_{113}=26.06$, $m_{114}=36.60$, whence $m_{11+}=8.14+15.06+26.06+36.60=85.86$ and $n_{11+}=8+14+25+39=86$. Also $\mathrm{Var}(n_{111})=(19\times9\times12\times2)/(21^2\times20)=0.4653$, $\mathrm{Var}(n_{112})=0.4832$, $\mathrm{Var}(n_{113})=1.1213$, $\mathrm{Var}(n_{114})=1.9096$, whence

$$\sum_s \mathrm{Var}(n_{11s}) = 0.4653 + 0.4832 + 1.1213 + 1.9096 = 3.9794,$$

so that

$$M^2 = (-0.1428)^2 / 3.9794 = 0.00512.$$

The extremely small value, 0.00512, means there is no evidence against the hypothesis of independence (P=0.943).

Output from the ANSM5 package in R can be obtained using `mantel.haenszel(ch13$drug, ch13$side.effects, ch13$age.group)`.

A continuity correction has been suggested whereby 0.5 is subtracted from $|m_{11+}-n_{11+}|$ before squaring. This may avoid over-weighting of evidence against H_0 in borderline cases.

The small differences between corresponding m_{11s} and n_{11s} point towards independence. However, as indicated above, situations arise where these differences may be large and with opposite signs in different strata, but the numerator in (13.13) remains small. It is preferable to use the Breslow–Day test with confidence intervals for a common θ if that hypothesis is accepted. For the data in this example, it can be shown that $M^2_{BD} = 6.497$ with an associated P=0.0898; there is no substantial evidence against equality of odds ratios, although there is room for a little doubt. The point estimator of the odds ratio is θ^*=1.0334 with a 95% confidence interval (0.4037, 2.6453). In light of these findings, it is not surprising that the Mantel–Haenszel test strongly supports independence. The Breslow–Day test, however, does raise a small doubt about the hypothesis of equality of odds ratios.

Output from the ANSM5 package in R can be obtained using `breslow.day(ch13$drug, ch13$side.effects, ch13$age.group)`.

13.2.5 Exact Tests for Equality of Odds Ratios in k 2 × 2 Tables

Zelen (1971) gives an exact permutation test for a common odds ratio. Gart (1970) gives a method for obtaining confidence intervals if a hypothesis of a common odds ratio is accepted. Unlike many other nonparametric tests, the asymptotic procedures described in the previous section are not based on the asymptotic properties of the statistics used in the exact test.

As computing facilities such as those provided by the ANSM5 package in R are needed to apply the exact methods, only their nature is sketched here.

Zelen's test for homogeneity of the ratios is based on a property of the hypergeometric probabilities arising in the Fisher exact test for 2×2 tables. Subject to the condition that all n_{11s} sum over s to the observed marginal total n_{11+}, the product of the hypergeometric probabilities over all possible k 2×2 tables satisfying this condition is a maximum when the observations strongly support a common odds ratio θ whatever the value of θ, and this product decreases steadily as that support weakens. Under the hypothesis that all odds ratios are equal, the actual probability P^* of observing any particular set of within-stratum outcomes in which the cell (1, 1) entries sum to the observed n_{11+} is given by the product of the k hypergeometric probabilities for those outcomes, divided by the sum of all such products over all possible cell (1, 1) values consistent with marginal totals for the relevant 2×2 tables, and which sum to n_{11+}. If the probability for the observed tables is denoted by P_0^*, the corresponding P-value is the sum of the probabilities P^* associated with all outcomes for which $P^* \le P_0^*$. If the test is applied to the data in Table 13.4, using the ANSM5 package in R and the command `zelen(ch13$drug, ch13$side.effects, ch13$age.group)` gives the probability associated with the observed outcome as $P_0^* = 0.008213$ and the relevant P=0.0689 compared with the asymptotic Breslow–Day value P=0.0898 quoted in Example 13.4.

The above command in R also gives the Breslow–Day 95% confidence interval for the common odds ratio to be (0.404, 2.645).

The procedure proposed by Gart for estimating the common odds ratio if the Zelen test supports that hypothesis uses a different approach. Suppose that in the asymptotic Breslow–Day procedure expected values are estimated under the assumption that the common odds ratio takes a particular value θ^*. If m_{11s} indicates the expected value in cell (1, 1) of stratum s, then in general $\sum_s m_{11s} \neq n_{11+}$. Large discrepancies indicate that such a value of θ^* is an unsatisfactory value for the common odds ratio. There may be no value of θ^* for which this discrepancy is not large if the hypothesis that there is a common odds ratio is not supported.

Gart's procedure for obtaining a confidence interval for a common odds ratio is in essence to compute a statistic T that is the sum of the observed n_{11s}. One then considers its position in the distribution of all possible values that this statistic might take if each observed n_{11s} were replaced by any possible value consistent with the marginal totals for the sth 2×2 table; i.e., the values are no longer constrained to sum to n_{11+}. Values θ^-, θ^+ may then be found for T such that they would just be accepted as hypothetical estimates of the common odds ratio in a test at the $100(\alpha/2)\%$ significance level. These provide $100(1-\alpha)\%$ confidence limits for θ. Modifications are needed if the data indicate acceptance of either a zero or an infinite common odds ratio.

For the data in Table 13.5 the exact 95% confidence interval given by Gart's method is (1.9149, 12.306), slightly longer than the asymptotic interval (2.0256, 10.9145) obtained in Example 13.3. The latter depends on the validity of certain variance estimates, involves a back-transformation from natural logarithms and assumes a reasonable rate of convergence to normality. From the practical viewpoint, the difference between the asymptotic and the exact permutation theory results should not cause alarm, bearing in mind that there are also questions of discontinuity in the distribution of Gart's T statistic, so the interval may not be an exact 95% interval.

13.2.6 The Linear-by-Linear Association Model

The linear-by-linear association model is a relatively simple, yet versatile, loglinear model with only one additive interaction term to represent association. For an $r\times c$ table, a score u_i is allocated to row i, $i=1, 2, \ldots, r$ subject to constraints $u_1 \le u_2 \le \ldots \le u_r$ and a score v_j to column j, $j=1, 2, \ldots, c$ subject to constraints $v_1 \le v_2 \le \ldots \le v_c$. If m_{ij} is the expected count in cell (i, j) the model may be written

$$\log m_{ij} = \text{independence terms} + \beta u_i v_j, \tag{13.14}$$

where β is a constant.

The model reduces to that for independence (no association) between row and column categories when $\beta=0$. It is sometimes convenient to use the modified form

$$\log m_{ij} = \text{independence terms} + \beta(u_i - m_u)(v_j - m_v), \tag{13.15}$$

where m_u, m_v are the means of the u_i and of the v_j, respectively. Note that (13.15) is equivalent to (13.14) if, in the latter, u_i, v_j are replaced by $u_i^* = u_i - m_u$ and $v_j^* = v_j - m_v$.

A number of standard nonparametric test statistics have exact permutation distributions based on this model. It is relevant to tables of counts where both row and column categories are ordered and when, if the classifications are not independent, they show a monotonic trend across row and column categories. When the scores are ranks, they are often referred to as *unit interval scores* because the difference between the scores associated with any pair of neighbouring rows, or of neighbouring columns, is in each case unity.

The parameter β may be estimated to reflect the extent of any relevant departure from independence. An indication of what is meant by a relevant departure is given in Example 13.5. Cells for which $|\beta(u_i - m_u)(v_j - m_v)|$ is large show the greatest departure from independence.

Example 13.5

For the case $r=5$, $c=7$ with rank scores, i.e., $u_i=i$, $v_j=j$, determine for each cell in the $r \times c$ table the contribution to $\log m_{ij}$ of the interaction term $\beta(u_i - m_u)(v_j - m_v)$ in (13.15).

Here $m_u=3$, $m_v=4$, so the interaction term in cell (i, j) in (13.15) is $\beta(i-3)(j-4)$. For any β, the interaction term in each cell is given in Table 13.6.

If $\beta>0$, $\log m_{ij}$ is greater than its value under independence for cells at the top left and bottom right of Table 13.6, and less for cells at the top right or bottom left. If $\beta<0$, the situation is reversed. If $\beta=0$, there is no association.

The name linear-by-linear association reflects the characteristic that within any row, say row i, for any chosen u_i, v_j, the interaction contribution across columns is a linear function of the v_j with slope $\beta(u_i - m_u)$. Similarly, within any column j the interaction contribution across rows is a linear function of the u_i with slope $\beta(v_j - m_v)$.

The relevance of this model to permutation theory is that many permutation tests depend upon a statistic, S, that may be written

$$S = \sum_{i,j} u_i v_j n_{ij} \tag{13.16}$$

with appropriate choices of u_i, v_j. Independence ($\beta=0$) corresponds to the null hypothesis; the null hypothesis H_0 is rejected if the observed outcome, expressed in the appropriate contingency table format, indicates a general pattern of the form exemplified for a particular case in Table 13.6 when $\beta \neq 0$. This corresponds to extreme values of S given by (13.16); intermediate values indicate lack of association.

TABLE 13.6

Interaction Terms in $\log m_{ij}$ for Rank Scores in a 5×7 Contingency Table

	Column Score						
Row Score	**−3**	**−2**	**−1**	**0**	**1**	**2**	**3**
−2	6β	4β	2β	0	-2β	-4β	-6β
−1	3β	2β	β	0	$-\beta$	-2β	-3β
0	0	0	0	0	0	0	0
1	-3β	-2β	$-\beta$	0	β	2β	3β
2	-6β	-4β	-2β	0	2β	4β	6β

TABLE 13.7

The Information in Table 12.2 with Times Replaced by Corresponding Ranks for Column Scores and 0, 1 Assigned as Row Scores

Time Ranks	1	2	3	4	5	6	7	8	9	10	11
Group A	1	1	1	0	1	1	0	1	1	0	0
Group B	0	0	0	1	0	0	1	0	0	1	1
Time Ranks	**12**	**13**	**14**	**15**	**16**	**17**	**18**	**19**	**20**	**21**	
Group A	1	0	1	0	1	0	0	0	0	0	
Group B	0	1	0	1	0	1	1	1	1	1	

For display purposes, the table is divided, but it should be read as a 2×21 table.

Example 13.6

In Example 6.1, the Wilcoxon–Mann–Whitney test was applied to a data set presented in a contingency table format in Table 12.2. Show that if, in that format, row scores 0, 1 are used to indicate Group A and Group B, respectively, and column scores correspond to the ranks of the observations in the combined groups, then S is a statistic for the Wilcoxon rank sum test.

In Table 12.2, the columns were ordered by increasing times taken to complete the task. The columns in that table are allocated the rank scores 1, 2, 3, … , 21 and the statistic S in (13.16) is computed using these scores.

Table 13.7 displays the information in Table 12.2 with the times replaced by their ranks. It follows immediately that S given by (13.16) is the sum of the ranks for Group B and here $S=155$. Hence, the Wilcoxon rank sum test is one for independence in a linear-by-linear association model if the above row and column scores are used.

If Groups A and B are interchanged, i.e., the rows are swapped but the row scores are retained, the statistic S becomes the sum of the Group A ranks, the alternative Wilcoxon rank sum statistic used in Example 6.1. Hence, in this two-sample situation, the ordering of the groups does not matter. There is an opposite association with the two different orderings.

In the context of the WMW test, independence, or lack of association, corresponds to a situation when there is no difference between group medians. It is associated with a broad scatter of zeros and ones in the cells of a table like Table 13.7. There is a strong linear-by-linear association if the ones (corresponding to an observation) tend to concentrate towards the left or right in the first row and towards the opposite extreme in the second row, a situation seen in a simpler case in Section 12.3.1.

The Spearman rank correlation coefficient provides a more informative illustration. If there are n paired observations without ties, the rank outcome pattern can be presented as an $n \times n$ contingency table with rows and columns representing x, y ranks, respectively. If an x-rank i, is paired with a y-rank j the value 1 is entered in cell (i, j). All other entries in row i or in column j are 0. The row and column totals are all 1. The pattern is illustrated for three cases when $n=5$.

Case A	x Ranks	1	2	3	4	5
	y Ranks	4	2	5	3	1
Case B	x Ranks	1	2	3	4	5
	y Ranks	2	1	3	5	4
Case C	x Ranks	1	2	3	4	5
	y Ranks	1	2	3	4	5

These are represented as 5×5 tables in (13.17), (13.18) and (13.19)

Case A	y **Ranks**	**1**	**2**	**3**	**4**	**5**
	x Ranks					
	1	0	0	0	1	0
	2	0	1	0	0	0
	3	0	0	0	0	1
	4	0	0	1	0	0
	5	1	0	0	0	0

(13.17)

Case B	y **Ranks**	**1**	**2**	**3**	**4**	**5**
	x Ranks					
	1	0	1	0	0	0
	2	1	0	0	0	0
	3	0	0	1	0	0
	4	0	0	0	0	1
	5	0	0	0	1	0

(13.18)

Case C	y **Ranks**	**1**	**2**	**3**	**4**	**5**
	x Ranks					
	1	1	0	0	0	0
	2	0	1	0	0	0
	3	0	0	1	0	0
	4	0	0	0	1	0
	5	0	0	0	0	1

(13.19)

In case A, the 1s are well scattered over rows and columns. In case B, there is a discernible tendency for the 1s to drift towards the top left and bottom right, hinting at some association between x- and y-ranks. Case C clearly represents the closest possible association and corresponds to $r_s=1$. The resemblance of these tables to scatter diagrams (with conventional axes rotated clockwise through 90°) is apparent if the 0s are deleted and the 1s replaced by dots.

It is implicit from the remarks made in Section 10.1.1 about permutation tests for the Pearson coefficient that the sum $\sum_i r_i s_i$ may be used here as a test statistic for the Spearman coefficient in an exact permutation test. It immediately follows that if the ordered x ranks and y ranks are taken as scores for a linear-by-linear association test in contingency tables of the form (13.17)–(13.19) that S given by (13.16) equals $\sum_i r_i s_i$. The relevant permutation distribution is over all such possible 5×5 tables with row and column sums of unity; there are 5! such tables in each case.

Similar arguments for n paired observations (x_i, y_i), with no ties in either variable, show that the statistic $S = \sum x_i y_i$ appropriate for an exact permutation test for the Pearson product moment correlation coefficient (Section 10.1) is equivalent to (13.16) with scores $u_i=x_{(i)}$, $v_j=y_{(j)}$. The relevant permutation is over all $n\times n$ tables with row and column sums all equaling unity.

The models extend easily to tied situations. For example, using mid-ranks for the Spearman coefficient with ties, consider the case:

x Ranks	1.5	1.5	3	4	5
y Ranks	1	4	4	2	4

The relevant contingency table with mid-rank scores is now:

y **Ranks**	**1**	**2**	**4**	**Total**
x **Ranks**				
1.5	1	0	1	2
3	0	0	1	1
4	0	1	0	1
5	0	0	1	1
Total	1	1	3	5

To generate the relevant permutation distribution S is calculated using mid-rank scores for all permutations of this table with the given marginal totals. Alternatively, the centralized scores $u_i - m_u$, $v_j - m_v$ may be used.

It is possible to compute relevant P-values for any linear-by-linear association model based on permutation or Monte Carlo simulation of such tables subject to fixed marginal totals for arbitrarily assigned scores. This is a powerful tool, but it is important to choose relevant and sensible scores. Such scores are considered in a reanalysis of the data in Example 12.7. The tests are essentially applications of the permutation test for a Pearson product moment correlation coefficient with many ties and scores (x, y values) chosen to reflect what may be regarded as appropriate measures of *distance* between categories.

Example 13.7

Given the data on side-effects of dose levels of a drug in Example 12.7, i.e.,

	Side-Effects			
Dose (mg)	**None**	**Slight**	**Moderate**	**Severe**
100	50	0	1	0
200	60	1	0	0
300	40	1	1	0
400	30	1	1	2

allocate appropriate scores and perform a linear-by-linear association test for evidence of association between dose level and side-effects.

One reasonable choice of scores would be row and column ranks. For rows, this corresponds to dose levels in 100 mg units. For columns, it represents an ordering of side-effects consistent with limited information. Another possibility would be to allocate column scores in a way a clinician might interpret the data. He or she might accept 1 and 2 as reasonable scores for no and slight side-effects, but regard moderate side-effects as 10 times as serious as slight ones, and severe side-effects as 100 times as serious as slight ones, giving logical column scores $v_1 = 1$, $v_2 = 2$, $v_3 = 20$, $v_4 = 200$.

The `linear.by.linear` function in this book's R package will perform Monte Carlo tests using either of these (or any other chosen) scoring systems. It will also give the asymptotic result available via the methods given below. This should be used with some reservations when, as here, there are many cells with small counts.

Using row and column rank scores gives a Monte Carlo one-tail $P=0.01099$ and doubling this gives a two-tail $P=0.02198$. With the alternative column scores 1, 2, 20, 200 the corresponding tail probabilities are $P=0.00715$ and $P=0.01430$.

Output from the ANSM5 package in R can be obtained using `linear.by.linear (ch13$dose, ch13$dose.side.effect)` and `linear.by.linear(ch13$dose, ch13 $dose.side.effect, v = c(1, 2, 20, 200))`.

A one-tail *P*-value is relevant as it is logical to expect side-effects to increase rather than decrease as dose increases. Whichever scoring system is used, there is strong evidence that side-effects increase with dose.

In Example 12.7, the Jonckheere–Terpstra test indicated strong evidence of association. As pointed out in Section 12.3.2, that test is equivalent to one using Kendall's tau. Using ranks for row and column scores is more like using Spearman's rho without taking ties into account, while other row and column scores give a test equivalent to that for a Pearson coefficient with those scores playing the roles of x, y values.

One might have performed a test exactly equivalent to that for Spearman's rho by allocating mid-rank scores for ties. Because there are 51 patients tied at the 100 mg dose, the score for row 1 would then be 26, that for row 2 would be $(52+112)/2=82$, and so on. A valid criticism of the use of these mid-ranks in this highly tied situation is that the mid-ranks depend heavily on the number of subjects allotted to each treatment (here dose levels). There must be some unease about such a scoring system, especially if the numbers allocated to each treatment differ substantially. There is a better case for basing scores on the nature of treatments or responses rather than upon the number of patients allocated to each treatment. Possible unsatisfactory influences of mid-rank scores with heavy tying and unbalanced row and column totals have been pointed out by Graubard and Korn (1987).

In a group of tests on the same data, each using different but plausible scores, the *P*-values often differ more substantially than they did in this example. An important caveat about any scoring system is that it should be chosen for its relevance to the problem at hand and not on a hint inspired by an inspection of the data that suggests that a particular scoring system may enhance the prospects of getting a small *P*-value. The speed of modern computers tempts one to try many analyses for the same data. The temptation is best avoided for linear-by-linear association tests by choosing a scoring system before data are obtained. Once again, the reader is referred to the final paragraph in Section 1.7.1.

It can be shown that for S given by (13.16)

$$\mathrm{E}(S)=\left[\left(\sum_i u_i n_{i+}\right)\left(\sum_j v_j n_{+j}\right)\right]/N$$

and

$$\mathrm{Var}(S)=\frac{\left[\sum_i u_i^2 n_{i+}-\frac{\left(\sum_i u_i n_{i+}\right)^2}{N}\right]\left[\sum_j v_j^2 n_{+j}-\frac{\left(\sum_j v_j n_{+j}\right)^2}{N}\right]}{(N-1)}$$

and asymptotically $Z=[S-\mathrm{E}(S)]/\sqrt{\mathrm{Var}(S)}$ has a standard normal distribution.

Example 13.8

Apply an asymptotic test to the data in Example 13.7 using rank scores for rows and columns.

E(S) and Var(S) are computed by the formulae given above. The value of Z is calculated.

The rank scores are $u_i=i$, $v_j=j$. Routine calculation gives $n_{1+}=51$, $n_{2+}=61$, $n_{3+}=42$, $n_{4+}=34$, $n_{+1}=180$, $n_{+2}=3$, $n_{+3}=3$, $n_{+4}=2$, $N=188$. In addition, $S=484$, $E(S)=469.7$ and $Var(S)=35.797$ whence $Z=2.388$, and $P=Pr(Z>2.388)=0.0085$. There is therefore strong evidence against H_0.

Output from the ANSM5 package in R can be obtained using `linear.by.linear(ch13$dose, ch13$dose.side.effect, alternative = "negative", do.asymp = TRUE)`.

This result is in line with that of the exact test. Despite many ties and many small cell counts, in this example, the asymptotic results are not misleading.

Many other tests already encountered may be formulated as linear-by-linear association tests. If the rank column scores in Example 13.6 are replaced by raw data scores, this leads to a Pitman-type test. Other alternatives might use the van der Waerden scores or, if appropriate, logrank scores.

13.2.7 The Chi-Squared Test for Trend

A test for trends in $r\times 2$ contingency tables referred to as the *Cochran–Armitage test* or more commonly as the *chi-squared test for trend* may be formulated as a linear-by-linear association test. One situation in which the test might be applied is where the rows represent explanatory variables and the columns two mutually exclusive and exhaustive outcomes. Typically, the rows are ordered explanatory variables such as age groups or increasing doses of a drug. It is assumed that columns represent binomial responses, the first for an event A, the second for an event B. For the ith row, $i=1, 2, \ldots, r$ for each of the n_{i+} units in the counts in that row, the probability of the event A is p_i and of the opposite event B is $q_i=1-p_i$.

The Cochran–Armitage test was proposed independently by Cochran (1954) and Armitage (1955). It is used to test the hypothesis H_0: $p_1=p_2=\ldots=p_r$ against H_1: $p_1\le p_2\le\ldots\le p_r$ (or a similar inequality with all signs reversed) where at least one inequality is strict. Effectively, Cochran and Armitage proposed a test which is a linear-by-linear association test with column scores 0, 1 and row scores 1, 2, 3, …, r. In Example 13.9 it is applied to a set of data given by Graubard and Korn (1987).

Example 13.9

Table 13.8 indicates whether or not congenital sex organ malformation was found among children born to mothers whose stated alcohol consumption (number of drinks per day) during pregnancy fell within various ranges. There is a slight indication that the probability that malformation is absent may decrease with increasing maternal alcohol consumption, since the maximum likelihood estimates of p_i, for each row are $p_1=17{,}066/(17{,}066+48)=0.9972$ and similarly $p_2=0.9974$, $p_3=0.9937$, $p_4=0.9921$, $p_5=0.9737$.

An alternative hypothesis to equality of the p_i is one of a decreasing trend. An exact Cochran–Armitage test is performed as a linear-by-linear association test with row scores 1, 2, 3, 4, 5 and column scores 0, 1. Using the ANSM5 package in R and the command `linear.by.linear(ch13$alcohol, ch13$malformation)`, the Monte Carlo one-tail $P=0.1025$. There is no strong evidence against H_0: all p_i are equal.

TABLE 13.8

Numbers of Children With or Without Malformations in Relation to Mother's Stated Alcohol Consumption (Graubard and Korn, 1987).

	Malformation	
Mother's Alcohol Intake (Drinks per Day)	**Absent**	**Present**
0	17,066	48
<1	14,466	38
1–2	788	5
3–5	126	1
≥6	37	1

A one-tailed test is appropriate on the grounds that there is no medical or other evidence to suggest a beneficial effect of increased alcohol uptake in this context. The situation may be different for other health issues. There is some evidence, for example, that moderate consumption of red wine may protect against the onset of heart disease.

The choice of row scores is arbitrary. There may be a temptation to use mid-rank scores for rows as an intuitively reasonable scoring system. The disadvantage in using these scores, pointed out in Example 13.7, is more marked in this example. It can be verified that the mid-rank scores for rows are 8,557.5, 24,365.5, 32,013, 32,473 and 32,555.5. The final three rows – where most of the changes in malformation probability, if any, would seem to take place – get almost identical scores. For this scoring system a Monte Carlo $P=0.2746$.

Output from the ANSM5 package in R can be obtained using `linear.by.linear(ch13$alcohol, ch13$malformation, u = c(8557.5, 24365.5, 32013, 32473, 32555.5))`.

The temptation to experiment with many scoring systems should be avoided. When no other system is obviously appropriate, the default scoring system 1, 2, 3, 4 … has much to commend it. However, if there are clinical grounds for adopting some other scoring system it is sensible to do this. For this example, Graubard and Korn suggest using the mean alcohol intake for row scores, i.e., 0, 0.5, 1.5, 4 and 7 where the choice of 7 is arbitrary but not unreasonable. This results in a Monte Carlo one-tail $P=0.0170$, suggesting that there is indeed some evidence against H_0. Sprent (1998, Section 13.5) suggests yet another scoring system that might well get clinical support. It gives a P-value not very different from that using the mean intake scores. This suggests that the analysis is not too sensitive to differences between scoring systems that may appear realistic under what most people would regard as rational assumptions.

Output from the ANSM5 package in R can be obtained using `linear.by.linear(ch13$alcohol, ch13$malformation, u = c(0, 0.5, 1.5, 4, 7))`.

Availability of programs like that in the ANSM5 package in R for exact and Monte Carlo linear-by-linear association tests with arbitrary scores adds flexibility to the Cochran–Armitage test as originally proposed with restricted row scores 1, 2, 3, 4, ….

The Cochran–Armitage test can also be applied in the analysis of questionnaire items for which the responses are ordered, e.g., in a survey of attitudes regarding the installation of solar panels for electricity generation in rural areas where respondents state whether they are happy, indifferent, or unhappy about a proposed development. Responses are recorded as 1, 2, 3, … , etc., with one extreme (e.g., highly satisfied) taking the lowest value and the other extreme (e.g., highly dissatisfied) taking the highest value.

TABLE 13.9

Reported Frequency of Receiving/Giving Feedback on PPI Comments

	Representatives	Researchers	Total
Frequency			
Never	12	4	16
Sometimes	40	17	57
Always	10	17	27
Total	62	38	100

Example 13.10

Table 13.9 shows the reported frequency of receiving/giving feedback in the patient and public involvement study of Mathie et al. (2018) (see Example 12.1). Is there any evidence of a trend between perceived feedback frequency and type of participant?

The responses are scored using the default scoring scheme with Never (1), Sometimes (2), Always (3). There is some indication of a trend as the proportions for each response that is given by representatives are $p_1=12/(12+4)=0.750$ and similarly $p_2=0.702$, $p_3=0.370$. Application of the Cochrane–Armitage test gives a Monte Carlo two-tail P-value of 0.00682. There is strong evidence for a trend in the direction observed.

Output from the ANSM5 package in R can be obtained using `linear.by.linear(ch13$frequency, ch13$person)`.

13.2.8 Capture–Recapture Techniques

Capture–recapture analysis has a long history (Goudie and Goudie, 2007). Dahl (1919) applied the reasoning involved to estimate the number of trout in a lake by taking a haul of fish using a net, tagging and releasing the fish, taking a further haul, and recording the proportion of fish in the second catch that had a tag.

To examine this process in detail, consider the estimation of the total number of pine martens in a wood. One might use a benign form of trapping to capture 25 animals, mark each for identification, and release them back into the wild. At a later date, a further 40 animals might be trapped, and these might include 5 marked, i.e., recaptured, animals, the remaining 35 being unmarked. If the unknown number of animals in the wood is N, the first sample marking procedure may be looked upon as dividing the population into two sub-populations – the first consisting of the 25 marked animals, and the second of the $N-25$ unmarked animals. If all animals have the same probability of being included in the first and second trapping, then the ratio of the number of marked animals to unmarked animals in the second sample is an intuitively reasonable estimate of the proportions in the population.

Equating these ratios, i.e., setting $25/(N-25)=5/35$, and solving gives $N=200$. It is easily verified that this is equivalent to using the ratio of marked to total populations, i.e., $25/N=5/40$. More generally, if the first, or capture, sample size is n_1 and the second, or recapture, sample size is n_2, and in this second sample there are b marked animals then N is estimated by equating the ratios $n_1/N=b/n_2$ giving the estimate $N=n_1n_2/b$.

The assumptions that all animals have an equal probability of being included in each sample and that the population size N remains constant between the capture and the recapture sample are crucial but may be unrealistic in practice. Some animals may be more inquisitive than others and thus more likely to be trapped. Sometimes the probability of a marked animal being trapped again may be decreased by some aspect of the marking process, such as being frightened by the initial trapping or marking procedure. Alternatively, it may be increased because the bait used was a desirable food item. These factors negate the equal probability assumption. Further, the population size may vary between samples due to births, deaths or migration.

It is shown below that when the model leading to the simple estimate given above is valid, approximate confidence limits for the true population size may be obtained by formulating the problem using an incomplete 2×2 contingency table. When the basic assumptions needed for the validity of the simple estimate break down, it is well known to users of capture–recapture methods that some of these can be dealt with by taking more than two samples. This leads to a multiple capture–recapture process, where in each succeeding sample, one identifies which animals have been captured previously, and at which sample or samples this happened. Analysis of such data is often based on loglinear models to describe interactions or associations.

Capture–recapture methods have been developed to include the use of covariates. The most straightforward situation is with a single categorical covariate. One suggested approach is to stratify the data according to the covariate (e.g., if gender is the covariate, stratify into males and females), estimate the total number of cases for each stratum and combine these to obtain an overall estimate (Hook and Regal, 1995). More sophisticated models that take into account continuous covariates are described by Tilling and Sterne (1998). The simple estimate of population size is biased if genuine covariate effects are ignored.

A detailed treatment is beyond the scope of this elementary text, but the basic ideas are outlined using a contingency table approach. The use of loglinear models for capture–recapture analysis is described by Bishop, Fienberg and Holland (1975), while Seber (1982) provides a comprehensive review of basic capture–recapture procedures. Cormack (1989) gives a detailed description of practical uses of loglinear models in this context.

For the two-sample procedure outlined at the start of this section, the results may be expressed in a 2×2 table such as Table 13.10, where rows represent numbers from the total population of unknown size N that are included, or are not included, in the first sample; columns represent similar information for the second sample.

Using the standard notation n_{ij} for the count in cell (i, j), the relevant numerical entries in Table 13.10 are $n_{11}=b$, $n_{12}=n_1-b$, $n_{21}=n_2-b$, and the first row and column totals are respectively n_1, n_2. In any particular application, these all have known numerical values. However, the value of n_{22} is unknown, which, in the example above, is the number of pine martens that are not observed in either sample. This means that the second row and column marginal totals are unknown, along with the grand total N. The unobservable number n_{22} is called a *structural zero*.

TABLE 13.10

A Contingency Table for a Simple Capture–Recapture Model

	Present in the Second Sample		
Present in the First Sample	**Yes**	**No**	**Total**
Yes	b	n_1-b	n_1
No	n_2-b	?	?
Total	n_2	?	?

Interrogation marks indicate an entry to be estimated.

Denoting the unknown population number by N, an intuitively reasonable estimate of N is obtained by choosing it so that the observed number in each cell n_{ij} equals the expected number $m_{ij}=n_{i+}n_{+j}/N$ under the assumption of independence.

Remember that in a 2×2 table, knowing m_{11} fixes all other m_{ij} to ensure correct marginal totals. The entry $n_{11}=b$ and from Table 13.10 it follows for any unknown total N that $m_{11}=n_1n_2/N$. Setting $m_{11}=n_{11}=b$ implies an estimate $\hat{N}=n_1n_2/b$, in line with the start of this section.

This approach may look like using a complex tool when a simpler alternative is available, but it opens up useful ways to obtain a confidence interval for N. It also has implications for developing loglinear models allowing for interactions when applied to more than two samples. Sekar and Deming (1949) derived an expression for the asymptotic variance $\hat{N}$ namely

$$\mathrm{Var}(\hat{N}) = \frac{n_{1+}n_{+1}n_{12}n_{21}}{n_{11}^3}, \tag{13.20}$$

and asymptotic 95% confidence limits for N are

$$\hat{N} \pm 1.96\sqrt{\mathrm{Var}(\hat{N})}$$

For populations not much bigger than the combined sample sizes, these asymptotic limits are sometimes unsatisfactory, giving a lower limit less than the total count of distinct units in the two samples.

Regal and Hook (1984) suggest an alternative approach that uses the likelihood ratio statistic G^2. If the capture–recapture estimate of the total number of cases is used in the 2×2 table, G^2 is zero, since the method effectively equates observed to expected numbers under independence. They suggested forming a 95% confidence interval that includes all N for which the fit of the independence model is adequate in a significance test at the 5% level. As G^2 is associated with a chi-squared distribution with one degree of freedom, the 95% confidence interval consists of the values of N giving a $G^2<3.84$. Unlike the interval calculated from the standard error, this gives a confidence interval that is asymmetric around $\hat{N}$ with the lower limit being closer to $\hat{N}$. It is also necessarily no less than the total observed distinct number of population members.

Capture–recapture methods play a valuable role in the needs assessment of the most marginalized members of society such as the homeless (Fisher et al., 1994) and drug abusers (Hickman et al., 2006). In addition, they can, as the following example shows, be used to correct for underestimation of the prevalence of a medical problem in human populations when several incomplete registers (each corresponding to a capture or a recapture sample of patients) exist (Hook and Regal, 1982; Smeeton et al., 1999).

Example 13.11

Guillain–Barré syndrome is a paralyzing neurological condition due to inflammation of the peripheral nerves. In the UK, patients are listed on various registers, none of which is complete. Rees et al. (1998) estimated the number of such cases in the southeast of England. Two lists were obtained, one from the British Neurological Surveillance Unit (BNSU) and another from hospital activity analysis (HAA). There were 23 patients on the BNSU list, 68 were on the HAA list and 17 were on both. Find an estimate of the total number of cases in the southeast of England, along with a 95% confidence interval.

The relevant data can be presented in a 2×2 table in the format of Table 13.9:

	Present in HAA List		
Present in BNSU	**Yes**	**No**	**Total**
Yes	17	6	23
No	51	?	?
Total	68	?	?

There are only two registers, so for the simple capture–recapture model, interactions cannot be investigated. The two lists are assumed to be independent. From the table, $n_{+1}=68$, $n_{1+}=23$ and $n_{11}=17$. The appropriate estimate of N is $\hat{N} = 68 \times 23/17 = 92$.

The asymptotic variance of this estimate obtained from (13.20) is

$$\text{Var}(\hat{N}) = (23 \times 68 \times 51 \times 6)/17^3 = 97.41.$$

This gives 95% confidence limits for N of $92 \pm 1.96\sqrt{97.41}$ that lead, after integer rounding towards the estimate $\hat{N}$, to the interval (73, 111). This is clearly unacceptable, since from the lists there are at least 68+23−17=74 cases.

Using the likelihood ratio statistic G^2 gives a lower limit for N of 79 and an upper limit of 121. These limits do not include the total number of known cases, and the corresponding interval is, in that respect, at least more appropriate than the earlier one.

In summary, the estimated total number of cases is 92. On the basis of the likelihood ratio and 95% confidence interval, the likely number of hidden cases of Guillain–Barré syndrome in this area of England is between 79−74=5 and 121−74=47.

With small samples, caution is needed when using the asymptotic standard error method to obtain a 95% confidence interval for N. The lower limit of the interval should always be checked against the number of known cases to see if it makes sense.

Confidence intervals for small samples will almost certainly be wide, and in this example may not shed much light on the number of cases missed by both lists. Sometimes the upper limit of the confidence interval may exceed any reasonable estimate expected by individuals working in the field of study.

If it was thought that the degree of overlap for the two registers might differ between males and females, gender could be used as a categorical covariate. The number of cases would be estimated for males and females separately, the overall total being the sum of the two estimates. In this example, little would be gained from this approach as the sample size is so small that the confidence intervals for the subgroups would be too wide to give any real information.

Additional registers are invaluable for improving the estimate of N. Using these extra samples, loglinear models allowing interaction terms can be investigated. Rees et al. had access to two further registers – a research database containing 22 cases and 5 death certificates giving this syndrome as a cause of death. This brought the total number of known cases to 79, with an estimate for N of 98. The new 95% confidence interval for N, based on the likelihood ratio, was (86, 120). As one would expect, further registers increased the number of known cases. Note that the 95% confidence interval is less wide. The upper limit for N is slightly smaller. An interaction term was required for an acceptable model. This casts doubt on the unavoidable assumption of independence in this example.

A problem arises if the count b of numbers in both samples is zero, for the estimate of N using the above approach is infinite. Adjustments are available in these circumstances, but it is usually better, when possible, to overcome the problem by taking larger samples.

The likelihood ratio statistic G^2 can be computed from the chi-squared test for 2×2 tables, and this provides a basis for a trial-and-error process for forming a confidence interval based on this statistic. An improved interval might be based on that for the Fisher exact test.

The contingency table approach generalizes to three or more samples in a multiple capture–recapture program. For three samples, the outcomes may be presented in an incomplete $2\times2\times2$ table, where the three dimensions refer to the respective samples and the categories in each are presence or absence. The observed cell count in cell (i, j, k) is written as n_{ijk}, where a subscript value 1 implies a population unit that is observed in the corresponding sample and a subscript 2 implies that it is not observed. Thus, for example, n_{121} is the number of animals, patients, or whatever observed in both samples 1 and 3, but not in sample 2. Similarly, n_{111} is the number observed in all three samples. Since the value of N is unknown, there is one unobservable or structural zero cell count corresponding to n_{222}, the number in the population that are never seen.

The expected numbers m_{ijk} in any cell, other than the one with a structural zero, can be modelled by a loglinear model that allows for possible interactions due to factors such as varying probabilities of capture for individual animals within or between samples, population changes due to births, deaths, immigration or emigration, etc. The detail is beyond the scope of this book, and the interested reader should refer to Cormack (1989).

In the rest of this section, some general remarks are made about the method that indicate the versatility of the approach.

The modelling process differs from that considered in Section 13.2.1 due to the lack of a specific count in cell (2, 2, 2). This means that in the three-sample case, it is not possible to make inferences about, or test for, a second-order interaction. However, it is possible to test hypotheses about first-order interactions. If samples are taken on k occasions, the results can be presented in a 2^k contingency table with one structural zero cell. Inferences may be made about interactions up to order $k-2$.

The loglinear models appropriate to such analyses are a special case of the important class of models known as *generalized linear models*. How well a model fits the data can be examined using the likelihood ratio test statistic G^2, which is often referred to as the *deviance*.

For the independence model with k samples, G^2 is associated with a chi-squared distribution having 2^k-k-2 degrees of freedom. If the P-value from this goodness-of-fit test is low, the model gives an inadequate fit to the data, so a more complex model is sought. If an extra interaction term is added to the model, the number of degrees of freedom is reduced by one. To create models that are as simple as possible, it is usual to accept the extra term only if the decrease in the deviance is statistically significant ($P<0.05$). In other words, the deviance has to decrease by at least 3.84. If a model provides a very poor fit, this is invariably the case. As extra terms are added, the P-value rises, i.e., the fit is better.

The estimate for N should be given with an appropriate confidence interval. Asymptotic formulae for calculating confidence intervals in the case of multiple samples have been developed (Bishop, Fienberg and Holland, 1975; Selvin, 1995). However, as with the situation of having only two samples, these may lead to confidence intervals that contain values less than the number of known cases. To avoid this problem, Regal and Hook (1984) suggest the calculation of $G^2_{\min}$, the minimum value of the likelihood ratio statistic for the loglinear model under consideration. This is obtained when using the capture–recapture estimate for N. Using similar reasoning to the two-sample case, the appropriate confidence interval for that particular model contains all those values of N for which G^2 is no more than $G^2_{\min}+3.84$.

In applying capture–recapture methods to real data, it is sometimes necessary to introduce a subjective element in the model selection process. A more complex model is sometimes preferred if this allows the *P*-value of the overall fit to rise above 0.05, even if the decrease in the deviance is modest. One should bear in mind, however, that additional terms in a model might lead to wider confidence intervals for *N*.

A more complex model may need to be rejected even if there is a statistically significant decrease in the deviance if the confidence interval contains implausible values from an ecological or clinical point of view (Hay, 1997).

Smeeton et al. (1999) used capture–recapture methods in order to estimate the prevalence of congenital malformation of the heart in the first year of life. All such infants are severely ill, and doctors would, in the light of experience, question a predicted rate of more than 20 cases per 1,000. In a practical study, a model with three interaction terms gave an estimate of 14.5 cases per 1,000 and a 95% confidence interval for the rate from 5.7 to 79.9 cases per 1,000. A clinician would not take these findings seriously. The removal of one of the interaction terms reduced the estimate to 6.8 cases per 1,000 and the upper limit of the confidence interval to a much more realistic rate of 13.3 cases per 1,000.

Regal and Hook (1999) give a further discussion of interval estimation of population size using overlapping lists or records.

Capture–recapture analysis of constant populations can be performed with various statistical packages, including specialist packages in R and the `loglin` command of the stats package. The estimate for *N* is calculated from the coefficients of the loglinear model. If *N* varies and birth, death and migration need to be taken into account, then modelling can be used in the way described in Cormack (1989).

SUMMARY POINTS:

- The loglinear model is used for the study of many aspects of association. It has close parallels to the linear model for treatment comparisons such as those arising in the analysis of factorial treatment structures in the analysis of variance (Section 13.2.1).
- A no-interaction model corresponds to independence and a model with interactions to patterns of association (Section 13.2.1).
- Sets of k tables each 2×2 are often associated with comparisons of binomial responses to each of two treatments at each of k levels of an explanatory variable (e.g., different age groups or different sources of a raw material). This explanatory variable is often called a covariate (Sections 13.2.2–13.2.5).
- Both asymptotic and exact test procedures are available to test whether the odds ratios may be supposed to be equal for all 2×2 tables, and if they are, whether or not they all take the common value of unity. The latter implies independence (Sections 13.2.2–13.2.5).
- For such tests, equality at a value other than unity implies a first-order interaction (Sections 13.2.2–13.2.5).
- The linear-by-linear association model is a loglinear model of special importance in nonparametric methods. Many well-known tests may be formulated as special cases (Section 13.2.6).

- The chi-squared test for trend (Cochran–Armitage test) is a modified form of the chi-squared test for contingency tables that takes the ordering of the categories for the explanatory and/or outcome variables into account (Section 13.2.7).
- The scoring of categories required in the Cochran–Armitage test should be considered carefully, as the scores used may have an impact on the findings (Section 13.2.7).
- In the straightforward capture–recapture model, it is assumed that all individuals involved have the same probability of capture; this probability is unchanged by capture experience, and the population size remains constant (Section 13.2.8).
- In order to investigate models that involve interactions, at least three capture samples are required (Section 13.2.8).
- Asymptotic confidence intervals for the size of the complete population can contain values that are less than the number of known individuals (Section 13.2.8).

13.3 Combining and Partitioning of Tables

Intuitively, one may feel that if several tables refer to the same row and column categories, then combining these into one table to get larger counts in cells may be helpful. It might, for example, increase the power of tests for detecting association. In other circumstances, one may feel that some restructuring or partitioning of tables, perhaps by combining certain rows and columns or breaking down into strata, may reveal information not easily discernible in the original table. Combining or partitioning should be done with caution. Detailed discussions are given by Agresti (2013, 2019). The treatment here covers only certain aspects of these topics to give some indication of when they can be useful and points out some of the pitfalls that may go with incautious use of the methods.

13.3.1 Simpson's Paradox and Combining Tables

Suppose the information in Table 13.11 is made available regarding numbers of individuals in urban and rural areas responding to a standard cough medicine and a new medicine where, for each person, it is recorded whether there is no effect or if it cures the cough.

Visual inspection of Table 13.11 shows that the proportion of cures in both urban and rural areas is higher with the new medicine than it is with the standard (26.05% versus

TABLE 13.11

Responses to Two Cough Medicines in Urban and Rural Areas

	Urban		Rural	
	No Effect	Cure	No Effect	Cure
Standard Medicine	498	103	339	341
New Medicine	1042	367	125	190

17.14% for urban and 60.32% versus 50.15% for rural). This is confirmed if the Fisher exact test is performed on each table separately, giving $P<0.0001$ for urban areas and $P=0.0017$ for rural areas. Both provide strong evidence against the hypothesis that the medicines are equally effective.

Output from the ANSM5 package in R can be obtained using `fishertest.ANSM (ch13$medicine[ch13$location == "Urban"], ch13$response[ch13$location == "Urban"], alternative = "greater")` for urban areas, with `ch13$location == "Urban"` replaced by `ch13$location == "Rural"` for rural areas.

Suppose that the split into urban and rural areas is now ignored to give the 2×2 table where the proportion of cures is worse for the new medicine than it is with the standard (32.31% versus 34.66%).

	No Effect	Cure
Standard Medicine	837	444
New Medicine	1,167	557

Applying the Fisher exact test to this table indicates $P=0.0946$ for the alternative hypothesis that the standard treatment is better than the new. In this example, combining the information does not enhance the evidence of association. Indeed, any association is reversed.

Output from the ANSM5 package in R can be obtained using `fishertest.ANSM(ch13$medicine, ch13$response, alternative = "less")`.

This is a situation where the individual tables indicate association, and here association is in the same direction. Yet the combined table does not indicate association but suggests that if there were any association, it might be in the opposite direction. This illustrates *Simpson's paradox*, a phenomenon reported by Pearson, Lee and Bramley-Moore (1899) and Yule (1903). This was first described formally by Simpson (1951). The paradox may arise when data for two strata are combined where different models hold, especially when, as here, the sample sizes in each stratum are different. In the above example, the models for urban and rural areas differ because the proportion of cures for each medicine in these areas is markedly different. Indeed, the percentages in each calculated from Table 13.11 are

	Urban	Rural
Standard Medicine	17.14	50.15
New Medicine	26.05	60.32

The moral is that tables should only be combined if it is reasonable to assume that the relevant identical distributional model is valid for each table. It is probably only worth attempting combinations if counts are small in corresponding cells in all tables that are to be combined.

The paradox has received a lot of attention. Entering *Simpson's paradox* in a web search engine will guide one to many examples.

13.3.2 Partitioning of Tables

As explained in Sections 12.2.2 and 12.2.3, both the Pearson X^2 statistic, and the likelihood ratio G^2 statistic have asymptotically a chi-squared distribution with $(r-1)(c-1)$ degrees of freedom. It is well known that a variate having a chi-squared distribution with ν degrees of freedom may be expressed as a sum of ν independent variates each with 1 degree of freedom. The G^2 statistic maybe partitioned similarly into components, each of which has asymptotically a chi-squared distribution with 1 degree of freedom. Rules for partitioning are complicated in the general $r \times c$ case. There is usually more than one possible partitioning. Necessary conditions for the single degree of freedom components to be additive are given by Agresti (2013, Section 3.3) and are:

- Subtable degrees of freedom must sum to the degrees of freedom of the original table.
- Each cell count in the original table must appear in one and only one subtable.
- Each marginal total of the original table must be a marginal total for one and only one subtable.

Lancaster (1949) suggested partitioning an $r \times c$ table into $(r-1)(c-1)$ separate 2×2 tables each with 1 degree of freedom having the form

$$\begin{array}{cc} \sum_{s<i}\sum_{t<j} n_{st} & \sum_{s<i} n_{sj} \\ \sum_{t<j} n_{it} & n_{ij} \end{array}$$

For example, for a 3×3 table

$$\begin{array}{ccc} n_{11} & n_{12} & n_{13} \\ n_{21} & n_{22} & n_{23} \\ n_{31} & n_{32} & n_{33} \end{array}$$

this partitioning gives the four tables

(i)		**(ii)**	
n_{11}	n_{12}	$n_{11}+n_{21}$	$n_{12}+n_{22}$
n_{21}	n_{22}	n_{31}	n_{32}
(iii)		**(iv)**	
$n_{11}+n_{12}$	n_{13}	$n_{11}+n_{12}+n_{21}+n_{22}$	$n_{13}+n_{23}$
$n_{21}+n_{22}$	n_{23}	$n_{31}+n_{32}$	n_{33}

It can be verified that this partitioning satisfies the above necessary conditions. The partitioning is not unique. For example, if rows and columns of the original table are permuted, the chi-squared statistic for the whole table is unchanged, but clearly the partitioning above will be altered numerically. Although, under the null hypothesis of independence, asymptotically the chi-squared statistics for subtables are independent and additive, the Pearson X^2 statistics computed for each of the four tables (i)–(iv) above do not, in general, add to the corresponding statistic for the original table.

However, if the likelihood ratio statistic G^2 is used, the components are additive.

Partial partitioning, where not all components correspond to single degrees of freedom, is also possible. In practice, one should choose a partitioning that helps to answer relevant questions about association.

Example 13.12

Two drugs used in chemotherapy are tested on 100 patients, 60 receiving drug A and 40 receiving drug B. Numbers treated with each drug who exhibit the presence or absence of two specific side-effects, hair loss (HL) and visual impairment (VI) are as follows:

	Side-Effect Status			
	HL	VI	HL+VI	None
Drug A	9	4	16	31
Drug B	3	16	2	19

Partition G^2 to examine association between single side-effects, between single and double side-effects, and between overall side-effect status for the two drugs.

The necessary conditions for additivity given above are satisfied by comparing column 1 with column 2, then columns 1+2 with column 3, and then columns 1+2+3 with column 4. These components represent appropriate tests for association (i) between single side-effects for each drug, (ii) between a single side-effect (either HL or VI) and a double side-effect (both HL and VI) and (iii) at least one side-effect and no side-effect.

Tests for each may be carried out using the G^2 statistic in either an exact or an asymptotic test if the latter is felt appropriate.

The relevant partitioned 2×2 tables are

(i)	HL	VI
Drug A	9	4
Drug B	3	16

(ii)	HL or VI	HL+VI
Drug A	13	16
Drug B	19	2

(iii)	Side-Effect	None
Drug A	29	31
Drug B	21	19

It can be verified that for the complete table $G^2=22.13$.

Output from the ANSM5 package in R can be obtained using `lik.ratio(ch13$chemo.drug, ch13$chemo.side.effect, do.asymp = TRUE)`.

From the partitioned 2×2 tables, for (i) $G^2=9.72$, for (ii) $G^2=12.24$ and for (iii) $G^2=0.17$. The sum of these components is $9.72+12.24+0.17=22.13$, the G^2 value for the complete 2×4 table. The ANSM5 package in R may be used to compute both Monte Carlo and asymptotic *P*-values for the original table and each of the 2×2 components and gives the following values:

	2×4 table	(i)	(ii)	(iii)
Monte Carlo *P*	0.0001	0.0035	0.0010	0.8384
Asymptotic *P*	0.0001	0.0018	0.0005	0.6830

The extremely low *P*-value for the full table indicates strong evidence of association between drugs and side-effect status. The low *P* for subtable (i) reflects the fact that if there is only one side-effect, it is more likely to be hair loss for drug A, and VI for drug B. The low P for subtable (ii) reflects the fact that the occurrence of both side-effects is more common with drug A, while the high P for subtable (iii) shows there is no evidence of association between drugs and the incidence of some side-effect, i.e., that both drugs seem equally likely to give rise to some side-effects.

The action a clinician might take in light of these findings depends upon the importance of each side-effect. If VI were slight and temporary, it might be rated less serious than hair loss and this would indicate a preference for drug B. If VI were severe and permanent, it would be more serious than hair loss. If both side-effects are moderate, two side-effects may be regarded as more serious than just one.

The above partitioning is not unique; the columns might be reordered; e.g., place column 4 first and then combine columns in the order used above. The order used here leads to a more logical interpretation of the components of G^2.

This example has used a particular case of a rule for partitioning G^2 into $c-1$ additive components each with 1 degree of freedom for the case of a $2\times c$ table. The rule is to form first the subtable with columns 1 and 2, next the subtable with column 1+2 as the first column and column 3 as the second column, next the subtable with the sum of columns 1, 2, 3 and column 4, and so on until the final table uses the sum of columns 1, 2, 3, … , $c-1$ and column c.

Although the Pearson X^2 statistic and G^2 have the same asymptotic distributions under H_0, if X^2 is partitioned as above, the component values of the Pearson statistic will not in general sum to X^2.

Best (1994) used a partitioning of chi-squared into components analogous to linear and quadratic components in a regression type analysis of variance. He was specifically interested in looking at location and dispersion differences for data obtained in tasting trials, but appropriate use of this partitioning extends to other types of data, and it can also be generalized to look at other features such as skewness. Best's approach uses partitioning proposed by Lancaster (1953). An account together with a comparison with some alternatives is given by Sprent (1998, Section 13.2). Partitioning of chi-squared is the main tool used in a wide range of nonparametric techniques discussed by Rayner and Best (2001).

13.3.3 Partitioning in a Different Kind of Contingency Table

To compare preferences for competing brands of a product like tomato soup, or for similar types of wine made from the same grape in different countries, or for different varieties of peas or beans, a common procedure is to ask a number of potential consumers, N, to rank each of k brands of the product from 1 to k in order of preference. In practice, the ranking is usually in descending order of preference, with the brand ranked 1 being the first choice. The results may be expressed in a $k\times N$ table like Table 10.7.

If there are no tied rankings, the results may also be expressed in a $k\times k$ table like Table 13.12 with each row corresponding to one of the k brands and each column to a rank, the ith column corresponding to rank i. Here the rows have no natural ordering, but the columns are ordered. The entry in cell (i, j) is the number among the N testers who gave rank j to brand i. A classic paper on this topic by Anderson (1959) gave data for a study in which 123 consumers were each asked to rank three varieties of snap beans in order of preference. Table 13.12 is a similar table for rankings by 90 potential purchasers for three models of broadly similar cars.

The basic difference between this contingency table and those considered earlier in the chapter is that the cell entries are not sampled independently from 270 units. Instead, each

TABLE 13.12

Numbers of Potential Purchasers Giving Various Rankings to Each of Three Models of Car

	Rank			
Model	**1**	**2**	**3**	**Total**
I	27	47	16	90
II	35	31	24	90
III	28	12	50	90
Total	90	90	90	270

of the 90 would-be purchasers, independently of each other, allocates the ranks 1, 2, 3 to each of the three models. Since each purchaser uses each rank, he or she is represented once in each column. This means one may validly carry out a goodness-of-fit test to see, for example, if the hypothesis that there is the same probability of each variety being ranked 1 is supported by considering only the data in column 1. This is because what happens in the allocations in the remaining columns is irrelevant to that question. The appropriate test is one for a uniform distribution, using the methods of Section 12.4 leading to an X^2 statistic that has asymptotically a chi-squared distribution with 2 degrees of freedom.

If there is no preferred first choice, the expected numbers in column 1 are each 90/3=30. A similar argument could be applied to column 2, or column 3, but these tests would not be independent of one another because, if they were, the components could be added to get a chi-squared with 6 degrees of freedom for the whole table. However, with fixed marginal totals, there are only 4 degrees of freedom associated with Table 13.12. A moment's reflection shows that if a conventional Pearson X^2 test is performed with these fixed marginal totals, the value of X^2 (with 4 degrees of freedom) would be the same as that obtained by adding the goodness-of-fit statistics for the no-preference goodness-of-fit test applied to each row with 6 associated degrees of freedom.

Anderson conjectured that if the usual X^2 statistic is reduced by a factor 2/3 (i.e., the ratio 4 to 6 of the degrees of freedom), one might obtain a statistic that had asymptotically a chi-squared distribution with 4 degrees of freedom under a hypothesis of independence. He proved this was so and that the result extended to the case of k items to be ranked in the case of a $k \times k$ table, where the conventional Pearson X^2 statistic is reduced by a factor $(k-1)/k$.

The modified statistic is $X'^2 = (k-1)X^2/k$.

The data in Table 13.12 give $X^2=42.8$. Since $k=3$, the chi-squared statistic for the overall test is $X'^2=2\times 42.8/3=28.53$ with 4 degrees of freedom. It can be verified that $P<0.0001$ indicating, not surprisingly in view of the data pattern, strong evidence of an association between models and purchaser preferences.

Anderson discusses partitioning X'^2 into single additive degrees of freedom but, as in the previous section, these are not unique. The important thing is to choose a partitioning that has a sensible interpretation. For a 3×3 table, one partitioning into contrasts proposed by Anderson, and the interpretation of each contrast is:

- $C_1: n_{21}-n_{23}$ Linear ranking effect for model II.
- $C_2: n_{11}-n_{13}-(n_{31}-n_{33})$ Difference between linear rankings for models I and III.
- $C_3: n_{21}-2n_{22}+n_{23}$ Quadratic ranking effect for model II.
- $C_4: n_{32}-n_{12}$ Difference between quadratic rankings for models I and III.

That this is the interpretation of C_4 becomes clear if it is written in an alternative form:

$$[(n_{11} - 2n_{12} + n_{13}) - (n_{31} - 2n_{32} + n_{33})] / 3.$$

Equivalence is established by noting that $n_{i1}+n_{i2}+n_{i3}=N$, $(i=1, 3)$.

Anderson also establishes that each single degree of freedom component of X'^2 takes the form $C_i^2 / \text{Var}(C_i)$. In determining $\text{Var}(C_i)$, one must take into account correlations between the various n_{ij} that arise because the same observers are involved in all counts. The calculations of the $\text{Var}(C_i)$ are explained in Anderson's original paper and outlined in Sprent (1998, Section 13.3). Here only the result is quoted, viz. $\text{Var}(C_1)=\text{Var}(C_4)=2N/3$ and $\text{Var}(C_2)=\text{Var}(C_3)=2N$.

An interesting feature of these contrasts is that if the components X'^2 for the first two contrasts are added, their sum is equal to the Friedman statistic T given by (7.5) based on the rank sums for each variety given by all 90 purchasers, equivalent to Kendall's coefficient of concordance. These points are illustrated in an example.

Example 13.13

For the data in Table 13.12, partition the statistic X'^2 into components corresponding to the contrasts C_1–C_4 described above. Verify the assertion that the addition of the first two components gives the Friedman statistic T in this case.

The necessary computations are based on the description of the components given above.

From Table 13.12 it is seen that $C_1=35-24=11$, and since $\text{Var}(C_1)=2N/3$ and $N=90$ it follows that the corresponding component of X'^2 is $11^2/[2\times90/3]=2.02$ and in like manner the remaining components are found to be 6.05, 0.05 and 20.42, summing to $X'^2=28.54$, the value quoted above apart from rounding error. To calculate the Friedman statistic for these data, note that the sum of the ranks given by all 90 purchasers for model I is $27+2\times47+3\times16=169$. Similarly, the sums for models II and III are 169 and 202, whence the Friedman statistic is found to be $T=8.07$, equal to the sum $2.02+6.05=8.07$ of the first two components of X'^2.

Hence, if an asymptotic chi-squared test is applied to each component, there is strong evidence of association between models and consumer preference for components C_2 and C_4, but not for C_1 or C_3.

Components C_3, C_4 are associated with departures from linearity of ranking. Linearity, broadly speaking, represents average rankings and departures therefrom represent differences in dispersion. Table 13.12 indicates a monotonic fall-off for rank preferences for model II, but a different pattern for models I and III. Model I is most frequently ranked 2, and least frequently ranked 3. On the other hand, model III is seldom ranked 2; a large number rank it 3, but it gets one more first ranking than does model I. Thus, model III shows a greater dispersion of ranks than do the other models.

Anderson (1959) suggests that in many practical situations, the experimenter may be more interested in model or varietal contrasts at the more favourable ranks, 1 and 2, in which case a more useful set of contrasts would be

$$D_1 : n_{31} - n_{11}, D_2 : n_{31} - 2n_{21} + n_{11},$$

$$D_3 : n_{12} - n_{13} - (n_{32} - n_{33}), D_4 : n_{23} - n_{22}.$$

The components D_1 and D_2 are linear and quadratic comparisons for the first rankings – of main interest in this approach – while the remaining two examine differences in the second and third rankings.

Problems similar to those briefly discussed in this section have been considered by Scholz and Stephens (1987), Best (1994), Best and Rayner (1996) and others.

SUMMARY POINTS:

- Combining and partitioning of contingency tables is often a useful tool. The former must be used with caution, as is illustrated by Simpson's paradox (Section 13.3.1).
- Partitioning into subtables each corresponding to a single degree of freedom in a statistic having an asymptotic chi-squared distribution is often useful in elucidating the structure of associations (Section 13.3.2).
- Most partitionings are not unique, and certain necessary conditions must be met to ensure the additivity of components (Section 13.3.2).
- In discrimination tests where panels of testers each state their order of preference for a range of similar products, the numbers allocating each rank to each product may be presented in a contingency table. Cell entries are the number of testers allocating each designated rank to each product (Section 13.3.3).
- The appropriate analysis for association, i.e., consistency between N testers assessing k products or no particular preferences (independence), requires a modification that takes into account the fact that although there are only N testers, the total number of counts (entries in all k^2 cells) is Nk. Partitioning of a chi-squared statistic associated with such tables is often informative about the nature of any association (Section 13.3.3).

13.4 A Legal Dilemma

A field of application where statistical notions such as P-values and the power of a test may be especially relevant is law and jurisprudence. This section illustrates some difficulties that may arise in the use of statistical notions in court proceedings, and also in some other fields, by a simple example using hypothetical data given by Gastwirth and Greenhouse (1995).

These authors give two sets of data for comparable groups of 100 employees for each of two companies, A and B, where employees are drawn from the same population of potential and eligible employees. For each company, the numbers are broken down by majority and minority ethnic groups and within each group by those promoted and those not promoted as follows:

	Company A		Company B	
	Promoted	Not Promoted	Promoted	Not Promoted
Minority	1	31	2	46
Majority	10	58	9	43

Gastwirth and Greenhouse indicate that if these data were presented in a U.S. legal action alleging discrimination on ethnic grounds in promotion policy, the courts may require a further explanation from a company to justify a claim that its policy is not discriminatory on those grounds if the statistical evidence indicates rejection at the 5% significance level of the hypothesis of no association between promotion and ethnic grouping. This further explanation might be that at the time of an offer for engagement, employees were notified that promotion would only be granted to those willing to work unsocial hours or to work for long periods away from home, and that relatively few of the minority group who accepted employment met that requirement.

For the Company A sample, the promotion rate for the minority group is 1/32, i.e., 3.125%, compared with an overall rate of 11% for all 100 employees. For Company B, the corresponding percentages are 4.167 and again 11%. It appears then that Company A shows slightly greater discrimination, but is the evidence sufficiently strong to support the 5% significance criterion in a Fisher exact test, which is appropriate here? Even if it is accepted that a one-tailed test is appropriate, the critical significance level is not achieved since the exact P=0.0765.

Output from the ANSM5 package in R can be obtained using `fishertest.ANSM(ch13$group[ch13$company == "A"], ch13$promoted[ch13$company == "A"], alternative = "less")`.

Intuitively, the lack of evidence regarding Company A should put Company B in a safe position because they promote 4.167% from the minority group compared with 3.125 for Company A. However, situations like this have been met before where intuition can be deceptive, e.g., that arising with Simpson's paradox, so the Fisher test should be applied to the Company B data. Here the relevant one-tailed test gives P=0.0351.

Given these data, if a significance level of 5% is to be used to indicate acceptable evidence of discrimination, the court is the one that is going to have to do some explaining to Company B! This may appear to a member of the general public, and to the legal profession, somewhat rough justice. The court may well want an explanation from a statistician as to how the apparent anomaly arises. One contributing factor is that Company A is promoting from a smaller pool of minority employees – 32% as opposed to 48% for Company B. Sample sizes do affect the behaviour of the Fisher exact test not only with regard to power; they may also affect discontinuities in possible P-values. It may be appropriate for a statistician to draw the attention of the court to this difficulty. The difference in the proportion of minority group employees in the two companies – 32% for Company A and 48% for Company B – is also of relevance. As Gastwirth and Greenhouse point out, this difference probably arises because each company has its own recruitment policy. They suggest that in this example in the workforce available to either company for employment, there may be 40% belonging to the minority group. A question that then arises is whether either company discriminates between ethnic groups in their recruitment policy. There is not acceptable evidence that either does, for assuming random recruitment, the number of minority individuals recruited per 100 employees has a B(100, 0.4) distribution, and using an exact test as, for example, the `binom.test` function from the ANSM5 package in R, recruitment of 32 has an associated two-tail P=0.125 while recruitment of 48 has a two-tail P=0.104. However, if it is assumed simply that each company is recruiting from a pool with some fixed, but not necessarily known, proportion in the minority ethnic group, recruitment numbers of 32 and 48 from that group do indicate a different employment policy, the relevant P being 0.030.

The effect of this is that for the company with the smaller number of minority group recruits, the Fisher exact test has lower power than it has when applied to data for the firm with the greater number of minority group recruits. There is a timely warning here about the need for care when using conditional tests when the marginal totals used for conditioning may themselves be conditional on another factor – such as employment policy in this example. This again, for statisticians, serves as a warning against placing too much faith in formal significance levels such as 5%.

Difficulties of the type outlined above in a legal context arise in other fields, especially medical statistics. For example, in a study of lung cancer in relation to smoking, the difficulty arises if one conditions upon events in the causal path such as patient status for other disorders such as emphysema or hypertension, which are often related to smoking.

The use of statistical arguments in courts and in related fields such as forensic medicine is a fast-growing area. Courts on the whole prefer straightforward and not too complicated statistical arguments, providing these are sound. There are many recorded cases where there have been miscarriages of justice because of crucial weaknesses in statistical evidence – evidence not necessarily given by statisticians. For example, in Great Britain there was a substantial miscarriage of justice because a nonstatistician gave evidence, quoting probabilities based on an assumption that two events were independent when there was a wealth of empirical evidence that in the relevant circumstances they were not (Sesardic, 2007).

Statisticians often complain of reluctance by lawyers and the judiciary to consider statistical or probabilistic arguments, finding this strange in view of the standard requirement of proof beyond reasonable doubt in criminal cases and the tendency to resolve civil disputes on a balance of probabilities.

Another factor may be ambiguity in interpretation, as indicated in the example on promotion rates in minority groups. Such difficulties are most apparent in adversarial systems when prosecution and defence experts may often present plausible analyses of the same data that lead to different conclusions.

Many examples of statistical problems in relation to justice and law are given by Gastwirth (1988) and Aitken, Taroni and Bozza (2021).

SUMMARY POINTS:

- If court rulings are based solely on a significance level of 5%, decisions can be made that might seem intuitively unfair.
- A common reason for miscarriage of justice is that event probabilities are multiplied together under the assumption of independence when in fact the events are associated.

13.5 Power

Increasing attention has been given in recent years to the power and sample size relationships for tests involving contingency tables. Exact power computations in practice are limited to fairly restricted and clearly defined models such as the comparison of binomial populations in $2 \times k$, or $k \times 2$ tables, particularly when there is ordering in the k categories.

It was shown in Section 6.7.1, where power calculations for the median test were considered, that exact power calculations were possible if the binomial probabilities relevant under H_0 could be determined along with a specific alternative of interest.

The latter binomial probability was, in that context, dependent upon what assumptions were made about the population distributions giving rise to the samples under consideration. However, exact power computations are feasible once that binomial probability, p_1, is ascertained. The procedure is then one for determining the power of the Fisher exact test of H_0: $p=0.5$ against H_1: $p=p_1$. The approach can be generalized to any p_0 specified in H_0.

Some computer packages may include a facility for exact power calculations for $2\times k$ tables where the columns correspond to k binomial samples, or the rows correspond to two multinomial samples each with k possible outcomes. These cover many potential situations of practical importance providing meaningful binomial or multinomial probabilities for the alternative hypothesis of interest can be specified. If this is possible, the programs may be applied to tests such as the Cochran–Armitage test and the extensions to that test with arbitrary scores given in Section 13.2.7, as well as to two independent sample permutation tests, including WMW, normal scores, logrank scores or even the Pitman raw data test.

An example of how relevant binomial probabilities may be assigned for power calculations for a Cochran–Armitage test in a clinical trial context using a logistic regression model is given in Mehta, Patel and Senchaudhuri (1998).

SUMMARY POINTS:

- Exact power calculations are limited to a few clearly defined models.
- Use of specialised statistical software is almost always essential.

13.6 Exercises

13.1 Prepare cross-sectional tables for the data in Table 13.1 to show

i. separately for each cholesterol level, presence or absence of CHD at each blood pressure level;

ii. separately for each blood pressure level, presence or absence of CHD for each cholesterol level.

13.2 Use an asymptotic test based on $\phi=\log\theta$ to determine for each of the 2×2 subtables in Table 13.2 whether there is evidence against the hypothesis of no association.

13.3 The study of young people reported by Brooks et al. (2014) presented eating breakfast every weekday (Yes, No) by gender and engagement in vigorous exercise during leisure activities (VEL) as:

	Boys		Girls	
	Less than Daily VEL	**Daily VEL**	**Less than Daily VEL**	**Daily VEL**
Breakfast	942	313	948	175
No Breakfast	447	156	811	113

Determine whether the data in this table are consistent with a model that specifies only first-order association.

13.4 Considering only the data for CHD in Table 13.1, and assuming that SBP levels and cholesterol levels are both in increasing orders, use an appropriate test to decide if there is evidence of an association between SBP and cholesterol levels. If you were told also that the blood pressure levels were I=normal or below normal; II=between 1% and 10% above normal; III=11%–20% above normal; IV=21%–50% above normal; V=more than 50% above normal, and that the cholesterol levels are A=normal or less; B=up to 50% above normal; C=over 50% above normal and including some patients as much as 200% above normal, and that the risk of a fatal stroke is positively associated with both SBP and cholesterol level, how might you take this information into account in your analysis?

13.5 Agresti (2002) quotes the following data on cross-classification of attitudes towards abortion and amounts of schooling based on the U.S. General Social Survey, 1972. Test these data for evidence of an association between attitudes and educational background.

	Attitude towards Abortion		
Schooling	**Disapprove**	**Neutral**	**Approve**
Less than High School	209	101	237
High School	151	126	426
More than High School	16	21	138

13.6 Town A is the only large community in an otherwise thinly populated area. During a particular year, of those who received medical attention for an injury from using a kitchen knife, 40 visited a general practice and 90 were seen at the hospital Accident and Emergency (A & E) department. In 15 cases, the individual attended both a general practice and hospital A & E. Use this information to estimate the total number of kitchen knife injury cases in Town A during that year, giving an approximate 95% confidence interval. Are the assumptions that you made appropriate?

13.7 Use a Pearson chi-squared test or a likelihood ratio test to determine whether either of the 2×2 subtables (urban, rural) in Table 13.11 indicates an association, and if so, whether it is reasonable to suppose a first-order association model is adequate.

13.8 Thirty adults are each asked to indicate their preferences for each of three brands of competing detergents by ranking them 1, 2, 3. The results are presented in the following tables of numbers, giving each rank:

	Pref. rank		
Brand	**1**	**2**	**3**
A	12	2	16
B	9	12	9
C	9	16	5

Determine the value of the Anderson statistic and calculate the contrasts C_1–C_4 as given in Section 13.3.3. Hence show that the corresponding components for these contrasts sum to the Anderson statistic.

13.9 The age distributions of the PPI representatives and researchers in the study of Mathie et al. (2018) were:

	Representatives	Researchers	Total
Age Group			
16–55	17	29	46
56–65	17	8	25
66–75	34	2	36
Total	68	39	107

Why might the chi-squared test for trend (Cochrane–Armitage test) be appropriate for investigating a possible association between type of participant and age group? Apply the test to these data and explain your findings.

13.10 When age discrimination is alleged in the compulsory layoff of employees, courts often compare layoff rates among those aged less than 40 and those aged 40 or over. Gastwirth (1997) for a case settled out of court before trial gives these data for layoffs.

	Age <40	Age 40+
Not Laid Off	787	1,113
Laid Off	18	50

If the court considers there is sufficient evidence of discrimination only if significance is established at the 5% level, using an appropriate test, what conclusion should they reach on this evidence?

13.11 Gastwirth (1997) reports that for the case considered in Exercise 13.10, a further breakdown of the data by age became available as follows:

	Age <40	40–49	50–59	60+
Not Laid Off	787	632	374	107
Laid Off	18	14	18	18

Should the court reach a different decision using this evidence than that reached using the information given in Exercise 13.10? If a different conclusion is reached, or the evidence is substantially stronger, what are the statistical factor or factors that explain that difference?

14

Robust Estimation

14.1 When Assumptions Break Down

In the application of nonparametric statistical methods, it is important to remember that distribution-free does not mean assumption-free. As observed with those methods described up to this point, sometimes it is assumed that samples are from populations whose cumulative distribution functions differ, if at all, only in their measures of location, e.g., in means or medians. In other cases, the only assumption made is that the population distributions are symmetric, while in others, samples came from populations where any differences are, in some sense, ordered. Independence, both between observations within a sample and between different samples, is often of major importance.

The most relaxed assumption is that associated with the sign test applied to a single sample, where the null hypothesis requires only that each observation comes from unspecified distributions (not necessarily all the same), providing they all have the same median under the null hypothesis. Some of the procedures encountered were more sensitive than others to breakdowns in assumptions.

The breakdown often takes the form of a few sample values being inconsistent with the pattern associated with the bulk of the observations. Such aberrant observations are often referred to as outliers. Outliers may arise in several ways. They may include the following:

- Values that are incorrectly measured or recorded.
- Measurements made on units that are atypical of the bulk of those in the population under study.
- Measurements made with less accuracy or precision than the bulk of the observations.

There are many causes of incorrect data. A measurement of 3.6 cm may be recorded as 33.6 cm or 36 cm. Repeating a digit is a common error in keyboard data entry, as is the omission of a decimal point. A temperature measurement might be incorrectly reported in degrees Fahrenheit when it should have been in degrees Centigrade. Such incorrectly reported data are not always easy to detect. For temperature measurements in a hot climate, a reading given in degrees Fahrenheit when it should have been in degrees Centigrade is readily picked up. This may not be the case in polar climates, however, where readings on either scale can be plausible.

It is sometimes hard to find the cause of correctly recorded but atypical data. One machine among many may produce poor-quality goods because it is operating at an abnormal temperature, a factor that may not be observed, or is thought to be irrelevant,

DOI: 10.1201/9780429326172-14

by the person collecting the data. The milk production of one cow in a herd may be exceptionally low because it suffers from a yet-to-be-diagnosed illness that affects yield. Remote sensors may not be subject to in-person monitoring, and, as a result, unanticipated conditions affecting the sensors, and thus the data reported, will not be observed.

A frequent complication when data from several sources are combined is that those from some sources may be less reliable than those from others. For example, if the calcium content of milk samples is measured in several laboratories, most of the laboratories may make very precise and accurate determinations while one or two may make less precise measurements (increasing the dispersion about the correct values). Another possibility is that one laboratory may return values that are consistently either too high or too low.

Higher dispersion around a similar mean could indicate the use of an incorrect scale. For instance, at polar research stations temperature measurements may be recorded using the Centigrade scale at all stations apart from one where the Fahrenheit scale is used. The Fahrenheit readings will have a greater spread of values, a factor that might draw attention to the different scales in use.

How to proceed in such circumstances depends to some extent on the objectives. Obvious measurement or recording errors should be corrected if possible, or else the observation should be discarded. Some such errors nearly always remain undetected, especially in large routine data collections. Although one might also try to reduce the other sources of outliers, such as atypical experimental units or poor laboratory performance, one often has to live at least in part with these. It is important to draw attention to anomalies and to state clearly what action, if any, is taken regarding them.

Recognition of these problems has led to several major statistical developments, especially since the 1970s. These include the use of robust methods. These aim to minimize the influence of outliers or other anomalous observations when these lead to a breakdown in basic assumptions, while at the same time performing almost as well as the optimum methods when relevant assumptions hold. Robust methods may also be highly efficient while requiring only minimal assumptions.

In this chapter, two approaches that have proved valuable in the context of possible outliers are described and some others are referred to briefly.

The first of these – known as *the bootstrap* – has the added benefit that it is also valuable in situations where existing theory is intractable or difficult to apply. It has both a parametric and a nonparametric form. Only the latter form is considered here. An interesting feature of the nonparametric version is that it often comes close to being assumption-free. The one key assumption always needed is that the sample cumulative distribution function, encountered in Section 2.5, is a reasonable approximation to the population cumulative distribution function for samples that are not too small.

Compared to some longer-established nonparametric methods, there is a price to pay for this relaxation of assumptions, but that price is often less than any that may arise from making unjustifiable assumptions.

The second main approach considered is known as *M-estimation* with the key property that the resulting estimators behave like conventional maximum likelihood estimators when the latter are appropriate while protecting against certain types of outliers by reducing the influence of these in determining the outcome of the analysis.

The basic ideas behind both the bootstrap and *M*-estimation are straightforward, but their practical application often requires care if less obvious pitfalls are to be avoided. A few basic principles are outlined here; practical aspects can be found in more detailed texts.

As a preliminary, a brief description of a characteristic called *influence* is given. This concept helps to clarify the relationship between outliers and estimation in the context of the detection of outliers.

SUMMARY POINTS

- Distribution-free statistical methods are not assumption-free.
- Outliers may be due to errors in measurement or recording but can be correctly recorded atypical values.
- Observations that are recorded incorrectly are not always easy to detect.
- Robust methods have been developed to minimize the influence of outliers.

14.2 Outliers and Influence

14.2.1 Nature and Detection of Outliers

This topic is discussed fully by Barnett and Lewis (1994). In broad terms, an outlier is an observation so remote from or out of line with other observations as to cause surprise. Whether a particular observation is surprising is a matter of subjectivity. It may depend upon what is known about the source of the data. For example, given the data set 0, 5, 9, 8, 3, 0, 125, 9, 17 without further information, most people would say that observation 125 was sufficiently remote from the others to cause surprise. However, an entomologist who knew these were counts of numbers of aphids on each of nine plants of the same rose species may not consider 125 to be odd, for in many situations involving insect populations heavy infestation on just one or two among a large number of plants is common. Barnett and Lewis (p. 16) quote even more extreme data from Fisher, Corbet and Williams (1943) for numbers of moths caught in a light trap, viz.,

11 54 5 7 4 15 560 18 120 24 3 51 3 12 84

Given the observations 2.7, 3.3, 3.5, 2.8, 4.1, 4.3, and no other information, none may seem surprising. But if we are told that these are weights in kg for one growing animal recorded at fortnightly intervals, the validity of the fourth observation 2.8 is questionable. A growing animal may suffer weight loss at some point but the decrease is unlikely to be as large as this. If a loss of this magnitude were recorded, the animal would probably be dead a fortnight later; it would certainly be unlikely to return to a weight consistent with a normal growth pattern within a further 2 weeks. This observation is not an outlier in the sense of being remote, but it is probably a contaminated observation, the contamination being a measurement error. One possibility is that a true weight of 3.8 was recorded as 2.8. If the weights were read from a digital display, it could be a careless misreading.

Outliers or contaminated observations are a nuisance that must, or should be, accounted for when making inferences. Criteria have been designed that give a basis for excluding outliers from an analysis. The aim here is often to make the remaining data more consistent with some parametric inference model.

Barnett and Lewis (1994, Section 6.3) list 48 tests for outliers in normal distributions alone, most of these being optimal only for fairly specific alternatives to the null hypothesis that all data belong to the same normal distribution where neither or one or both parameters may be known. Difficulties are often caused by a masking effect, whereby the power of a test for one outlier is reduced by the presence of other outliers.

As mentioned already, if an outlier can be shown to be an error, it should be corrected if possible; if it cannot be corrected, it should be rejected. When there is no clear indication that an outlier is a measuring or recording error, the appropriate action is less clear. It depends upon the population of interest and on what questions are being asked about that population. For example, in an experiment to test a drug for reducing blood pressure, it may be known that it is ineffective if the recipient drinks alcohol. All recipients might be instructed not to drink alcohol while taking the drug. If, in a group of 25 patients receiving the drug, the decreases in systolic blood pressure in mm Hg are

−3	0	2	5	6	21	23	23	27	30	32	35	37
39	41	43	47	47	49	52	54	57	59	60	64	

a clinician might suspect that the first five readings were for patients who had almost certainly ignored the alcohol instruction. How firmly the clinician held that view would be a matter of experience. If it were known, for instance, that the drug is ineffective for about 1 person in 50 whether or not they drink alcohol, it would be reasonable to suppose that one or two of the first five readings might be for such cases, but rather unlikely for all five. If, on the other hand, there were strong grounds for believing it was almost certain that a substantial reduction in blood pressure would take place if alcohol was not consumed, and what was of interest was the mean or median reduction in blood pressure in such cases, it makes sense to omit the five lowest readings. To assume these five had all ignored the instruction regarding the use of alcohol might be unfair. Participants might be asked to self-report their alcohol intake. A dose-response relationship may mean that a little alcohol does not have the same impact as a larger intake. The accuracy of self-reported data could, of course, be disputed.

If there were indications that between 5% and 10% of the population might not respond positively to the drug, the above results are consistent with that hypothesis. In that case, if the whole population is of interest, no observation should be rejected when making inferences. If interest is only in the population of those who responded positively, the five observations should be rejected. The purpose of this hypothetical example is to show that there may be no easy answers to dealing with outliers. It is, of course, important, and ethically sound, to indicate clearly what has been done – and why – if outliers are present or suspected.

14.2.2 A Test for Outliers

Many tests for outliers lack robustness. Some are notoriously bad at detecting more than one outlier in the same tail; others tend to miss a pair of outliers in opposite tails. A simple and reasonably robust test (Huber, 1981) is to classify any observation x^* as an outlier if

$$|x^* - \text{med}(x_i)| / \{\text{med}[|x_i - \text{med}(x_i)|]\} > 5$$

Here, med(x_i) is the median of all observations in the sample. The denominator is a measure of spread called the *median absolute deviation*, often abbreviated to MAD. The choice of 5 as a critical value is motivated by the reasoning that if the observations other than outliers have an approximately normal distribution, the test then picks up as an outlier any observations more than about three standard deviations from the mean.

Example 14.1

Use the above test to detect any outliers in the data set

8.9 6.2 7.2 5.4 3.7 2.8 22.2 12.7 6.9 3.1 29.8

It is easiest to determine the median and MAD after ordering the observations. First test the observation furthest from the median. If this is not detected as an outlier, the process is terminated. If the test indicates that this observation is an outlier, the next most extreme in either tail is taken. The process continues until an observation that is not an outlier is found.

On ordering, the observations are

2.8 3.1 3.7 5.4 6.2 6.9 7.2 8.9 12.7 22.2 29.8

The median is 6.9 and the absolute deviation of the observation 2.8 from this median is $|2.8 - 6.9| = 4.1$. Similarly, the remaining absolute deviations are 3.8, 3.2, 1.5, 0.7, 0.0, 0.3, 2.0, 5.8, 15.3 and 22.9. Ordering these deviations shows that their median, the MAD, is 3.2. Setting $x^* = 29.8$, the left-hand side of the test statistic is $22.9/3.2 = 7.16$, indicating that 29.8 is an outlier. Setting $x^* = 22.2$ gives a test value of $15.3/3.2 = 4.78$, so neither 22.2 nor any other observation is classified as an outlier. The data set therefore contains one outlier, $x^* = 29.8$.

Having classed 29.8 as an outlier, it is necessary to decide what to do about it. An obvious line to follow is to check (i) whether it may be an error, and if it is whether it can be corrected. If it is not obviously an error then (ii) is there anything peculiar about the experimental unit giving rise to that value?

The data look likely to have come from a skew rather than a normal distribution. The test applied here should not be used for assessing normality. Lilliefors' test for normality might be more appropriate.

14.2.3 Influence and Robustness

In robustness studies, a key role is played by *influence functions*. The idea is due to Hampel who describes their role in robust estimation in Hampel (1974). Here, two simple influence functions are described briefly. A detailed treatment of these functions, including applications in various fields, is given in Barnett and Lewis (1994, Section 3.1.3).

First, the effect of a single outlier in relation to the mean is considered. Suppose there are n uncontaminated observations $x_1, x_2, \ldots, x_n$ regarded here as being *good*. There is also one contaminated observation z, which is an outlier in the sense that it takes a value greater than any good observation. Let $\bar{x} = \left(\sum x_i\right)/n$ be the mean of the good observations and $\bar{x}_a$ the mean of the augmented set that includes z. It can be shown that

$$\bar{x}_a - \bar{x} = \frac{n\bar{x} + z}{n+1} - \bar{x} = \frac{z - \bar{x}}{n+1} \tag{14.1}$$

whence it follows that the effect of z on the sample mean $\bar{x}$ for the good observations is a linear function of z that tends to infinity as $z \to \infty$. In other words, the effect of just one contaminated observation on the sample mean may be infinite! While (14.1) is a measure of the influence of one contaminated observation, the influence function $I_n(z)$ for a sample of n is obtained by multiplying the difference between the two means by $n+1$, i.e.,

$$I_n(z) = (n+1)(\bar{x}_a - \bar{x}) = z - \bar{x}.$$

As $n \to \infty, \bar{x} \to \mu$, the population mean for the good observations, and the limiting function

$$I(z) = z - \mu \tag{14.2}$$

is the asymptotic influence function, but it is often referred to simply as the *influence function* as it is the form of greatest interest. Broadly similar and easily modified arguments apply if z is an outlier in the lower tail, i.e., it has a value markedly less than all good observations.

In contrast to the situation with the mean where one outlier may have an appreciable, indeed unbounded, influence, the effect of a single outlier upon the median is usually small. It is indeed bounded providing the number of good observations $n \geq 2$. The situation is slightly but not essentially different depending upon whether the number of good data is odd or even.

Consider first the case of an even number of good observations $n=2m$. If the observations are arranged in ascending order and $x_{(i)}$ denotes the ith largest sample value, then the median of the good observations is $(x_{(m)}+x_{(m+1)})/2$. If an outlier z has a value greater than $x_{(2m)}$, the effect is to shift the median of the combined sample only to $x_{(m+1)}$. Similarly, an outlier below $x_{(1)}$ only shifts the median to $x_{(m)}$.

If the number of good observations is odd and $n=2m+1$ giving a median $x_{(m+1)}$, then an outlier $z>x_{(2m+1)}$ only shifts the median to $(x_{(m+1)}+x_{(m+2)})/2$. Similarly, any $z<x_{(1)}$ shifts the median to $(x_{(m)}+x_{(m+1)})/2$.

Because the underlying population distribution influences the values of the order statistics, a different approach to that used for the mean is needed to determine the asymptotic influence function. Details are given in Barnett and Lewis (1994, Chapter 4). They show that an appropriate expression is

$$I(z) = \frac{\text{sgn}(z-m)}{2f(m)} \tag{14.3}$$

where m is the median for the good observations that are distributed with a frequency function $f(x)$ and $\text{sgn}(z-m)=+1$, 0 or -1 depending upon whether z is greater than, equal to, or less than m. The function $I(z)$ given by (14.3) is bounded for a continuous distribution and the *supremum* or greatest value of $|I(z)|$ is $1/2f(m)$. This supremum is called the *gross error sensitivity*. For a large sample from a $N(\mu, \sigma^2)$ distribution, if just one contaminated observation is added, then the possible effect upon the sample mean is unbounded, while the maximum effect upon the sample median cannot exceed $1/[2f(\mu)]=\sigma\sqrt{(2\pi)}/2=1.253\sigma$, since the median of the normal distribution of the good values is μ.

If there are $n=2m+1$ good observations and two contaminated observations are added then, whatever their magnitude, the greatest effect upon the adulterated sample is that the median shifts from $x_{(m+1)}$ to either $x_{(m)}$ or $x_{(m+2)}$. The argument easily extends to four contaminants with bounds at most $x_{(m-1)}$ or $x_{(m+3)}$. Proceeding in this way, when $2m$ contaminated observations are added, the bounds become $x_{(1)}$ and $x_{(2m+1)}$. Only when there are at least $2m+1$ contaminants is it possible for the median to include some of the contaminated observations and thus become unbounded if some of the contaminated values are unbounded. Thus, at least 50% of the observations must be contaminated in a way that makes them outliers before the median becomes unbounded. A broadly similar argument may be used for an even number of good observations.

The number of contaminated observations needed to make a mean or median unbounded determines what is called the *breakdown point*. This is sometimes expressed as a percentage, i.e., 50% for the median, but more usefully as a fraction, i.e., 1/2 for the median. As already seen, the breakdown point is $1/(n+1)$ for the mean, since only one contaminated observation needs to be added to a sample of n to make the sample mean potentially unbounded. The breakdown point is an important measure of robustness because, in many practical situations, there is a strong suspicion, or even direct evidence, that an appreciable proportion of observations may be contaminated.

It should now be intuitively clear why in earlier chapters estimators based on medians were found to be more robust against outliers than those based on means.

SUMMARY POINTS

- To decide whether an outlier is a genuine observation, it is important to know the context in which it was observed (Section 14.2.1).
- A simple and reasonably robust test for outliers is based on the median absolute deviation (MAD) (Section 14.2.2).
- Influence studies indicate both the extent to which outliers may influence sample characteristics such as the mean or median and also explain why some estimators (e.g., the median) will be less affected by substantial numbers of outliers than others (e.g., the mean) (Section 14.2.3).

14.3 The Bootstrap

14.3.1 Motivation

As previously indicated, the motivation for the bootstrap is that the sample cumulative distribution function $S(x)$ introduced in Section 4.2 should, for all but small samples, reflect many characteristics of the population cumulative distribution function, $F(x)$, for the population from which the random sample was obtained. This was implicit in the Kolmogorov test introduced in that section, where $S(x)$ was used in deriving a statistic to determine whether a sample was consistent with some given $F(x)$. For the bootstrap, there is often limited concern with hypotheses about any particular $F(x)$. More often the focus is on making inferences simply on the assumption that $S(x)$ is a good approximator to some $F(x)$ without being specific about the precise form of $F(x)$. The vaguer the assumptions about population distribution, the more useful the bootstrap becomes.

The method is based on repeated resampling of data to discover more about the characteristics of the population from which the data are a random sample. Resampling, as seen as early as Example 2.3 and in numerous other examples of permutation test procedures throughout this book, is at the heart of permutation or randomization test theory. Permutation tests are based on resampling without replacement and, where appropriate, lead to exact tests and estimation procedures. Monte Carlo alternatives to permutation tests are also based on resampling without replacement. At each stage of a Monte Carlo procedure, a random permutation of the original data is used to create the test statistic of interest, and, with repeated random permutations, an empirical distribution for the test statistic is obtained. A Monte Carlo *P*-value is then obtained by comparing the test statistic obtained from the original data with this empirical distribution. By contrast, bootstrapping uses sampling with replacement and leads to approximate tests and estimation procedures; however, it is also applicable in many situations where no permutation test is available or appropriate.

Bootstrapping can be used without prior specification of any distributional model, but the procedure itself can only be justified by fairly complicated mathematics. The main practical requirement for its application is suitable computing facilities. Although bootstrapping results are by their nature usually only approximate, they are often more reliable and informative than those obtained by fitting a wrong model, e.g., assuming normality when that is not valid, or when it is not justified by some asymptotic result like the central limit theorem.

There is seldom a unique or best solution to a bootstrap problem. In practice, the method is used mainly either because there is no tractable analytic solution, when a permutation test or the facility to carry it out is not available, or when there is doubt about whether conditions needed for a particular analytic solution or permutation test actually hold.

The approach is intuitively reasonable in many applications, and confidence in the method is enhanced because, when it is correctly used, inferences are usually similar to those given by analytic or permutation solutions when these exist or are relevant. Important applications include those to complicated data structures or ones that involve inferences about concepts such as correlation or ratios of variables, where analytic results are not readily available except under very restrictive distributional assumptions.

The use of the bootstrap stems largely from the work by Efron (1979). A straightforward introduction to it and related techniques is given by Efron and Gong (1983). Full treatments at the elementary and intermediate level, with many examples, are given by Chernick (2008), Dikta and Scheer (2021), Efron and Tibshirani (1993) and Davison and Hinkley (1997).

14.3.2 Bootstrap Samples

Given a random sample of n observations $x_1, x_2, \ldots, x_n$ from some population, a bootstrap sample is a random sample of size n obtained from these data by sampling with replacement. This means some x_i may occur more than once, others not at all, in a bootstrap sample.

It is notionally possible to determine the distribution of all possible bootstrap samples and the distribution for many associated statistics, such as the means or medians of the bootstrap samples. Such distributions are called *true bootstrap distributions* to distinguish them from estimates based on a random subset of all possible bootstrap samples. The latter are important in practice because determining true bootstrap distributions analytically is a formidable task for all but small n, except for a few statistics where some general analytic results hold. In this latter case, inferences can usually be made easily without bootstrapping.

The complete bootstrap distribution for the mean and median for a sample of three observations is now derived. This is an interesting illustration, but trivial in practice because so small a sample is not very informative about the distribution associated with a large population. All possible samples of three formed with replacement from three distinct data x_1, x_2, x_3 are considered. Because of replacement, the probability of any x_i being selected is always 1/3. In a bootstrap sample of three, any x_i may appear r_i times, where r_i=0, 1, 2 or 3, subject to the constraint $r_1+r_2+r_3=3$, where the probability of each possible sample is given by the multinomial distribution. Writing $n(x_i)$ for the number of times each x_i occurs, the probability of a sample where x_1 occurs r_1 times, x_2 occurs r_2 times and x_3 occurs r_3 times is

$$\Pr\left[n(x_1)=r_1,\ n(x_2)=r_2, n(x_3)=r_3\right]=\frac{3!}{r_1!r_2!r_3!}\left(\frac{1}{3}\right)^3. \tag{14.4}$$

It is straightforward to record all possible bootstrap samples and the associated probabilities given by (14.4). Table 14.1 gives these together with each sample mean and median, where it is assumed, without loss of generality, that $x_1<x_2<x_3$.

For each sample, there is an associated probability p_i given by (14.4). These are shown in the second column of Table 14.1, so the probability distribution of the bootstrap samples is known. These probabilities also apply to the sample means y_i in column 3. Using the standard formulae $E(Y)=\sum p_i y_i$ and $\mathrm{Var}(Y)=\sum p_i y_i^2-\left(\sum p_i y_i\right)^2$, it can be shown that $E(Y)=\bar{x}=(x_1+x_2+x_3)/3$, the mean of the original sample of 3. Rather more complex algebra shows that the variance of the bootstrap sample means is given by $\mathrm{Var}(Y)=\sum(x_i-\bar{x})^2/9$.

These results generalize to samples of size n, where once again the mean of the bootstrap means is the overall sample mean, and now the variance of the sample means is s^2/n, where $s^2=\sum(x_i-\bar{x})^2/n$. This is similar to the classical estimator of variance of the mean and is indeed equal to the maximum likelihood estimator for samples from a normal distribution. Thus, bootstrapping, here as a method of determining the standard error of the mean, shows no advantage over classical methods. Any advantage shows for estimators other than the mean, where standard errors are not easy to obtain analytically except under restrictive assumptions about the population distribution.

For the small sample of three, bootstrap estimation of the standard error of the median is easy to illustrate. The bootstrap medians in column 4 of Table 14.1 take only three possible values x_1, x_2, x_3. For example, x_1 is the median of each of the three samples:

TABLE 14.1

Bootstrap Samples for $n=3$ and Sample Means and Medians

Sample	Probability	Sample Mean	Sample Median
x_1, x_1, x_1	1/27	x_1	x_1
x_1, x_1, x_2	3/27	$(2x_1+x_2)/3$	x_1
x_1, x_1, x_3	3/27	$(2x_1+x_3)/3$	x_1
x_1, x_2, x_2	3/27	$(x_1+2x_2)/3$	x_2
x_1, x_2, x_3	6/27	$(x_1+x_2+x_3)/3$	x_2
x_1, x_3, x_3	3/27	$(x_1+2x_3)/3$	x_3
x_2, x_2, x_2	1/27	x_2	x_2
x_2, x_2, x_3	3/27	$(2x_2+x_3)/3$	x_2
x_2, x_3, x_3	3/27	$(x_2+2x_3)/3$	x_3
x_3, x_3, x_3	1/27	x_3	x_3

(i) x_1, x_1, x_1, (ii) x_1, x_1, x_2, (iii) x_1, x_1, x_3, and the total associated probability calculated from column 2 is 1/27+3/27+3/27=7/27. Similarly, the median x_2 has a total probability of 13/27, and the median x_3 has a total probability of 7/27, whence, if Y represents the bootstrap median

$$\mathrm{E}(Y) = \frac{7x_1 + 13x_2 + 7x_3}{27}$$

and

$$\mathrm{Var}(Y) = \frac{7x_1^2 + 13x_2^2 + 7x_3^2}{27} - \frac{(7x_1 + 13x_2 + 7x_3)^2}{729}.$$

The bootstrap standard error of the median is the square root of this variance.

One must not expect to make useful inferences about characteristics such as the mean or median of a large population from a sample of three, by bootstrapping or by any other method. Even with realistic but still relatively small samples of, say, 10 or 12 observations, there is only limited evidence about the nature of the population from such scant data. In particular, when sampling from a large population, the sample cumulative distribution function (cdf) of a small sample may not be a good approximation to the population cdf.

The number of possible different bootstrap samples when $n=3$ is only the ten given in Table 14.1. The number of possible samples increases rapidly as n increases. When $n=4$, it is 35, with the probabilities associated with each ranging from 1/256 to 24/256, while for $n=10$ there are 92,378 distinct bootstrap samples.

In practice, bootstrap inferences are nearly always based on Monte Carlo sampling to generate a predetermined fixed number, B, say, of bootstrap samples. Here, a typical bootstrap sample is denoted by

$$x_1^*, x_2^*, \ldots, x_n^*$$

where each x_i^* is equal to one of the original observations. As already indicated, because sampling is with replacement some of the original sample values (x_j) may not appear, while others may appear more than once.

For example, if $n=9$ and the observations are

2.5, 3.1, 4.2, 5.1, 5.3, 5.9, 6.7, 7.2, 10.5,

a typical bootstrap sample might be

$x_1^*=4.2$ $x_2^*=6.7$ $x_3^*=5.1$ $x_4^*=6.7$ $x_5^*=7.2$

$x_6^*=6.7$ $x_7^*=2.5$ $x_8^*=10.5$ $x_9^*=5.1$

If B bootstrap samples are generated, the bth sample may be written

$$x_1^{*b}, x_2^{*b}, \ldots, x_n^{*b}.$$

Vector notation provides a convenient shorthand by writing $\mathbf{x}^*$ for any bootstrap sample and $\mathbf{x}^{*b}$ for the bth sample. For the numerical example above

$$\mathbf{x}^* = (4.2,\ 6.7,\ 5.1,\ 6.7,\ 7.2,\ 6.7,\ 2.5,\ 10.5,\ 5.1).$$

In bivariate or multivariate situations, such as those in correlation or regression, each x_i may itself be a vector. For each bootstrap sample $\mathbf{x}^*$, statistics such as the sample mean, median or sample variance are computed and used to estimate the corresponding population distribution characteristic. Denote the parameter or other population characteristic of interest by θ and the statistic used to estimate it from the bootstrap sample $\mathbf{x}^*$ by $s(\mathbf{x}^*)$. The estimate for θ can then be written as $\theta^* = s(\mathbf{x}^*)$.

Because sampling is with replacement, the numerical value of $s(\mathbf{x}^*)$ changes from sample to sample, and so the statistic has a distribution. As the number of bootstrap samples, B, tends to infinity, the mean of any statistic $s(\mathbf{x}^*)$ will tend to the mean computed for the true bootstrap sampling distribution. However, as pointed out above, this is only known in a few special cases or can only be worked out for general cases for small values of n, as in the illustration given earlier for $n=3$.

Practical experience shows that in many but not in all situations, even for small n, the mean of B bootstrap samples converges rapidly to the limiting value that holds when $B \to \infty$.

If B bootstrap samples are generated and the mean of the $s(\mathbf{x}^{*b})$ is denoted by $s(.^*)$, i.e., $s(.^*) = \sum_b \left[s(\mathbf{x}^{*b}) \right] / b$, then the appropriate estimator of the true bootstrap standard error of θ^*, denoted here by $\mathrm{se}_B(s)$, is

$$\mathrm{se}_B(s) = \left\{ \sum_b \left[s(\mathbf{x}^{*b}) - s(.^*) \right]^2 / (B-1) \right\}^{1/2}. \tag{14.5}$$

This is the usual estimator of a population standard deviation based upon a random sample of B from some population and tends to the true bootstrap standard deviation for the statistic s as $B \to \infty$. In practice, the approximation is often good for B as low as 20, provided n is not too small. Even for small n reasonable estimates may be obtained with $B=100$, although again indiscriminate use of the bootstrap with very small samples is cautioned, because then the sample distribution may not truly reflect a population characteristic of interest.

Pitfalls to watch out for when using the bootstrap with small samples include difficulties with bias. This may carry over to moderate or large samples in some contexts. Bias is discussed for a range of applications by Efron and Tibshirani (1993). It is important to realize that in applications using only a finite number, B, of bootstrap samples, there are two sources of error.

The first is the usual *sampling error* applicable to all sampling-based inferences about a population, no matter what method of inference is used. For example, in parametric inference about the mean of a normal population based on a sample of n, the sample mean $\bar{x}$ will, in general, not equal the population mean μ. The sampling 'error' in taking $\bar{x}$ as an estimator of μ is measured by the standard error.

The second source of error specific to bootstrap sampling is that made when the true bootstrap standard error is approximated using only a finite number B of bootstrap samples. It is not proved here, but it can be shown that the true bootstrap standard deviation is approximately equal to the estimated population standard deviation for the corresponding estimator. Example 14.2 sheds some light on the way sampling variation is reflected in bootstrap estimates.

Example 14.2

For the following sample data, use the bootstrap to estimate the population median.

0.04	0.06	0.27	0.32	0.33	0.40	0.50	0.63	0.69	0.92
1.09	1.10	1.35	1.61	1.66	1.69	1.71	1.80	1.98	2.65
2.83	3.50	3.72	3.75	3.99	5.16	5.49	6.31	7.05	16.05

The data are a random sample of 30 from a known distribution, but they are not unlike some that may arise in practice. They might, for example, be the percentage of a pollutant in various water sources, a situation where it is not uncommon to find small levels of pollution in many of the sources but quite high and wide-ranging levels in others. The data are arranged in ascending order to indicate more clearly the skewness with a long upper tail including the value 16.05, which might suggest itself as an outlier. A boxplot would be a useful exploratory data analysis (EDA) tool to show these characteristics. The sample median is (1.66+1.69)/2=1.675. The information available is explored by generating $B=40$ bootstrap samples and obtaining the median of each for use in estimating the bootstrap standard error of this statistic.

The bs function from the ANSM5 package in R was used to generate the 40 bootstrap samples. The function computes the median for each sample, which ranged from 0.805 to 3.5. The estimated bootstrap standard error given by (14.5) was 0.4615.

Output from the ANSM5 package in R can be obtained using the following.

```
bs.sample <- bs(ch14$example14.2, nsims.bs = 40)
range(bs.sample$vector)
sd(bs.sample$vector)
```

To help interpret these results, it is now disclosed that the data are a computer-generated random sample from an exponential distribution with mean $\mu=3$. The theoretical median of that distribution is $\theta=2.079$. A crude rule-of-thumb indication used in bootstrapping is that a hypothetical value of a parameter θ should be accepted if the sample equivalent estimator (here the sample median) lies within two standard deviations of that hypothesized value. Taking the estimated bootstrap standard error as a reasonable estimate of this standard deviation in this example, the difference 2.079 – 1.675=0.404 is less than one estimated standard error, which here is 0.4615.

There may be some unease as to whether 40 bootstrap samples give a reliable estimate of the true bootstrap standard error. That it does so is also an implicit requirement for the validity of the crude rule of thumb used above. This rule requires the assumption that the true bootstrap standard error is reasonably close to the best estimated standard error where this is known.

In this example, where sampling was from a known distribution, there is an analytic expression for the standard error of a median estimator from a sample of n, namely,

$$\text{se}(\text{median}) = 1/\left(4nf^2\right)^{1/2} \tag{14.6}$$

where f is the ordinate of the probability density function at the median. For an exponential distribution with mean 3, it can be shown that $f=0.1667$ and when $n=30$ that se(median) = 0.5476. In this example, the estimate of the bootstrap standard error, 0.4615, underestimates the true standard error. This is a bias introduced by the fact that the chosen sample has a median $m = 1.675$, which is appreciably below the population median $\theta = 2.079$. In fact, 19 of the 30 sample values are less than the population median. This is not an alarming departure from the expected number 15. The situation is not very different from that with many familiar tests. Recall that when sampling from a normal distribution, the usual sample estimator s^2 of σ^2 may differ appreciably from the latter in many samples. That is why, when σ^2 is unknown, the *t*-test is invoked to allow for such uncertainty.

It is interesting to note that if it is assumed that the population median is equal to the sample median of 1.675, then the standard error of the median estimator assuming the sample was from an exponential distribution with this median would now be 0.4412, close to the bootstrap estimate of 0.4615 obtained above.

The bootstrap may also be used to obtain an estimated confidence interval for a parameter such as the median. There are a number of practical difficulties in obtaining such intervals, but one approximate procedure is briefly outlined in Section 14.3.4.

In this example, a sample from a known distribution was considered and so it was possible to invoke some theoretical results for the standard error of the median because f was known in (14.6). Had f been unknown or if it was suspected the sample was from a mixture of distributions or that some observations were outliers, the bootstrap might be a serious competitor to, say, sign test procedures. This point is explored further in Example 14.5.

14.3.3 Bootstrapping versus Permutation Procedures

The bootstrap is particularly useful when no simple analytical procedure exists. It is also useful when no exact test based on permutations is available; it may still provide an approximation to a permutation procedure. Its application in these circumstances is, in general, not as straightforward as that of exact procedures, or even that for Monte Carlo approximations in permutation tests. As mentioned previously, a key difference between a permutation test and a bootstrap procedure is that the former is based on sampling *without* replacement, while the latter is based on sampling *with* replacement. The consequences of this apparently small difference are far reaching, some of which will be illustrated for two-independent-sample problems.

The Pitman permutation procedure for two independent samples from continuous distributions was not considered in much detail in Section 6.1.7 because it lacks robustness. For hypothesis testing, it is in theory straightforward, and, as in the case of most permutation tests, there are many equivalent statistics. Given samples

$$x_1, x_2, \ldots, x_m$$

$$y_1, y_2, \ldots, y_n$$

for the Pitman or raw data permutation test equivalent permutation test statistics for the hypothesis H_0: *the population means are identical* against the usual one- or two-tail alternatives under the assumption that the samples are from populations with otherwise identical distributions include the usual *t*-statistic, the first sample sum $\sum_i x_i$, the second sample sum $\sum_j y_j$, the first sample mean $\bar{x}$, the second sample mean $\bar{y}$, and the difference between sample means $\bar{x}-\bar{y}$. This equivalence follows from the fact that under permutation both the denominator in the usual t-statistic and the sum of the combined sample values remain constant, so that all the above statistics are linear functions of one another, have the same ordering, and the same P-value is associated with each corresponding value in that ordering. Remember, though, that if this permutation t-statistic is used, it does not have the tabulated t-distribution values relevant to the normal theory parametric test.

The above equivalences all assume sampling is without replacement. They do not carry over to the bootstrap, where sampling is with replacement. Then the sum of all $m+n$ values will change between bootstrap samples depending upon which of the original sample values do not appear at all, appear only once, or appear more than once in a given bootstrap sample. Thus, different conclusions will be reached if bootstrap inferences are based on each of the above possible permutation distribution statistics. As pointed out already, there is, in general, no unique and best bootstrap for a particular situation. What is important is to use a statistic that is intuitively reasonable in a particular case. If it is reasonable to assume that the samples come from identical distributions that differ only in mean, it makes sense to base bootstrap inferences on the sample mean difference $\bar{x}-\bar{y}$. How this might be done is shown in Examples 14.3 and 14.4.

Example 14.3

In Example 6.1, the times taken by two groups of students to perform a list of calculations were analysed. The observations were as follows:

Group A	23	18	17	25	22	19	31	26	29	33	
Group B	21	28	32	30	41	24	35	34	27	39	36

Assuming that the samples are from populations with distributions that differ if at all only in their means μ_1 and μ_2, use a bootstrap analysis to assess the strength of evidence against the hypothesis H_0: $\mu_1=\mu_2$ when the alternative is H_1: $\mu_1 \neq \mu_2$.

The analysis in Example 6.1 was based on the Wilcoxon rank sum test. However, here the analysis is based on the raw data, so that the bootstrap may be regarded as an approximation to a Pitman permutation test. Having taken 1,000 bootstrap samples, for each compute $\bar{x}^*-\bar{y}^*$ in the way described below. The relevant P-value is estimated as the proportion of the $\bar{x}^*-\bar{y}^*$ that exceed the observed $\bar{x}-\bar{y}$ in magnitude.

In a Pitman permutation test, where sampling is without replacement, each permutation sample of size m is obtained by drawing a sample of m observations from the combined sample of $m+n$ observations. The n observations not selected automatically form the complementary permutation sample of size n. In this example, $m=10$ and $n=11$. To obtain an analogous bootstrap sample, first select a sample of $m+n=21$ with replacement from the combined Group A and Group B data. The first ten values chosen for this sample are designated as a bootstrap sample of size m. The remaining n sample values constitute the bootstrap sample of size n. For each such bootstrap sample, compute $\bar{x}^*-\bar{y}^*$ and compare its magnitude with that of the observed $\bar{x}-\bar{y}=-7.245$ for the given data.

Using this procedure, one sample of 21 that could be obtained by sampling with replacement from the given data is

17 22 22 27 31 25 41 34 26 17 30 23 28 21 33 21 27 34 23 33 19

leading to bootstrap samples of 10 and 11 values, respectively,

17	22	22	27	31	25	41	34	26	17	
30	23	28	21	33	21	27	34	23	33	19

The difference between the sample means is $\bar{x}^* - \bar{y}^* = 26.2 - 26.545 = -0.345$.

An estimate of the two-tail test *P*-value based on *B* bootstrap samples is then given by $P^* =$ (number of samples for which $|\bar{x}^* - \bar{y}^*| \geq 7.245)/B$. Using the commands in R given below, for 1,000 such samples it was found that $|\bar{x}^* - \bar{y}^*| \geq 7.245$ in 11 cases implying $P^*=0.011$. A second sampling again with $B=1{,}000$ found $P^*=0.017$. Hence, there is strong evidence against the hypothesis that the means are equal.

R commands that will carry out the above analysis are below:

```
n.A <- length(ch6$groupA); n.B <- length(ch6$groupB)
bs.A.mean <- NULL; bs.B.mean <- NULL
for (i in 1:1000){
   bs.sample <- sample(c(ch6$groupA, ch6$groupB),
                       n.A + n.B, replace = TRUE)
   bs.A.mean[i] <- mean(bs.sample[1:n.A])
   bs.B.mean[i] <- mean(bs.sample[(n.A + 1):(n.A + n.B)])
}
sum(abs(bs.A.mean - bs.B.mean) > 7.245) / 1000
```

For these data, a Pitman permutation test gives the exact $P=0.014$, in close agreement with the bootstrap result obtained above.

The Pitman test usually gives similar results to the *t*-test even when assumptions relevant to the latter do not hold. For the above data, few statisticians would have reservations about the use of the parametric *t*-test and indeed for these data, the two-tail *P*-value given by that test is $P=0.011$, again in broad agreement. Also, in Example 6.1, the application of the exact Wilcoxon-Mann-Whitney (WMW) test to these data gave $P=0.016$, so there is good agreement between all tests considered here.

Using the bootstrap statistic $\bar{x}^* - \bar{y}^*$ is intuitively reasonable under the assumption that the only possible population distributional differences involve the means. A different approach is appropriate if the assumption of identical distributions under the null hypothesis is dropped. Although not pursued further here, this problem is discussed fully, together with some alternatives to the approach used in this example by Efron and Tibshirani (1993, Chapters 15 and 16). Even in simple parametric situations, choice of appropriate procedures depends strongly on relevant assumptions. For instance, validity of the *t*-test depends not only on an assumption of normality but also upon one of equal population variances.

Nearly all major statistical packages have a facility for random sampling with replacement and even if no direct bootstrapping program is included for the illustrative example considered here, it is usually straightforward to write a macro to form many bootstrap samples quickly. In R, the `sample` function can be used, noting that its default is to undertake sampling without replacement and the argument `replace = TRUE` must be used for sampling with replacement.

Bootstrapping, following the broad pattern in Example 14.3, may also be applied in situations where WMW methods are used. With one possible modification, the situation is essentially the same as that for the raw data, noting that the observations are now replaced by their ranks. The modification is that when bootstrapping in this manner, it is not strictly true that a WMW procedure is used. Why this is so can be illustrated by a trivial example.

Consider the situation in Example 2.3 where sailors were ranked 1–12 according to their health, two of them having received citrus fruits and the remaining ten other diets. What was performed in that example was essentially a WMW test (see Section 6.1.1). If the procedure in Example 14.3 is followed for combined sample raw data 1, 2, 3, 4, 5, 6, 7, 8, 9, 10, 11, 12, a combined bootstrap sample, of say, 10, 5, 4, 7, 2, 9, 2, 8, 1, 3, 3, 11 could be obtained. Continuing as in Example 14.3, this sample would split into bootstrap samples 10, 5 and 4, 7, 2, 9, 2, 8, 1, 3, 3, 11 of sizes 2 and 10, respectively. To do so is not unreasonable, but proceeding in this way is not performing a strict WMW bootstrap because the combined sample bootstrap data have not been reranked to allow for ties induced by the resampling. Doing so, using appropriate mid-ranks would replace 10, 5, 4, 7, 2, 9, 2, 8, 1, 3, 3, 11 by 11, 7, 6, 8, 2.5, 10, 2.5, 9, 1, 4.5, 4.5, 12 giving bootstrap samples 11, 7 and 6, 8, 2.5, 10, 2.5, 9, 1, 4.5, 4.5, 12. Although the difference is clear in this example, in practice, this tends to be small for samples of reasonable size.

The bootstrap is not recommended for small samples for a reason already explained – namely that a small sample may not be a good representation of an underlying distribution. This is in itself a limitation in any form of inference, but for bootstrapping, it is compounded by the bootstrap sampling error.

Example 14.4

For the data in Example 14.3, explore the use of a bootstrap in the estimation of a *P*-value using ranked data.

Bootstrap sampling may be used to obtain an approximation to the exact $P=0.016$ for a two-tailed test obtained in Example 6.1 using the WMW procedure. The process used follows broadly the pattern in Example 14.3 and is illustrated by our own findings.

For the combined samples, the rank sums are $S_m=76$ and $S_n=155$ (Example 6.1). These statistics are inappropriate for bootstrapping. This is because, for bootstrap samples, the corresponding S_m^* and S_n^* no longer satisfy the condition $S_m^*+S_n^*=(m+n)(m+n+1)/2$ in general. In an approach like that in Example 14.3, we used the difference between the means of the bootstrap samples as the appropriate statistic, comparing these with the difference between the observed mean ranks for the two groups of (155/11) – (76/10) or 6.49. We proceeded by taking 1,000 samples in the manner described in Example 14.3 using (i) the data obtained by resampling ranks directly and (ii) that obtained by modification by reranking the resampled data to allow for ties introduced by resampling with replacement, using the R commands given below. The respective two-tail *P*-values obtained in the way described in Example 14.3 were $P=0.015$ for (i) and $P=0.012$ for (ii). Both of these results are close to the exact WMW result. Hence, there is strong evidence against H_0.

R commands that will carry out the above analysis for (i) are shown below:

```
n.A <- length(ch6$groupA); n.B <- length(ch6$groupB)
bs.meanrnk.A <- NULL; bs.meanrnk.B <- NULL
for (i in 1:1000){
   bs.ranks <- sample(rank(c(ch6$groupA, ch6$groupB)),
                          n.A + n.B, replace = TRUE)
   bs.meanrnk.A[i] <- mean(bs.ranks[1:n.A])
   bs.meanrnk.B[i] <- mean(bs.ranks[(n.A + 1):(n.A + n.B)])
}
sum(abs(bs.meanrnk.A - bs.meanrnk.B) > 6.49) / 1000
```

For (ii), the command immediately beneath the for statement is replaced by the following.

```
bs.ranks <- rank(sample(c(ch6$groupA, ch6$groupB),
                n.A + n.B, replace = TRUE))
```

14.3.4 Bootstrap Confidence Intervals

Bootstrapping is often used to obtain approximate confidence intervals. Simplistic approximations suffer from two defects. The first is one of bias, which is covered briefly in Section 14.3.5. The other is that they often tend to provide less than the claimed coverage, e.g., an interval computed in a manner that purports to give a 95% coverage may only give about 91% or 92% coverage. At a basic level, Efron and Tibshirani (1993, Section 14.2) list five different ways to compute confidence intervals using the bootstrap. Each works tolerably well when appropriate, but it is hard to know which is appropriate if little is known about distributional properties of the population from which the data are obtained – a situation when statisticians are particularly likely to want to use the bootstrap. Another approach to bootstrap confidence intervals is given by Hall (1992).

Before proceeding further, it may be helpful to recall the well-known result that, given a sample of n from a normal distribution with variance σ^2, the 95% confidence limits for the population mean μ are $\bar{x} \pm 1.96\sigma/(\sqrt{n})$, where $\bar{x}$ is the sample mean. If σ is unknown, $\sigma/(\sqrt{n})$ is replaced by its usual sample estimate $s/(\sqrt{n})$ and 1.96 by the appropriate quantile of the t-distribution with $n-1$ degrees of freedom, where $s = \sqrt{\left\{\left[\sum_i (x_i - \bar{x})^2\right] / (n-1)\right\}}$.

For samples of more than 30 observations, the t-distribution quantiles closely approach those of the standard normal distribution. Modifications for other confidence levels are straightforward. In practice, the above limits are widely used even if there is little evidence that the sample is from a normal distribution, faith in the outcome relying on the central limit theorem holding, despite its asymptotic nature, even for a moderate sample size.

This fundamental result for parametric inference stimulated two approaches to forming bootstrap confidence intervals. The first is that for reasonably large samples, approximate 95% confidence limits for a parameter θ may be based on a sample estimator $\hat{\theta}$ of θ and are given by $\hat{\theta} \pm 2\text{se}^*(\hat{\theta})$, where $\text{se}^*(\hat{\theta})$ is an estimate of the true bootstrap standard error based on B bootstrap samples. For samples from a normal population, this works for the mean because it can be shown that in that case, the bootstrap standard error is a good approximation to the usual estimated standard error (although it shows a small bias). However, as emphasized earlier, the bootstrap is most useful when analytic theory is nonexistent or is highly distribution dependent and therefore any parametric theory may not hold for a particular sample. Experience has shown that this bootstrap method does not translate well to such situations. This is partly due to potential bias in bootstrap standard errors but more importantly because confidence intervals based on symmetry about an estimator $\hat{\theta}$ are inappropriate if that estimator has a skew distribution.

A more fruitful and relatively easy to apply approach to bootstrap confidence intervals is based on the distribution of appropriate bootstrap estimators, $s(\mathbf{x}^{*b})$ of each of B bootstrap samples. When B is large, then providing the $s(\mathbf{x}^{*b})$ are reasonably free from bias, the distribution of $s(\mathbf{x}^{*b})$ is likely to approach the sampling distribution of the estimator of the parameter θ of interest. An obvious way to estimate a 95% confidence interval is to take the 0.025 and 0.975 quantiles of the distribution of the B computed $s(\mathbf{x}^{*b})$. Thus, if $B=1{,}000$, for a

95% confidence interval, the limits are the 25th and 975th largest values of the $s(\mathbf{x}^{*b})$. More generally, for a (1–2α)100% interval with B bootstrap samples, the limits are the αBth and the $(1-\alpha)B$th largest sample value of $s(\mathbf{x}^{*b})$. For nonintegral values of such quantiles, it usually suffices to round to the nearest integer. Confidence intervals formed in this way are called *quantile-based intervals*. The main defects of these intervals are that they may be misleading due to bias and tend to give less than the nominal coverage. The effect of bias is that the true probabilities associated with each tail are not equal to each other, and which is the greater depends upon the direction of bias. Efron and Tibshirani (1993, Chapter 14) present two commonly used corrections to remedy these weaknesses, but these are not discussed here.

Example 14.5

For the data in Example 14.2, obtain 95% and 99% quantile-based confidence intervals for the population median using 2,000 bootstrap samples.

For each of 2,000 bootstrap samples, the median is obtained and these are arranged in order and the appropriate limits are obtained in the way described below.

We obtained 2,000 medians of bootstrap samples using a computer program and arranged these in ascending order. The 95% limits were the 50th and 1,950th largest. Using the ANSM5 package in R and the command `bs(ch14$example14.2, nsims.bs=2000, CI.width=0.95)$CI` gives limits of 1.005 and 2.74. Similarly, 99% limits are given by the 10th and 1,990th largest, and using the command `bs(ch14$example14.2, nsims.bs=2000, CI.width=0.99)$CI` turned out to be 0.69–3.50.

Estimated quantile-based 95% and 99% bootstrap confidence intervals for the population median are therefore (1.005, 2.74) and (0.69, 3.50).

For our 2,000 samples, the estimated standard error of the bootstrap median turned out to be 0.4144, in close agreement with the estimate of 0.4615 obtained in Example 14.1 from only 40 samples. It is common experience with bootstrapping that while samples of 100 or less give good estimates of standard errors, samples of 1,000 or more are needed for satisfactory quantile-based or related estimates of confidence intervals.

Because of the asymmetric sample (it has been disclosed that the data came from an asymmetric distribution (see Example 14.2)), approximate confidence intervals based on the estimated bootstrap standard error are not appropriate, even if that standard error estimate is itself good. In this example, using this approach, approximate 95% confidence limits based around the sample median are $1.675 \pm 2 \times 0.4144$, giving the unsatisfactory interval (0.85, 2.50).

When a statistic of interest is calculated for a large number of bootstrap samples, it is often useful to examine the distribution of these bootstrap statistics informally, perhaps presenting them in a histogram or as a box and whisker plot. Although the data set is too small to give bootstrap results likely to be relevant to a larger population, some bootstrap samples will be considered using the data in Example 11.2 as illustrations.

Example 14.6

In Example 11.2, the relationship between the amount of time spent by adolescents playing video games during weekends and the attitudes of their parents towards video gaming was depicted using a scatter diagram (Figure 11.2). The data given were as follows:

Upper Limit Stated by Parents (x)	0	1	2	3	4	5	6
Reported Gaming Time (y)	2.5	3.1	3.4	4.0	4.6	5.1	11.1

In that example and in Example 11.4, respectively, least-squares and Theil–Kendall estimates of the regression line slope β were obtained. Use a bootstrap with 1,000 samples to obtain approximate 95% inter-quantile-based confidence intervals for each of these estimates. Use histograms to illustrate the main characteristics of bootstrap estimates of β in each case. Explain the main characteristics of and differences between the forms of these histograms.

One, but not the only, basis for bootstrapping in a bivariate regression problem with n observed data points (x_i, y_i) is to take samples of n of these points with replacement. Using this approach for least-squares bootstrap estimates b^* of β, the usual least-squares estimate is obtained for each bootstrap sample. Similarly, for the Theil–Kendall method, the bootstrap estimator for each sample is the median of the pairwise b_{ij} (as defined in Section 11.4.2 for that sample).

For each bootstrap sample of data points, the bootstrap least-squares or Theil–Kendall estimator of β is computed in the way described above. For 1,000 samples, the least-squares estimators in one run gave a quantile-based 95% confidence interval for β as (0.49, 1.95), in reasonable agreement with the classic least-squares interval (0.23, 1.98). The Theil–Kendall estimator for a different sample of 1,000 gave an interval (0.50, 1.925), identical to the theory-based interval (0.50, 1.925) obtained in Example 11.4.

Output from R can be obtained using the following for the least squares.

```
n <- length(ch11$reportedtime)
bs.beta <- NULL
for (i in 1:1000){
   ids <- sample(n, n, replace = TRUE)
bs.beta[i] <- lm(ch11$reportedtime[ids] ~
                 ch11$parentlimit[ids])$coefficients[2]
}
quantile(bs.beta, c(0.025, 0.975))
```

For the Theil–Kendall estimator, the command starting 'bs.beta[i] <-' is replaced by

```
bs.beta[i] <- theil.kendall(ch11$reportedtime[ids],
                            ch11$parentlimit[ids])$stat
```

Figures 14.1 and 14.2 show histograms of the 1,000 bootstrap estimators of β obtained for each method using a class interval of 0.1. The class interval for 0.8–0.9, for example, contains all b^* such that $0.8 < b^* \leq 0.9$, with similar inequalities applying to other intervals.

Owing to appreciable discontinuities in possible values of the b^* (especially in the case of the Theil–Kendall estimator), not too much reliance should be placed on confidence intervals for so small a sample size. The most interesting feature of the histograms is the evident bimodality (or in the case of the Theil–Kendall estimator perhaps multimodality). A moment's reflection would suggest that nearly all bootstrap samples that do not include the aberrant point (6, 11.1) give a $b^* \leq 0.6$ and the breaks in the histograms between 0.6 and

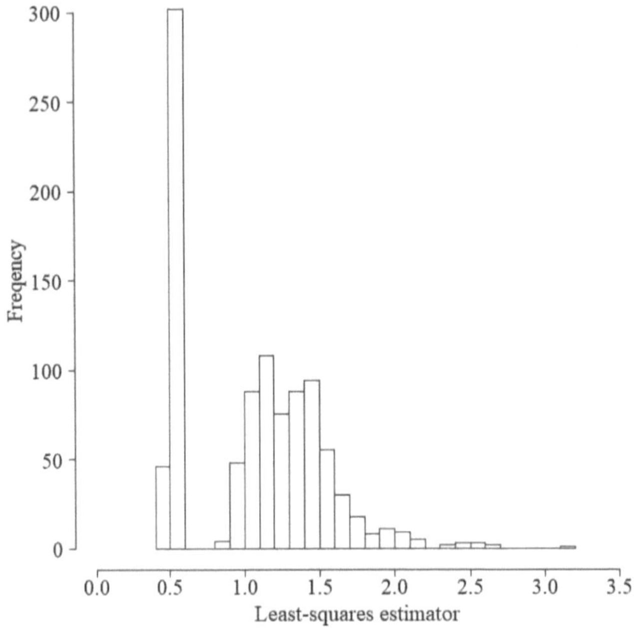

FIGURE 14.1
Histogram for 1,000 bootstrap least-squares slope estimators for the data in Example 14.6.

0.8 in both Figures 14.1 and 14.2 suggest that values of $b^* > 0.8$ arise from samples where the point (6, 11.1) occurs at least once. The fact that $b^* \leq 0.6$ for only 34.8% of the samples for least squares, while $b^* \leq 0.6$ for 64.9% of all samples using Theil–Kendall reflects the more robust nature of the Theil–Kendall estimator. If the point (6, 11.1) occurs only once, then for most samples, this aberrant point is likely to have little influence upon the median of the b_{ij}, whereas only one occurrence of this point in a sample will influence the least-squares estimator appreciably. In Section 11.5, it was pointed out that the least-squares estimator can be expressed as a weighted mean of the b_{ij}, and the robustness properties noted here are broadly in line with the findings on influence in Section 14.2.2.

The number of b_{ij} in many bootstrap samples will be less than the total of 21 possible pairings of points for the original sample because of repeated values, and also many of the pairwise estimates will be identical if there are several repeated values. These factors may also influence robustness in a rather complicated manner. With either estimator in an extreme case, the bootstrap sample may consist of seven replicates of the same point in which case the estimate of β is undefined for both least-squares and the Theil–Kendall method.

The results discussed here are based on one sample of 1,000 for least-squares estimators and on another sample of 1,000 for Theil–Kendall estimators. It would, of course, be feasible and indeed interesting to compare both estimators for the same sample. The discussion above referred to discontinuities in the Theil–Kendall estimators. Table 14.2 indicates the extent of these for the particular sample used in this example. These arise because a variety of samples give the same sampling median (which must be an observed b_{ij} or the mean of two observed b_{ij}).

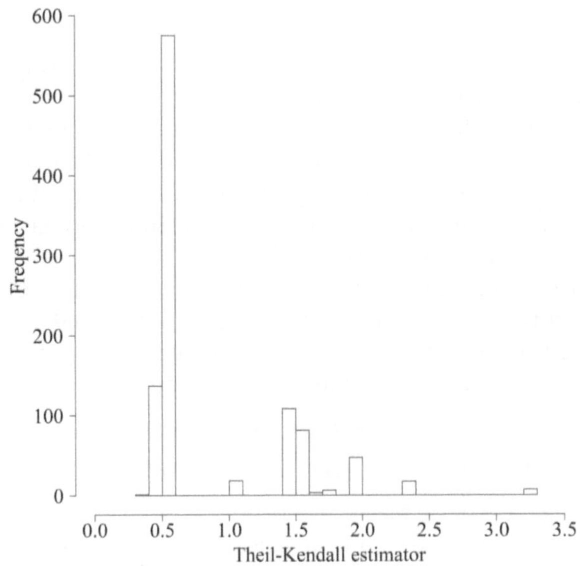

FIGURE 14.2
Histogram for 1,000 bootstrap Theil–Kendall slope estimators for the data in Example 14.6.

TABLE 14.2
Values of 1,000 Bootstrap Slope Estimates b^* and the Number of Times (r) That Each Occurs Using the Theil–Kendall Estimator for the Example 14.6 Data

b^*	r	b^*	r	b^*	r	b^*	r
0.3	1	0.525	69	0.567	82	1.517	9
0.45	17	0.535	1	0.575	4	1.6	72
0.5	120	0.537	8	0.583	11	1.679	3
0.51	2	0.546	11	0.6	215	1.762	6
0.512	3	0.55	55	1.017	15	1.925	47
0.52	73	0.558	25	1.1	3	2.367	17
0.522	11	0.562	5	1.433	108	3.25	7

These simple illustrations of bootstrapping convey only the flavour of this approach and indicate in an intuitive way some of its strengths and weaknesses. As pointed out earlier, the method is most useful in situations where either there is no simple analytic method or the data are samples from essentially unknown distributions. Correlation problems and inferences based on ratios of two variables are situations often met where theory is generally intractable unless very simple distributional assumptions can be made. Several detailed examples of its use in problems involving both correlation coefficients and ratios are given by Efron and Tibshirani (1993) and by Davison and Hinkley (1997). Efron (1981) and Robinson (1983) applied the bootstrap to censored data, while Hall and Hart (1990) used the bootstrap to test for differences between means expressed as very general regression functions.

14.3.5 Related Techniques

Although it was not apparent in the examples above, the bootstrap frequently leads to biased estimators. In a few cases, theoretical adjustments may be made to remove bias, and bootstrap methods exist for estimating that bias. In many practical applications, bias is more easily estimated by a related but older technique called the *jackknife*. The idea was introduced specifically for bias estimation by Quenouille (1949), but the name jackknife is due to Tukey (1958). The jackknife will not always work; in particular, it will not work well when the median is used as an estimator.

Many commonly used estimators that are sample analogues of their population counterparts are biased except perhaps when rather strong distributional assumptions are made. Indeed, one of the few universally unbiased estimators in common use is the sample mean, which is unbiased when used as an estimator of the population mean. A simple example of a biased estimator is the sample variance $s^2 = \sum (x_i - \bar{x})^2 / n$ as an estimator of a population mean σ^2. Here the bias is easily estimated, it being well known that $E(s^2) = (n-1)\sigma^2/n = \sigma^2 - \sigma^2/n$, so the bias is $-\sigma^2/n$.

The jackknife works this way. Given a sample of n, for any parameter θ, for which an estimator is the sample analogue of the parameter being estimated and is denoted by $\hat{\theta}$, n further estimators are formed by replacing the original sample with a set of samples each identical to the original except that one observation is omitted from each in turn. These are called *jackknife samples*. Thus, the ith jackknife sample is

$$x_1, x_2, \ldots x_{i-1}, x_{i+1}, \ldots, x_n. \quad i = 1, 2, \ldots, n.$$

The estimator of θ based on the sample with x_i omitted is conventionally denoted by $\hat{\theta}_{(i)}$, and the mean of the $\hat{\theta}_{(i)}$ by $\hat{\theta}_{(.)}$, i.e., $\hat{\theta}_{(.)} = \sum_i \hat{\theta}_{(i)} / n$. Finally, the jackknife estimator $\theta^{\dagger}$ of θ is

$$\theta^{\dagger} = n\hat{\theta} - (n-1)\hat{\theta}_{(.)} \tag{14.7}$$

and the jackknife estimator of bias is

$$\text{bias}(\hat{\theta}) = (n-1)(\hat{\theta}_{(.)} - \hat{\theta}) \tag{14.8}$$

The motivation for (14.7) and (14.8) is that, for both the sample mean and the sample variance as estimators of the corresponding population parameters, (14.7) gives unbiased estimators of these parameters and (14.8) correctly estimates the bias of the sample mean as estimator of the population mean as zero and that of the sample variance as estimator of the population variance as $-\sigma^2/n$. These results can be verified by calculating expectations. In general, (14.8) does not reduce the bias of every estimator to zero nor does (14.7) always provide an unbiased estimator of any parameter. However, the latter will usually reduce any bias in $\hat{\theta}$.

The jackknife and its use in association with the bootstrap to reduce bias and make other improvements to bootstrap estimation are discussed in detail both by Efron and Tibshirani (1993, especially Chapter 11) and by Davison and Hinkley (1997).

Another widely used technique that bears some resemblance to the bootstrap is *cross-validation*. In fields like regression and classification methods, parameter estimation is often a part of the model building process to produce a model that will be used to make predictions when new data become available. One source of concern is how good these predictions will be when applied to the new data. Ideally, all available data should be used to fit a model, the process then being followed by further sampling from the original population to see how well the fitted model acts as a predictor. This may be costly or impossible, in which case an alternative is cross-validation. For this, the data are divided into two or more portions and a model is fitted in turn to all but one of these portions. The prediction error of the fitted model when it is fitted to the omitted portion is then calculated. Each portion is omitted in turn and a combined estimated prediction error is obtained.

An extreme case especially suited to relatively small data sets is the *leave-one-out* cross-validation procedure where each of n observations is omitted in turn and a model is fitted to the remaining $n - 1$ data, a predicted value is then obtained for the omitted observation using that model. For example, in a regression context if, when the ith observation is omitted, the predicted value of y_i is $y^*_{(i)}$. A measure of cross-validation prediction error is $\sum_i \left(y_i - y^*_{(i)}\right)/n$. The computational procedure is similar to that for the jackknife, but the objective is different. A more detailed description of cross-validation and its relationship to the bootstrap is given by Efron and Tibshirani (1993, Chapter 17).

SUMMARY POINTS

- Bootstrapping is based on the repeated resampling of data with replacement (Section 14.3.1).
- The distribution of all possible bootstrap samples is called the true bootstrap distribution (Section 14.3.2).
- Bootstrapping can provide an approximation to a permutation procedure (Section 14.3.3).
- Bootstrap confidence intervals can be constructed based on the distribution of appropriate bootstrap estimators. These are known as quantile-based intervals (Section 14.3.4).
- Problems with bias that are associated with certain bootstrap estimators may be eased by the use of the jackknife (Section 14.3.5).
- Cross-validation is useful for exploring the predictive ability of a fitted model (Section 14.3.5).

14.4 *M*-Estimators and Other Robust Estimators

Throughout this book, the estimators described have shown varying degrees of tolerance to breakdowns in assumptions especially those associated with a few observations being out of line with the general patterns. As indicated in Section 14.2 and elsewhere, the median is generally more robust than the mean in the presence of a few rogue observations – behaviour explained by the influence functions. The price to be paid for this robustness may be an increase in the standard error of the estimator or perhaps the introduction of bias in procedures such as the bootstrap. There are several compromise estimators that retain some of the desirable properties of the mean as an estimator but either downweigh, or eliminate, the more extreme observations.

One such estimator is the *trimmed mean*, in which a proportion of the observations (typically the 10% most extreme) are eliminated, and inferences are based on the mean of the remaining observations. The median is an extreme case of the trimmed mean, in which all but one observation is trimmed if n is odd, and all but two are trimmed if n is even. Another estimator is the *Winsorized mean*, where extreme observations are shrunk to the value of the largest remaining observation. These methods are not discussed in detail here, but they, along with related ones, are fully covered in Barnett and Lewis (1994) in a range of contexts.

The mean, median, trimmed mean and Winsorized mean are members of a class of estimators sometimes called *L-estimators*. Here, L stands for *linear*, the name being given because if the ith ordered sample value is written as $x_{(i)}$, then any of these estimators take the form

$$\mu^* = \sum_i w_i x_{(i)}$$

where w_i are weights.

Putting $w_i=1/n$ for all i gives μ^* as the sample mean. If n is odd, say $n=2m+1$ and put $w_{m+1}=1$ and $w_i=0$ if $i \neq m+1$, then μ^* is the sample median. For $n=2m$, the median is the mean of the middle pair of values, so the weights are $w_m=w_{m+1}=0.5$ and $w_i=0$ otherwise. For a trimmed mean, where the k largest and k smallest observations are trimmed, the weights are $w_i=0$ if $i \leq k$ or if $i \geq n-k+1$, and $w_i=1/(n-2k)$ otherwise. For Winsorization, if k values in each tail are shrunk to $x_{(k+1)}$ and $x_{(n-k)}$, respectively, then $w_i=0$ if $i \leq k$ or if $i \geq n-k+1$, while $w_{k+1}=w_{n-k}=(k+1)/n$, otherwise $w_i=1/n$.

Since they include the mean, L-estimators are not necessarily robust. They tend to gain robustness when extreme order statistics are downweighted.

Inevitably, there are often doubts about whether certain observations are or are not in some way distorted relative to the bulk of the observations. Even if there are reasonable models to cover such discrepancies, there may still be either one or two gross departures from the general pattern or alternatively a large number of small perturbations, these often arising as a consequence of rounding or using measuring devices of limited accuracy. This has led to the development of procedures that are almost as good as the best available procedures when there are no such disturbances and are little affected by just a few major or many minor disturbances.

The Huber–Hampel *M-estimators* are one such class. The M indicates that they are like optimal maximum likelihood estimators when these are appropriate. They are little disturbed by either a few grossly aberrant observations or by small perturbations in many observations. For all but trivial problems, their use requires adequate computer programs. Here the basic ideas for estimation of the mean of a symmetric distribution where the maximum likelihood estimator is the sample mean is demonstrated. In particular, given a sample of n observations $x_1, x_2, \ldots, x_n$ from a normal distribution, this is equivalent to least-squares estimation where the estimator $\hat{\mu}$ is the value of μ that minimizes

$$U(x,\mu) = \sum_i (x_i - \mu)^2, \tag{14.9}$$

i.e., the sum of squares of deviations of the x_i from μ. The influence function (14.2) shows that for any one observation, $\hat{\mu}$ is linearly dependent on the distance of that observation from the true mean. Thus, an outlier z lying six standard deviations from the mean will have twice the influence of one lying three standard deviations from the mean.

Huber (1972) and others developed *M*-estimators to cope with the possible presence of a few outliers, where one really wanted inferences applicable to the remaining data in situations where specific maximum likelihood estimators were available for that remaining data. The proposed estimators have a built-in mechanism for reducing the effect of any outliers. Apart from this, they are exactly, or almost, identical to maximum likelihood estimators.

The quantity $(x - \mu)^2$ in (14.9) is an example of what is called a *distance measure* because it measures (the square) of the distances of the x_i from μ. The complete function in (14.9) is called a *distance function*. Another example of a distance function is the function $V(x,\mu) = \sum_i |x_i - \mu|$, i.e., the sum of the absolute deviations. This is minimized by setting $\mu^* = \text{med}(x_i)$. It is well known that in the case of U in (14.9), the minimum is found by differentiating with respect to μ and equating that derivative to zero leading to the normal or estimating equation

$$\sum_i [-2(x_i - \mu)] = 0$$

with solution $\hat{\mu} = \left(\sum_i x_i\right)/n = \bar{x}$, the sample mean.

A distance measure is defined as a function $d(t)$ such that for any t

i. $d(t) \geq 0$,
ii. $d(t) = d(-t)$,
iii. the derivative $\psi(t) = d'(t)$ is a nondecreasing function of t for all t.

For robust estimation of the mean μ of a symmetric distribution, Huber (1972) proposed a function $d(t)$ such that for some fixed $k>0$,

$$\begin{aligned} d(t) &= 1/2t^2 && \text{if } |t| \leq k \\ &= k|t^2/2| - k^2/2 && \text{if } |t| > k. \end{aligned}$$

The derivative $\psi(t) = d'(t)$ is

$$\begin{aligned} \psi(t) &= -k && \text{if } t < -k, \\ &= t && \text{if } |t| \leq k, \\ &= k && \text{if } t > k. \end{aligned}$$

The value of μ is estimated by the solution μ^* of the normal equation

$$\sum_i \psi(x_i - \mu) = 0. \tag{14.10}$$

If $k \to \infty$, then for all x, $\psi(x_i - \mu) = x_i - \mu$, and (14.10) gives the sample mean as the appropriate estimator, equivalent to the maximum likelihood estimator in the normal case. For finite k, the form of $d(t)$ implies that for all x_i for which $|x_i - \mu| \leq k$ a function equivalent to that for least squares is minimized. For $|x_i - \mu| > k$, a linear function of absolute differences is minimized.

A suitable value for k needs to be chosen. In this simple problem, practical experience has indicated that a useful choice is one such that the interval with endpoints given by med(x_i)±k contains between 70% and 90% of all observations. In general, (14.10) must be solved iteratively. When k is chosen, the procedure starts by rewriting that equation in the form

$$\sum_i \frac{\psi(x_i - \mu)}{x_i - \mu}(x_i - \mu) = 0. \tag{14.11}$$

Setting $w_i = \psi(x_i - \mu)/(x_i - \mu)$, (14.11) becomes $\sum_i w_i(x_i - \mu) = 0$ with solution

$$\mu^* = \left[\sum_i (w_i x_i)\right] / \left(\sum_i w_i\right). \tag{14.12}$$

The weights are functions of the unknown μ^*, so an estimate μ_0 of μ^* is needed to calculate initial weights w_0 and use these in (14.12) to calculate a new estimate μ_1. This procedure is repeated until convergence, which is usually achieved in a few iterations. For even moderate sample sizes, computation is tedious without a suitable computer program, but the steps for three small data sets are illustrated here.

Example 14.7

Obtain and compare Huber M-estimators of the mean for the three data sets

0,	1.2,	2.3,	3.8,	5.2,	7.1,	7.9,	8.5
0,	1.2,	2.3,	3.8,	5.2,	7.1,	7.9,	19.7
0,	1.2,	2.3,	3.8,	5.2,	7.1,	7.9,	115.5

A suitable choice of k is required after which the iterative procedure outlined above is carried out for each data set.

The procedure is illustrated for the second data set; similar steps are needed for the other sets. The median of the second set is (3.8+5.2)/2=4.5. If this is taken as the first estimate of μ and k=4 is chosen, the interval 4.5±4 (i.e., 0.5 to 8.5) contains 1.2, 2.3, 3.8, 5.2, 7.1 and 7.9, representing 75% of all observations. Similarly, if k=5 is chosen, the interval 4.5±5 (i.e., –0.5 to 9.5) includes 87.5%. Both inclusion percentages are within the suggested range 70–90.

For k=5, by setting $\psi(x_i - \mu) = (x_i - \mu)$ for observations in the interval (–0.5, 9.5), it follows that w_i=1 for all but the observation x_8=19.7. For that observation $\psi(x_i - \mu) = k = 5$, so that w_8=5/(19.7 – 4.5)=0.3289, whence the new estimate μ_1 is

$$\mu_1 = \frac{0+1.2+3.8+5.2+7.1+7.9+0.3289\times 19.7}{1+1+1+1+1+1+1+0.3289} = 33.3733/7.3289 = 4.554$$

For the next iteration, the same k=5 is retained. This gives full weight w=1 to all observations in the interval 4.554±5, i.e., (–0.446, 9.554). The weight for x_8=19.7 is now w_8=5/(19.7 – 4.554)=0.3301, whence a similar calculation to that above gives a new estimate μ_2=4.639.

Iterating again, only x_8 has a weight less than 1, which is found to be w_8=5/(19.7–4.639)=0.3320, and with this weight, the next estimate μ_3=4.643. The new adjusted weight for x_8 is w_8=5/(19.7 – 4.643)=0.3321, leading to the new estimate μ_4=4.643. To this degree of accuracy, there is no improvement on the previous iteration, so it can be concluded that μ^*=4.643.

With the same choice of sample median and k for the first data set, there are no observations for which $w_i \neq 1$ and the estimator is thus the sample mean, i.e., $\mu^*=4.5$. For the third data set, similar calculations to that for the second set, with x_8 there replaced by $x_8=115.5$, lead to the estimate $\mu^*=4.643$ after several iterations.

For the three data sets, here sample means are, respectively, 4.5, 5.9 and 17.875, indicating the strong influence of the outliers. The above *M*-estimation procedure leads to 4.5, 4.643 and 4.643 as estimates of the population mean, indicating a striking effect of downweighting the outliers.

Other possible estimators in this case include trimmed means, Winsorization or the median. The sample median for each set is 4.5, while a trimmed mean excluding $k=1$ observation from each tail leads to estimators 4.58 in each case. Similarly, in each case, the Winsorization estimates are 4.575.

Different choices of k, or different amounts of trimming or Winsorization, will give different estimates, but in practice, if estimators are robust against the outliers in the data, there should be little difference between rational choices, i.e., choices that downweigh rogue observations substantially or even eliminate them from the computation while directly or indirectly giving strong support to good observations. The median may appear not to obey these criteria, its computation taking into account at most two observations. However, all observations do have a role in determining the median, because the ordering of the observations determines which observation, or pair of observations, make a nonzero contribution to its determination.

Only if the observations are sampled from a symmetric distribution, however, will the sample median be a sensible estimator of the population mean (and then only if the latter exists). In many practical situations, even in the absence of any rogue observations, the median estimator will, however, have a greater standard error than that for the mean.

Several practical considerations influence the choice of k when using *M*-estimators. In particular, if k is too small only a few observations get full weight, and the estimator may be strongly influenced by rounding or grouping effects in those fully weighted observations. Also, there may be unsatisfactory features relating to change of scale that can be overcome by using different weights to those used above. Numerous alternative distance functions have been proposed, but details are omitted here. Many aspects of these procedures are discussed by Andrews et al. (1972).

Andrews (1974) introduced an *M*-estimator for robust linear multiple regression, and Härdle and Gasser (1984) applied a similar type of estimator for fitting nonlinear curves.

Thomas (2000) describes the use of the bootstrap to determine the precision of robust estimates of location, illustrating the use of appropriate methods for ten data sets that are either highly skewed or contain outliers.

SUMMARY POINTS

- Robust *L*-estimators (e.g., the trimmed mean, Winsorized mean, median) give reduced or no weight to the most extreme values.
- *M*-estimators, which are based on distance functions, are robust to either a few extreme values or many less obvious outliers.
- Use of *M*-estimators is particularly appropriate when the status of outliers is uncertain, but inferences applicable to the population from which the bulk of the data was sampled are sought.

14.5 Exercises

14.1 Use the test $|x^* - \text{med}(x_i)|/\{\text{med}[|x_i - \text{med}(x_i)|]\} > 5$ (*see* Section 14.2.2) to check for outliers in each of the following data sets:

Set A	3.2	4.9	1.3	9.7	12.9	–6.3	4.2	22.5	0.3	–7.1	
Set B	1	35	71	13	45	18	91	34	777	29	452

14.2 Use the outlier test in Exercise 14.1 to examine the data set:

0.04	0.06	0.27	0.32	0.33	0.40	0.50	0.63	0.69	0.92
1.09	1.10	1.35	1.61	1.66	1.69	1.71	1.80	1.98	2.65
2.83	3.50	3.72	3.75	3.99	5.16	5.49	6.31	7.05	16.05

14.3 Smith and Naylor (1987) give the following data for strengths of 15 cm lengths of glass fibre and suggest that the two smallest observations may be outliers. Does the test proposed in Section 14.2.2 confirm this?

0.37	0.40	0.70	0.75	0.80	0.81	0.83	0.86	0.92	0.92	0.94	0.95
0.98	1.03	1.06	1.06	1.08	1.09	1.10	1.10	1.13	1.14	1.15	1.17
1.20	1.20	1.21	1.22	1.25	1.28	1.28	1.29	1.29	1.30	1.35	1.35
1.37	1.37	1.38	1.40	1.40	1.42	1.43	1.51	1.53	1.61		

14.4 A sample of paired observations of X and Y are

X	1	2	3	5	8
Y	3	7	9	7	12

Use a bootstrap based on 1,000 samples to obtain an estimate of the bootstrap standard error of the estimator $\bar{x}/\bar{y}$ for the ratio of population means, as well as the estimator itself. In addition, obtain 95% quantile-based confidence intervals for the population ratio.

14.5 For the data set:

X	1	2	3	5	8
Y	3	7	9	7	12

use the jackknife to estimate the bias in the estimator $\bar{x}/\bar{y}$.

14.6 Calculate the mean of the data set:

0.04	0.06	0.27	0.32	0.33	0.40	0.50	0.63	0.69	0.92
1.09	1.10	1.35	1.61	1.66	1.69	1.71	1.80	1.98	2.65
2.83	3.50	3.72	3.75	3.99	5.16	5.49	6.31	7.05	16.05

and also the trimmed means removing the top and bottom 10% (deciles) and 25% (quartiles) of all observations. Explain any difference between these estimates of the population mean. Which do you prefer and why?

14.7 For the data set:

0 1.2 2.3 3.8 5.2 7.1 7.9 115.5

find the Huber M-estimator of the mean when $k=4$.

14.8 The data below give the age of first stroke (years) for ten patients in a hospital:

0 34 52 57 61 65 68 76 81 85

Use the method described in Section 14.2.2 to test for outliers. Can you think of possible reasons why outliers might arise in this type of data?

14.9 Records at a medical centre indicate for nine patients the following numbers of doctor consultations in a 12-month period:

0 1 1 1 1 2 3 6 12

Calculate the mean and median, and compare these values with the trimmed mean and Winsorized mean based on an assumption that the smallest and largest values may be outliers.

15

Nonparametric Methods in Action

15.1 Introduction

This chapter addresses how nonparametric methods can be applied in the context of a published medical research study on the surgical removal of the gallbladder, although the issues discussed are not limited to studies from this field. It examines various questions that can be investigated using the techniques described in this book. The methods selected are not exhaustive, and the strengths and weaknesses of those chosen are discussed. At the end of this chapter, extension questions arising from this study are given for further exploration of the data.

Understanding the context in which statistical tools are to be applied is necessary for their practical use. In this case study, statisticians and those from other nonclinical fields, e.g., social scientists and economists, may only require background information on gallstone disease whereas medical students and practitioners, especially those working in the specialty of this research, will find the study of greatest value with an in-depth understanding of this disorder. A concise account of the formation and management of gallstones can be found in the Cochrane systematic review as mentioned below. Literature aimed at a clinical audience can be found in the references of the systematic review and the article on which this case study is based.

Note that the narrative apart from the description of the data is fictional. This case study is not intended to reflect the views of the staff, the delivery of healthcare, and the service management at the hospital where the research was conducted. The same caveat should be borne in mind regarding the views of the authors of this book.

The data were originally supplied by Mike Lewis to assist with a systematic review of modified dietary fat intake and gallstone disease. We acknowledge his permission to extend the use of these data to this case study.

SUMMARY POINTS:

- Researchers should obtain knowledge about the field of the study before commencing data analysis.
- The level of knowledge should be appropriate for the role of the researcher.
- References found in general articles can lead to more detailed sources of information.

DOI: 10.1201/9780429326172-15

15.2 Intake of Dietary Fat and Gallstone Surgery

15.2.1 Dietary Fat and Gallstone Disease

Gallstone disease, also known as cholelithiasis, can result in hard deposits or stones in the gallbladder and biliary tract. The gallbladder stores bile, which is released for digestion, following either the consumption of solid food or liquids or cephalic stimulation. Bile contains cholesterol, produced by the body from metabolic processes and/or from dietary intake, and bilirubin from the breaking down of old red blood cells. If bile concentrations of cholesterol and/or bilirubin pigments increase excessively or if the gallbladder is not functioning as it should or if the release of bile is impaired, gallstones are able to form. The cholesterol saturation index (CSI), which is the ratio of the level of cholesterol in a bile sample compared to that sample's maximum cholesterol-carrying capacity as determined under laboratory conditions, is critically important. If this value exceeds 1, supersaturation occurs, leading to precipitation of cholesterol and increased risk of the formation of gallstones.

A Cochrane systematic review of randomized controlled trials on the effect of modified dietary fat on the management of gallstone disease identified several studies conducted in the 1980s in both the United States and Britain (Madden et al., 2024). These studies focussed on the CSI rather than the formation or symptomatic management of gallstones. Evidence from randomized trials regarding the management of gallstones and diet is extremely scant. This systematic review identified a single study that examined the effect of diet on the dissolution of gallstones (Lefkof et al., 1986), which showed no evidence for a dietary effect. Despite the lack of clinical evidence, low-fat diets have been advised for people with gallstones but it is unclear whether this was to prevent further gallstone formation or to manage gallstone symptoms including abdominal pain.

Interest in the role that diet plays in gallstone disease has switched to the treatment of hard deposits in gallbladder bile referred to as cholecystolithiasis. Gallstones can remain dormant in the gallbladder for many years, but when symptoms arise they can involve a significant level of pain. Currently, the standard treatment for symptomatic cholecystolithiasis is cholecystectomy, which is the surgical removal of the gallbladder. This procedure is performed either via laparoscopy or open surgery. The research on which this case study is based investigated whether a pre-operative low-fat diet to reduce patient weight can ease surgical removal.

15.2.2 The Laparoscopic Cholecystectomy Study

To minimize the time that patients spend under anaesthetic, optimize patient flow, and increase cost effectiveness, the length of this type of operation should be as short as safely possible and discharge on the day of the operation should be routine. It is known that obesity is associated with a longer operative time (OT). To investigate this issue further, a randomized trial was conducted on the effect of a very low calorie diet (VLCD) in the 2 weeks leading up to the operation on obese patients undergoing laparoscopic surgery at a hospital in the UK (Burnand et al., 2016). The primary outcome variable, operative time, was defined as the length of the surgical procedure in minutes measured from the first incision to the end of skin closure. Other outcomes of interest were whether a patient could be discharged on the day of surgery and the perceived difficulty of the procedure as reported by the operating surgeon. Members of the surgical team were blind to the allocated group for each patient.

Patients with a body mass index (BMI) of more than 30 kg/m^2, who gave their consent and were suitable for recruitment based on the inclusion and exclusion criteria were allocated by simple randomization to either a VLCD or continuation of their normal diet (diet as normal) at a pre-operative clinic 2 weeks before their operation. All study participants kept a detailed food diary. Explanatory variables included age, gender, weight, and BMI as recorded at the pre-operative clinic.

In total, 77 patients were invited to participate. Of these, 46 patients gave their consent and were recruited of which 21 were randomized to the VLCD group and 25 to the diet as normal group. The protocol was adhered to well, patient engagement in the study was high, and measurements were recorded carefully, resulting in complete data for almost all variables.

SUMMARY POINTS:

- The previous work on a research question can be surprisingly scant (Section 15.2.1).
- In designing a study, a primary outcome variable should be stated (Section 15.2.2).
- In practical studies, the findings should lead to identifiable benefits such as better use of resources (Section 15.2.2).
- Not all individuals invited to participate in a study will give their consent (Section 15.2.2).

15.3 Questions Arising from the Cholecystectomy Study

Note that the analyses and comments on each question are not intended to be comprehensive. Exploration of further issues raised by these questions will give additional insight.

To draw meaningful conclusions using the methods described below, two overriding assumptions need to be made. These are required for almost all the commonly encountered statistical techniques, nonparametric or otherwise. The first is that observations are independent of each other. In studies of cholecystectomy, this is generally a reasonable assumption to make. However, if the data include repetitions of the procedure on the same individual, as could occur with hip replacement surgery, the independence assumption will not be valid. So called clustered data require more sophisticated statistical modelling.

The second assumption concerns the generalization of the study findings. In order to do this, the observations in the study are assumed to be representative of the population under consideration. For this study, it might be reasonable to generalize the findings to other hospitals within the UK. However, they may not be applicable in countries where, for instance, the normal diet in the population contains less dietary fat than the UK.

15.3.1 BMI at Entry to Study

A reviewer states that a 60% recruitment rate is quite low and asks whether it is possible that patients with a very high BMI at a pre-operative clinic might have a different participation rate than others. If in the general population, the median BMI for obese patients is around 36 kg/m^2, is there any evidence either way from this study?

BMI values recorded at the pre-operative clinic are shown below:

30.18	30.20	30.29	30.43	30.74	30.82	31.08	31.18	31.33	31.33
31.61	31.70	31.80	31.88	31.91	32.28	33.13	33.38	33.39	33.51
33.72	33.78	33.87	33.90	34.03	34.63	34.75	34.88	35.44	35.85
36.11	36.15	36.23	36.34	36.62	36.88	37.07	37.51	37.53	37.74
39.34	39.71	40.15	40.43	42.63	43.25				

A plot of these data is given in Figure 15.1. The population median of 36 kg/m^2 is around the middle of the distribution. The sample median is 33.88 kg/m^2, which suggests that those with higher BMI may be underrepresented in the sample.

The null hypothesis to be tested is H_0: *Median*=36 against H_1: *Median*≠36. The test is two-sided as the original proposition concerns the participation rate for those with very high BMI being *different* to that of others, rather than specifying whether that would be lower or higher.

We consider the following three nonparametric tests: the sign test, the Wilcoxon signed-rank test (which would be the most widely used in this situation), and the Pitman test. A test based on normal distribution theory would be a one-sample *t*-test, testing the mean rather than the median. Figure 15.1 does not give much support for the assumption that the data come from a normal distribution but we report the results below nevertheless for comparison purposes. Table 15.1 shows the results of applying these tests and associated 95% confidence intervals.

The null hypothesis is rejected by all the tests at the 5% level of significance (marginally) except the sign test, reflecting the lack of power that it has. The Pitman test performs well

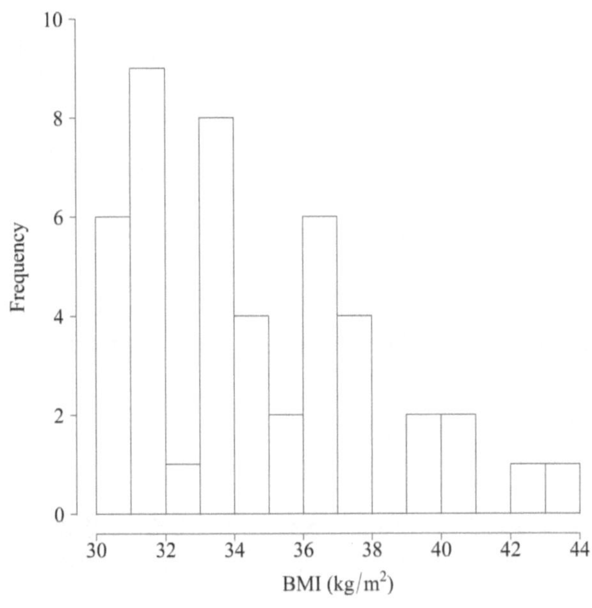

FIGURE 15.1
Distribution of body mass index values (in kg/m^2) for the combined dietary groups.

TABLE 15.1

P-values and confidence intervals for testing body mass index.

Test	Asymptotic *P*-Value	Asymptotic 95% CI	Exact *P*-Value	Exact 95% CI	Bootstrap 95% CI
Sign test	0.0553	(33.13, 35.85)	0.0541	(32.28, 36.11)	
Wilcoxon signed-rank test	0.0052	(33.38, 35.46)	0.0045	(33.38, 35.46)	
Pitman test	0.0079		0.0070		(33.13, 35.85)
t-test	0.0065	(33.58, 35.58)			

despite its lack of robustness. The confidence intervals associated with the sign test are also wider than those associated with the Wilcoxon signed-rank test and *t*-test, although identical to the bootstrap interval associated with the Pitman test. It is notable that the *t*-test, although being applied to data which do not exhibit much sign of normality as shown in Figure 15.1, produces very similar results to the Wilcoxon signed-rank test. It is also of note that whilst the asymptotic *P*-value for the sign test does not reject the null hypothesis at the 5% level of significance, the associated asymptotic 95% confidence interval does not include the null hypothesis value of 36. The reason for this is that the calculation of the *P*-value includes the use of a continuity correction (by default) that is not present in the calculation of the confidence interval. If the continuity correction is deliberately omitted (e.g., by using the `cont.corr = FALSE` argument in the R function), the *P*-value produced is 0.039, rejecting the null hypothesis at the 5% level of significance.

The results in Table 15.1, other than those for the sign test, show that there is strong evidence against the null hypothesis. The values contained in the 95% confidence intervals, being less than 36, provide evidence that those with higher BMI may be less likely to participate. Note that the upper limits of the 95% confidence intervals are very close to 36, showing that the true difference from the median under the null hypothesis might not be of practical importance. Both statistical significance and practical importance should be considered in the interpretation of the results.

The analyses described above can be performed in R using the following commands:

```
sign.test(ch15$BMI, H0 = 36, do.asymp = TRUE)
wilcoxon.signedrank(ch15$BMI, H0 = 36, do.asymp = TRUE)
pitman(ch15$BMI, H0 = 36, do.asymp = TRUE)
t.test(ch15$BMI, mu = 36)
```

15.3.2 Dietary Group and Patient Weight on the Day of the Operation

Patients in the VLCD group might be expected to have a lower weight than those in the diet as normal group by the day of their operation. Do the data (in kg) provide any evidence?

VLCD Group

85.80	83.10	86.50	83.70	85.60	83.50	84.90	95.25	100.00	78.20
81.10	88.45	96.40	98.80	102.60	72.50	130.45	79.80	81.05	102.65
79.20									

Diet as Normal Group

109.90	90.00	90.50	92.30	105.40	113.40	95.50	91.20	92.30	81.00
107.00	89.30	83.70	104.25	114.50	99.15	114.70	73.20	95.90	87.40
83.35	85.15	79.20	72.60	83.10					

A plot of these data is given in Figure 15.2. It is difficult to see any clear difference between the median weights for the two groups, although if one ignores the single large value in the VLCD group, the weights in the diet as normal group may be more spread out. The sample median for the VLCD group is 85.6 kg and for the diet as normal group is 90.2 kg, which suggests that the VLCD group may, in fact, have a lower weight than those in the other group.

The null hypothesis to be tested is H_0: *median weight for the VLCD group = median weight for the diet as normal group* against H_1: *median weight for the VLCD group < median weight for the diet as normal group.* The test is one-sided.

We consider the following three nonparametric tests: the Wilcoxon–Mann–Whitney test (which would be the most widely used in this situation), the median test, and the normal scores test. A test based on normal distribution theory would be a two-sample *t*-test, testing the mean rather than the median. Figure 15.2 gives some support for the assumption that the data come from a normal distribution. Table 15.2 shows the results of applying these tests and associated 95% confidence intervals.

The null hypothesis is rejected at the 5% level of significance by the median test (the bluntness of the test looking only at whether values are above or below the median results in the several VLCD weights just below the median of 88.875 providing stronger evidence against H_0 than might be warranted) and at the 10% level of significance by the other nonparametric tests (the effectiveness of the normal scores test being aided by the lack of

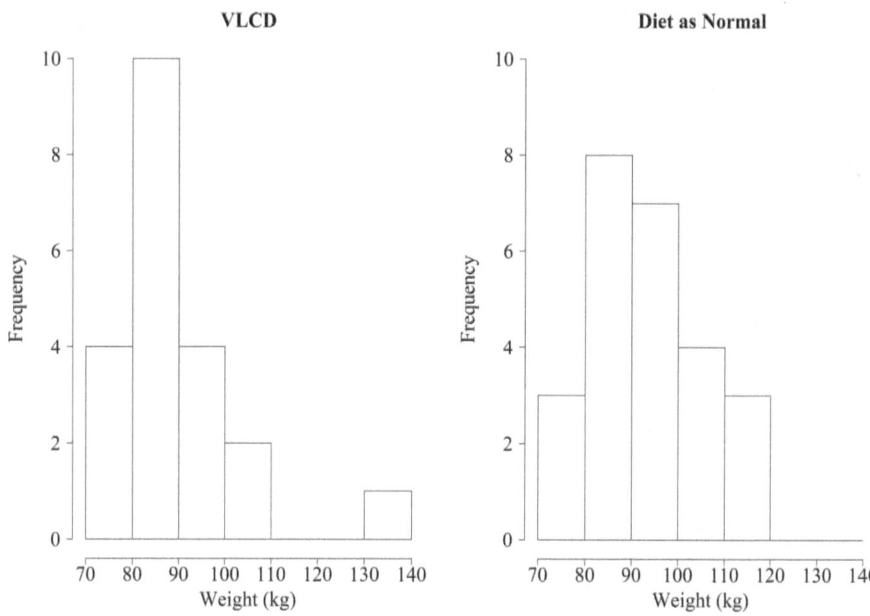

FIGURE 15.2
Distribution of weight (in kg) by dietary group.

TABLE 15.2

P-values and confidence intervals for testing differences in weights.

Test	Asymptotic *P*-value	Asymptotic 95% CI	Exact/Monte Carlo *P*-value	Exact/bootstrap 95% CI
Wilcoxon–Mann–Whitney test	0.0875	(−11.20, 2.50)	0.0881 (Monte Carlo)	(−13.35, 4.75) (Bootstrap)
Median test	0.03786		0.0373 (Exact)	(−12.85, 7.85) (Exact)
Normal scores test	0.0985			
t-test	0.1504	(−11.40, 3.61)		

values far into either tail), but not even at the 15% level of significance by the two-sample *t*-test. The exact/bootstrap confidence intervals are wider than the asymptotic intervals. The exact interval associated with the median test is particularly wide at least partially due to it having a width of 98.309% (an exact 95% interval not being available). It is notable that the *t*-test, although being applied to data for which the normal distribution seems reasonable in Figure 15.1, yields different results from the nonparametric tests. This may well be due to the single large weight in the VLCD group having a greater effect on the calculation of the test statistic in its raw form than as a rank. In summary, evidence against the null hypothesis is weak.

The analyses described above can be performed in R using the following commands:

```
wilcoxon.mann.whitney(ch15$wgt.VLCD, ch15$wgt.norm,
                      alternative = "less", do.asymp = TRUE)
med.test(ch15$wgt.VLCD, ch15$wgt.norm,
         alternative = "less", do.asymp = TRUE)
normal.scores.test(ch15$wgt.VLCD, ch15$wgt.norm,
                   alternative = "less")
t.test(ch15$wgt.VLCD, ch15$wgt.norm, alternative = "less")
t.test(ch15$wgt.VLCD, ch15$wgt.norm)$conf.int
```

15.3.3 Variation in OT Within the VLCD and Diet as Normal Groups

If a VLCD does reduce OT (presented below as minutes:seconds), there may be more extremely high values in the diet as normal group. These extreme values could increase the dispersion of the observations in that group. Does the study suggest this?

VLCD Group

33:00 24:10 35:00 25:51 25:00 22:55 26:40 23:00 32:04 20:36
21:28 27:16 25:00 18:00 21:30 40:00 40:50 25:27 18:15 22:14
28:00

Diet as Normal Group

66:27 64:00 55:57 29:16 34:46 24:50 28:33 36:30 27:51 31:00
42:00 42:00 31:00 73:00 45:00 26:45 29:10 20:42 23:30 40:00
22:30 20:10 26:30 22:50 170:34

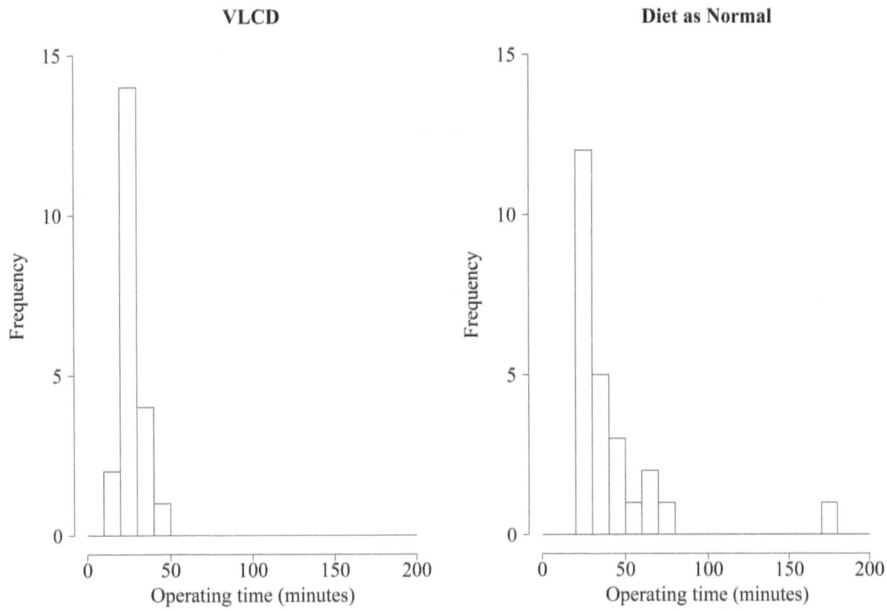

FIGURE 15.3
Distribution of operative time (OT, in minutes) by dietary group.

These data are clearer to work with if converted to minutes so that, for example, the first row of the *VLCD* group has the following format:

33.00 24.17 35.00 25.85 25.00 22.92 26.67 23.00 32.07 20.60

Although for the purpose of illustration the converted observations are expressed to just two decimal places, the maximum accuracy available should be used in analyses.

A plot of these data is shown in Figure 15.3. It is clear that the OTs associated with the diet as normal group are more dispersed than those for the VLCD group, with a number of times exceeding 50 minutes and one extreme value. The standard deviation for the VLCD group is 6.39 whilst that for the diet as normal group is 30.68.

In the light of the clear impression from the standard deviations and the plots, it is tempting not to conduct a formal hypothesis test. The appropriate null hypothesis to be tested is H_0: *Variance for the VLCD group = Variance for the diet as normal group* against H_1: *Variance for the VLCD group < Variance for the diet as normal group.* The test is one-sided as the likely direction of any difference has a straightforward explanation.

We consider three nonparametric tests: the Siegel–Tukey test (perhaps the most commonly used in this situation), the Ansari–Bradley test, and the Conover Squared-Rank test. A test based on normal distribution theory would be an *F*-test. Figure 15.3 shows that the OT for the VLCD group might be normally distributed but that for the diet as normal group has considerable skew. Despite this we report the normal distribution theory results below for comparison purposes. Table 15.3 shows the results of applying these tests.

In all cases the null hypothesis is rejected at the 5% level of significance which is not surprising given the summary statistics and plots shown above. The Siegel–Tukey and Ansari–Bradley tests are in close agreement, but the Conover Squared-Rank test gives a higher *P*-value. The *F*-test gives a very small *P*-value and will have been influenced heavily by the single extreme value in the diet as normal group. All analyses show strong evidence of the observations from the diet as normal group having greater dispersion.

TABLE 15.3

P-values and confidence intervals for testing dispersion in operative times (OTs).

Test	Asymptotic *P*-value	Monte Carlo *P*-value
Siegel–Tukey test		0.0048
Ansari–Bradley test	0.0056	
Conover squared-rank test		0.0102
F-test	<0.0001	

The analyses described above can be performed in R using the following commands:

```
siegel.tukey(ch15$optime.VLCD, ch15$optime.norm,
             alternative = "less")
ansari.bradley(ch15$optime.VLCD, ch15$optime.norm,
               alternative = "less")
conover(ch15$optime.VLCD, ch15$optime.norm,
        alternative = "less")
var.test(ch15$optime.VLCD, ch15$optime.norm,
         alternative = "less")
```

15.3.4 Association between OT and Length of Stay (LoS)

It is reasonable to expect the LoS to be greater if the OT is high. Investigate a possible association between the two variables (OT in minutes:seconds, LoS in hours).

VLCD Group

OT:	33:00	24:10	35:00	25:51	25:00	22:55	26:40	23:00	32:04	20:36
LoS:	6.00	7.00	6.50	3.75	9.66	9.75	28.00	10.00	10.00	7.50
OT:	21:28	27:16	25:00	18:00	21:30	40:00	40:50	25:27	18:15	22:14
LoS:	10.80	11.00	14.00	9.00	11.00	11.00	26.00	6.00	8.50	7.75
OT:	28:00									
LoS:	7:75									

Diet as Normal Group

OT:	66:27	64:00	55:57	29:16	34:46	24:50	28:33	36:30	27:51	170:34
LoS:	6.50	9.00	9.00	6.50	7.50	9.00	8.50	5.00	6.00	146.00
OT:	31:00	42:00	42:00	31:00	73:00	45:00	26:45	29:10	20:42	23:30
LoS:	8.00	8.00	10.00	4.66	5.75	10.00	7.25	11.00	8.00	9.00
OT:	40:00	22:30	20:10	26:30	22:50					
LoS:	12.00	11.00	11.50	6.50	8.00					

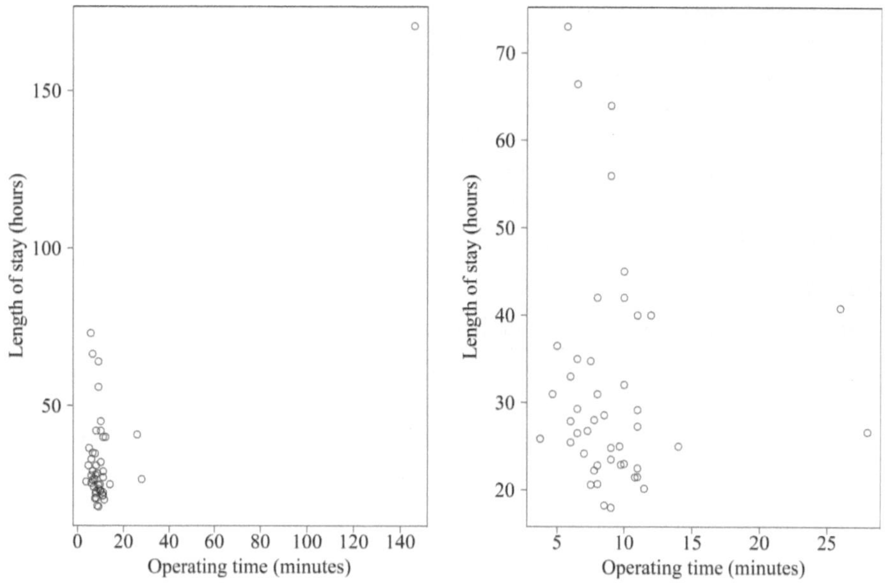

FIGURE 15.4
Scatter diagrams for length of stay (LoS, in hours) against OT (in minutes) with and without extreme case.

A plot of these data is given in Figure 15.4, the first panel showing all the data and the second excluding the single extreme case. Although it might be tempting to conclude from the first panel that there is a positive association between LoS and OT, the second panel shows very little evidence of such a relationship.

The null hypothesis to be tested is H_0: *No relationship between LoS and OT* against H_1: *There is a relationship between LoS and OT.*

We consider three nonparametric tests: one based on Spearman's rank correlation coefficient, one based on Kendall's tau and one based on Blomqvist's coefficient. Spearman and Kendall's tau would perhaps be the most commonly used tests in this situation. A test based on normal distribution theory comes from the Pearson correlation coefficient. Figure 15.4 shows that the distribution of LoS is skewed with many small values and a few large values and, coupled with the apparent lack of normality of LoS in Figure 15.3, we cannot be confident that the assumptions required for a normal distribution theory approach are satisfied. Nevertheless, we report the normal distribution theory results below for comparison purposes. Table 15.4 shows the results of applying these tests with, and without the extreme case present.

Considering first the results when all cases are used, we see that the Spearman's rank correlation, Kendall's tau and Blomqvist's coefficient all give very similar results, but the Pearson correlation is extremely different. This is due to the extreme case, having a very large LoS and OT, resulting in an enormous influence on the calculation of the Pearson correlation. When this extreme case is removed, we see all the methods giving very similar results.

The analyses described above can be performed in R using the following commands:

```
spearman(ch15$los, ch15$optime)
kendall.tau(ch15$los, ch15$optime)
blomqvist(ch15$los, ch15$optime)
pearson(ch15$los, ch15$optime)
```

```
spearman(ch15$los[ch15$los < 100],
         ch15$optime[ch15$optime < 100])
kendall.tau(ch15$los[ch15$los < 100],
            ch15$optime[ch15$optime < 100])
blomqvist(ch15$los[ch15$los < 100],
          ch15$optime[ch15$optime < 100])
pearson(ch15$los[ch15$los < 100],
        ch15$optime[ch15$optime < 100])
```

TABLE 15.4

Coefficients and *P*-values for testing relationship between length of stay and OT.

	Using all cases		With extreme case removed	
Coefficient/test	**Coefficient**	**Exact/Monte Carlo *P*-value**	**Coefficient**	**Exact/Monte Carlo *P*-value**
Spearman's rank correlation	–0.0750	0.6198 (Monte Carlo)	–0.1485	0.3296 (Monte Carlo)
Kendall's tau	–0.0603	0.5696 (Monte Carlo)	–0.1098	0.3038 (Monte Carlo)
Blomqvist's coefficient	–0.1304	0.5559 (Exact)	–0.1429	0.5377 (Exact)
Pearson's correlation	0.8286	0.0128 (Monte Carlo)	–0.0508	0.7280 (Monte Carlo)

15.3.5 Dietary Group and Operative Difficulty

The literature on the impact of obesity and surgical difficulty has found a difference between the VLCD and the diet as normal groups regarding the reported difficulty of the operation. This is because a high level of body fat can make surgical access to the gallbladder more difficult. Does this study support what has been found by others?

Reported Difficulty of Surgery

	Extremely Difficult	Difficult	Normal	Easy	Total
VLCD	1	2	4	14	21
Diet as Normal	2	1	14	8	25
Total	3	3	18	22	46

A visual inspection of the table shows that the VLCD group are clustered in the "Easy" category (14/21=66.7%) whereas for the diet as normal group they are clustered in the "Normal" category (14/25=56%). This would suggest that there is a clear difference between the groups regarding the difficulty of the operation.

The null hypothesis to be tested is H_0: *No relationship between operating difficulty and dietary group* against H_1: *There is a relationship between operating difficulty and dietary group.*

The tests we consider here are the chi-squared test, Fisher's exact test (the two most commonly used tests in this situation), the likelihood ratio test and the Wilcoxon–Mann–Whitney test which allows for the ordering of the categories for the operating difficulty. Table 15.5 shows the results of applying these tests. Because the difficulty of surgery is categorical, we effectively have many tied values when applying the Wilcoxon–Mann–Whitney test. This would cause difficulties when calculating an exact *P*-value but here a Monte Carlo *P*-value is produced.

TABLE 15.5

P-values for testing relationship between operating difficulty and dietary group.

	With "Extremely Difficult" and "Difficult" separate		With "Extremely Difficult" and "Difficult" combined	
Test	**Exact *P*-value**	**Monte Carlo *P*-value**	**Exact *P*-value**	**Monte Carlo *P*-value**
Chi-squared test		0.0456		0.0269
Fisher's exact test	0.0275		0.0242	
Likelihood ratio test		0.0869		0.0426
Wilcoxon–Mann–Whitney test		0.0697		0.0724

Considering the data with "Extremely difficult" and "Difficult" as separate categories and applying conventional *P*-value benchmarks produces mixed findings. Whilst all the results have a *P*-value of less than 10%, only the chi-squared test and Fisher's exact test give a *P*-value of less than 5%. Examining the exact *P*-values is more enlightening, as all of these provide some evidence against the null hypothesis. Although from examination of the table we saw distinct differences between the dietary groups, these are being diluted in the tests by the fact that they have similar proportions of cases in the "Extremely Difficult" and "Difficult" categories. When these two categories are combined, the chi-squared test and likelihood ratio test then produce results that would reject the null hypothesis at the 5% level of significance whilst Fisher's exact test and the Wilcoxon–Mann–Whitney test give essentially unchanged results. Once again, all the exact *P*-values indicate some evidence against the null hypothesis. These observations highlight the importance of choices being made about the data and the tests chosen and caution against the temptation to "fish" for tests which give desired results.

The analyses described above can be performed in R using the following commands:

```
chisqtest.ANSM(ch15$diet, ch15$opdiff)
fishertest.ANSM(ch15$diet, ch15$opdiff)
lik.ratio(ch15$diet, ch15$opdiff)
wilcoxon.mann.whitney(ch15$opdiff[ch15$diet == "VLCD"],
                      ch15$opdiff[ch15$diet == "Normal diet"])
opdiff2 <- ch15$opdiff
opdiff2[opdiff2 == "Extremely difficult"] <- "Difficult"
opdiff2 <- droplevels(opdiff2)
chisqtest.ANSM(ch15$diet, opdiff2)
fishertest.ANSM(ch15$diet, opdiff2)
lik.ratio(ch15$diet, opdiff2)
wilcoxon.mann.whitney(opdiff2[ch15$diet == "VLCD"],
                      opdiff2[ch15$diet == "Normal diet"])
```

15.3.6 Difference in the Regression Coefficients of the Dietary Groups for the Relationship between OT and LoS

In Section 15.3.4, the strength of the relationship between LoS and OT was explored, without taking the dietary group into account. In this section, we examine whether the form of the relationship between LoS and OT is the same in each of the dietary groups. We exclude the extreme case from the diet as normal group due to the undue influence it would have.

VLCD Group

OT:	33.00	24.10	35.00	25.51	25.00	22.55	26.40	23.00	32.04	20.36
LoS:	6.00	7.00	6.50	3.75	9.66	9.75	28.00	10.00	10.00	7.50
OT:	21.28	27.16	25.00	18.00	21.30	40.00	40.50	25.27	18.15	22.14
LoS:	10.80	11.00	14.00	9.00	11.00	11.00	26.00	6.00	8.50	7.75
OT:	28.00									
LoS:	7.75									

Diet as Normal Group

OT:	66.27	64.00	55.57	29.16	34.46	24.50	28.33	36.30	27.51	31.00
LoS:	6.50	9.00	9.00	6.50	7.50	9.00	8.50	5.00	6.00	8.00
OT:	42.00	42.00	31.00	73.00	45.00	26.45	29.10	20.42	23.30	40.00
LoS:	8.00	10.00	4.66	5.75	10.00	7.25	11.00	8.00	9.00	12.00
OT:	22.30	20.10	26.30	22.50						
LoS:	11.00	11.50	6.50	8.00						

The plots of these data are shown in Figure 15.5. From these plots, it is reasonable to suggest that the two dietary groups have a different relationship between LoS and OT.

The null hypothesis to be tested is H_0: *The dietary groups have the same relationship between LoS and OT* against H_1: *The dietary groups have different relationships between LoS and OT.*

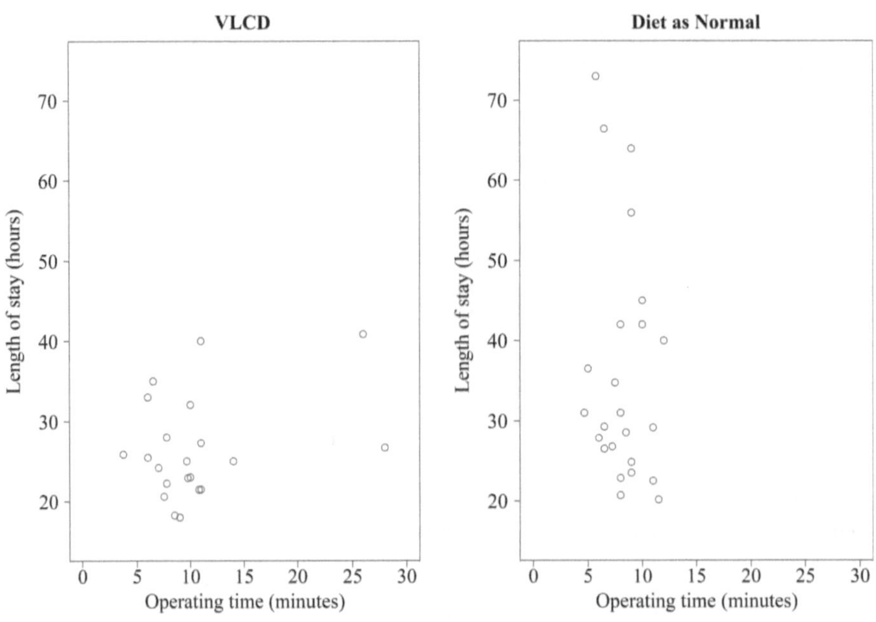

FIGURE 15.5
Scatter diagrams for LoS (in hours) against OT (in minutes) by dietary group.

We will use the ordinary least squares regression coefficient as a measure of the relationship between LoS and OT for each group and thus more precisely define the hypotheses to be H_0: *The dietary groups have the same regression coefficient relating LoS and OT*; H_1: *The dietary groups have different regression coefficients relating LoS and OT.* Bootstrapping will be used to create a 95% confidence interval for the difference between regression coefficients for the dietary groups. The lack of normality discussed above in Sections 15.3.4 and 15.3.5 for these data is not relevant when considering bootstrapping.

Using 1,000 bootstrap samples, we obtained a 95% confidence interval of (−0.1427, 0.7313). As zero is clearly within this interval, we conclude that we have insufficient evidence to reject the null hypothesis.

The analyses described above can be performed in R using the following commands:

```
n.VLCD <- length(ch15$los.VLCD)
n.norm <- length(ch15$los.norm[ch15$los.norm < 100])
bs.beta.diff <- NULL
for (i in 1:1000){
  bs.sample.VLCD <- sample(1:n.VLCD, n.VLCD, replace = TRUE)
  bs.VLCD.beta <-
    lm(ch15$los.VLCD[bs.sample.VLCD] ~
         ch15$optime.VLCD[bs.sample.VLCD])$coefficients[2]
  bs.sample.norm <- sample(1:n.norm, n.norm, replace = TRUE)
  bs.norm.beta <-
    lm(ch15$los.norm[ch15$los.norm < 100][bs.sample.norm] ~
         ch15$optime.norm[ch15$optime.norm < 100]
       [bs.sample.norm])$coefficients[2]
  bs.beta.diff[i] <- bs.VLCD.beta - bs.norm.beta
}
quantile(bs.beta.diff, c(0.025, 0.975))
```

SUMMARY POINTS:

- Depending on the data being analysed, nonparametric tests may give very similar results (Section 15.3.1) or may give results that differ to some degree (Section 15.3.2).
- When there is very strong evidence to reject the null hypothesis, the choice of test is rather open to question (Section 15.3.3).
- Coefficients summarizing relationships between variables can vary depending on the methods used to calculate them (Section 15.3.4).
- To avoid a search for "interesting" results, statistical tests should be decided before analysis starts (Section 15.3.5).
- Bootstrapping can be an effective method for investigating differences between groups when standard measures are not available (Section 15.3.6).

15.4 Going Forward

In this section, we seek to place nonparametric statistical methods in the context of modern applied statistics. We discuss issues regarding criticisms of hypothesis testing in the context of the replication crisis and look forward to what advances in computing power and artificial intelligence (AI) may mean for nonparametric statistical methods.

15.4.1 Nonparametric Statistical Methods and the Replication Crisis

The replication crisis (Ioannidis, 2005) is concerned with the publication of scientific results that cannot be replicated when the experiment or study is repeated. Frequently, hypothesis testing is accused of being a contributory factor, with researchers having available to them the ability to carry out all manner of tests on all manner of relationships and, potentially, pick and choose which to report.

Nonparametric tests, being part of the wider family of hypothesis tests, are implicated in this crisis, and there is an additional risk involved due to there being a tendency for researchers to regard nonparametric tests as "less risky" because they incorrectly believe that assumptions do not have to be made. As a result, they may be at greater risk of being used in "*P*-hacking" operations where multiple tests are carried out in the hope of finding "interesting" results.

15.4.2 Nonparametric Statistical Methods and Computing Power

In the examples used in this book, the ANSM5 package in R has been used to perform analyses, producing asymptotic, exact and Monte Carlo *P*-values, and confidence intervals. These three types of *P*-value have a story to tell concerning the history of nonparametric methods and computing power. Probability calculations, sometimes based on complete enumeration of all possibilities, were the basis of the first tests and, preceding the innovation of asymptotic results and Monte Carlo simulation, were, of course, what we now call "exact". It was, however, impossible to carry out complete enumerations for many problems, a situation which Fisher recognized in 1935 (Fisher, 1935, p. 63) and which still exists today for larger problems. As a result, asymptotic theory was developed, enabling data with larger numbers of cases to be handled. The advent of calculating machines and what we now know as computers meant that greater numbers of exact calculations could be carried out, and asymptotic tests could handle much larger datasets. As computing power increased, Monte Carlo methods became feasible, using simulation to approximate exact tests without having to rely on assumptions made by asymptotic alternatives. A more detailed background of the history of nonparametric methods can be found in Sen (1991) and Hollander, Wolfe and Chicken (2014).

What does the future hold for nonparametric statistical methods in relation to computing power? We can be confident that computing power will continue to increase in coming years and quantum computing is likely to provide a further step change in the speed with which some types of analysis can be carried out. As a result, we can predict that it will become possible to calculate exact *P*-values and confidence intervals for increasingly large and complex problems within useful timeframes.The increase in power will mean the number of Monte Carlo simulations that can be carried out in a set time will grow and even greater accuracy can be obtained.

But how much will really be gained from this increased ability to conduct exact tests and gain increased accuracy from Monte Carlo simulations? The effort expended to add reliable decimal places to *P*-values, beyond the first few figures, has little practical effect on conclusions drawn from tests. The ability to conduct a larger multitude of tests in a set amount of time may encourage *P*-hacking activities discussed in Section 15.4.1. As we write this edition in the mid-2020s, we are living through a climate crisis and becoming increasingly aware of our need as a species to reduce the resources we consume. Although the impact of computing for nonparametric tests may be small compared to other human activities, researchers are likely to wish to justify the resources they use. Lannelongue et al. (2023) suggest principles for environmentally sustainable computational science which embed environmental sustainability as a key element of any research. Consideration of these issues may see resurgence in the popularity of asymptotic tests that, as we have seen in preceding chapters, can perform very well in favourable circumstances and use a small fraction of the computing power compared to exact and Monte Carlo calculations.

We might also raise a question concerning the threat that AI may pose to the future of hypothesis testing, not from the point of view of an apocalyptic suggestion that AI might wipe out humankind (although science fiction and conspiracy theories are silent when it comes to predicting whether "machines" might still use the Wilcoxon–Mann–Whitney test without us), but whether the predictions that can be made using AI will render hypothesis testing unnecessary. This is moving well beyond the scope of this book, but humans will always want to ask questions and explore hypotheses, and despite its acknowledged shortcomings, hypothesis testing is a way of doing this. Nonparametric statistical methods, with their ability to help understand data whilst making minimal assumptions, are likely to remain an important tool in the armoury of data analysis.

SUMMARY POINTS:

- Nonparametric statistical methods are implicated in the replication crisis, alongside other hypothesis tests (Section 15.4.1).
- The computing power needed for some nonparametric tests may present challenges for sustainable computational science (Section 15.4.2).

15.5 Extension Questions

Note that these problems may require knowledge of statistical methods described in this edition of ANSM but not covered in the examples described in this chapter. The reader is also encouraged to consider methods beyond the scope of this book that may be useful for answering the question posed, such as charts and parametric analyses (if appropriate), as nonparametric methods should be considered part of the analytical armoury rather than its entirety.

15.1 Investigate whether the median participant age in the population might be 50 years.

Participant ages are as follows:

21	22	23	26	27	28	29	34	35	35
35	36	36	38	40	41	41	42	43	45
45	45	46	47	47	48	48	48	48	49
49	51	52	52	53	53	54	56	59	61
63	63	64	66	67	69				

15.2 Do the participants from this study demonstrate an association between height (m) and weight (kg)?

Height:	1.49	1.53	1.54	1.55	1.56	1.56	1.59	1.59	1.59	1.59
Weight:	73.2	83.35	78.2	72.5	85.15	72.6	90.0	85.8	95.5	79.2
Height:	1.59	1.59	1.59	1.60	1.60	1.61	1.62	1.62	1.63	1.63
Weight:	86.5	83.7	83.5	81.0	100.0	79.2	114.5	96.4	90.5	84.9
Height:	1.64	1.64	1.64	1.65	1.65	1.65	1.65	1.65	1.65	1.66
Weight:	104.25	95.25	102.65	92.3	83.7	83.1	83.1	81.1	79.8	99.15
Height:	1.66	1.66	1.67	1.67	1.68	1.69	1.69	1.70	1.70	1.72
Weight:	85.6	98.0	89.3	88.45	87.4	102.6	81.05	113.4	92.3	109.9
Height:	1.73	1.74	1.78	1.79	1.80	1.83				
Weight:	91.2	105.4	107.0	130.45	95.9	114.7				

15.3 Laparoscopic cholecystectomy involves what is known as the dissection of Calot's triangle. The difficulty of this stage of the surgery was reported separately.
Reported difficulty of performing dissection of Calot's triangle is shown below:

	Extremely difficult	Difficult	Normal	Easy	Total
VLCD	0	4	7	10	21
Diet as normal	2	9	7	5	23
Total	2	13	14	15	44

Is there evidence of an association between dietary group and reported degree of difficulty?

15.4 Ordered by the date and time of operation, the lengths of the procedure for the diet as normal patients were:

66:27	64:00	170:34	55:57	29:16	34:46	24:50	28:33	36:30	27:51	31:00
42:00	42:00	31:00	73:00	45:00	26:45	29:10	20:42	23:30	40:00	22:30
20:10	26:30	22:50								

Is there any evidence for a trend in the time taken to perform the procedure?

Appendix

Badenscallie Burial Data

Several examples in this book use data on the age at death of male members of four Scottish clans in the burial ground at Badenscallie in the Coigach district of Wester Ross, Scotland. The data were collected in June 1987. Clan names have been changed, but the records are as complete as possible for four clans. There were a few missing values because names or dates were unreadable on a few headstones, and several headstones appeared to be missing.

Minor spelling variations, especially those of M', Mc and Mac, were ignored. Ages are given for complete years, e.g., 0 means before first birthday and 79 means on or after 79th but before 80th birthday, according to the information on the tombstone. Ages are given in ascending order within each clan.

McAlpha (59 members)

0	0	1	2	3	9	14	22	23	29	33	41	41	42	44
52	56	57	58	58	60	62	63	64	65	69	72	72	73	74
74	75	75	75	77	77	78	78	79	79	80	81	81	81	81
82	82	83	84	84	85	86	87	87	88	90	92	93	95	

McBeta (24 members)

0	19	22	30	31	37	55	56	66	66	67	67	68	71
73	75	75	78	79	82	83	83	88	96				

McGamma (21 members)

13	13	22	26	33	33	59	72	72	72	77	78	78	80	81
82	85	85	85	86	88									

McDelta (13 members)

1	11	13	13	16	34	65	68	74	77	83	83	87

References

Chapter 1

Bury, K. (1999) *Statistical Distributions in Engineering*. Cambridge: Cambridge University Press.
Cannistra, R. (2004) The ethics of early stopping rules: who is protecting whom? *Journal of Clinical Oncology*, **22**, 1542–1545.
Easterbrook, P.J., Berlin, J.A., Gopalan, R. and Matthews, D.R. (1991) Publication bias in clinical research. *Lancet*, **337**, 867–872.
Fisher, R.A. (1925) *Statistical Methods for Research Workers*. 1st edn. Edinburgh: Oliver & Boyd.
Herring, J. (2022) *Medical Law and Ethics*. 9th edn. Oxford: Oxford University Press.
Hutton, J.L. (1995) Statistics is essential for professional ethics. *Journal of Applied Philosophy*, **12**, 253–261.
Krantz, D.H. (1999) The null hypothesis testing controversy in psychology. *Journal of the American Statistical Association*, **94**, 1372–1381.
Mathie, E., Wythe, H., Munday, D., Millac, P., Rhodes, G., Roberts, N., Smeeton, N., Poland, F. and Jones, J. (2018) Reciprocal relationships and the importance of feedback in patient and public involvement: a mixed methods study. *Health Expectations*, **21**, 899–908.
Nelder, J.A. (1999) From statistics to statistical science (with comment). *The Statistician*, **48**, 257–270.
Pearson, K. (1900) On a criterion that a given system of deviations from the probable in the case of a correlated system of variables is such that it can reasonably be supposed to have arisen in random sampling. *Philosophical Magazine*, **50**, 157–175.
Popper, K. (1980) *The Logic of Scientific Discovery*. 4th edn.-revised. London: Hutchinson.
Ruidavets, J.-B., Ducimetière, P., Montaye, M., Haas, B., Bingham, A., Yarnell, J., Amouyel, P., Arveiler, D., Kee, F., Bongard, V. and Ferrières, J. (2010) Patterns of alcohol consumption and ischaemic heart disease in culturally divergent countries: the Prospective Epidemiological Study of Myocardial Infarction (PRIME). *BMJ*, **341**, c6077.
Schulz, K.F., Altman, D.G. and Moher, D. (2010) CONSORT 2010 Statement: updated guidelines for reporting parallel group randomised trials. *BMC Medicine*, **8**, 18.
Spiegel, K.M. and Stephens, L.J. (2017) *Schaum's Outline of Statistics (Schaum's Outlines)*. 6th edn. New York: McGraw-Hill.
Williamson, P., Hutton, J.L., Bliss, J., Blunt, J., Campbell, M.J. and Nicholson, R. (2000) Statistical review by research ethics committees. *Journal of the Royal Statistical Society Series A*, **163**, 5–13.

Chapter 2

Agresti, A. (2010) *Analysis of Ordinal Categorical Data*. 2nd edn. Hoboken, NJ: John Wiley & Sons.
Agresti, A. (2013) *Categorical Data Analysis*. 3rd edn. Hoboken, NJ: John Wiley & Sons.
Agresti, A. (2019) *An Introduction to Categorical Data Analysis*. 3rd edn. Hoboken, NJ: John Wiley & Sons.
Arbuthnot, J. (1710) An argument for divine providence, taken from the constant regularity observ'd in the births of both sexes. *Philosophical Transactions of the Royal Society of London*, **27**, 186–190.
Bahadur, R.R. (1967) Rates of convergence of estimates and test statistics. *The Annals of Mathematical Statistics*, **38**, 303–324.
Berry, K.J., Johnston, J.E. and Mielke, P.W. (2014) *A Chronicle of Permutation Statistical Methods: 1920–2000, and Beyond*. New York: Springer.

Bury, K. (1999) *Statistical Distributions in Engineering*. Cambridge: Cambridge University Press.

Chernick, M.R. and LaBudde, R.A. (2011) *An Introduction to Bootstrap Methods with Applications to R*. Hoboken, NJ: John Wiley and Sons.

Conover, W.J. (1999) *Practical Nonparametric Statistics*. 3rd edn. New York: John Wiley & Sons.

Corder, G.W. and Foreman, D.I. (2014) *Nonparametric Statistics: A Step-by-Step Approach*. 2nd edn. Hoboken, NJ: John Wiley & Sons.

Daniel, W.W. (2000) *Applied Nonparametric Statistics*. 2nd edn. Belmont, CA: Duxbury.

Davison, A.C. and Hinkley, D.V. (1997) *Bootstrap Methods and Their Application*. Cambridge: Cambridge University Press.

Desu, M.M. and Raghavarao, D. (2004) *Nonparametric Statistical Methods for Complete and Censored Data*. Boca Raton, FL: Chapman & Hall/ CRC.

Dickhaus, T. (2018) *Theory of Nonparametric Tests*. Cham: Springer.

Edgington, E.S. and Onghena, P. (2007) *Randomization Tests*. 4th edn. Boca Raton, FL: Chapman & Hall/ CRC.

Efron, B. and Hastie, T. (2016) *Computer Age Statistical Inference: Algorithms, Evidence and Data Science*. New York: Cambridge University Press.

Efron, B. and Tibshirani, R.J. (1993) *An Introduction to the Bootstrap*. London: Chapman & Hall.

Everitt, B.S. (1992) *The Analysis of Contingency Tables*. 2nd edn. London: Chapman & Hall.

Gibbons, J.D. and Chakraborti, S. (2020) *Nonparametric Statistical Inference*. 6th edn. Boca Raton, FL: Chapman & Hall/CRC.

Good, P. (2005) *Permutation, Parametric and Bootstrap Tests of Hypotheses*. New York: Springer.

Hájek, J., Šidák, Z. and Sen, P.K. (1999) *Theory of Rank Tests*. 2nd edn. San Diego, CA: Academic Press.

Hettmansperger, T.P. and McKean, J.W. (2011) *Robust Nonparametric Statistical Methods*. 2nd edn. Boca Raton, FL: Chapman & Hall/CRC.

Higgins, J.J. (2004) *Introduction to Modern Nonparametric Statistics*. Belmont, CA: Duxbury.

Hollander, M., Wolfe, D.A. and Chicken, E. (2014) *Nonparametric Statistical Methods*. 3rd edn. Hoboken, NJ: John Wiley & Sons.

Inchley, J., Currie, D., Budisavljevic, S., Torsheim, T., Jåstad, A., Cosma, A., Kelly, C., Már Arnasson, Á. and Samdal, O. (eds) (2020) Spotlight on Adolescent Health and Well-Being. Findings from the 2017/2018 Health Behaviour in School-Aged Children (HBSC) Study. International Report. Volume 2. Key Data. Copenhagen: WHO Health Organization.

Kendall, M.G. and Gibbons, J.D. (1990) *Rank Correlation Methods*. 5th edn. London: Edward Arnold.

Leach, C. (1979) *Introduction to Statistics. A Nonparametric Approach for the Social Sciences*. Chichester: John Wiley & Sons.

Lehmann, E.L. (2006) *Nonparametrics: Statistical Methods Based on Ranks*. Revised edn. Berlin: Springer.

Li, Q. and Racine, J.S. (2006) *Nonparametric Econometrics: Theory and Practice*. Princeton, NJ: Princeton University Press.

Lind, J. (1753) *A Treatise of the Scurvy. In Three Parts. Containing an Inquiry into the Nature, Causes and Cure, of that Disease. Together with a Critical and Chronological View of What Has Been Published on the Subject*. Edinburgh: Printed by Sands, Murray and Cochran for A. Kincaid and A. Donaldson.

Ling, H., Morris, H.R., Neal, J.W., Lees, A.J., Hardy, J., Holton, J.L., Revesz, T. and Williams, D.D.R. (2017) Mixed pathologies including chronic traumatic encephalopathy account for dementia in retired association football (soccer) players. *Acta Neuropathologica*, **133**, 337–352.

Manly, B.F.J. (2006) *Randomization, Bootstrap and Monte Carlo Methods in Biology*. 3rd edn. Boca Raton, FL: Chapman & Hall/CRC.

Maritz, J.S. (1995) *Distribution-Free Statistical Methods*. 2nd edn. London: Chapman & Hall.

Mood, A.M. (1954) On the asymptotic efficiency of certain nonparametric two-sample tests. *The Annals of Mathematical Statistics*, **25**, 514–522.

Noether, G.E. (1991) *Introduction to Statistics: The Nonparametric Way*. New York: Springer–Verlag.

Pitman, E.J.G. (1949) Notes on Non-Parametric Statistical Inference. Institute of Statistics Mimeo Series 27. North Carolina State University. Department of Statistics: Rayleigh, NC. https://repository.lib.ncsu.edu/items/fa11f321-aa76-47d7-a061-8fff12d2e547 (accessed 16 December 2024).

Randles, R.H. and Wolfe, D.A. (1991) *Introduction to the Theory of Nonparametric Statistics*. New York: John Wiley & Sons.

Rayner, J.C.W. and Best, D.J. (2001) *A Contingency Table Approach to Nonparametric Testing*. Boca Raton, FL: Chapman & Hall/CRC.

Siegel, S. and Castellan, N.J. (1988) *Nonparametric Statistics for the Behavioural Sciences*, 2nd edn. New York: McGraw-Hill.

Thas, O. (2010) *Comparing Distributions*. New York: Springer.

Wasserman, L. (2006) *All of Nonparametric Statistics*. New York: Springer.

Yates, F. (1984) Tests of significance for 2×2 contingency tables. *Journal of the Royal Statistical Society Series A*, **147**, 426–463.

Chapter 3

Braun, A., Kendall, S., Cole, C., Smeeton, N. and Angus-Leppan, H. (2017) Development of the Epilepsy Risk Awareness scale (ERA scale) for people with epilepsy. *Seizure: European Journal of Epilepsy*, **46**, 13–18.

Conover, W.J. (1999) *Practical Nonparametric Statistics*, 3rd edn. New York: John Wiley & Sons.

Desu, M.M. and Raghavarao, D. (2004) *Nonparametric Statistical Methods for Complete and Censored Data*. Boca Raton, FL: Chapman & Hall/ CRC.

Duncan, J.S. and Sander, J.W. (1991). The Chalfont seizure severity scale. *Journal of Neurology, Neurosurgery and Psychiatry*, **54**, 873–876.

Fisher, R.A. (1935) *The Design of Experiments*. Edinburgh: Oliver & Boyd.

Fisher, R.A. and Yates, F. (1963) *Statistical Tables for Biological, Agricultural and Medical Research*, 6th edn. Edinburgh: Oliver & Boyd.

Gibbons, J.D. and Chakraborti, S. (2020) *Nonparametric Statistical Inference*, 6th edn. Boca Raton, FL: Chapman & Hall/ CRC.

Gumpertz, M.L., Graham, J.M. and Ristiano, J.B. (1997) Autologistic model of spatial pattern of phytophthora epidemic in bell pepper: effects of soil variables on disease presence. *Journal of Agricultural, Biological and Environmental Statistics*, **2**, 131–156.

Hodges, J.L. Jr. and Lehmann, E.L. (1963) Estimates of location based on rank tests. *The Annals of Mathematical Statistics*, **34**, 598–611.

Hollander, M., Wolfe, D.A. and Chicken, E. (2014) *Nonparametric Statistical Methods*, 3rd edn. Hoboken, NJ: John Wiley & Sons.

Holt, C.A. and Sullivan, S.P. (2023) Permutation tests for experimental data. *Experimental Economics*, **26**, 775–812. https://doi.org/10.1007/s10683-023-09799-6.

Kimura, D.K. and Chikuni, S. (1987) Mixtures of empirical distributions: an iterative application of the age-length key. *Biometrics*, **43**, 23–35.

Knapp, T.R. (1982) The birthday problem: some empirical data and some approximations. *Teaching Statistics*, **4**, 10–14.

Lehmann, E.L. (2006) *Nonparametrics: Statistical Methods Based on Ranks*. Revised edn. Berlin: Springer.

Pitman, E.J.G. (1949) Notes on Non-Parametric Statistical Inference. Institute of Statistics Mimeo Series 27. North Carolina State University. Department of Statistics: Rayleigh, NC. https://repository.lib.ncsu.edu/items/fa11f321-aa76-47d7-a061-8fff12d2e547 (accessed 16 December 2024).

Randles, R.H., Fligner, M.A., Policello, G.E. and Wolfe, D.A. (1980) An asymptotically distribution-free test for symmetry versus asymmetry. *Journal of the American Statistical Association*, **75**, 168–172.

Rayner, J.C.W. and Best, D.J. (2001) *A Contingency Table Approach to Nonparametric Testing*. Boca Raton, FL: Chapman & Hall/CRC.

Siegel, S. and Castellan, N.J. (1988) *Nonparametric Statistics for the Behavioural Sciences*, 2nd edn. New York: McGraw-Hill.

van der Waerden, B.L. (1952) Order tests for the two-sample problem and their power, I. *Proceedings of the Koninklijke Nederlandse Akademie van Wetenschappen A*, **55**, 453–458. Correction, **56**, 80.

van der Waerden, B.L. (1953) Order tests for the two-sample problem and their power; II, III. *Proceedings of the Koninklijke Nederlandse Akademie van Wetenschappen A*, **56**, 303–310, 311–316.

Walsh, J.E. (1949a) Applications of some significance tests for the median which are valid under very general conditions. *Journal of the American Statistical Association*, **44**, 342–355.

Walsh, J.E. (1949b) Some significance tests for the median which are valid under very general conditions. *The Annals of Mathematical Statistics*, **20**, 64–81.

Wilcoxon, F. (1945) Individual comparisons by ranking methods. *Biometrics*, **1**, 80–83.

Chapter 4

Ajne, B. (1968) A simple test for uniformity of a circular distribution. *Biometrika*, **55**, 343–354.

Balakrishnan, N. and Koutras, M.V. (2002) *Runs and Scans with Applications*. New York: John Wiley & Sons.

Banerjee, A.T. and Shah, B.R. (2018) Differences in prevalence of diabetes among immigrants to Canada from South Asian countries. *Diabetic Medicine*, **35**, 937–943.

Barton, D.E. and David, F.N. (1957) Multiple runs. *Biometrika*, **44**, 168–178.

Bradley, J.V. (1968) *Distribution-Free Statistical Tests*. Englewood Cliffs, NJ: Prentice-Hall.

Conover, W.J. (1999) *Practical Nonparametric Statistical Methods*, 3rd edn. New York: John Wiley & Sons.

Cox, D.R. and Stuart, A. (1955) Some quick tests for trend in location and dispersion. *Biometrika*, **42**, 80–95.

Fisher, N.I. (1993) *Statistical Analysis of Circular Data*. Cambridge: Cambridge University Press.

Fisher, R.A. (1929) Tests of significance in harmonic analysis. *Proceedings of the Royal Society of London*, **A125**, 54–59.

Gibbons, J.D. and Chakraborti, S. (2020) *Nonparametric Statistical Inference*, 6th edn. Boca Raton, FL: Chapman & Hall/CRC.

Hodges, J.L. Jr. (1955) A bivariate sign test. *The Annals of Mathematical Statistics*, **26**, 523–527.

Jeandron, A., Saidi, J.M., Kapama, A., Burhole, M., Birembano, F., Vandevelde, T., Gasparrini, A., Armstrong, B., Cairncross, S. and Ensink, J.H. (2015) Water supply interruptions and suspected cholera incidence: a time-series regression in the Democratic Republic of the Congo. *PLoS Medicine*, **12**, e10001893. https://doi.org/10.1371/journal.pmed.1001893.

Kolmogorov, A.N. (1933) Sulla determinazione empirica di una legge di distribuzione. *Giornale dell'Istituto Italiano degli Attuari*, **4**, 83–91.

Kolmogorov, A.N. (1941) Confidence limits for an unknown distribution function. *The Annals of Mathematical Statistics*, **12**, 461–463.

Leslie, J.R., Stephens, M.A. and Fotopoulos, S. (1986). Asymptotic distribution of the Shapiro-Wilk W for testing normality. *The Annals of Statistics*, **14**, 1497–1506.

Lilliefors, H.W. (1967) On the Kolmogorov-Smirnov test for normality with mean and variance unknown. *Journal of the American Statistical Association*, **62**, 399–402.

Mardia, K.V. (1972) *Statistics of Directional Data*. London: Academic Press.

Mardia, K.V. and Jupp, P.E. (2000) *Directional Statistics*. Chichester: John Wiley & Sons.

Mood, A.M. (1940) The distribution theory of runs. *The Annals of Mathematical Statistics*, **11**, 367– 392.

Neave, H.R. (1981) *Elementary Statistical Tables*. London: George Allen & Unwin.

Ross, G.J.S. (1990) *Nonlinear Estimation*. New York: Springer–Verlag.

Sackett, D.L., Haynes, R.B., Guyatt, G.H. and Tugwell, P. (1991) *Clinical Epidemiology: A Basic Science for Clinical Medicine*. 2nd edn. Boston, MA: Little Brown.

Schuster, E.F. and Gu, X. (1997) On the conditional and unconditional distributions of the number of runs in a sample from a multisymbol alphabet. *Communications in Statistics: Simulation and Computation*, **26**, 423–442.

Shapiro, S.S. and Wilk, M.B. (1965) An analysis of variance test for normality (complete samples). *Biometrika*, **52**, 591–611.

Shapiro, S.S., Wilk, M.B. and Chen, H.J. (1968) A comparative study of various tests for normality. *Journal of the American Statistical Association*, **63**, 1343–1372.

Shaughnessy, P.W. (1981) Multiple runs distributions: recurrences and critical values. *Journal of the American Statistical Association*, **76**, 732–736.

Siegel, S. and Castellan, N.J. (1988) *Nonparametric Statistics for the Behavioral Sciences*. 2nd edn. New York: McGraw-Hill.

Smeeton, N. (2021) Early statistical findings and authorship misattribution: an unsystematic review of the literature. *Chance*, **34**, 39–43.

Smeeton, N.C. and Cox, N.J. (2003) Do-it-yourself shuffling and the number of runs under randomness. *The Stata Journal*, **3**, 270–277.

Smeeton, N.C. and Cox, N.J. (2006) Software update: do-it-yourself shuffling and the number of runs under randomness. *The Stata Journal*, **6**, 597.

Smeeton, N. and Wilkinson, G. (1988) The detection of annual clusters in individual patterns of parasuicide. *Journal of Applied Statistics*, **15**, 179–182.

Sprent, P. (1998) *Data Driven Statistical Methods*. London: Chapman & Hall.

Swed, F.S. and Eisenhart, C. (1943) Tables for testing randomness of grouping in a series of alternatives. *The Annals of Mathematical Statistics*, **14**, 83–86.

Whitworth, W.A. (1886) *Choice and Chance*. Cambridge: Deighton Bell.

Whitworth, W.A. (1897) *DCC Exercises: Including Hints for the Solution of All the Questions in Choice and Chance*. Cambridge: Deighton Bell.

Chapter 5

Arnold, H.J. (1965) Small sample power for the one sample Wilcoxon test for nonnormal shift alternatives. *The Annals of Mathematical Statistics*, **36**, 1767–1778.

Conover, W.J. (1999) *Practical Nonparametric Statistics*. 3rd edn. New York: John Wiley & Sons.

Geffen, G., Bradshaw, J.L. and Nettleton, N.C. (1973) Attention and hemispheric differences in reaction time during simultaneous audio-visual tasks. *Quarterly Journal of Experimental Psychology*, **25**, 404–412.

Hollander, M., Wolfe, D.A. and Chicken, E. (2014) *Nonparametric Statistical Methods*, 3rd edn. Hoboken, NJ: John Wiley & Sons.

Jeandron, A., Saidi, J.M., Kapama, A., Burhole, M., Birembano, F., Vandevelde, T., Gasparrini, A., Armstrong, B., Cairncross, S. and Ensink, J.H. (2015) Water supply interruptions and suspected cholera incidence: a time-series regression in the Democratic Republic of the Congo. *PLoS Medicine*, **12**, e10001893. https://doi.org/10.1371/journal.pmed.1001893.

Jones, B. and Kenward, M.G. (2014) *Design and Analysis of Cross-Over Trials*. 3rd edn. London: Chapman and Hall/CRC.

Klotz, J. (1963) Small sample power and efficiency for the one-sample Wilcoxon and normal scores tests. *The Annals of Mathematical Statistics*, **34**, 624–632.

Lehmann, E.L. (2006) *Nonparametrics: Statistical Methods Based on Ranks*. Revised edn. Berlin: Springer.

McNemar, Q. (1947) Note on the sampling error of the difference between correlated proportions or percentages. *Psychometrika*, **12**, 153–157.

Noether, G.E. (1987) Sample size determination for some common nonparametric tests. *Journal of the American Statistical Association*, **82**, 645–647.

Pearson, J.C.G. and Sprent, P. (1968) Trends in hearing loss associated with age or exposure to noise. *Journal of the Royal Statistical Society Series C (Applied Statistics)*, **17**, 205–215.

Scott, A.J., Smith, T.M.F. and Jones, R.G. (1977) The application of time series methods to the analysis of repeated surveys. *International Statistical Review*, **45**, 13–28.

Spiegel, M. and Wrede, R. (2010) *Schaum's Outline of Advanced Calculus (Schaum's Outlines)*. 3rd edn. New York: McGraw-Hill.

Chapter 6

Ansari, A.R. and Bradley, R.A. (1960) Rank sum tests for dispersion. *The Annals of Mathematical Statistics*, **31**, 1174–1189.

Armitage, P. (1984) Editor's report, 1979–1984. *Biometrics*, **40**, 1171–1172.

Brooks, F.M., Smeeton, N.C., Chester, K., Spencer, N. and Klemera, E. (2014) Association between physical activity in adolescence and health behaviours, well-being, family and social relations. *International Journal of Health Promotion and Education*, **52**, 271–282.

Brown, G.W. and Mood, A.M. (1948) Homogeneity of several samples. *The American Statistician*, **2**, 22.

Capon, J. (1961) Asymptotic efficiency of certain locally most powerful rank tests. *The Annals of Mathematical Statistics*, **32**, 88–100.

Carter, E.M. and Hubert, J.J. (1985) Analysis of parallel line assays with multivariate responses. *Biometrics*, **41**, 703–710.

Chandra, M., Singpurwalla, N.D. and Stephens, M.A. (1981) Kolmogorov statistics for test of fit for the extreme value and Weibull distributions. *Journal of the American Statistical Association*, **76**, 729–731.

Conover, W.J. (1980) *Practical Nonparametric Statistics.* 2nd edn. New York: John Wiley & Sons.

Conover, W.J. (1999) *Practical Nonparametric Statistics.* 3rd edn. New York: John Wiley & Sons.

Cramér, H. (1928) On the composition of elementary errors. *Skandinavisk Aktuarietidskrift*, **11**, 13–74, 141–180.

Darwin, J.H. (1959) Note on the comparison of several realizations of a Markoff chain. *Biometrika*, **46**, 412–419.

David, F.N. and Barton, D.E. (1958) A test for birth-order effects. *Annuals of Human Eugenics*, **22**, 250–257.

Davis, T.P. and Lawrance, A.J. (1989) The likelihood for competing risk survival analysis. *Scandinavian Journal of Statistics*, **16**, 23–28.

Deuchler, G. (1914) Über die methoden der korrelationsrechnung in der pädagogik und psychologie. *Zeitschrift für Pädagogische Psychologie und Jugendkunde*, **15**, 229–242.

Eplett, W.J.R. (1982) The distribution of Smirnov type two-sample rank tests for discontinuous distribution functions. *Journal of the Royal Statistical Society Series B*, **44**, 361–369.

Flanders, N. (1970) *Analyzing Teacher Behaviour*. London: Harper and Row.

Fligner, M.A. and Rust, S.W. (1982) A modification of Mood's median test for the generalized Behrens–Fisher problem. *Biometrika*, **69**, 221–226.

Freidlin, B. and Gastwirth, J.L. (2000) Should the median test be retired from general use? *The American Statistician*, **54**, 161–164.

Freund, J.E. and Ansari, A.R. (1957) *Two Way Rank Sum Tests for Variance.* Blacksburg, VA: Virginia Polytechnic Institute Report to Office of Ordnance Research and National Science Foundation, **34**.

Gart, J.J. (1963) A median test with sequential application. *Biometrika*, **50**, 55–62.

Gastwirth, J.L. (1965) Percentile modifications of two sample rank sum tests. *Journal of the American Statistical Association*, **60**, 1127–1141.

Gibbons, J.D. and Chakraborti, S. (2020) *Nonparametric Statistical Inference.* 6th edn. Boca Raton, FL: Chapman & Hall/CRC.

Hill, N.S. and Padmanabhan, A.L. (1984) Robust comparison of two regression lines and biomedical applications. *Biometrics*, **40**, 985–994.

Hollander, M., Wolfe, D.A. and Chicken, E. (2014) *Nonparametric Statistical Methods*, 3rd edn. Hoboken, NJ: John Wiley & Sons.

Klotz, J. (1962) Nonparametric tests for scale. *The Annals of Mathematical Statistics*, **33**, 498–512.

Kruskal, W. (1957) Historical notes on the Wilcoxon unpaired two-sample test. *Journal of the American Statistical Association*, **52**, 356–360.

Lehmann, E.L. (2006) *Nonparametrics: Statistical Methods Based on Ranks*. Revised edn. Berlin: Springer.

Mann, H.B. and Whitney, D.R. (1947) On a test of whether one of two random variables is stochastically larger than the other. *The Annals of Mathematical Statistics*, **18**, 50–60.

Mathisen, H.C. (1943) A method of testing the hypothesis that two samples are from the same population. *The Annals of Mathematical Statistics*, **14**, 188–194.

Miller, S.M. (1995) Monitoring versus blunting styles of coping with cancer influence the information patients want and need about their disease. Implications for cancer screening and management. *Cancer*, **76**, 167–177.

Mood, A.M. (1954) On the asymptotic efficiency of certain nonparametric two sample tests. *The Annals of Mathematical Statistics*, **25**, 514–522.

Moses, L.E. (1952) Nonparametric statistics for psychological research. *Psychological Bulletin*, **49**, 122–143.

Neave, H.R. (1981) *Elementary Statistical Tables*. London: George Allen & Unwin.

Newson, R.B. (2017) Asymptotic Distributions of Two-Sample Rank Statistics for Continuous Outcomes. https://www.rogernewsonresources.org.uk/miscdocs/asy2con1.pdf.

Noether, G.E. (1987) Sample size determination for some common nonparametric tests. *Journal of the American Statistical Association*, **82**, 645–647.

Pitman, E.J.G. (1937a) Significance tests that may be applied to samples from any population. *Supplement to the Journal of the Royal Statistical Society*, **4**, 119–130.

Pothoff, R.F. (1963) Use of the Wilcoxon statistic for a generalized Behrens-Fisher problem. *The Annals of Mathematical Statistics*, **34**, 1596–1599.

Saunders, R. and Laud, P. (1980) The multidimensional Kolmogorov goodness-of-fit test. *Biometrika*, **67**, 237.

Siegel, S. and Tukey, J.W. (1960) A nonparametric sum of ranks procedure for relative spread in unpaired samples. *Journal of the American Statistical Association*, **55**, 429–444.

Smirnov, N.V. (1939) On the estimation of discrepancy between empirical curves of distribution for two independent samples. (In Russian). *Moscow University Mathematics Bulletin*, **2**, 3–16.

Smirnov, N.V. (1948) Tables for estimating the goodness of fit of empirical distributions. *The Annals of Mathematical Statistics*, **19**, 279–281.

Sprent, P. (1998) *Data Driven Statistical Methods*. London: Chapman & Hall.

Sukhatme, B.V. (1957) On certain two sample nonparametric tests for variances. *The Annals of Mathematical Statistics*, **28**, 188–194.

von Mises, R. (1931) *Wahrscheinlichkeitrechnung und ihre Anwendung in der Statistik und Theoretischen Physik*. Leipzig: F. Deuticke.

Wald, A. and Wolfowitz, J. (1940) On a test whether two samples are from the same population. *The Annals of Mathematical Statistics*, **11**, 147–162.

Wilcoxon, F. (1945) Individual comparisons by ranking methods. *Biometrics*, **1**, 80–83.

Chapter 7

Barton, D.E. and David, F.N. (1957) Multiple runs. *Biometrika*, **44**, 168–178.

Berry, D.A. (1987) Logarithmic transformations in ANOVA. *Biometrics*, **43**, 439–456.

Biggins, J.D., Loynes, R.M. and Walker, A.M. (1987) Combining examination results. *British Journal of Mathematical and Statistical Psychology*, **39**, 150–167.

Bland, M (2015) *An Introduction to Medical Statistics*, 4th edn. Oxford: Oxford University Press.

Boos, D.D. (1986) Comparing *K* populations with linear rank statistics. *Journal of the American Statistical Association*, **81**, 1018–1025.

Bünning, H. and Kössler, W. (1999) The asymptotic power of Jonckheere-type tests for ordered alternatives. *Australian & New Zealand Journal of Statistics*, **41**, 67–72.

Cochran, W.G. (1950) The comparison of percentages in matched samples. *Biometrika*, **37**, 256–266.

Cohen, A. (1983) Seasonal daily effect on the number of births in Israel. *Journal of the Royal Statistical Society Series C (Applied Statistics)*, **32**, 228–235.

Durbin, J. (1951) Incomplete blocks in ranking experiments. *British Journal of Statistical Psychology*, **4**, 85–90.

Friedman, M. (1937) The use of ranks to avoid the assumptions of normality implicit in the analysis of variance. *Journal of the American Statistical Association*, **32**, 675–701.

Gibbons, J.D. and Chakraborti, S. (2020) *Nonparametric Statistical Inference*, 6th edn. Boca Raton, FL: Chapman & Hall/CRC.

Gill, A.N. and Mehta, G.P. (1989) Selection procedures for scale parameters using two-sample statistics. *Sankhya, Series B*, **51**, 149–157.

Gill, A.N. and Mehta, G.P. (1991) Selection procedures for scalar parameters using two-sample *U*-statistics. *Australian Journal of Statistics*, **33**, 347–362.

Hinkley, D.V. (1989) Modified profile likelihood in transformed linear models. *Journal of the Royal Statistical Society Series C (Applied Statistics)*, **38**, 495–506.

Hollander, M. and Wolfe, D.A. (1999) *Nonparametric Statistical Methods*. 2nd edn. New York: John Wiley & Sons.

Hollander, M., Wolfe, D.A. and Chicken, E. (2014) *Nonparametric Statistical Methods*, 3rd edn. Hoboken, NJ: John Wiley & Sons.

House, D.E. (1986) A nonparametric version of Williams' test for a randomized block design. *Biometrics*, **42**, 187–190.

Iman, R.L. and Davenport, J.M. (1980) Approximations of the critical region of the Friedman statistic. *Communications in Statistics*, **A9**, 571–595.

Jonckheere, A.R. (1954) A distribution free *k*-sample test against ordered alternatives. *Biometrika*, **41**, 133–145.

Kruskal, W.H. and Wallis, W.A. (1952) Use of ranks in one-criterion variance analysis. *Journal of the American Statistical Association*, **47**, 583–621.

Lehmann, E.L. (2006) *Nonparametrics: Statistical Methods Based on Ranks*. Revised edn. Berlin: Springer.

Lind, J. (1753) *A Treatise of the Scurvy in Three Parts. Containing an Inquiry into the Nature, Causes and Cure, of That Disease. Together with a Critical and Chronological View of What Has Been Published on the Subject*. Edinburgh: Printed by Sands, Murray and Cochran for A Kincaid and A Donaldson.

Ling, H., Morris, H.R., Neal, J.W., Lees, A.J., Hardy, J., Holton, J.L., Revesz, T. and Williams, D.D.R. (2017) Mixed pathologies including chronic traumatic encephalopathy account for dementia in retired association football (soccer) players. *Acta Neuropathologica*, **133**, 337–352.

Lubischew, A.A. (1962) On the use of discriminant functions in taxonomy. *Biometrics*, **18**, 455–477.

Mack, G.A. and Wolfe, D.A. (1981) *K*-sample rank test for umbrella alternatives. *Journal of the American Statistical Association*, **76**, 175–181.

Meyer, J.P. and Seaman, M.A. (2014) A comparison of the exact Kruskal-Wallis distribution to asymptotic approximations for all sample sizes up to 105. *Journal of Experimental Education*, **81**, 139–156.

Neave, H.R. (1981) *Elementary Statistical Tables*. London: George Allen & Unwin.

Pearce, S.C. (1965) *Biological Statistics: An Introduction*. New York: McGraw–Hill.

Shirley, E. (1977) A nonparametric equivalent of Williams' test for contrasting existing dose levels of a treatment. *Biometrics*, **33**, 386–389.

Snee, R.D. (1985) Graphical display of results of three treatment randomized block experiments. *Journal of the Royal Statistical Society Series C (Applied Statistics)*, **34**, 71–77.

Spiegel, K.M. and Stephens, L.J. (2017) *Schaum's Outline of Statistics (Schaum's Outlines)*. 6th edn. New York: McGraw-Hill.

Terpstra, T.J. (1952) The asymptotic normality and consistency of Kendall's test against trend when ties are present in one ranking. *Indagationes Mathematicae*, **14**, 327–333.

Willemain, T.R. (1980) Estimating the population median by nomination sampling. *Journal of the American Statistical Association*, **75**, 908–911.

Williams, D.A. (1986) A note on Shirley's nonparametric test for comparing several dose levels with a zero-dose control. *Biometrics*, **42**, 183–186.

Chapter 8

Akritas, M.G. and Arnold, S.F. (1994) Fully nonparametric hypotheses for factorial designs. I. Multivariate repeated measures designs. *Journal of the American Statistical Association*, **89**, 336–343.

Akritas, M.G., Arnold, S.F. and Brunner, E. (1997) Nonparametric hypotheses and rank statistics for unbalanced factorial designs. *Journal of the American Statistical Association*, **92**, 256–265.

Ben Abdelouahed, H., Gharbi, F., Roumié, M., Baccouche, S., Ben Romdhane, K., Nsouli, B. and Trabelsi, A. (2010) PIXE analysis of medieval silver coins. *Material Characterization*, **61**, 59–64.

Bland, J.M. and Altman, D.G. (1995) Multiple significance tests: the Bonferroni method. *British Medical Journal*, **310**, 170.

Brunner, E. and Puri, M.L. (1996) Nonparametric methods in design and analysis of experiments. In *Handbook of Statistics.* Eds S. Ghosh and C.R. Rao, vol 13, 631–702. New York: Elsevier.

Brunner, E. and Puri, M.L. (2001) Nonparametric methods in factorial designs. *Statistical Papers*, **42**, 1–52.

Conover, W.J. (1999) *Practical Nonparametric Statistics.* 3rd edn. New York: John Wiley & Sons.

Conover, W.J. and Iman, R.L. (1976) On some alternative procedures using ranks for the analysis of experimental designs. *Communications in Statistics: Theory and Methods*, **A5**, 1349–1368.

Conover, W.J. and Iman, R.L. (1981) Rank transformation as a bridge between parametric and non-parametric statistics. *The American Statistician*, **35**, 124–133.

Cox, D.R. (1984) Interaction. *International Statistical Review*, **52**, 1–31.

Duncan, J.S. and Sander, J.W. (1991) The Chalfont seizure severity scale. *Journal of Neurology, Neurosurgery and Psychiatry*, **54**, 873–876.

Emerson, J.D. and Hoaglin, D.C. (1983) Analysis of two-way tables by medians. In *Understanding Robust and Exploratory Data Analysis.* Eds D.C. Hoaglin, F. Mosteller and J.W. Tukey, 166–207. New York: John Wiley & Sons.

Hettmansperger, T.P. and Elmore, R. (2002) Tests for interaction in a two-way layout: should they be included in a nonparametrics course? *Proceedings of the 6th International Conference on the Teaching of Statistics, Holiday Inn Cape Town, Cape Town, South Africa*, July, 2002.

Hettmansperger, T.P. and McKean, J.W. (2011) *Robust Nonparametric Statistical Methods*, 2nd edn. Boca Raton, FL: Chapman and Hall/CRC.

Hollander, M., Wolfe, D.A. and Chicken, E. (2014) *Nonparametric Statistical Methods*, 3rd edn. Hoboken, NJ: John Wiley & Sons.

Higgins, J.J. (2004) *Introduction to Modern Nonparametric Statistics*. Belmont, CA: Duxbury.

Jones, B. and Kenward, M.G. (2014) *Design and Analysis of Cross-Over Trials*, 3rd edn. London: Chapman and Hall/CRC.

Koch, G.G. (1972) The use of nonparametric methods in the statistical analysis of the two-period change-over design. *Biometrics*, **28**, 577–594.

Leach, C. (1979) *Introduction to Statistics. A Nonparametric Approach for the Social Sciences.* Chichester: John Wiley & Sons.

Ling, H., Morris, H.R., Neal, J.W., Lees, A.J., Hardy, J., Holton, J.L., Revesz, T. and Williams, D.D.R. (2017) Mixed pathologies including chronic traumatic encephalopathy account for dementia in retired association football (soccer) players. *Acta Neuropathologica*, **133**, 337–352.

Pearce, S.C. (1965) *Biological Statistics – An Introduction*. New York: McGraw–Hill.

Putt, M.E. and Chinchilli, V.M. (2004) Nonparametric approaches to the analysis of crossover studies. *Statistical Science*, **19**, 712–719.

Rosenthal, I. and Ferguson, T.S. (1965) An asymptotic distribution-free multiple comparison method with application to the problem of *n* rankings of *m* objects. *British Journal of Mathematical and Statistical Psychology*, **18**, 243–254.

Satterthwaite, F.E. (1946) An approximate distribution of estimates of variance components. *Biometrics Bulletin*, **2**, 110–114.

Schwarze, A., Ehrenpfordt, I. and Eggert, F. (2014). Workload of younger and elderly drivers in different infrastructural situations. *Transportation Research, Part F*, **26**, 102–115.

Shirley, E.A.C. (1987) Applications of ranking methods to multiple comparison procedures and factorial experiments. *Journal of the Royal Statistical Society Series C (Applied Statistics)*, **36**, 205–213.
Spiegel, K.M. and Stephens, L.J. (2017) *Schaum's Outline of Statistics (Schaum's Outlines)*. 6th edn. New York: McGraw-Hill.

Chapter 9

Akritas, M.G. (2004) Nonparametric survival analysis. *Statistical Science*, **19**, 615–623.
Akritas, M.G. and Brunner, E. (1997) Nonparametric methods for factorial designs with censored data. *Journal of the American Statistical Association*, **92**, 568–576.
Altshuler, B. (1970) Theory for the measurement of competing risks in animal experiments. *Mathematical Biosciences*, **6**, 1–11.
Andersen, P.K., Borgan, O., Gill, R.D. and Keiding, N. (1982) Linear nonparametric tests for the comparison of counting processes with application to censored survival data. *International Statistical Review*, **50**, 219–258.
Cox, D.R. (1972) Regression models and life tables. *Journal of the Royal Statistical Society Series B*, **34**, 187–220.
Davis, T.P. and Lawrance, A.J. (1989) The likelihood for competing risk survival analysis. *Scandinavian Journal of Statistics*, **16**, 23–28.
Desu, M.M. and Raghavarao, D. (2004) *Nonparametric Statistical Methods for Complete and Censored Data*. Boca Raton, FL: Chapman & Hall/CRC.
Dinse, G.E. (1982) Nonparametric estimation for partially-complete time and type of failure data. *Biometrics*, **38**, 417–431.
Freidlin, B., Podgor, M.J. and Gastwirth, J.L. (1999) Efficiency robust tests for survival or ordered categorical data. *Biometrics*, **55**, 883–886.
Gankam-Kengue, F., Ayers, C., Khera, A., de Lemos, J. and Maalouf, N.M. (2013) Mild hyponatremia is associated with an increased risk of death in an ambulatory setting. *Kidney International*, **83**, 700–706.
Gastwirth, J.L. (1968) The first-median test: a two-sided version of the control median test. *Journal of the American Statistical Association*, **63**, 692–706.
Gastwirth, J.L. (1985) The use of maximum efficiency robust tests in combining contingency tables and survival analysis. *Journal of the American Statistical Association*, **80**, 380–384.
Gehan, E.A. (1965a) A generalized Wilcoxon test for comparing arbitrarily singly censored samples. *Biometrika*, **52**, 203–223.
Gehan, E.A. (1965b) A generalized two-sample Wilcoxon test for doubly censored data. *Biometrika*, **52**, 650–653.
Graunt, J. (1665) *Natural and Political Observations Made upon the Bills of Mortality*. 3rd edn. London: Royal Society.
Innes, N.P.T., Stirrups, D.R., Evans, D.J.P., Hall, N. and Leggate, M. (2006) A novel technique using preformed metal crowns for managing carious primary molars in general practice—a retrospective analysis. *British Dental Journal*, **200**, 451–454.
Kalbfleisch, J.D. and Prentice, R.L. (2002) *The Statistical Analysis of Failure Time Data*. 2nd edn. Hoboken, NJ: John Wiley & Sons.
Kaplan, E.L. and Meier, P. (1958) Nonparametric estimation from incomplete observations. *Journal of the American Statistical Association*, **53**, 457–81.
Kleinbaum, D.G. and Klein, M. (2011) *Survival Analysis: A Self-Learning Text*. 3rd edn. (Statistics for Biology and Health). New York: Springer.
Ogle, W. (1887) Summary of several male life tables. *Journal of the Royal Statistical Society*, **50**, 648–652.
Opdam, N.J.M., Bronkhorst, E.M., Roeters, J.M. and Loomans, B.A.C. (2007) A retrospective clinical study on longevity of posterior composite and amalgam restorations. *Dental Materials*, **23**, 2–8.

Peto, R. and Peto, J. (1972) Asymptotically efficient rank-invariant test procedures. *Journal of the Royal Statistical Society Series A*, **135**, 185–206.

Peto, R., Pike, M.C., Armitage, P., Breslow, N.E.D., Cox, D.R., Howard, S.V., Mantel, N., McPherson, K., Peto, J. and Smith, P.G. (1977) Design and analysis of randomized clinical trials requiring prolonged observation of each patient. II Analysis and examples. *British Journal of Cancer*, **35**, 1–39.

Promislow, S.D. (2015) *Fundamentals of Actuarial Mathematics*. 3rd edn. Chichester: Wiley.

Savage, I.R. (1956) Contributions to the theory of rank order statistics: the two-sample case. *The Annals of Mathematical Statistics*, **27**, 590–615.

Tarone, R. and Ware, J. (1977) On distribution-free tests for equality of survival distributions. *Biometrika*, **64**, 156–160.

Thompson, J.R. (2011) *Empirical Model Building*. 2nd edn. Hoboken, NJ: John Wiley & Sons.

Tsiatis, A. (1975) A nonidentifiability aspect of the problem of competing risks. *Proceedings of the National Academy of Sciences*, **72**, 20–22.

Vinovskis, M.A. (1971) The 1789 life table of Edward Wigglesworth. *The Journal of Economic History*, **31**, 570–590.

Walford, C. (1885) History of life assurance in the United Kingdom. *Journal of the Institute of Actuaries*, **25**, 114–133.

Wolfe, C.D.A., Smeeton, N.C., Coshall, C., Tilling, K. and Rudd, A.G. (2005). Survival differences after stroke in a multiethnic population: follow-up study with the south London stroke register. *British Medical Journal*, **331**, 431–433.

Chapter 10

Blomqvist, N. (1950) On a measure of dependence between two random variables. *The Annals of Mathematical Statistics*, **21**, 593–600.

Blomqvist, N. (1951) Some tests based on dichotomization. *The Annals of Mathematical Statistics*, **22**, 362–371.

Borkowf, C.B., Gail, M.H., Carroll, R.J. and Gill, R.D. (1997) Analyzing bivariate continuous data grouped into categories defined by empirical quantiles of marginal distributions. *Biometrics*, **53**, 1054–1069.

Braun, A., Kendall, S., Cole, C., Smeeton, N. and Angus-Leppan, H. (2017) Development of the Epilepsy Risk Awareness scale (ERA scale) for people with epilepsy. *Seizure: European Journal of Epilepsy*, **46**, 13–18. https://doi.org/10.1016/j.seizure.2017.02.005

Cantor, A.B. (1996) Sample-size calculations for Cohen's kappa. *Psychological Methods*, **1**, 150–153.

Cohen, J. (1960) A coefficient of agreement for nominal scales. *Educational and Psychological Measurement*, **20**, 37–46.

Cohen, J. (1968) Weighted kappa: nominal scale agreement with provision for scaled disagreement or partial credit. *Psychological Bulletin*, **70**, 213–220.

Conger, A.J. (1980) Integration and generalization of kappa for multiple raters. *Psychological Bulletin*, **88**, 322–328.

Crane, M., Mallol, J., Beasley, R., Stewart, A. and Asher, M.I. (2003) Agreement between written and video questions for comparing asthma symptoms in ISAAC. *European Respiratory Journal*, **21**, 455–461.

Daniels, H.E. and Kendall, M.G. (1947) The significance of rank correlation where parental correlation exists. *Biometrika*, **34**, 197–208.

Dietz, E.J. and Killeen, T.J. (1981) A nonparametric multivariate test for monotone trend with pharmaceutical applications. *Journal of the American Statistical Association*, **76**, 169–174.

Donner, A., Eliasziw, M. and Klar, N. (1996). Testing the homogeneity of kappa statistics. *Biometrics*, **52**, 176–183.

Doolittle, M.H. (1888) Association ratios. *Bulletin of the Philosophical Society of Washington*, **10**, 83–87, 94–96.

Fleiss, J.L. (1978) Measuring nominal scale agreement among many raters. *Psychological Bulletin*, **76**, 378–382.

Fleiss, J.L., Levin, B. and Paik, M.C. (2003) *Statistical Methods for Rates and Proportions*. 3rd edn. Hoboken, NJ: John Wiley & Sons.

Fleiss, J.L. and Cohen, J. (1973) The equivalence of weighted kappa and the intraclass correlation coefficient as measures of reliability. *Educational and Psychological Measurement*, **33**, 613–619.

Fleiss, J.L., Lee, J.C.M. and Landis, J.R. (1979) The large sample variance of kappa in the case of different sets of raters. *Psychological Bulletin*, **86**, 974–977.

Galton, F. (1892) *Fingerprints*. London: Macmillan.

Gibbons, J.D. and Chakraborti, S. (2020) *Nonparametric Statistical Inference*, 6th edn. Boca Raton, FL: Chapman & Hall/ CRC.

Gideon, R.A. and Hollister, R.A. (1987) A rank correlation coefficient resistant to outliers. *Journal of the American Statistical Association*, **82**, 656–666.

Hollander, M., Wolfe, D.A. and Chicken, E. (2014) *Nonparametric Statistical Methods*, 3rd edn. Hoboken, NJ: John Wiley & Sons.

Kendall, M.G. (1938) A new measure of rank correlation. *Biometrika*, **30**, 81–93.

Kendall, M.G. and Gibbons, J.D. (1990) *Rank Correlation Methods*. 5th edn. London: Edward Arnold.

Kerridge, D. (1975) The interpretation of rank correlations. *Journal of the Royal Statistical Society Series C (Applied Statistics)*, **24**, 257–258.

Landis, J.R. and Koch, G.G. (1977) The measurement of observer agreement for categorical data. *Biometrics*, **33**, 159–174.

Light, R.J. (1971) Measures of response agreement for qualitative data: some generalizations and alternatives. *Psychological Bulletin*, **76**, 365–377.

Neave, H.R. (1981) *Elementary Statistical Tables*. London: George Allen & Unwin.

Noether, G.E. (1987) Sample size determination for some common nonparametric tests. *Journal of the American Statistical Association*, **82**, 645–647.

Paul, S.R. (1979) Models and estimation procedures for the calibration of examiners. *British Journal of Mathematical and Statistical Psychology*, **32**, 242–251.

Pearson, K. (1895) Note on regression and inheritance in the case of two parents. *Proceedings of the Royal Society of London*, **58**, 240–242.

Pitman, E.J.G. (1937b) Significance tests that may be applied to samples from any population, II: The correlation coefficient test. *Supplement to the Journal of the Royal Statistical Society*, **4**, 225–232.

Posner, K.L., Sampson, P.D., Caplan, R.A., Ward, R.J. and Chenly, F.W. (1990) Measuring interrater reliability among multiple raters: an example of methods for nominal data. *Statistics in Medicine*, **9**, 1103–1116.

Rae, G. (1988) The equivalence of multiple rater kappa statistics and intraclass correlation coefficients. *Educational and Psychological Measurement*, **48**, 921–933.

Smeeton, N.C., Rona, R.J., Oyarzun, M. and Diaz, P.V. (2006) Agreement between responses to a standardized asthma questionnaire and a questionnaire following a demonstration of asthma symptoms in adults. *American Journal of Epidemiology*, **163**, 384–391.

Spearman, C. (1904) The proof and measurement of association between two things. *The American Journal of Psychology*, **15**, 72–101.

Sprent, P. (1998) *Data Driven Statistical Methods*. London: Chapman & Hall.

Thomas, G.E. (1989) A note on the correcting of ties with Spearman's ρ. *Journal of Statistical Computation and Simulation*, **31**, 37–40.

Chapter 11

Adichie, J.N. (1967) Estimates of regression parameters based on rank tests. *The Annals of Mathematical Statistics*, **38**, 894–904.

Adichie, J.N. (1984) Rank tests in linear models. In *Handbook of Statistics.* Eds Krishnaiah, P.R. and Sen, P.K., vol 4, 229–257. Amsterdam: Elsevier.

Atkinson, A.C. (1985) *Plots, Transformations and Regression. An Introduction to Graphical Methods of Diagnostic Regression Analysis*. Oxford: Clarendon Press.

Braun, A., Kendall, S., Cole, C., Smeeton, N. and Angus-Leppan, H. (2017) Development of the Epilepsy Risk Awareness scale (ERA scale) for people with epilepsy. *Seizure: European Journal of Epilepsy*, **46**, 13–18. https://doi.org/10.1016/j.seizure.2017.02.005.

Cook, R.D. and Weisberg, S. (1982) *Residuals and Influence in Regression*. London: Chapman & Hall.

Gat, J.R. and Nissenbaum, A. (1976) Limnology and ecology of the Dead Sea. *National Geographic Society Research Reports*, **17**, 413–418.

Hollander, M., Wolfe, D.A. and Chicken, E. (2014) *Nonparametric Statistical Methods*, 3rd edn. Hoboken, NJ: John Wiley & Sons.

Hussain, S.S. and Sprent, P. (1983) Nonparametric regression. *Journal of the Royal Statistical Society Series A*, **146**, 182–191.

Jaeckel, L.A. (1972) Estimating regression equations by minimizing the dispersion of residuals. *The Annals of Mathematical Statistics*, **43**, 144–158.

Jeffree, E.P. (1960) Some long-term means from *The Phenological Reports* (1891–1948) of the Royal Meteorological Society. *Quarterly Journal of the Royal Meteorological Society*, **86**, 95–103.

Katti, S.K. (1965) Multivariate covariance analysis. *Biometrics*, **21**, 957–974.

Kimura, D.K. and Chikuni, S. (1987) Mixtures of empirical distributions: an iterative application of the age-length key. *Biometrics*, **43**, 23–35.

Maritz, J.S. (1995) *Distribution-Free Statistical Methods*. 2nd edn. London: Chapman & Hall.

McKean, J.W., Sheather, S.J. and Hettmansperger, T.P. (1990) Regression diagnostics for rank based methods. *Journal of the American Statistical Association*, **85**, 1018–1028.

Sen, P.K. (1968) Estimates of the regression coefficient based on Kendall's tau. *Journal of the American Statistical Association*, **63**, 1379–1389.

Sen, P.K. (1969) On a class of rank order tests for the parallelism of several regression lines. *The Annals of Mathematical Statistics*, **40**, 1668–1683.

Sprent, P. (1998) *Data Driven Statistical Methods*. London: Chapman & Hall.

Theil, H. (1950) A rank invariant method of linear and polynomial regression analysis, I, II, III. *Proceedings of the Koninklijke Nederlandse Akademie van Wetenschappen A*, **53**, 386–392, 521–525, 1397–1412.

Chapter 12

Agresti, A. (1992) A survey of exact inferences for contingency tables. *Statistical Science*, **7**, 131–153.

Agresti, A. (2010) *Analysis of Ordinal Categorical Data*. 2nd edn. Hoboken, NJ: John Wiley & Sons.

Agresti, A. (2013) *Categorical Data Analysis*. 3rd edn. Hoboken, NJ: John Wiley & Sons.

Agresti, A. (2019) *An Introduction to Categorical Data Analysis*. 3rd edn. Hoboken, NJ: John Wiley & Sons.

Armitage, P., Berry, G. and Matthews, J.N.S. (2001) *Statistical Methods in Medical Research*. 4th edn. New York: Wiley Blackwell.

Bassendine, M.F., Collins, J.D., Stephenson, J., Saunders, P. and James, O.W.F. (1985) Platelet associated immunoglobulins in primary biliary cirrhosis: a cause of thrombocytopenia? *Gut*, **26**, 1074–1079.

Bishop, Y.M.M., Fienberg, S.E. and Holland, P. (1975) *Discrete Multivariate Analysis: Theory and Practice*. Cambridge, MA: MIT Press.

Bowker, A.H. (1948) A test for symmetry in contingency tables. *Journal of the American Statistical Association*, **43**, 572–574.

Breslow, N.E. and Day, N.E. (1980) *Statistical Methods in Cancer Research. I. The Analysis of Case-Control Studies*. Lyons: WHO/IARC Scientific Publications, No. 32.

Christensen, R.A. (1997) *Log-Linear Models and Logistic Regression*. 2nd edn. New York: Springer-Verlag.

Cox, P.R. (1978) *Demography*. 5th edn. Cambridge: Cambridge University Press.

Everitt, B.S. (1992) *The Analysis of Contingency* Tables . 2nd edn. London: Chapman & Hall.

Fisher, R.A. (1934) *Statistical Methods for Research Workers*. 5th edn. Edinburgh: Oliver & Boyd.

Freeman, G.H. and Halton, J.H. (1951) Note on an exact treatment of contingency, goodness of fit and other problems of significance. *Biometrika*, **38**, 141–149.

Freidlin, B. and Gastwirth, J.L. (1999) Unconditional versions of several tests commonly used in the analysis of contingency tables. *Biometrics*, **55**, 264–267.

Graubard, B.I. and Korn, E.L. (1987) Choice of column scores for testing independence in ordered $2 \times K$ contingency tables. *Biometrics*, **43**, 471–476.

Hornbrook, M.C., Hurtado, A.V. and Johnson, R.E. (1985) Health care episodes: definition, measurement and use. *Medical Care Review*, **42**, 163–218.

Kendall, M.G. and Gibbons, J.D. (1990) *Rank Correlation Methods*. 5th edn. London: Edward Arnold.

Kim, B., Kim, J., Choi, S. and Yoon, Y. (2018) Understanding the characteristics of recurrent visits to the emergency department by paediatric patients: a retrospective observational study conducted at three tertiary hospitals in Korea. *BMJ Open*, **8**, e018208. https://doi.org/10.1136/bmjopen-2017–018208.

Kimber, A.C. (1987) When is a χ^2 not a χ^2? *Teaching Statistics*, **9**, 74–77.

Lehmann, E.L. (1975) *Nonparametrics: Statistical Methods Based on Ranks*. San Francisco, CA: Holden Day, Inc.

Lehmann, E.L. (2006) *Nonparametrics: Statistical Methods Based on Ranks*. Revised edn. Berlin: Springer.

Ling, H., Morris, H.R., Neal, J.W., Lees, A.J., Hardy, J., Holton, J.L., Revesz, T. and Williams, D.D. (2017) Mixed pathologies including chronic traumatic encephalopathy account for dementia in retired association football (soccer) players. *Acta Neuropathologica*, **133**, 337–352.

Lloyd, C.J. (1999) *Statistical Analysis of Categorical Data*. New York: John Wiley & Sons.

Marascuilo, L.A. and McSweeney, M. (1977) *Nonparametric and Distribution-free Methods for the Social Sciences*. Monterey, CA: Brooks/Cole Publishing Company.

Marascuilo, L.A. and Serlin, R.C. (1979) Tests and contrasts for comparing change parameters for a multiple sample McNemar data model. *British Journal of Mathematical and Statistical Psychology*, **32**, 105–112.

Mathie, E., Wythe, H., Munday, D., Millac, P., Rhodes, G., Roberts, N., Smeeton, N., Poland, F. and Jones, J. (2018) Reciprocal relationships and the importance of feedback in patient and public involvement: a mixed methods study. *Health Expectations*, **21**, 899–908.

Muehlbauer, T. (2010) Relationship between starting and finishing position in Formula One car races. *International Journal of Performance Analysis in Sport*, **10**, 98–102.

Newcombe, L.P. and Newcombe, R.G. (2001) Adult language learning: the effect of background, motivation and practice on perseverance. *International Journal of Bilingual Education and Bilingualism*, **4**, 332–354.

Pearson, K. (1900) On a criterion that a given system of deviations from the probable in the case of a correlated system of variables is such that it can reasonably be supposed to have arisen in random sampling. *Philosophical Magazine*, **50**, 157–175.

Pearson, K. (1904) *Mathematical contributions to the theory of evolution. XIII. On the theory of contingency and its relation to association and normal correlation. Drapers' Company Research Memoirs. Biometric Series 1*. London: Dulau.

Plackett, R.L. (1981) *The Analysis of Categorical Data*. 2nd edn. London: Charles Griffin & Co.

Shigematsu, K., Watanabe, Y. and Nakano, H. (2015) Weekly variations of stroke occurrence: an observational cohort study based on the Kyoto Stroke Registry, Japan. *BMJ Open*, **5**, e006294. https://doi.org/10.1136/bmjopen-2014-006294.

Siegel, S. and Castellan, N.J. (1988) *Nonparametric Statistics for the Behavioral Sciences*. 2nd edn. New York: McGraw-Hill.

Simonoff, J.S. (2003) *Analyzing Categorical Data*. New York: Springer–Verlag.

Smeeton, N.C. (1986) Modelling episodes of mental illness: some results from the Second U.K. National Morbidity Survey. *The Statistician*, **35**, 55–63. https://doi.org/10.2307/2988299.

Somers, R.H. (1962) A new asymmetric measure of association for ordinal variables. *American Sociological Review*, **27**, 799–811.
Stuart, A. (1955) A test for homogeneity of the marginal distributions in a two-way classification. *Biometrika*, **42**, 412–416.
Stuart, A. (1957) The comparison of frequencies in matched samples. *British Journal of Statistical Psychology*, **10**, 29–32.
Upton, G.J.G. (1992) Fisher's exact test. *Journal of the Royal Statistical Society Series A*, **155**, 395–402.
Yates, F. (1934) Contingency tables involving small numbers and the χ^2 test. *Supplement to the Journal of the Royal Statistical Society*, **1**, 217–235.
Yates, F. (1984) Tests of significance for 2×2 contingency tables. *Journal of the Royal Statistical Society Series A*, **147**, 426–463.

Chapter 13

Agresti, A. (2002) *Categorical Data Analysis*, 2nd edn. New York: John Wiley & Sons.
Agresti, A. (2010) *Analysis of Ordinal Categorical Data*. 2nd edn. Hoboken, NJ: John Wiley & Sons.
Agresti, A. (2013) *Categorical Data Analysis*, 3rd edn. Hoboken, NJ: John Wiley & Sons.
Agresti, A. (2019) *An Introduction to Categorical Data Analysis*. 3rd edn. Hoboken, NJ: John Wiley & Sons.
Aitken, C., Taroni, F. and Bozza, S. (2021) *Statistics and the Evaluation of Evidence for Forensic Scientists*. 3rd edn. Hoboken, NJ: John Wiley & Sons.
Anderson, R.L. (1959) Use of contingency tables in the analysis of consumer preference studies. *Biometrics*, **15**, 582–590.
Armitage, P. (1955) Tests for linear trends in proportions and frequencies. *Biometrics*, **11**, 375–386.
Best, D.J. (1994) Nonparametric comparison of two histograms. *Biometrics*, **50**, 538–541.
Best, D.J. and Rayner, J.C.W. (1996) Nonparametric analysis for doubly ordered two-way contingency tables. *Biometrics*, **52**, 1153–1156.
Bishop, Y.M.M., Fienberg, S.E. and Holland, P. (1975) *Discrete Multivariate Analysis: Theory and Practice*. Cambridge, MA: MIT Press.
Breslow, N.E. and Day, N.E. (1980) *Statistical Methods in Cancer Research. I. The Analysis of Case-Control Studies*. Lyons: WHO/IARC Scientific Publications, No. 32.
Brooks, F.M., Smeeton, N.C., Chester, K., Spencer, N. and Klemera, E. (2014) Associations between physical activity in adolescence and health behaviours, well being, family and social relations. *International Journal of Health Promotion and Education*, **52**, 271–282.
Cochran, W.G. (1954) Some methods for strengthening the common $\chi 2$ tests. *Biometrics*, **10**, 417–454.
Cormack, R.M. (1989) Log-linear models for capture-recapture. *Biometrics*, **45**, 395– 413.
Cox, D.R. and Snell, E.J. (1989) *Analysis of Binary Data*. 2nd edn. London: Chapman & Hall.
Dahl, K. (1919) Studies of trout and trout-waters in Norway. *Salmon and Trout Magazine*, **18**, 16–33.
Dobson, A.J. and Barnett, A.G. (2018) *An Introduction to Generalized Linear Models*. 4th edn. Boca Raton, FL: Chapman & Hall/CRC.
Everitt, B.S. (1992) *The Analysis of Contingency* Tables. 2nd edn. London: Chapman & Hall.
Fienberg, S.E. (1980) *The Analysis of Cross Classified Categorical Data*. 2nd edn. Cambridge, MA: MIT Press.
Fisher, N., Turner, S.W., Pugh, R. and Taylor, C. (1994) Estimating numbers of homeless and homeless mentally ill people in north east Westminster by using capture-recapture analysis. *BMJ*, **308**, 27–30.
Gart, J. (1970) Point and interval estimation of the common odds ratio in the combination of 2×2 tables with fixed marginals. *Biometrika*, **57**, 471–475.
Gastwirth, J.L. (1988) *Statistical Reasoning in Law and Public Policy: Statistical Concepts and Issues of Fairness*. San Diego, CA: Academic Press.

Gastwirth, J.L. (1997) Statistical evidence in discrimination cases. *Journal of the Royal Statistical Society Series A*, **160**, 289–303.

Gastwirth, J.L. and Greenhouse, S.W. (1995) Biostatistical concepts and methods in the legal setting. *Statistics in Medicine*, **14**, 1641–1653.

Goudie, I.B.T. and Goudie, M. (2007) Who captures the marks for the Petersen estimator? *Journal of the Royal Statistical Society Series A*, **170**, 825–839.

Graubard, B.I. and Korn, E.L. (1987) Choice of column scores for testing independence in ordered $2{\times}k$ contingency tables. *Biometrics*, **43**, 471–476.

Hay, G. (1997) The selection from multiple data sources in epidemiological capture-recapture studies. *The Statistician*, **46**, 515–520.

Hickman, M., Hope, V., Platt, L., Higgins, V., Bellis, M., Rhodes, T., Taylor, C. and Tilling, K. (2006) Estimating prevalence of injecting drug use: a comparison of multiplier and capture-recapture methods in cities in England and Russia. *Drug and Alcohol Review*, **25**, 131–140.

Hook, E.B. and Regal, R.R. (1982) Validity of Bernoulli census, log-linear, and truncated binomial models for correcting for underestimates in prevalence studies. *American Journal of Epidemiology*, **116**, 168–176.

Hook, E.B. and Regal, R.R. (1995) Capture-recapture methods in epidemiology: methods and limitations. *Epidemiologic Reviews*, **17**, 243–264.

Lancaster, H.O. (1949) The derivation and partitioning of χ^2 in certain discrete distributions. *Biometrika*, **36**, 117–129.

Lancaster, H.O. (1953) A reconciliation of χ^2, considering metrical and enumerative aspects. *Sankhya*, **13**, 1–9.

McCullagh, P. and Nelder, J.A. (1989) Generalized Linear Models. 2nd edn. London: Chapman & Hall.

Mantel, N. and Haenszel, W. (1959) Statistical aspects of the analysis of data from retrospective studies of disease. *Journal of the National Cancer Institute*, **22**, 719–748.

Mathie, E., Wythe, H., Munday, D., Millac, P., Rhodes, G., Roberts, N., Smeeton, N., Poland, F. and Jones, J. (2018) Reciprocal relationships and the importance of feedback in patient and public involvement: a mixed methods study. *Health Expectations*, **21**, 899–908.

Mehta, C.R., Patel, N.R. and Senchaudhuri, P. (1998) Exact power and sample size computations for the Cochran-Armitage trend test. *Biometrics*, **54**, 1615–1621.

Pearson, K., Lee, A. and Bramley-Moore, L. (1899) Genetic (reproductive) selection: inheritance of fertility in man, and of fecundity in thoroughbred racehorses. *Philosophical Transactions of the Royal Society* A, **192**, 257–330.

Plackett, R.L. (1981) *The Analysis of Categorical Data*. 2nd edn. London: Charles Griffin & Co.

Rayner, J.C.W. and Best, D.J. (2001) *A Contingency Table Approach to Nonparametric Testing*. Boca Raton, FL: Chapman & Hall/CRC.

Rees, J.H., Thompson, R.D., Smeeton, N.C. and Hughes, R.A.C. (1998) Epidemiological study of Guillain-Barré syndrome in south east England. *Journal of Neurology, Neurosurgery & Psychiatry*, **64**, 74–77.

Regal, R.R. and Hook, E.B. (1984) Goodness-of-fit based confidence intervals for estimates of the size of a closed population. *Statistics in Medicine*, **3**, 287–291.

Regal, R.R. and Hook, E.B. (1999) An exact test for all-way interactions in a 2^M contingency table: application to interval capture-recapture estimation of population size. *Biometrics*, **55**, 1241–1246.

Robins, J., Breslow, N. and Greenland, S. (1986) Estimation of the Mantel–Haenszel variance consistent in both sparse data and large-strata limiting models. *Biometrics*, **42**, 311–323.

Scholz, F.W. and Stephens, M.A. (1987) *K*-sample Anderson–Darling tests. *Journal of the American Statistical Association*, **82**, 918–924.

Seber, G.A.F. (1982) *The Estimation of Animal Abundance and Related Parameters*. 2nd edn. London: Charles Griffin and Company.

Sekar, C.C. and Deming, W.E. (1949) On a method of estimating birth and death rates and the extent of registration. *Journal of the American Statistical Association*, **44**, 101–115.

Selvin, S. (1995) *Practical Biostatistical Methods*. Belmont, CA: Wadsworth Publishing Company.

Sesardic, N. (2007) Sudden infant death or murder? A royal confusion about probabilities. *British Journal for the Philosophy of Science*, **58**, 299–329.

Simpson, E.H. (1951) The interpretation of interaction in contingency tables. *Journal of the Royal Statistical Society Series B*, **13**, 238–241.

Smeeton, N.C., Rona, R.J., Sharland, G., Botting, B.J., Barnett, A. and Dundas, R. (1999) Estimating the prevalence of malformation of the heart in the first year of life using capture-recapture methods. *American Journal of Epidemiology*, **150**, 778–785.

Sprent, P. (1998) *Data Driven Statistical Methods*. London: Chapman & Hall.

Tilling, K. and Sterne, J.A.C (1998) Capture-recapture methods including covariate effects. *American Journal of Epidemiology*, **149**, 392–400.

Yule, G.U. (1903) Notes on the theory of association of attributes in statistics. *Biometrika*, **2**, 121–134.

Zelen, M. (1971) The analysis of several 2×2 contingency tables. *Biometrika*, **58**, 129–137.

Chapter 14

Andrews, D.F. (1974) A robust method for multiple linear regression. *Technometrics*, **16**, 523–531.

Andrews, D.F., Bickel, P.J., Hampel, F.R., Huber, P.J., Rogers, W.H. and Tukey, J.W. (1972) *Robust Estimates of Location: Survey and Advances*. Princeton, NJ: Princeton University Press.

Barnett, V.D. and Lewis, T. (1994) *Outliers in Statistical Data*. 3rd edn. Chichester: John Wiley & Sons.

Chernick, M.R. (2008) *Bootstrap Methods: A Guide for Practitioners and Researchers*. 2nd edn. Hoboken, NJ: John Wiley & Sons.

Davison, A.C. and Hinkley, D.V. (1997) *Bootstrap Methods and Their Application*. Cambridge: Cambridge University Press.

Dikta, G. and Scheer, M. (2021) *Bootstrap Methods: With Applications in R*. Switzerland: Springer Nature.

Efron, B. (1979) Bootstrap methods: another look at the Jackknife. *The Annals of Statistics*, **7**, 1–26.

Efron, B. (1981) Censored data and the bootstrap. *Journal of the American Statistical Association*, **76**, 312–319.

Efron, B. and Gong, G. (1983) A leisurely look at the bootstrap, the jackknife and cross-validation. *The American Statistician*, **37**, 36–48.

Efron, B. and Tibshirani, R.J. (1993) *An Introduction to the Bootstrap*. London: Chapman & Hall.

Fisher, R.A., Corbet, A.S. and Williams, C.B. (1943) The relation between the number of species and the number of individuals in a random sample of an animal population. *The Journal of Animal Ecology*, **12**, 42–57.

Hall, P. (1992) *The Bootstrap and Edgeworth Expansion*. New York: Springer–Verlag.

Hall, P. and Hart, J.D. (1990) Bootstrap test for difference between means in nonparametric regression. *Journal of the American Statistical Association*, **85**, 1039–1049.

Hampel, F.R. (1974) The influence curve and its role in robust estimation. *Journal of the American Statistical Association*, **69**, 383–393.

Härdle, W. and Gasser, T. (1984) Robust nonparametric function fitting. *Journal of the Royal Statistical Society Series B*, **46**, 42–51.

Huber, P.J. (1972) The 1972 Wald Lecture. Robust statistics: a review. The Annals of Mathematical Statistics, 43, 1041-1067.

Huber, P.J. (1981) *Robust Statistics*. New York: John Wiley & Sons.

Quenouille, M.H. (1949) Approximate tests of correlation in time series. *Journal of the Royal Statistical Society Series B*, **11**, 18–84.

Robinson, J.A. (1983) Bootstrap confidence intervals in location–scale models with progressive censoring. *Technometrics*, **25**, 179–187.

Smith, R.L. and Naylor, J.C. (1987) A comparison of maximum likelihood and Bayesian estimators for the three-parameter Weibull distribution. *Journal of the Royal Statistical Society Series C (Applied Statistics)*, **36**, 358–369.

Thomas, G.E. (2000) Use of the bootstrap in robust estimation of location. *Journal of the Royal Statistical Society Series D (The Statistician)*, **49**, 63–77.
Tukey, J.W. (1958) Bias and confidence in not quite large samples (Abstract). *The Annals of Mathematical Statistics*, **29**, 614.

Chapter 15

Burnand, K.M., Lahiri, R.P., Burr, N., van Rensburg, L.J. and Lewis, M.P.N. (2016) A randomised, single blinded trial, assessing the effect of a two week preoperative very low calorie diet on laparoscopic cholecystectomy in obese patients. *HPB*, **18**, 456–461.
Fisher, R.A. (1935) *The Design of Experiments*. Edinburgh: Oliver & Boyd.
Hollander, M., Wolfe, D.A. and Chicken, E. (2014) *Nonparametric Statistical Methods*. 3rd edn. Hoboken, NJ: John Wiley & Sons.
Ioannidis, J.P.A. (2005) Why most published research findings are false. *PLoS Medicine*, **2**, e124.
Lannelongue, L., Aronson, H.-E.G., Bateman, A., Birney, E., Caplan, T., Juckes, M., McEntyre, J., Morris, A.D., Reilly, G. and Inouye, M, (2023) GREENER principles for environmentally sustainable computational science. *Nature Computational Science*, **3**, 514–521.
Lefkof, I.R., Frenkiel, P.G., Lee, D.W.T., Cohen, H., Bonorris, G.G., Gilmore, C.J., Hamlin, J.A., Komaiko, M., Resser, K., Marks, J.W. and Schoenfield, L.J. (1986) Effect of diet on dissolution of gallstones by ursodeoxycholic acid, including a comparison between ultrasonography and cholecystography. *The Mount Sinai Journal of Medicine*, **53**, 241–249.
Madden, A.M., Smeeton, N.C., Culkin, A. and Trivedi, D. (2024) Modified dietary fat intake for treatment of gallstone disease in people of any age. *Cochrane Database of Systematic Reviews*, **2**, CD012608. https://doi.org/10.1002/14651858.CD012608.pub3.
Sen, P.K. (1991) Nonparametrics: retrospectives and perspectives. *Journal of Nonparametric Statistics*, **1**, 1–32.

Index